EU LAW AND THE DEVELOPMENT OF A SUSTAINABLE, COMPETITIVE AND SECURE ENERGY POLICY

EU LAW AND THE DEVELOPMENT OF A SUSTAINABLE, COMPETITIVE AND SECURE ENERGY POLICY

Opportunities and Shortcomings

Bram Delvaux

intersentia
Cambridge – Antwerp – Portland

Intersentia Publishing Ltd.
Trinity House | Cambridge Business Park | Cowley Road
Cambridge | CB4 0WZ | United Kingdom
Tel.: +44 1223 393 753 | Email: mail@intersentia.co.uk

Distribution for the UK:
NBN International
Airport Business Centre, 10 Thornbury Road
Plymouth, PL6 7 PP
United Kingdom
Tel.: +44 1752 202 301 | Fax: +44 1752 202 331
Email: orders@nbninternational.com

Distribution for the USA and Canada:
International Specialized Book Services
920 NE 58th Ave. Suite 300
Portland, OR 97213
USA
Tel.: +1 800 944 6190 (toll free)
Email: info@isbs.com

Distribution for Austria:
Neuer Wissenschaftlicher Verlag
Argentinierstraße 42/6
1040 Wien
Austria
Tel.: +43 1 535 61 03 24
Email: office@nwv.at

Distribution for other countries:
Intersentia Publishing nv
Groenstraat 31
2640 Mortsel
Belgium
Tel.: +32 3 680 15 50
Email: mail@intersentia.be

EU Law and the Development of a Sustainable, Competitive and Secure Energy Policy. Opportunities and Shortcomings
Bram Delvaux

ISBN 978-1-78068-064-4
D/2013/7849/9
NUR 828

British Library Cataloguing in Publication Data. A catalogue record for this book is available from the British Library.

ACKNOWLEDGEMENTS

This book is an edited, updated and revised version of my doctoral dissertation, defended at the Faculty of Law, *Collegium Falconis*, KU Leuven on 30 September 2011. It is the result of several years of research and experience in the field of European energy law and policy. It signals the end of a journey of discovery, hard labour, patience, challenges, self-doubt and inspired insights. I would never have reached the end of this journey without the continuous inspiration, involvement, help and support of colleagues, friends and family. I would like to express my deepest gratitude to all the people here who made the end of this journey possible, admitting that the list is much longer in reality.

First and foremost, I would like to extend my gratitude to my promotor, Prof. dr. Kurt Deketelaere, for giving me the opportunity to conduct this research project. He has been a source of inspiration from the very beginning to the end of the writing of my book. You were always there for me even at short notice. You believed in me and encouraged me. You have guided me throughout this journey with great commitment, enthusiasm and insight. For me it was a great honour to write this thesis by your side. I hope there are many more years of cooperation to come.

Secondly, I would like to express my gratefulness to my co-promotor, Prof. dr. Ronnie Belmans, whom I have had the pleasure to work with at ELECTA. He introduced me to the world of the technical aspects of energy and energy markets. I would like to thank you for the motivating cooperation of the past years. I could not have completed this book without your support and enthusiasm.

Thirdly, I particularly would like to thank dr. Tom Vanden Borre for his support throughout the writing of this book. I am very grateful that you followed up on its developments, for being part of the jury and your invaluable suggestions and comments. Thank you for comforting me when I was losing faith and all the encouraging conversations we have had in the past years.

Prof. dr. Leigh Hancher, Prof. dr. Wouter Devroe and Prof. dr. Marc Boes, it is my great honour that you all accepted to be a member of my jury. I am very grateful for taking the time and making the effort to read this book thoroughly and providing me with invaluable suggestions and comments.

Many thanks are due to my colleagues at the Faculty of Law and Faculty of Engineering of the KU Leuven and the KU Leuven Energie-instituut. Thank you all for the continuously encouraging discussions and shared inspiration. I am greatly thankful to my colleagues of the IDO project, who supported me throughout the writing of this book. Especially, I would like to thank, Prof. dr. William D'haeseleer, Leonardo, Erik, David and Gerd. I also would like to thank the following current and former colleagues for their support, Prof. dr. Geert Van Calster, Kristof, Leen, Hans, Patrick, Cedric, Peter, Dimitris, Karolien and the mysterious mister Ener Guy. Sincere thanks go to Ann, Katleen, Kathleen, Nathalie, Natacha, Veerle and Marina who took care of all my administrative tasks.

I would also like to express my sincere gratitude to the friends of the Energy Law Research Forum (ELRF) for the support and inspiring reflections and discussions. I would like to thank Kim, Michaël, Henrik, Catherine, Anatole and Nicole in particular.

I am deeply grateful to my friends and their support during the writing of my book. In particular, I would like to thank Gert, Pieter, Davey, Emely and Didier.

I owe special thanks to Ann Groffils for all the efforts she has put in the layout of the book! Thank you very much! Since English is not my mother tongue, I have put the proofreading into the hands of nonkel Johny. For your language revisions, many thanks to you, nonkel Johny!

A heartfelt thank you goes to my parents, for providing me with love, support, encouragement and every opportunity I could ever want. Without them I would not have written this book. I hope that my efforts and accomplishments here reflect how very much my parents have done for me and how dear they are to me. I also would like to thank my sister, parents-in-law, sister-in-law, brothers-in-law and family for all their support during this journey.

Last, but by no means least, my greatest debt and deepest thanks are reserved for my wife Marijke. When the heavy demands of research induced a premature grumpiness, you digested it with a smile. To you go my heartfelt thanks for your patience, love and understanding throughout all these years. I dedicate this work to you and our greatest gifts, our daughter Roos and son Alexander.

Bram Delvaux – December 2012

CONTENTS

Acknowledgements v

Chapter 1. General introduction 1

Chapter 2. Objectives of the European energy policy 9

A. Introduction 9
B. European energy policy: historical overview 10
C. Recent developments in the European energy policy 15
  1. The European challenges of the new energy era 15
  2. The European response to this new energy era 17
    (a) Green Energy Paper of 2006 17
    (b) 2006 and 2011 Energy Efficiency Action Plans 18
    (c) Energy sector inquiry 20
    (d) The New Energy Policy for Europe 21
    (e) Strategic Energy Technology Plan 23
    (f) Energy and Climate Package 24
    (g) Second Strategic Energy Review 25
    (h) European Energy Programme for Recovery 25
    (i) Energy infrastructure priorities for 2020 and beyond 26
    (j) A roadmap for moving to a competitive low-carbon economy in 2050 28
    (k) Energy roadmap 2050 29
D. The three key objectives of the European energy policy 29
  1. Introduction 29
  2. Liberalisation process of the internal energy market 30
  3. Sustainability and climate change challenge 31
  4. Security of energy supply 35
E. An insufficient, incoherent and fragmented EU energy policy 38
F. Concluding remarks 46

Chapter 3. Key EU secondary legislation in the energy field 49

A. Introduction 49
B. Internal market for electricity and gas 50
  1. Introduction 50
  2. Electricity and Gas Directives and Regulations 50

(a) Electricity and Gas Directives 50
(b) Electricity and Gas Regulations. 54
3. ACER Regulation 57
4. Regulation on data and information on investment projects in energy infrastructure 59
C. Sustainability and climate change 60
1. Introduction 60
2. Measures dealing with the supply of energy 60
(a) Renewable Energy Directives. 60
(b) Directive on combined heat and power 63
3. Measures dealing with energy demand 64
(a) Directive on energy end-use efficiency and energy services 65
(b) Directive on the energy performance of buildings. 67
(c) Directive on Eco-design for energy-using appliances 68
(d) Directives on efficiency in energy using products 70
i) Household appliances. 70
ii) Office appliances 71
iii) Tyre labelling 73
4. Emission Trading Directive. 73
5. Union framework for the taxation of energy products and electricity .. 76
D. Security of energy supply 77
1. Introduction 77
2. Security of energy supply directives and regulations 78
(a) Directives on minimum stocks of crude oil and/or petroleum 78
(b) Security of electricity supply and infrastructure directive and security of natural gas supply directive 79
3. External security of energy supply 82
(a) Energy Charter Treaty. 83
(b) Energy Community Treaty. 83
E. Problems faced with EU secondary legislation in the energy field 85
F. Conclusion 89

Chapter 4. Principles of EU law and legal bases in the TFEU allowing Union action in the field of energy prior to the Lisbon Treaty. 91

A. Introduction 91
B. Energy and the Treaty context: historical overview. 92
1. Introduction 92
2. Treaty establishing the European Coal and Steel Community 93
3. Treaty establishing the European Atomic Energy Community 94
4. Treaty establishing the European Economic Community 96
(a) Adoption of the Single European Act. 97
(b) Adoption of the European Union Treaty 99

(c) Treaty framework after the adoption of the Treaty of Amsterdam 101
(d) Developments with the adoption of the Treaty of Nice 102
(e) Constitutional Treaty of Europe 102
5. Conclusion 105
C. Division of competences between the Union and its Member States in the energy field 106
1. Introduction 106
2. The principle of attribution of powers 107
(a) Meaning of the principle of attribution of powers on a European level 107
(b) Energy and the principle of attribution of powers 109
3. The principle of subsidiarity 112
(a) Introduction 112
(b) Meaning of the principle of subsidiarity on a European level 113
i) Principle of subsidiarity: not without reason 113
ii) Present form and meaning of the principle of subsidiarity 115
iii) Subsidiarity test 116
iv) Conclusion 119
(c) Energy and the principle of subsidiarity 121
4. The principle of proportionality 123
(a) Introduction 123
(b) Meaning of the principle of proportionality on a European level 124
(c) Energy and the principle of proportionality 128
5. Conclusion 130
D. The provisions of the TFEU allowing Union action in the field of energy 131
1. Introduction 131
2. The general TFEU provisions and their application in the energy sector 132
(a) Introduction 132
(b) Electricity and gas markets and the limits of the general TFEU provisions 134
i) The application of Article 106 of the TFEU 135
ii) The application of Article 345 of the TFEU 160
(c) The application of the rules on the free movement of goods 170
i) Fiscal barriers to trade or measures having an effect equivalent to customs duties 171
ii) Non-fiscal barriers to trade 181
3. Internal market provisions 222
(a) Meaning of Articles 114 and 115 of the TFEU 222

(b) Case law of the Court of Justice interpreting Article 114 of the TFEU ... 223
(c) Article 114 and energy ... 228
(d) Conclusion ... 228
4. Article 113 of TFEU ... 229
(a) Meaning of Article 113 of the TFEU ... 229
(b) Article 113 of the TFEU and energy ... 229
(c) Conclusion ... 230
5. The environmental law provisions in the TFEU ... 230
(a) Introduction ... 230
(b) Present environmental competences of the Union ... 231
i) Objectives of the environmental policy ... 231
ii) Principles of the environmental policy ... 242
(c) Legal basis in the TFEU for environmental measures ... 255
i) Provisions primarily of a fiscal nature ... 256
ii) Measures affecting town and country planning, quantitative management of water resources and land use ... 257
iii) Measures significantly affecting a Member State's choice between different energy sources and the general structure of its energy supply ... 258
(d) Derogation clauses of the TFEU in the environmental field ... 261
(e) Renewable Directive 2009/28/EC and its legal bases Articles 192 and 114 of the TFEU ... 266
i) Introduction ... 267
ii) Legal basis of the Renewable Energy Directive 2009/28/EC ... 268
iii) Conclusion ... 275
6. Loophole Article 352(1) of the TFEU ... 276
(a) Meaning of Article 352(1) of the TFEU ... 276
(b) Article 352 of the TFEU and energy ... 277
7. The trans-European networks provision ... 277
(a) Meaning of Articles 170–172 of the TFEU ... 277
(b) Secondary legislation on the basis of Articles 170–172 of the TFEU ... 279
(c) Conclusion ... 283
8. Meaning of Article 122 of the TFEU ... 283
9. Impact of case law of the Court of Justice on the adoption of secondary legislation in the energy field ... 284
(a) Background ... 284
(b) The *Meroni* doctrine and ACER ... 287
(c) The adopted text on ACER ... 290
i) The Agency's set of tasks ... 291
ii) Analysis and conclusion ... 296
E. Conclusion ... 297

Chapter 5. Union competences in the energy field after the adoption of the Lisbon Treaty 301

A. Introduction 301
B. Relevant changes to provisions of the TFEU 302
1. Categories of Union competence 302
(a) Introduction 302
(b) Exclusive competences 302
(c) Shared competences 304
(d) Competence to support, coordinate or supplement 310
(e) Since Lisbon, energy is a shared competence 311
2. Revised Protocol on the application of the principles of subsidiarity and proportionality 313
(a) Role of the national Parliaments 313
(b) Compliance with subsidiarity under the revised Protocol 319
3. Article 122 of the TFEU and the Lisbon Treaty 321
4. Climate change inserted in Article 191 of the TFEU 322
5. Foreign direct investment under the Lisbon Treaty 323
6. Changes to Article 352(1) of the TFEU 326
C. The new Article 194 of the TFEU 328
1. Article 194 of the TFEU and the objectives of European Energy Policy 329
(a) Meaning of Union policy on energy 330
(b) Meaning of "*in the context of the establishment and functioning of the internal market and with regard for the need to preserve and improve the environment*" 332
(c) In a spirit of solidarity between Member States 333
(d) Ensure the functioning of the energy market 336
(e) Ensure security of energy supply in the Union 336
(f) Promote energy efficiency and energy saving and the development of new and renewable forms of energy 340
(g) Promote the interconnection of energy networks 341
(h) Conclusion 342
2. The legal basis of European Energy Policy 344
(a) Paragraph 2 of Article 194 TFEU 344
i) Member State's right to determine the conditions for exploiting its energy resources 345
ii) Choice between different energy sources and the general structure of its energy supply, without prejudice to Article 192(2)(c) 345
(b) Third paragraph of Article 194 of the TFEU 347
3. Declaration 35 on Article 194 of the TFEU 348
4. Gas Regulation No 994/2010 and Article 194 of the TFEU 349

(a) Background 349
(b) Main features of the Gas Regulation No 994/2010 350
(c) Critical assessment of the Gas Security Regulation No 994/2010 . 352
5. Concluding remarks with respect to the new energy title 354
D. Article 194 of the TFEU and/or other provisions of the TFEU 355
1. Choice between Article 192 of the TFEU and Article 114 of the TFEU 355
2. Choice between Article 194 of the TFEU and other provisions of the TFEU 358
(a) General remarks 358
(b) Identifying the most appropriate legal basis for energy measures 362
E. Concluding remarks 367

Chapter 6. Legal and institutional options for a European energy framework 369

A. Introduction 369
B. Suggestions in legal doctrine and current practices 370
1. Enhanced cooperation 370
2. Reinforced cooperation or in-built closer cooperation – the Schengen model 372
3. Functional or regional cooperation 375
C. Amendments to the provisions of the Treaties 376
1. Procedure for amending the Treaties 376
2. Treaty amendments with respect to energy 378
(a) Article 3 of the TFEU 378
(b) Article 122 of the TFEU 378
(c) Articles 191 and 192 of the TFEU 379
(d) Articles 170–172 of the TFEU 380
(e) Article 194 of the TFEU 380
(f) Article 288 of the TFEU 382
3. Conclusion 386
D. A new energy Treaty on a European level 386
E. Conclusion 391

Chapter 7. General conclusion 395

Bibliography 403
Index 447

CHAPTER 1
GENERAL INTRODUCTION

1. For several years, various policy documents on a European and international level have highlighted the need to establish secure and sustainable energy policies and strategies. This is done with the objective, from an economic perspective, to address the problems of security and diversification of energy supply and, from an environmental perspective, to reduce the increase of greenhouse gases. The underlying reason of the need for such energy policies and strategies is the energy demand which continues to rise in virtually all regions of the world.

2. The well-being of European citizens, industry and economy depends on affordable, secure and sustainable energy. Moreover, in the EU, energy accounts for 80% of all greenhouse gas emissions. The days of cheap energy for Europe are over; energy prices are rising on an international level.

3. All EU Member States are facing these challenges of rising energy prices, climate change and increasing import dependence. Higher energy prices and the increasing dependence on energy imports jeopardise the EU's competitiveness and security. Not only is the dependence on energy imports increasing, but also the interdependence of the Member States with respect to energy. Consequently, the energy and climate change challenge is at the moment, and in the longer run, one of the greatest tests which the EU and its Member States are facing. It will take decades to steer our energy system into a more secure and sustainable system. However, the decisions needed to set the EU on the right path should be taken now.[1]

4. Since the adoption of the Treaty establishing the European Coal and Steel Community, signed in Paris in 1951, and the Treaty establishing the European

---

1 Communication from the Commission of 10 November 2010 to the Council, the European Parliament, the European Economic and Social Committee and the Committee of the Regions, *Energy 2020 A Strategy for competitive, sustainable and secure energy*, COM(2010)639, 20 p.; Green Paper, *A European Strategy for Sustainable, Competitive and Secure Energy*, COM(2006)105, 20 p. and Communication from the Commission to the European Council and the European Parliament of 10 January 2007, *An energy policy for Europe*, COM(2007)1, 28 p.

Atomic Energy Community, signed in Rome on 25 March 1957, the energy markets and geopolitical considerations have changed significantly. The three core objectives of the EU energy policy to date and in the long term are a sustainable, competitive and secure energy supply. These objectives are complementary goals and therefore, each aspect of the integrated energy policy needs to contribute to all three goals.

Europe is indeed in a new energy era which is characterised by the challenges of climate change and security of supply. To ensure that today's customers and those of tomorrow are supplied with competitively priced, secure and sustainable supplies of electricity, the European energy market requires an over-arching EU legal regulatory framework.[2] The need to deliver secure, sustainable and competitive energy is stronger than ever and therefore, the EU and its Member States must act now. Moreover, the price of failure is too high.[3]

5. The European Union and its Member States have common concerns *viz.* security of energy supply which could lead to a growing willingness to establish a more coordinated, developed and integrated EU energy policy. Such a European energy policy needs both political agreement of all Member States (including a potential transfer of more powers to the EU), and a reliable legal regulatory framework.

6. The overall objective of this book is to analyse whether the current EU regulatory framework for energy law and policy suffices to address the triple objectives of the EU's energy policy (a competitive, secure and sustainable energy supply), or by contrast, whether changes to that framework are needed. An efficient EU regulatory framework for energy law and policy will not only allow the EU and its Member States to achieve the triple objectives, but will also generate the right investment signals for potential investors. It should provide all the market players, such as generators and network operators, with the real and effective incentives to invest in the EU energy market. Moreover, this framework should support a transition into an energy economy which improves service, choice and security for households and industry.

7. Why a stronger energy policy on an EU level and more regulatory harmonisation instead of regulatory competition? In various regulatory fields

2 CEER response to the Energy Green Paper of 11 July 2006, p. 3 and 4. According to the CEER paper such an over-arching EU legal regulatory framework should be based on four principles: European obligations on network companies; improved operational co-ordination of the networks; regulators with the necessary powers and effective unbundling.

3 Communication from the Commission of 10 November 2010 to the Council, the European Parliament, the European Economic and Social Committee and the Committee of the Regions, *Energy 2020 A Strategy for competitive, sustainable and secure energy*, COM(2010)639, p. 2.

this question is being asked on a global, regional and national level. One author argued that the proper question is *"to what extent should regulators cooperate with regulators in other jurisdictions"* or the harder question *"whether cooperation on ultimate regulatory goals or targets is beneficial"*.[4] The answer to the latter question, according to this author, is highly context-specific.[5]

Today, the academic debate with respect to regulatory competition is focussed on the direction of the races between states. There are two main approaches in this debate. Firstly, there are those who are worried that inter-jurisdictional competition could lead to a welfare-reduction if states fail to set minimum standards. They predict a "race to the bottom" between states with a systematic lowering of regulatory standards in order not to endanger the states' competiveness. The second approach, on the other hand, predicts a "race to the top". They argue that, based on an extension of the principles of competition and free market, regulatory competition stimulates innovation, differentiation, efficiency and experimentation which results in more efficient laws and enhanced social welfare. Recently, a middle way emerged referred to as "regulatory co-opetition". This third approach argues that the optimal governance level is a flexible mix of competition and cooperation between states.[6]

8. For many years Member States have considered energy policy as an issue of national security which should remain within the competence of individual states and therefore they were unwilling to lose parts of their sovereignty in this area. Moreover, there are a number of reasons why a common approach on an EU level in the energy field was difficult to achieve. In the 27 Member States there are different political and social conditions, foreign policy, levels of economic development and growth, etc. which should somehow be coordinated on an EU level. Member States have a different energy mix and therefore, a different level of import dependence, suppliers and diplomatic relations with major energy producing countries. Another factor is that Member States have a different structure of energy sectors (for example some are interested in nuclear energy development, others in coal industry, and others in renewable energy sources).

9. Despite these differences, Member States have in common their reliance on fossil fuels and their import dependence for oil and natural gas. All Member

---

4 A. O. SYKES, "Regulatory Competition or regulatory harmonization? A Silly Question" *Journal of International Economic Law* (2000), p. 257.

5 *Ibid.*

6 A. M. SACHDEVA, "Regulatory competition in European company law" *European Journal of Law & Economics* (2010), p. 137–139; D. C. ESTY, "Regulatory competition in focus" *Journal of International Economic Law* (2000), p. 215 and 216 and G. VAN CALSTER, "Against Harmonisation – Regulatory Competition in Climate Change Law" *Carbon & Climate Law Review* (2008), p. 89 and 90.

States are facing, not only the increasing import dependence, but also the challenges of climate change and other environmental constraints, rising demand for energy and higher energy prices. Furthermore, given the increasing interdependence of Member States in energy, a power failure in one Member State has immediate effects in other Member States. Access to sustainable, competitive and secure energy is of the utmost importance for all Member States.[7]

10. Taking into account these common challenges of the Member States, an approach based on 27 individual energy policies competing with each other is likely to be insufficient. Member States are not strong enough to cope with the increasing competition for energy resources on an international level. A combined weight of the EU will be more effective than dispersed actions of the Member States.[8] The European energy markets will function better when all actors operate on the same conditions and within a harmonised regulatory framework. An EU-wide energy regulatory framework should be developed to provide a robust basis for energy trade and markets across borders which promote investment, competition and security of supply on an EU-level.

Given the challenges in the energy sphere and the specific characteristics of energy[9], regulatory competition is, in the author's view, not the best track to be followed in meeting these urgent challenges. The book therefore argues in favour of a more centralised EU level approach to deal with the energy challenges of the Member States.

11. With respect to the methodology, the research method in this book of primary and secondary law provisions is based upon traditional sources of legal and policy documents, case law and legal doctrine. The main means to reflect on this topic is an analysis of the legal and policy documents, mainly on a European level, as well as a relevant study of the legal doctrine in this field.

12. A significant source of interpretation for the primary law provisions in this book is the existing case law of the Court of Justice of the European Union and to a lesser extent the General Court. With respect to the provisions of secondary law, the interpretation was done in the light of the objectives, the wording and preamble of the measure concerned. Moreover, legal doctrine, proposals of the European Commission, various reports and preparatory documents have extensively been analysed and used in order to interpret the primary and secondary law provisions in this book.

---

7 A. GANOVA, *European Union Energy Supply Policy: Diversified in Unity?* (Institut Européen des Hautes Etudes Internationales, 2007), p. 35–45.

8 *Ibid.*

9 See *Infra.*

13. The book contains seven chapters. Following this introduction, chapter 2 discusses the objectives of the Union's energy policy. For many years, the Union has been adopting various documents dealing with the objectives of its energy policy. Chapter 2 therefore starts with a historical overview of the European objectives in the energy field followed by the more recent developments in the energy policy area, and identifies the European challenges of the new energy era. It highlights that the three core objectives of the EU energy policy to date and in the long term are a sustainable, competitive and secure energy supply. This chapter concludes that the current European energy policy framework, being very ambitious, will not achieve its three central goals. Thus, it is undeniable that urgent action is required.

14. On the basis of the adopted energy policy objectives, chapter 3 provides an overview of the different provisions of key secondary legislation on a European level implementing these objectives into the EU energy law framework. The points of departure for the analysis of the EU energy law framework in this book are the current provisions of secondary legislation. The choice to start with secondary law instead of primary law was a resolute one. It will enable us to determine whether the current legal framework of secondary law provisions in the energy field on a European level is sufficient to achieve the central goals of the EU energy policy, or by contrast, whether changes are needed to these provisions, but also to the primary law provisions. This starting point enables us to analyse in the next chapter whether an insufficient legal framework of secondary law is attributable to a lack of adequate legal basis in the primary law provisions. In the light of the foregoing, chapter 3 will discuss the current legal framework which adopted several specific instruments on a European level in order (a) to achieve the full opening of the internal market for electricity and/or gas, (b) to address security of supply crises leading to supply disruption and (c) to tackle the challenge of climate change.

15. For each of the three central goals, chapter 3 will provide an overview of the most relevant secondary legislation adopted over the past years. It highlights, that the Union currently has a relatively well-developed set of measures in order to achieve these three central goals of its energy policy. Moreover, the on-going attempts on a European level to improve the existing legal framework of secondary legislation cannot be denied. Nonetheless, this chapter concludes that the common EU energy policy remains both insufficient in its results and incoherent in its aims for achieving the three central and ambitious goals.

16. Before the adoption of the Lisbon Treaty, the Union, even without any specific articles, chapter or title on energy, adopted various measures of secondary legislation in the energy field. Thus, in the energy sphere, the Union intervened, on the one hand, by enforcing the general Treaty provisions and, on the other hand, by adopting secondary legislation on the basis of its competences regarding the internal market, tax, infrastructure, environment, etc. These general provisions and legal bases for the adoption of secondary legislation in the energy sphere are the subject of chapter 4.

17. It should be highlighted that, although chapter 4 analyses the primary Treaty provisions before the adoption of the Lisbon Treaty, it applies the new numbering of the provision of the TFEU in order to facilitate its reading. Chapter 4 first provides an historical overview of the Treaty context with respect to energy. Since the post-World War reconstruction and the great need for energy sources, Europe recognises the necessity of coordinated action in the energy sector. This is reflected in the fact that in the European Treaties of the 1950s, two of these three Treaties are specifically directed at the energy sector, namely the ECSC Treaty and the EURATOM Treaty. In addition to a brief discussion of these Treaties, this chapter gives an overview of the energy-related amendments to the Treaty over the past years.

18. This part of chapter 4 is followed by an analysis of the principles of attribution of competences, subsidiarity and proportionality which govern the existence and exercise of Union powers and their application to the energy field. It allows us to answer the question who (Union and/or Member States), on the basis of the fundamental principles (conferral, subsidiarity and proportionality), is competent to act in the energy field and to which extent. Moreover, it answers the question whether these fundamental principles constitute a barrier to the adoption of measures in the energy field.

19. Chapter 4 continues with a detailed discussion of the most relevant general provisions of the TFEU applicable to the energy sector such as Articles 106 and 345 of the TFEU, and the rules on the free movement of goods. The purpose of this analysis is to determine whether and to which extent these primary law provisions could constitute a barrier to the development of an adequate European legal framework in the energy field. It highlights the possibilities for Member States to adopt deviating provisions on the basis of primary law which could hamper the adoption of an efficient legal framework in the energy field on a European level.

It should be highlighted that this book contains, given its scope, an analysis of the relevant competition law provisions of the TFEU and their application to the energy sector. In the early 1990s the European Commission has realised that Member States were not willing to adopt national measures leading towards a liberalisation process of the telecom and energy sectors. It has therefore started to use and apply the Treaty provisions relating to competition law and has tried to force Member States to liberalise these sectors. This was a very important step in the liberalisation process of the energy markets since it brought the Member

State to the negotiating table to discuss secondary legislation requiring the progressive opening of the electricity and gas markets. Although, this initiative of the European Commission, on the basis of its exclusive competence in competition law, has resulted in the past years in an extensive package of secondary legislation, competition law, today, is still of vital importance to the energy sector.

This was confirmed by the energy sector inquiry of 2007 in which the European Commission has emphasised that a full and combined use of its powers under antitrust rules, merger and State aid control is required in order to maximise the impact of its enforcement action in the energy sector. The European Commission furthermore confirmed that competition law enforcement can make a significant contribution. However, it cannot by itself open the electricity and gas markets and resolve all the shortcomings and therefore a number of regulatory measures and remedies are needed on a European level. These regulatory measures and remedies and in particular their proper legal basis have mainly been analysed in the book. But in this context the relevant competition law provisions of the TFEU have also been analysed.

20. Beside the general TFEU provisions, the next part examines the primary law provisions, such as Article 114 of the TFEU, Title XX of the TFEU, comprising Articles 191–193, Article 113, Articles 170–172 of the TFEU, etc. which establish a legal basis for the adoption of energy-related legislation. On the basis of the analyses of these Treaty provisions, chapter 4 discusses the Treaty bases of the Renewable Directive 2009/28/EC and Directive 2008/101/EC which includes aviation in the scheme for greenhouse gas emission allowance trading within the Union.

21. This chapter 4 not only discusses the Union principles and Treaty provisions of the TFEU, it also illustrates, with the example of the Agency for the Cooperation of Energy Regulators, that case law of the Court of Justice could hamper the adoption of secondary legislation in the energy field. Chapter 4 concludes that, due to the recourse to more general legal provisions of the TFEU, the European legislator was, to a certain extent, unable to address the specific issues of the energy sector.

22. Chapter 5 examines the Treaty developments in the energy field after the adoption of the Lisbon Treaty. Before the analysis of the new energy title in the TFEU, chapter 5 provides an overview of the most relevant changes in the TFEU for the energy field. This part discusses, for example, the explicit recognition of energy, since the adoption of the Lisbon Treaty, in Article 4(2), (i) of the TFEU as a new policy area in the general category of shared competences. Thus, it has become undeniable that both the Union and the Member States are competent to regulate in the energy field. However, since energy has been categorised as a shared competence in primary law, this does not mean that the EU can exercise this competence unconditionally. The Union, when it proposes measures in the energy field, still needs to comply with the principles of attribution of competences, subsidiarity and proportionality which govern the existence and exercise of Union powers.

23. The chapter continues with a detailed analysis of the new energy title of the TFEU. After some 40 years of the Union's legislative activities in the energy policy area, an energy title XXI has been inserted into the TFEU which compromises the new Article 194 of the TFEU. These provisions establish a legal basis for the further development of the European energy policy. Thus, finally a *"proper"* legal basis for the adoption of measures in the energy sphere has been inserted into the Treaty. This chapter discusses in detail the scope of these new energy provisions in the TFEU.

24. Beside the detailed discussion of Article 194 of the TFEU, chapter 5 identifies the most appropriate legal basis for the adoption of measures of secondary legislation in the energy field. With respect to the choice between Article 194 of the TFEU and other provisions of the TFEU such as Articles 113, 114, 192, etc., this chapter argues that, merely because a proposed measure concerns the energy sector, this does not automatically mean that the given measure must be based on Article 194 of the TFEU. It must be asserted on a case-by-case basis which Treaty article (energy, environmental, fiscal, etc. title) matches the most with the main aim, objective and content of a given measure on a European level.

25. Chapter 5 concludes that, although the adoption of the new energy title has been welcomed warmly, Article 194 of the TFEU can be seen as a mere codification of the existing practice in the energy field. Consequently, the existing gaps, flaws and limitations remain in place.

26. To overcome the current limitations relating to Article 194 of the TFEU, chapter 6 provides a number of institutional and legal amendments to the current European energy framework which should remedy the current deficiencies and allow the Union to adopt a specific, common and effective European energy policy and legislation. It highlights that urgent, decisive and immediate action is required. Hence, chapter 6 of this book argues that there are two options in this respect: on the one hand, different amendments to the TFEU, on the other hand, the adoption of a new energy Treaty on a European level. These legal and institutional options should allow the European energy policy to achieve its three central goals in the coming years. The book ends with a general conclusion in chapter 7.

CHAPTER 2
OBJECTIVES OF THE EUROPEAN ENERGY POLICY

A. INTRODUCTION

27. Since the post-World War reconstruction and the great need for energy sources, Europe has recognised the necessity of coordinated action in the energy sector. The primary concern of both the Communities and their Member States was the issue of energy supply. The supply of a sufficient amount of energy at a reasonable price became rapidly an essential condition for a well-functioning and growing economy.[10] Energy has been considered as *"a production factor of fundamental and universal importance to economic activity, which cannot be ignored or downgraded in the construction of an economic union"*.[11]

28. Energy policies did vary over time and differ from one country to another. Such variations are mainly caused due to technological developments and local conditions. It is important to bear these differences in mind when developing energy policies since the choices made today must not prejudice the development of future energy policies.[12]

---

10 J. SCHWARZE, "European Energy Policy in Community Law" in E. J. MESTMÄCKER (ed.), *Natural Gas in the Internal Market – A Review of Energy Policy* (London, Graham and Trotman, 1993), p. 153 and 154. See furthermore, E. F. SCHUMACHER, "The Struggle for a European Energy Policy", *Journal of Common Market Studies* (1964), p. 199 *et seq.*; P. O. LAPIE, "La Politique énergétique commune", *Revue Marché Commun* (1964), p. 134–141; W. MÜLLER-MICHAELIS, *Die Integration der westeuropäischen Energiemärkte* (Göttingen, Vandenhoeck u. Ruprecht, 1968); S. ANADUCCI, "Towards a Comprehensive Community Energy Policy: Some Legal Considerations", *Diritto comunitario e degli scambi internazionali* (1976), p. 247–281; M. BYWATER, "La Politique énergétique commune", *Revue Marché Commun* (1977), p. 409 *et seq.*; H. MICHAELIS, *Energiemarkt und Energiepolitik in einer Europäischen Union* (Frankfurt am Main, A. Metzner, 1976), 96 p. and N. GREEN, "The legal basis of a Community Energy Policy", *European Law Review* (1983), p. 52 *et seq.*

11 T. DAINTITH and S. F. WILLIAMS, *The legal integration of energy markets* (Berlin, W. De Gruyter, 1987), p. 5.

12 Policy Proposal by Jacques Delors, *Towards a European Energy Community: A Policy proposal*, study by S. ANDOURA, L. HANCHER and M. VAN DER WOUDE, p. 3.

29. Over the years, the European Union adopted numerous documents dealing with the objectives of its energy policy. The aim of this chapter is to provide an overview of these objectives in the energy field, and to introduce the uninitiated into the current energy policies on a European level.

B. EUROPEAN ENERGY POLICY: HISTORICAL OVERVIEW

30. In the fifties, the European energy policy, with the exception of the coal and nuclear sector, was characterised by a self-restriction of the Union's institutions and an undistorted development of the national energy markets. Because of the unexpected change into a *surplus* of hard coal at the end of the 1950s there was a need for intervention in this sector. In 1952, most of the Member States' energy consumption was coal-based and produced in the Union.[13]

31. In the sixties the consumption of coal went down to 60% and oil already contributed 30% of the Member States' energy consumption. The High Authority under the Treaty establishing the European Coal and Steel Community (hereinafter ECSC Treaty) adopted several decisions which permitted the Member States to grant financial support to mining companies – in spite of Article 4, (c) of the ECSC Treaty which prohibits subsidies or aids granted by the Member States within the Union – in order to delay the decline in this sector.[14] Furthermore, the three European Community (hereinafter EC) institutions adopted a memorandum in 1963 formulating the priorities of a mineral oil policy, *inter alia*, transparency of prices, stock policy and information concerning future investments. In this period, the demand for energy has increased by 6% per annum with a growing dependence on oil imports which in 1968 had a share of 54%.[15]

---

13 For further analysis of this period see S. S. HAGHIGHI, *Energy Security The External Legal Relations of the European Union with Major Oil- and Gas-Supplying Countries* (Oxford, Hart Publishing, 2007), p. 38–46; N. J. D. LUCAS, *Energy and the European Communities* (London, Europa Publications, 1977), 175 p. and W. G. JENSEN, *Energy in Europe: 1945–1980* (London, GT Foulis, 1967), 224 p.

14 The Council has recently decided to allow EU Member States to maintain, on certain conditions, state aid for uncompetitive coal mines until December 2018. See the press release of the Council of the European Union of 10 December 2010, *Council extends state aid regime for the coal sector*, PRESSE 344, 17766/10 and the Report of the Committee on Economic and Monetary Affairs of 12 November 2010 on the proposal for a Council regulation on State aid to facilitate the closure of uncompetitive coal mines, A7–0324/2010 and the press release of the European Parliament of 23 November 2010, *Parliament asks for state aid for coal mines to be allowed until 2018*.

15 U. IMMENGA, "The Development of European Energy Policy: From the ECSC Treaty to the Internal market" in E. J. MESTMÄCKER (ed.), *Natural Gas in the Internal Market – A Review*

32. Due to this development, the Commission became very concerned, and prepared the study *"Further orientation for a common energy policy"* for the Council of 18 December 1968. It defined the urgent problems in the energy sector such as dependence on oil and energy saving and contained recommendations which expressed the need for a European energy policy. This study appeared to be one of the first studies which contained an overall analysis addressing a comprehensive energy market instead of a narrow sectorial approach concerning a single energy source. Unfortunately, with the exception of some measures stabilising the energy supply, no further initiatives were taken on the basis of this study.[16]

33. Following the first oil crisis of 1973, the Member States and the Union reconsidered the need for an energy policy, which resulted in particular policies and legislative initiatives.[17, 18] An energy committee consisting of leading public servants was set up by the Council on 30 January 1974. The committee's task was to advise the Commission on all important issues concerning energy policy.[19] To ensure a continuous overview of the energy supply, stock and consumption, the Council also agreed that comprehensive energy balances be drawn up every three months.[20]

34. On 17 September 1974, the Council adopted the Resolution concerning a new strategy on energy policy for the Union, which confirmed the importance of a common energy policy due to the developments on world energy markets. This Resolution envisaged a reduction of energy consumption in the Union and, through a diversification of energy sources, a secure energy supply. A few months

---

of Energy Policy (London, Graham and Trotman, 1993), p. 51, quoted from H. VON DER GROEBEN, H. VON BOECKH, J. THIESING, C.D. EHLERMANN (eds.), *Kommentar zum EWG-Vertrag* (Baden-Baden, Nomos, 1983), p. 1505.

16 U. IMMENGA, "The Development of European Energy Policy: From the ECSC Treaty to the Internal market" in E. J. MESTMÄCKER (ed.), *Natural Gas in the Internal Market – A Review of Energy Policy* (London, Graham and Trotman, 1993), p. 50 and 51. For further details see H. VON DER GROEBEN, H. VON BOECKH, J. THIESING, C.D. EHLERMANN (eds.), *Kommentar zum EWG-Vertrag* (Baden-Baden, Nomos, 1983), p. 1507 and S. S. HAGHIGHI, *Energy Security The External Legal Relations of the European Union with Major Oil- and Gas Supplying Countries* (Oxford, Hart Publishing, 2007), p. 46–53.

17 For example: Directive 75/339/EEC of 20 May 1975 obliging the Member States to maintain minimum stocks of fossil fuel at thermal power stations, *OJ* L 153/35; Directive 73/238/EEC of 24 July 1973 on measures to mitigate the effects of difficulties in the supply of crude oil and petroleum products, *OJ* L 228/1; Decision 77/706/EEC on the setting of a community target for a reduction in the consumption of primary energy sources of energy in the event difficulties in the supply of crude oil and petroleum products, *OJ* L 292/9.

18 For further details on the oil crisis see S. S. HAGHIGHI, *Energy Security The External Legal Relations of the European Union with Major Oil- and Gas Supplying Countries* (Oxford, Hart Publishing, 2007), p. 53–62.

19 H. VON DER GROEBEN, H. VON BOECKH, J. THIESING, C.D. EHLERMANN (eds.), *Kommentar zum EWG-Vertrag* (Baden-Baden, Nomos, 1983), p. 1509.

20 Regulation No 293/74 of the Council of 30 January 1974 on information for the establishment of comprehensive energy balance sheets for the Community, *OJ* L 32/1.

later, on 17 December 1974, the Council adopted a second Resolution concerning objectives of a common energy policy for 1985. On 13 May 1980, the Resolution was updated to the year 1990 and completed by the Resolution *"Evolution of an energy strategy for the Community"*.[21]

35. The Resolution of 17 December 1974 led to a number of analyses containing proposals to save energy in households, transport and industry and some suggestions to optimise energy production. In order to realise the objectives of a common energy policy, the Union institutions adopted different recommendations and directives on the efficient use of energy (such as the improvement of the conditions of car driving, improvement of public transport services, heat insulation and heating systems in buildings) between 1976 and 1980. Moreover, financial incentives were given for demonstration projects aiming at better exploitation of traditional energy resources and the development of new alternatives. As a result of these measures, the 62% share of imported oil of the gross energy consumption of the Union in 1973 was reduced significantly in 1985.[22]

36. The main strategy of the Union's energy policy in the eighties was, on the one hand, the reduction of dependency on energy imports by saving energy and, on the other, strengthening the exploitation of nuclear energy and other energy sources in the Union. This strategy was pursued through five operational priorities: (a) ensuring favourable investment conditions for substitute energies and energy saving; (b) consensus among the Member States concerning price policy; (c) taking measures to avoid distortions on energy markets; (d) enforcement of common policies in the field of research and development and (e) developing common concepts and initiatives concerning external relations.[23]

37. In June 1987, the Energy Council advised the Commission to draw up an inventory of the existing obstacles and their possible remedies to create an internal energy market. This led to the adoption of the 1988 Working Document in which the Commission suggested, on the one hand, an application of the existing EC law and, on the other, the submission of specific directives to address existing obstacles.[24] As the name of the 1988 Working Document suggests, it does not provide a blueprint or a comprehensive programme in order to create a single internal energy market.

---

21 H. VON DER GROEBEN, H. VON BOECKH, J. THIESING, C.D. EHLERMANN (eds.), *Kommentar zum EWG-Vertrag* (Baden-Baden, Nomos, 1983), p. 1510.

22 T. OPPERMANN, *Europarecht* (Munchen, Beck, 1991), p. 472 and 473.

23 U. IMMENGA, "The Development of European Energy Policy: From the ECSC Treaty to the Internal market" in E. J. MESTMÄCKER (ed.), *Natural Gas in the Internal Market – A Review of Energy Policy* (London, Graham and Trotman, 1993), p. 51–53.

24 Commission working document of 2 May 1988, *The internal energy market*, COM(1988)238, 90 p.

38. Unfortunately, the 1988 Working Document contained various shortcomings. It was not clear which type of competition had to be pursued. Was the desired objective the promotion of inter-fuel competition (e.g. gas *versus* electricity) or intra-fuel competition (which means increased competition within the separate fuel sectors e.g. within the electricity sector)? It did not contain a clear commitment towards a truly unified energy market. In the 1988 Working Document there is confusion on the scope of the available instruments to obtain the desired objectives. It did not clarify what combination of competition and intervention is deemed appropriate and who will adopt the desired measures (only Member States or only the Union legislator or both)?[25]

39. Moreover, due to the resistance of certain Member States, further progress towards an internal energy market could not be achieved.[26] Consequently, Member States remained essentially free to regulate and organise their national energy markets. Hence, the Commission was left with the basic Treaty provisions (competition and free movement) as its principal instruments to establish the internal energy market.[27]

40. Shortly before the adoption of the First Electricity Directive[28] in late 1996 and the First Gas Directive[29] in 1998, the Commission adopted a Green Paper[30] in 1994 which was followed by a White Paper on an Energy Policy for the European Union and External Security of Energy Supply.[31] The latter highlighted that the external dimension should be considered as the most important vehicle for action in order to secure the energy supply of the Union. It stated furthermore that *"the conditions of access of energy products into the EC market on the basis of its trade competences, the access of energy companies to third countries' markets, and the organisation of dialogue, assistance, cooperation and emergency aid are the international responsibilities of the Community"*.[32] This White Paper

25 L. HANCHER, "A single European Energy Market – rhetoric or reality?", *Energy Law Journal* (1990), p. 222 and 223.

26 P. CAMERON, *Competition in Energy Markets, Law and Regulation in the European Union*, (Oxford, Oxford University Press, 2007), p. 53 and 54 and D. S. M. DOUGALL, "European Community Energy law in Perspective" in D. S. M. DOUGALL and T. W. WÄLDE (eds.), *European Community Energy Law – selected topics, International Energy and Resources law & Policy Series* (Nijhoff, London, Graham & Trotman/Martinus, 1994), p. 1 and 2.

27 L. HANCHER, "A single European Energy Market – rhetoric or reality?", *Energy Law Journal* (1990), p. 223–225.

28 Directive 96/92/EC of the European Parliament and the Council of 19 December 1996 concerning common rules for the internal market in electricity, *OJ* L 27/10 (hereinafter First Electricity Directive).

29 Directive 98/30/EC of the European Parliament and the Council of 22 June 1998 concerning common rules for the internal market in natural gas, *OJ* L 204/1 (hereinafter First Gas Directive).

30 Green Paper, *For a European Union Energy Policy*, COM(94)659, 128 p.

31 White Paper on an Energy Policy for the European Union, COM(95)682, 45 p.

32 *Ibid.*, p. 3 and 26.

established a strong link between the promotion of international cooperation and ensuring the security of the energy supply of the Union.[33]

41. In 2000, the Commission adopted a Green Paper entitled "*Towards a European Strategy for Security of Energy Supply*" (hereinafter the Green Paper of 2000) in which various security issues were discussed and some solutions were put in place.[34] It acknowledged that the Union's energy production is insufficient to cover its energy demand. As a result, the energy import dependence was deemed to increase constantly. One of the main consequences of this increasing import dependency highlighted in the Green Paper of 2000, are the increasing costs. According to the Green Paper of 2000, in the year 1999, 6% of the total imports into the Union were already dedicated to energy which amounted to 240 billion euros. Moreover, the enlargement of the EU to 27 Member States raised concerns. Hence, the Green Paper of 2000 sets the goals of guaranteeing "*uninterrupted physical availability of energy products on the market, at a price which is affordable for all consumers while respecting environmental concerns and looking towards sustainable development*". This Green Paper of 2000 also rightfully acknowledges that "*the European Union has very limited scope to influence the energy supply side*" and therefore, the EU should concentrate its energy policies on the demand side, in particular by promoting energy savings in buildings and transport.[35]

42. Beside promoting energy savings, the Green Paper of 2000 highlighted that the increased use of new and renewable energies could significantly reinforce sustainable and secure energy supply. Although, an increased and efficient use of these new and renewable energies will require substantial investments. To realise these substantial investments, the Green Paper of 2000 hopes for a policy which "*ranges from drastic fiscal measures in favour of renewable sources or the obligation on the part of electricity producers and distributors to purchase a minimum percentage of electricity produced from renewable sources to aid to research or to financing mechanisms*".[36] However, at that moment such policy was yet not taken seriously by most Member States.[37]

43. The Green paper of 2000 gave rise to various opinions from the Council, Member States, European Parliament, Economic and Social Committee and the

---

[33] S. S. HAGHIGHI, *Energy Security – The External Legal Relations of the European Union with Major Oil- and Gas-Supplying Countries* (Oxford, Hart Publishing, 2007), p. 156–159.

[34] Green Paper, *Towards a European Strategy for the Security of Energy Supply*, COM(2000)769.

[35] S. S. HAGHIGHI, *Energy Security – The External Legal Relations of the European Union with Major Oil- and Gas-Supplying Countries* (Oxford, Hart Publishing, 2007), p. 159–162.

[36] Green Paper, *Towards a European Strategy for the Security of Energy Supply*, COM(2000)769, p. 50.

[37] S. S. HAGHIGHI, *Energy Security – The External Legal Relations of the European Union with Major Oil- and Gas-Supplying Countries* (Oxford, Hart Publishing, 2007), p. 163.

Committee of the Regions. The common denominator of these opinions was that in the debate with respect to the security of energy supply, the EU should aim at strongly reducing energy consumption and therefore, demand management should be at the forefront. However, the adoption of only an efficient framework for demand management is insufficient.

44. In that period the externalities of a secure energy supply were not considered vital by many. For example the Resolution of the European Parliament on the Green Paper of 2000[38] revealed the attitude of the Parliament with respect to the externalities of security of energy supply. It made clear that there was a disagreement on whether or not the Union should be granted external competences in the energy field. This resulted in a lack of a proper framework for managing the import dependency of the EU.[39]

45. In 2005, the Commission adopted a report on the progress of the Green Paper of 2000 that outlined which efforts were still required to reach the objectives enumerated in the Green Paper of 2000. It stated that in the areas of energy efficiency, renewable energy sources, the use of clean coal and diversification of energy sources, additional measure and guidelines were still needed. This report also emphasised that, in order to ensure stable energy prices, dialogues should be pursued with major energy-producing countries.[40] These concerns led to the adoption of the Green Paper: *"A European Strategy for Sustainable, Competitive and Secure Energy"*[41] which will be elaborated in the next part.

C. RECENT DEVELOPMENTS IN THE EUROPEAN ENERGY POLICY

1. THE EUROPEAN CHALLENGES OF THE NEW ENERGY ERA

46. For many years we took energy for granted, assuming that it would always be there, affordable and easily accessible. However a number of events, such as the large black-out in Italy in 2003, the need to limit the forthcoming rise of global temperatures, a tripling of oil prices in 2008 and the gas disputes between

---

38 European Parliament Resolution on the Commission Green Paper Towards a European Strategy of Energy Supply, COM(2000)769 – C5–0145/2001–2001/2071(COS).

39 S. S. HAGHIGHI, *Energy Security – The External Legal Relations of the European Union with Major Oil- and Gas-Supplying Countries* (Oxford, Hart Publishing, 2007), p. 166 and 167.

40 *Ibid.*, p. 167 and 168.

41 Green Paper, *A European Strategy for Sustainable, Competitive and Secure Energy*, COM(2006)105, 20 p.

Ukraine and Russia in 2006–2009, made clear that this is no longer the case. Therefore, various documents on an international, European and national level currently emphasise the need to establish secure and sustainable energy policies and strategies. These energy policies and strategies should, from an economic perspective, address the problems of security and diversification of energy supply, the sufficient investment for secure energy delivery and, from an environmental perspective, reduce the increase of greenhouse gases.[42]

47. Since 1970, world energy consumption has almost doubled and the World Energy Outlook of the International Energy Agency foresees that the world primary energy demand will increase by 36% between 2008 and 2035.[43] The International Energy Agency expects that this growth will be fossil fuel-based of which oil and gas take one third each of this increase. Also renewables will contribute to meet the higher world energy demand, but on the other hand the share of nuclear energy will decline.[44]

48. The 2009 update of the energy trends for 2030[45], foresees in a slight growth of primary energy consumption in the European Union between today and 2030 according to the so-called baseline scenario. In the overall energy mix, the share of oil and coal declines, while the gas and nuclear demand remain largely stable. The share of renewables is set to increase significantly. The import dependency of the European Union will continue to be high for both oil and coal and will increase mainly due to the depletion of indigenous resources, to reach about 73–79% of consumption by 2020 and 81–89% by 2030.[46]

49. During the informal summit of 27 September 2005 at Hampton Court, the EU Heads of State and Government discussed the call of the British Prime Minister Tony Blair to create a better coordinated energy policy on a European level to tackle climate change, improve security of supply and to react to oil price increases. Subsequent summits in October and December 2005, then fully recognised the need for an integrated long-term energy policy.

---

42 Http://ec.europa.eu/energy/index_en.htm.

43 This means from around 12300 million tonnes of oil equivalent (hereinafter Mtoe) to over 16700 Mtoe or 1.2% per year on average. See IEA, *World Energy Outlook 2010*, November 2010, p. 77–101.

44 Commission staff working document, *Annex to the Green Paper, A European strategy for sustainable, competitive and secure energy – what is at stake – background document*, COM(2006)105, p. 3–6.

45 Http://ec.europa.eu/energy/observatory/trends_2030/doc/trends_to_2030_update_2009.pdf.

46 Communication from the Commission of 17 November 2010 to the European Parliament, the Council, the European Economic and Social Committee and the Committee of the Regions, *Energy infrastructure priorities for 2020 and beyond – A Blueprint for an integrated European energy network*, COM(2010)677, p. 19–21.

50. In March 2006, this led to the adoption by the Commission of the Green Paper: "*A European Strategy for Sustainable, Competitive and Secure Energy* (hereinafter: "Green Energy Paper of 2006")". The Green Energy Paper of 2006 has set out on the one hand, the new energy realities facing the EU and, on the other, the basis of an integrated European energy policy.[47]

2. THE EUROPEAN RESPONSE TO THIS NEW ENERGY ERA

(a) *Green Energy Paper of 2006*

51. The three core objectives of the EU energy policy to date and in the long term are a sustainable, competitive and secure energy supply. These objectives are complementary goals and therefore, each aspect of the integrated energy policy needs to contribute to all three. In order to achieve these objectives, the European Commission has defined six priority areas and concrete measures for each of these priority areas in the Green Energy Paper of 2006[48]:

(a) *completing the internal European electricity and gas markets* with concrete measures such as creating a single European grid for both electricity and gas, a priority interconnection plan, a level playing field, boosting the competitiveness of European industry and investments in generation capacity;
(b) *ensuring the security of supply and the solidarity between Member States* with concrete measures such as the establishment of a European Energy Supply Observatory, improving the network security through increased collaboration and exchange of information between transmission system operators, improving the physical security of infrastructure, etc.;
(c) *towards a more sustainable, efficient and diverse energy mix* with concrete measures such as a Strategic EU Energy Review which should allow a transparent and objective debate on the various energy sources;
(d) *an integrated approach to tackle climate change* with concrete measures such as the adoption of an Action Plan on Energy Efficiency to save 20% of the EU primary energy use by 2020 compared to the BAU scenario, to bring forward a Renewable Energy Road Map and to stimulate carbon capture and geological storage;
(e) *encouraging innovation* through the adoption of a strategic European energy technology plan and
(f) *towards a coherent external energy policy* with concrete measures such as the adoption of a clear policy on securing and diversifying energy supplies,

---

47 Green Paper, *A European Strategy for Sustainable, Competitive and Secure Energy*, COM(2006)105, 20 p.

48 *Ibid.*

establishing dialogues with major energy producers and suppliers, reacting effectively to external energy crisis situations, etc.

52. With the adoption of these six priority areas, the European Commission started its negotiations on a European level in order to establish a common European energy strategy based on three pillars, sustainability, competitiveness and security of supply.

(b) 2006 and 2011 Energy Efficiency Action Plans

53. On 19 October 2006, the European Commission adopted, next to the Green Energy Paper of 2006, an Energy Efficiency Action Plan.[49] This action plan contains various measures in order to put the EU on the path to achieve a 20% energy efficiency objective, equivalent to some 390 Mtoe, compared to a business-as-usual-scenario, by 2020. If it is successful, by 2020 the CO_2 emissions should be reduced each year by 780 Mt CO_2 with respect to the baseline scenario and lead to fuel savings of more than € 100 billions annually. The action plan lists a range of priority actions to be initiated immediately and other actions to be initiated gradually during the six year period of the plan.[50]

54. In order to realise this energy efficiency potential a significant shift in energy consumption is required. The behavioural patterns will have to change, using less primary energy while enjoying the same quality of life. Hence, the EU should

---

49 Communication from the Commission of 19 October 2006, *Action Plan for Energy Efficiency: Realising the Potential*, COM(2006)545, 25 p. (hereinafter Energy Efficiency Action Plan). This Energy Efficiency Action Plan and other policy initiatives on a European level has led to the adoption of for example: Directive 2010/31/EU of the European Parliament and of the Council of 19 May 2010 on the energy performance of buildings, *OJ* L 153/13; Directive 2010/30/EU of the European Parliament and of the Council of 19 May 2010 on the indication by labelling and standard product information of the consumption of energy and other resources by energy-related products, *OJ* L 153/1; Regulation (EC) No 1222/2009 of the European Parliament and of the Council of 25 November 2009 on the labelling of tyres with respect to fuel efficiency and other essential parameters, *OJ* L 342/46; Regulation (EC) No 106/2008 of the European Parliament and of the Council of 15 January 2008 on a Community energy-efficiency labelling programme for office equipment, *OJ* L 39/1; Council Decision of 18 December 2006 concerning conclusion of the Agreement between the Government of the United States of America and the European Community on the coordination of energy-efficiency labelling programmes for office equipment, *OJ* L 381/24; Commission Regulation (EU) No 1016/2010 of 10 November 2010 implementing Directive 2009/125/EC of the European Parliament and of the Council with regard to ecodesign requirements for household dishwashers, *OJ* L 293/31; Commission Regulation (EU) No 1015/2010 of 10 November 2010 implementing Directive 2009/125/EC of the European Parliament and of the Council with regard to ecodesign requirements for household washing machines, *OJ* L 293/21; Commission Regulation (EC) No 859/2009 of 18 September 2009 amending Regulation (EC) No 244/2009 as regards the ecodesign requirements on ultraviolet radiation of non-directional household lamps, *OJ* L 247/3, etc.

50 Communication from the Commission of 19 October 2006, *Action Plan for Energy Efficiency: Realising the Potential*, COM(2006)545, p. 3 and 4.

encourage producers to develop more energy-efficient products and technologies, and consumers need stronger incentives to buy such energy-efficient products.

55. To reach the full potential by 2020, further action is required. The European Commission scheduled for 2009 a mid-term review of the 2006 Energy Efficiency Action Plan.[51]

56. On 8 March 2011, the Commission adopted its Energy Efficiency Plan 2011[52] which highlighted that energy efficiency is one of the most cost-effective ways to enhance security of energy supply and to reduce emissions of greenhouse gases and other pollutants. It acknowledged that substantial steps have been taken in order to achieve a 20% energy efficiency objective by 2020. However, recent Commission estimations suggest that the EU is on course to achieve only half of this 20% objective. This plan therefore will try to reinforce the EU's efforts in the energy efficiency field and will focus on the building, industry and transport sectors.[53] The latter is addressed in the White Paper on transport in which the Commission adopted a roadmap of 40 initiatives to build a competitive transport system for the next decade. This roadmap will, *inter alia*, increase mobility, employment, but also reduce Europe's dependence on imported oil dramatically and cut greenhouse gases by at least 60% in transport by 2050.[54]

57. The Energy Efficiency Plan 2001 has outlined various measures which ensure the full achievement of the 20% objective, *inter alia*, high standards of energy efficiency should systematically be applied when public authorities purchase goods, services and works; legislative proposals in order to overcome the ambiguities experienced with the deployment of energy performance contracting; the creation of a network of smart cities to pioneer new technologies on a the municipal level; address heat consumption in buildings by exploring the range of available solutions and the promotion of the use of district heating; a European training strategy for energy efficiency workforce; exploring ways to tackle the effective recovery of heat losses from electricity and industrial processes; improving national and European financial support for energy efficiency investments; promoting energy-and resource-efficient appliances; etc.

---

51 See also the European Council conclusions of 4 February 2011, p. 3. According to these conclusions, the Council is invited to examine the upcoming proposal for a new Energy Efficiency Plan of the European Commission. Moreover, the Council will review the implementation of the EU energy efficiency target by 2013 and consider whether further measures are necessary.

52 Communication of the Commission to the European Parliament, the Council, the European Economic and Social Committee and the Committee of the Regions of 8 March 2011, Energy Efficiency Plan 2011, COM(2011)109, 15 p.

53 *Ibid.*, p. 2 and 3.

54 White Paper of 28 March 2011, *Roadmap to a Single European Transport Area – Towards a competitive and resource transport efficient transport system*, COM(2011)144, 30 p.

58. The revised plan acknowledged that targets for energy efficiency are an effective way to create political willingness and trigger action. Contrary to the request of the Members of the European Parliament's Industry Committee (hereinafter ITRE) to adopt a legally binding target on energy efficiency in its revised plan[55], the Commission proposed a two-step approach to target setting. As a first step, Member States are currently setting indicative national energy efficiency targets and programmes. The Commission will closely monitor their implementation. In 2013, it will assess whether these indicative targets and programmes will deliver the 20% objective. If the reviews in 2013 show that the overall EU target is unlikely to be achieved, then, as second step, the Commission will propose legally binding national energy efficiency targets for 2020.

59. This approach of the Commission is not surprising since it is technically more difficult to administer an energy efficiency target than, for example, a renewables target because it is not clear how efficiency can be measured. Moreover, the approach of the Commission is in accordance with the decision taken by the European Council at its meeting of 4 February 2011. The Council will review the implementation of the EU energy efficiency target by 2013.[56]

(c) Energy sector inquiry

60. Beside the adoption of the Green Energy Paper and the 2006 and 2011 Energy Efficiency Action Plans, the Commission has launched an inquiry into competition of the gas and electricity markets in June 2005, pursuant to Article 17 of the Regulation 1/2003/EC.[57, 58]

61. The energy sector inquiry responded to concerns voiced by consumers and new entrants on dramatic rises in electricity and gas wholesale prices, which could not be fully explained by higher primary fuel costs and environmental obligations, and persistent complaints about barriers to entry and limited choice for consumers. With this inquiry, the Commission aimed at assessing the competition conditions in the gas and electricity markets, and establishing whether the indications of market malfunctions result from breaches of competition law provisions. The energy sector inquiry was launched on 17 June 2005 and led to various initial findings presented in the Issues Paper[59] on 15 November 2005.

---

[55] Press Release ITRE of 9 November 2010, *Energy Efficiency Action Plan: focusing on efficient buildings and eco-design.*

[56] European Council conclusions of 4 February 2011, p. 3.

[57] Regulation (EC) No 1/2003 of 16 December 2002 on the implementation of the rules on competition laid down in Articles 81 and 82 of the Treaty, *OJ* L1/1 (hereinafter Regulation 1/2003/EC) and see for further details with respect to the energy sector inquiry http://ec.europa.eu/competition/sectors/energy/inquiry/index.html.

[58] Http://ec.europa.eu/comm/competition/sectors/energy/inquiry/index.html.

[59] European Commission Competition DG, *Energy sector inquiry – issues paper non-confidential version of 15 November 2005* http://ec.europa.eu/comm/competition/sectors/energy/inquiry/issues_paper 15112005.pdf.

62. On 10 January 2007 the Commission has published its final report which identified a number of serious shortcomings such as too much market concentration in most national markets, vertical foreclosure, too little integration between Member States' markets, lack of transparency, lack of competition on a retail level through long-term contracts, balancing markets which favour incumbents, lack of effective and transparent time information.

63. To address these shortcomings in the electricity and gas markets, the Commission suggested in its report to apply both competition (merger control, anti-trust and state aids) and regulatory-based remedies.[60] The Commission correctly argued that the existing competition law provisions can make a significant contribution with respect to the liberalisation process. However, these provisions cannot by itself open the electricity and gas markets and resolve all the shortcomings identified in both markets by the energy sector inquiry. Therefore, various regulatory measures are essential in order to open up the European gas and electricity markets and to create a single European energy market.[61] Beside the adoption of the final report of the inquiry, on 10 January 2007 the Commission adopted a comprehensive package of measures which propose a new Energy Policy for Europe.

(d) The New Energy Policy for Europe

64. In its Communication of 10 January 2007 entitled *"An Energy Policy for Europe"*, the Commission has put forward an integrated package of measures underpinned by:

(a) an EU objective in international negotiations of 30% reduction of greenhouse gas emissions by developed countries by 2020 compared to 1990. In addition, 2050 global greenhouse gas emissions must be reduced by up to 50% compared to 1990, implying reductions in industrialised countries of 60–80% by 2050 and
(b) an EU commitment to achieve, in any event, at least a 20% reduction of greenhouse gases by 2020 compared to 1990.

65. In order to achieve this strategic objective, Europe must be transformed into a highly energy-efficient and low CO_2 energy economy. This strategic objective and the integrated package of measures in the Communication of 10 January 2007 represent the core of the new European Energy Policy.[62]

---

60 Communication from the Commission of 10 January 2007, *Inquiry pursuant to Article 17 of Regulation (EC) No 1/2003 into the European gas and electricity sectors (Final Report)*, COM(2006)851, 15 p.

61 *Ibid.*, p. 9.

62 Communication from the Commission to the European Council and the European Parliament of 10 January 2007, *An energy policy for Europe*, COM(2007)1 and http://ec.europa.eu/energy/energy_policy/index_en. htm.

66. The European Spring Council of 8 and 9 March 2007 agreed on this new European Energy Policy and invited the Commission to come forward with concrete measures such as the effective separation of supply and generation activities from network operation, the further harmonisation of the powers and enhanced independence of the national regulators, greater transparency in energy market operations and also to achieve the following targets:

(a) an independent EU commitment to achieve at least a 20% reduction of greenhouse gases by 2020 compared to 1990 levels and an objective for a 30% reduction by 2020 subject to the conclusion of a comprehensive international climate change agreement and
(b) a mandatory EU target of 20% renewable energy by 2020 including a 10% biofuels target.[63]

67. Not only the European Spring Council, but also the European Parliament in its Resolution of 10 July 2007 on "*Prospects for the internal gas and electricity market*" expressed a clear and strong political support in favour of a common energy policy. The European Parliament stated in particular that "*transmission ownership unbundling is the most effective tool to promote investments in infrastructures in a non-discriminatory way, fair access to the grid for new entrants and transparency in the market*". It furthermore called for enhancing "*cooperation between national regulators at EU level, through an EU entity, as a way to promote a more European approach to regulation on cross-border issues*" and emphasised that "*increased convergence and harmonisation of the competences of national regulators is essential to overcome technical and regulatory differences that impose serious barriers to cross-border trade and interconnections*".

68. The European Parliament also underlined that different measures on a European level are necessary with respect to consumer protection, interconnections, electricity and gas networks, etc. It also calls on Member States to end the application of generalised regulated tariffs gradually, to facilitate an increase in grid capacity in order to allow the incorporation of new massive onshore and offshore renewable energy generation, to facilitate and limit the duration of authorisation procedures for the building of interconnection lines, etc.[64]

69. As a response to these calls of the European Spring Council and European Parliament, the Commission adopted the Third Package of Legislative Proposals on 19 September 2007 amending the existing gas and electricity directives and

---

63 Council of the European Union of Brussels, Brussels European Council of 8/9 March 2007, 2 May 2007, 7224/1/07 REV 1 and www.consilium.europa.eu/ueDocs/cms_Data/docs/pressData/en/ec/93135.pdf.

64 European Parliament resolution of 10 July 2007 on prospects for the internal gas and electricity market (2007/2089(INI)).

regulations, and a draft regulation establishing an agency for the cooperation of energy regulators.[65] These legislative proposals contained a number of important provisions. However, the central features of the Third Package of Legislative Proposals concerned further unbundling of network operations from supply and generation activities, coordination of regulators on an EU level, enhancing the role of national regulators, increasing TSO coordination, consumer protection and increasing transparency.[66] The Third Package was adopted in a second reading by the European Parliament on 21 April 2009. Subsequently, after final adoption by the Council, the 2 new directives and 3 regulations were adopted on 13 July 2009.[67, 68]

(e) *Strategic Energy Technology Plan*

70. In November 2007 the European Commission also presented its Strategic Energy Technology (hereinafter SET) Plan. This Plan, adopted in March 2008,

---

65 Proposal for a Directive of the European Parliament and of the Council amending Directive 2003/54/EC of the European Parliament and of the Council of 26 June 2003 concerning common rules for the internal market in electricity, COM(2007)528; Proposal for a Directive of the European Parliament and of the Council amending Directive 2003/55/EC of the European Parliament and of the Council of 26 June 2003 concerning common rules for the internal market in natural gas, COM(2007)529; Proposal for a Regulation of the European Parliament and of the Council establishing an Agency for the Cooperation of Energy Regulators, COM(2007)530; Proposal for a Regulation of the European Parliament and of the Council amending Regulation (EC) No 1228/30 on conditions for access to the network for cross-border exchanges in electricity, COM(2007)531 and Proposal for a Regulation of the European Parliament and of the Council amending Regulation (EC) No 1775/2005 on conditions for access to the natural gas transmission networks, COM(2007)532.

66 M. HUNT, "EU Energy Policy in 2007–2008" in N. HUNT and K. TALUS (eds.), *The EU Energy Directory* (Rixensart, Euroconfidentiel, 2008), p. 11–21.

67 Directive 2009/72/EC of the European Parliament and of the Council of 13 July 2009 concerning common rules for the internal market in electricity and repealing Directive 2003/54/EC, *OJ* L 211/55; Directive 2009/73/EC of the European Parliament and of the Council of 13 July 2009 concerning common rules for the internal market in natural gas and repealing Directive 2003/55/EC, *OJ* L 211/94; Regulation (EC) No 713/2009 of the European Parliament and of the Council of 13 July 2009 establishing an Agency for the Cooperation of Energy Regulators, *OJ* L 211/1; Regulation (EC) No 714/2009 of the European Parliament and of the Council of 13 July 2009 on conditions for access to the network for cross-border exchanges in electricity and repealing Regulation (EC) No 1228/2003, *OJ* L 211/15 and Regulation (EC) No 715/2009 of the European Parliament and of the Council of 13 July 2009 on conditions for access to the natural gas transmission networks and repealing Regulation (EC) No 1775/2005, *OJ* L 211/36.

68 For a further analysis of these provisions see B. DELVAUX, "Recente ontwikkelingen in het Europees energierecht en –beleid – Is de Sagrada Familia eindelijk af?" in K. DEKETELAERE en B. DELVAUX (eds.), *Jaarboek Energierecht 2007* (Antwerpen, Intersentia, 2008), p. 1–33; B. DELVAUX en T. VANDEN BORRE, "Recente ontwikkelingen in het Europese Energierecht en –beleid – Heel wat lekkers in de Europese pijplijn" in K. DEKETELAERE en B. DELVAUX (eds.), *Jaarboek Energierecht 2008* (Antwerpen, Intersentia, 2009), p. 1–44; B. DELVAUX, M. HUNT and K. TALUS (eds.), *EU Energy Law and Policy Issues* (Rixensart, Euroconfidentiel, 2008), 298 p. and B. DELVAUX, M. HUNT and K. TALUS (eds.), *EU Energy Law and Policy Issues* (Rixensart, Euroconfidentiel, 2009), 360 p.

contains the new Union approach to accelerate the development and deployment of cost-effective low-carbon technologies which are crucial to achieve the 20–20–20 goals by 2020.[69]

(f) Energy and Climate Package

71. On 23 January 2008, the Commission issued its Energy and Climate Package[70] which in pursuance of the European Council's request provided for a proposal for a directive on the promotion of renewable energy[71], a proposal for a directive amending the EU Emissions Trading Directive[72] and a proposal for a decision on the effort of Member States to reduce their greenhouse gas emissions in sectors not covered by the EU emission trading scheme such as transport, buildings, services, smaller industrial installations, agriculture and waste.[73] The European Parliament adopted this package on 17 December 2008, after which the Council gave its final approval on 6 April 2009. As a result, a number of legislative acts were adopted on a European level.[74]

---

69 Communication from the Commission of 22 November 2007 to the Council, the European Parliament, the European Economic and Social Committee and the Committee of the Regions, *A European strategic energy technology plan (SET-plan) – Towards a low carbon future*, COM(2007)723.

70 Communication from the Commission to the European Parliament, the Council, the European Economic and Social Committee and the Committee of the Regions – *2020 by 2020 – Europe's climate change opportunity*, COM(2008)30.

71 Proposal for a Directive of the European Parliament and of the Council on the promotion of the use of energy from renewable sources, COM(2008)19.

72 Proposal for a Directive of the European Parliament and of the Council amending Directive 2003/87/EC so as to improve and extend the greenhouse gas emission allowance trading system of the Community, COM(2008)16.

73 Proposal for a Decision of the European Parliament and of the Council on the effort of Member States to reduce their greenhouse gas emissions to meet the Community's greenhouse gas emission reduction commitments up to 2020, COM(2008)17. See also the Communication from the Commission to the Council and the European Parliament on a first assessment of national energy efficiency action plans as required by Directive 2006/32/EC on energy end-use efficiency and energy services – Moving forward together on energy efficiency, COM(2008)11, 16 p.; Communication from the Commission to the European Parliament, the Council, the European Economic and Social Committee and the Committee of the Regions – *Supporting early demonstration of sustainable power generation from fossil fuels*, COM(2008)13, 11 p. and Community guidelines of 1 April 2008 on State aid for environmental protection, *OJ* C 82/1.

74 Directive 2009/28/EC of the European Parliament and of the Council of 23 April 2009 on the promotion of the use of energy from renewable sources and amending and subsequently repealing Directives 2001/77/EC and 2003/30/EC (hereinafter Renewable Directive 2009/28/EC), *OJ* L 140/16; Directive 2009/29/EC of the European Parliament and of the Council of 23 April 2009 amending Directive 2003/87/EC so as to improve and extend the greenhouse gas emission allowance trading scheme of the Community, *OJ* L 140/63; Directive 2009/31/EC of the European Parliament and of the Council of 23 April 2009 on the geological storage of carbon dioxide and amending Council Directive 85/337/EEC, European Parliament and Council Directives 2000/60/EC, 2001/80/EC, 2004/35/EC, 2006/12/EC, 2008/1/EC and Regulation (EC) No 1013/2006, *OJ* L 140/114 and Decision No 406/2009/EC of the European

(g) Second Strategic Energy Review

72. These measures were not the only energy policy initiatives on a European level. In addition, the European Commission adopted in its second Strategic Energy Review an action plan to strengthen the EU energy security and solidarity.[75, 76] This action plan focuses on five points among which the promotion of infrastructure essential to the EU's energy needs and diversification of energy supplies, a greater focus on energy in the EU's international relations, improved oil and gas stocks and crises response mechanisms, a new impetus on energy efficiency and using the EU's indigenous energy reserves much better. In addition, this Strategic Energy Review contains the first steps towards a vision for 2050.

73. The EU has currently set out the essential first steps towards the 2020 goals. However, if the EU chooses for radical changes such as zero carbon electricity generation or a fundamental technology shift (e.g. breaking the dependency on oil in its transport sector), such changes require more time and choices to be made by policymakers, investors, scientists and educational institutions. Therefore, the Commission has decided to start with the charting of a policy agenda for 2030 and a vision for 2050.[77]

(h) European Energy Programme for Recovery

74. Another recent initiative on a European level is the agreement of July 2009 on the European Energy Programme for Recovery.[78] This Recovery Programme should speed up and secure investments in infrastructure and technology projects in the energy sector. This programme has allocated € 3.98 billion to

Parliament and of the Council of 23 April 2009 on the effort of Member States to reduce their greenhouse gas emissions to meet the Community's greenhouse gas emission reduction commitments up to 2020, *OJ* L 140/136.

75 Communication from the Commission of 13 November 2008 to the European Parliament, the Council, the European Economic and Social Committee and the Committee of the Regions, *Second Strategic Energy Review: an EU energy security and solidarity action plan*, COM(2008)781, 20 p.

76 B. DELVAUX en T. VANDEN BORRE, "Recente ontwikkelingen in het Europese Energierecht en -beleid – Heel wat lekkers in de Europese pijplijn" in K. DEKETELAERE en B. DELVAUX (eds.), *Jaarboek Energierecht 2008* (Antwerpen, Intersentia, 2009), p. 26–40.

77 Communication from the Commission of 13 November 2008 to the European Parliament, the Council, the European Economic and Social Committee and the Committee of the Regions, *Second Strategic Energy Review: an EU energy security and solidarity action plan*, COM(2008)781, p. 2–18.

78 Communication from the Commission to the European Council – *A European Economic Recovery Plan*, COM(2008)800, 19 p. and Report from the Commission to the Council and the European Parliament on the implementation of the European Energy Programme for Recovery, COM(2010)191, 9 p.

finance mature energy infrastructure and technology projects (wind energy and CCS) during 2010 and 2011.[79]

75. Its main objective was to help overcome possible financial difficulties faced by mature projects due to the economic and financial crises.[80] In this context, reference can also be made to the amended Regulation on the European Regional Development Fund which extended its scope for sustainable energy investments in buildings. Not only energy efficiency investments in public and commercial buildings can be financed by this fund, but also energy efficiency investments in the residential sector.[81]

(i) Energy infrastructure priorities for 2020 and beyond

76. Other energy policy initiatives of the Commission are those on the Energy 2020 and the energy infrastructure priorities for 2020 and beyond. The Union's energy infrastructure is the central nervous system of our economy. The Commission stressed that without a major shift in the way the European infrastructure is developed, the EU energy goals will not be achieved. Not only for the EU energy goals, but also for the EU's economic strategy it's a crucial prerequisite that our energy infrastructures and networks are adequate, integrated and reliable.[82]

---

79 Regulation (EC) No 663/2009 of the European Parliament and of the Council of 13 July 2009 establishing a programme to aid economic recovery by granting Community financial assistance to projects in the field of energy, *OJ* L 200/31.

80 Commission staff working document of 17 November 2010 impact assessment accompanying document to the Communication from the Commission to the European Parliament, the Council, the European Economic and Social Committee and the Committee of the Regions, *Energy infrastructure priorities for 2020 and beyond – A Blueprint for an integrated European energy network*, SEC(2010)1395, p. 10.

81 Regulation (EU) No 539/2010 of the European Parliament and of the Council of 16 June 2010 amending Council Regulation (EC) No 1083/2006 laying down general provisions on the European Regional Development Fund, the European Social Fund and the Cohesion Fund as regards simplification of certain requirements and as regards certain provisions relating to financial management, *OJ* L 158/1. This amendment means that up to 4% of the national European Regional Development Fund allocations are available for energy efficiency investments in the residential sector. Commission staff working document of 10 November 2010, *State of play in the EU energy policy* accompanying document in the Communication from the Commission to the Council, the European Parliament, the European Economic and Social Committee and the Committee of the Regions, *Energy 2020 A Strategy for competitive, sustainable and secure energy*, SEC(2010)1346, p. 2 and 3.

82 Communication from the Commission of 17 November 2010 to the European Parliament, the Council, the European Economic and Social Committee and the Committee of the Regions, *Energy infrastructure priorities for 2020 and beyond – A Blueprint for an integrated European energy network*, COM(2010)677, p. 4 and 5. See also the Commission staff working document of 17 November 2010 impact assessment accompanying document to the Communication from the Commission to the European Parliament, the Council, the European Economic and Social Committee and the Committee of the Regions, *Energy*

77. The development of the European energy infrastructure and networks[83] will not only enable the Union to realise its properly functioning internal electricity and gas market, but also enhance security of supply, increase energy efficiency, enable the integration of renewable energy sources and allow consumers to benefit from the liberalisation process of the electricity and gas markets, new technologies and intelligent energy use.

78. To enable the EU to reap the benefits of an integrated European grid, the Commission adopted a blueprint which aims to provide the EU with a vision of what is needed to make our energy infrastructure and networks efficient. In addition, to identify the obstacles and issues which need to be addressed such as long and uncertain permitting procedures, the financing of energy networks, etc., the Commission proposed short term and longer term priorities to make our energy infrastructure suitable for the 21st century[84], and a toolbox to speed up implementation.[85]

---

infrastructure priorities for 2020 and beyond – A Blueprint for an integrated European energy network, SEC(2010)1395, p. 57.

83 According to the Primes model calculations around one trillion euros must be invested in our energy system between today and 2020 in order to meet the 2020 goals. Half of it will be required for networks, including electricity and gas distribution and transmission, storage and smart grids. For the energy transmission networks about 200 bn Euros will be needed of which 50% will be taken up by the market by 2020. Consequently, there will be a gap of about 100 bn euros. For further details see the Commission staff working document of 17 November 2010 impact assessment accompanying document to the Communication from the Commission to the European Parliament, the Council, the European Economic and Social Committee and the Committee of the Regions, *Energy infrastructure priorities for 2020 and beyond – A Blueprint for an integrated European energy network*, SEC(2010)1395, 57 p.

84 The short-term priorities for electricity, gas and oil are for example the offshore grid in the Northern Seas and connection as well as central Europe; interconnections in southwestern Europe, Connections in central eastern and southeastern Europe; completion of the Baltic energy market interconnection plan; the Southern corridor to diversify sources on an EU level further and to bring gas to the Union from the Caspian Basin, Central Asia and the Middle East; linking the Baltic, Black, Adriatic and Aegean Seas; reinforcing the interoperability of the central eastern European pipeline network; provide the necessary framework and initial incentives for rapid investments in a new intelligent network infrastructure. The longer-term priorities are for example the accommodating of ever-increasing wind surplus generation in and around the Northern and Baltic seas and increasing renewable generation in the east and south of Europe and also North Africa; connecting new generation hubs with major storage capacities in Nordic countries and the Alps; examination and agreement on the technical and practical modalities of a future CO_2 transport infrastructure. For further details see the Communication from the Commission of 17 November 2010 to the European Parliament, the Council, the European Economic and Social Committee and the Committee of the Regions, *Energy infrastructure priorities for 2020 and beyond – A Blueprint for an integrated European energy network*, COM(2010)677, p. 10–13 and in particular its Annex, p. 25–45.

85 The toolbox contains measures such as reinforcing the existing Regional Initiatives; faster and more transparent permit granting procedures; better methods and information for decision-makers and citizens and a stable framework for financing.

79. On the basis of this blueprint and the views expressed by the energy sector and institutions on this blueprint, the Commission intends to prepare appropriate initiatives (e.g. through energy security and infrastructure instruments and the mainstreaming of energy priorities in different programmes) in 2011. These initiatives will address the current regulatory and financial issues faced by our energy infrastructure and networks.[86]

(j) A roadmap for moving to a competitive low-carbon economy in 2050

80. The Communication of the Commission of 8 March 2011 on a roadmap for moving to a competitive low-carbon economy in 2050[87] presents the possible action to be taken up to 2050 which could enable the Union to deliver greenhouse gas emission reductions in line with the 80 to 95% objective. This objective of reducing greenhouse gas emissions by 80 to 95% has been reconfirmed by European Council at its meeting of 4 February 2011. Together with the Energy Efficiency Plan 2011[88] and the white paper on transport[89] this communication is a key deliverable under the Europe 2020 flagship initiative for a resource-efficient Europe.[90]

81. The roadmap indicates that in order to be in line with the 80 to 95% overall greenhouse gas emission reductions by 2050, a 40% domestic reduction compared to 1990 is required as a milestone for 2030 and 80% for 2050. This communication states that if the EU wants to move in this direction, it should start working on appropriate strategies now. Not only the EU but also the Member States should develop national low-carbon roadmaps soon, if it has not already been done.

82. For key sectors such as power, industry, transport, residence, services and agriculture, it highlights ranges for emission reductions for 2030 and for 2050. Moreover, the roadmap emphasises that electricity will play a central role in achieving a low-carbon economy. In both the transport and heating

---

86 Communication from the Commission of 17 November 2010 to the European Parliament, the Council, the European Economic and Social Committee and the Committee of the Regions, *Energy infrastructure priorities for 2020 and beyond – A Blueprint for an integrated European energy network*, COM(2010)677, p. 4–17.

87 Communication from the Commission of 17 November 2010 to the European Parliament, the Council, the European Economic and Social Committee and the Committee of the Regions, *A Roadmap for moving to a competitive low-carbon economy in 2050*, COM(2011)112, 15 p.

88 See *Supra*.

89 White Paper of 28 March 2011, *Roadmap to a Single European Transport Area – Towards a competitive and resource transport efficient transport system*, COM(2011)144, 30 p.

90 Communication from the Commission to the European Parliament, the Council, the European Economic and Social Committee of the Regions, *A Resource-efficient Europe – Flagship initiative under the Europe 2020 Strategy*, COM(2011)21, 16 p.

sector, electricity will increasingly be used, however its overall consumption would only have to continue to increase at historic growth race due to the continuous improvements in efficiency. This roadmap will be used as a basis for developing sector-specific initiatives and roadmaps in the coming months and years.[91]

(k) Energy roadmap 2050

83. In order to achieve the goal of cutting emissions by over 80% by 2050, discussed above, the Energy roadmap 2050 of 15 December 2011[92] explores the challenges posed by delivering the EU's decarbonisation objective while at the same time ensuring security of energy supply and competiveness. It describes on the basis of a set of scenarios the consequences of a carbon free energy system and the long-term European policy and technology-neutral framework which is needed. It shows, according to the Commission, that the decarbonisation objective is feasible. On the basis of this roadmap the Commission will continue to bring forward initiatives in the energy field.[93]

D. THE THREE KEY OBJECTIVES OF THE EUROPEAN ENERGY POLICY

1. INTRODUCTION

84. As from the moment states started to impose conditions on the extraction and use of fossil fuels, the need for specific energy policies gradually arose. In the early nineteenth century, the energy-related policies of states mainly concerned the granting of concessions, taxation, but also in some cases, pollution. With the increased use of fossil fuels, safety and environmental considerations also became more important.

85. Over the past ten years, a common EU energy policy has evolved around the common objective to ensure for all consumers, private and industrial, the uninterrupted physical availability of energy products and services on the market at affordable prices, while contributing to the Union's security of supply and sustainability goals. As discussed above, the Union has a relatively well-developed set of policy documents at this moment which suggests

---

91 *Ibid.*, p. 4–15.

92 Communication from the Commission to the European Parliament, the Council, the European Economic and Social Committee and the Committee of the Regions, *Energy Roadmap 2050*, COM(2011)885, 19 p.

93 *Ibid.*

various measures to achieve EU energy policy goals.[94] The purpose of the next part is to have a closer look at the three central goals of the current EU energy policy.

2. LIBERALISATION PROCESS OF THE INTERNAL ENERGY MARKET

86. Since the second half of the twentieth century *"access to energy at reasonable prices"* became a policy objective. Access to electricity and gas in a non-discriminatory manner for all Europeans became one of the basic features of the European energy policy.

87. One of the means to achieve this objective is the creation of an EU internal market in electricity and gas. For a considerable amount of time, the EU has been promoting free trade and competition in a wide European energy market assuming that the achievement of a fully liberalised single market for electricity and gas will exert downward pressure on prices and ensure security of supply. This aim of competitiveness has been pursued through the enforcement of the primary Treaty provisions on free movement and competition as well as with the adoption of various acts of secondary legislation.[95]

88. Energy companies have grown beyond national borders, but their development remains hampered by a host of various national practices and measures. Consequently, the EU is far from being able to guarantee that any energy company has the right to sell gas and electricity in any Member State on equal terms with the existing national companies, without discrimination or disadvantage.

89. A recent study on the functioning of retail electricity markets for EU consumers published by the European Commission concluded that EU consumers do not make full use of the saving opportunities of energy market liberalisation.[96] The study shows that EU consumers could save about € 13 billion in total if they switched to the cheapest electricity tariff they could find. Consumer protection has significantly improved in the past. However, different issues such as lack of information, billing information, green tariffs and peak/

---

[94] Policy Proposal by Jacques Delors, *Towards a European Energy Community: A Policy proposal*, study by S. ANDOURA, L. HANCHER and M. VAN DER WOUDE, p. 27–60.

[95] *Ibid.*, p. 17–28.

[96] Study of November 2010, *The functioning of the retail electricity markets for consumers in the European Union*, ECME Consortium, EAHC/FWC/2009 86 01, 560 p.

off-peak tariffs, dissatisfaction with complaint handling, etc. remain outstanding and need to be solved.[97]

90. The financial and economic crisis also had a significant impact on the internal market. This crisis has put pressure on planned investments due to possible financing difficulties and uncertainties affecting the supply side and resulted in a fall in the consumption of electricity and gas which is more pronounced for gas than for electricity.[98]

91. The positive element of the crisis was that it created new opportunities for competition because on liquid hubs more gas[99] came available at lower prices. Moreover, the situation with excess supply in gas markets paved the way for final consumers to see gas prices that are based on the supply and demand for gas rather than on the price of oil. This lower consumption had an impact on international oil and gas prices and on end user prices. Although the fall witnessed in wholesale energy costs has not been entirely reflected in end user prices of electricity and gas.[100]

3. SUSTAINABILITY AND CLIMATE CHANGE CHALLENGE

92. The most recent key issue of European energy policy is the sustainability challenge and in particular, the *"climate change challenge"*, the environmental dimension of which has been high on the political agenda for some years now. The climate change challenge is certainly an environmental challenge, however, it

97 Communication from the Commission of 10 November 2010 to the Council, the European Parliament, the European Economic and Social Committee and the Committee of the Regions, *Energy 2020 A Strategy for competitive, sustainable and secure energy*, COM(2010)639, p. 2 and see for further details Commission staff working document of 11 November 2010, *The functioning of the retail electricity markets for consumers in the European Union*, SEC(2010)1409, 38 p. and Commission staff working paper of 11 November 2010, *An energy policy for consumers*, SEC(2010)1407, 22 p.

98 For detailed analyses see the quarterly reports on electricity and gas markets of the Market Observatory for Energy of the European Commission accessible at the following address: http://ec.europa.eu/energy/observatory/gas/gas_en.htm.

99 It should be noted that not only the financial and economic crisis led to more gas in that period other reasons are the increased supply of unconventional gas to Europe and the launch of a new LNG terminal project in Qatar.

100 Communication from the Commission of 11 March 2010 to the Council and the European Parliament, *Report on progress in creating the internal gas and electricity market*, COM(2010)84, 14 p.; Commission staff working document of 10 November 2010, *State of play in the EU energy policy* accompanying document in the Communication from the Commission to the Council, the European Parliament, the European Economic and Social Committee and the Committee of the Regions, *Energy 2020 A Strategy for competitive, sustainable and secure energy*, SEC(2010)1346, p. 7 and 8.

is closely linked to the volume of energy and the way energy is produced (nuclear, fossil or renewable sources), transported (grids, pipes, cables and vessels) and consumed (energy savings and efficient use of energy). Indeed, 80% of all the greenhouse gases emitted in the EU come from energy production and use.

93. The cornerstone on an international level of the battle against climate change is the adoption of the Kyoto Protocol of 11 December 1997 in which the acceding industrialised countries undertake to reduce their greenhouse gas emissions with at least 5% during the period 2008–2012 compared to the 1990 levels. On 25 April 2002, the Union acceded to this Protocol and thus, it committed itself to reduce its greenhouse gas emissions by 8% in 2012.[101]

94. With the adoption of the Kyoto Protocol[102] on an international level, governments have agreed on legally binding reduction targets for greenhouse gases[103] within fixed timetables for the industrialised or 'Annex I' countries.[104] In addition to these fixed levels, the Kyoto Protocol has also introduced three international flexibility mechanisms, namely: international emissions trading, joint implementation (hereinafter JI) and clean development mechanism (hereinafter CDM).

95. To achieve the Protocol's targets, the contracting Parties need to implement national climate change policies and measures. To this end, Article 2 of the Kyoto Protocol lays down a list of policies and measures, which might help to mitigate climate change and to promote sustainable development. As this list is indicative, the Protocol leaves the national governments with absolute discretion to implement any particular policy. The indicative list includes enhancing energy efficiency, promoting sustainable agriculture, promoting renewable energy, carbon sequestration and other environmentally-friendly technologies.[105] Energy-related measures are thus prominently present as one of the ways in which climate change can be mitigated. This is not surprising, knowing that about three-quarters of the world's energy is provided by fossil

---

101 Council Decision 2002/358/EC of 25 April 2002 concerning the approval, on behalf of the European Community, of the Kyoto Protocol to the United Nations Framework Convention on Climate Change and the joint fulfilment of commitments thereunder, *OJ* L 130/1.

102 For a detailed analysis of the Kyoto Protocol, see R. LEAL ARCAS, "Is the Kyoto Protocol an adequate environmental agreement to resolve the climate change problem', *EELR* (2002), p. 282-294.

103 The Kyoto Protocol emission targets cover six of the main greenhouse gases: Carbon dioxide (CO_2), Methane (CH_4), Nitrous oxide (N_2O), Hydrofluorocarbons (HFCs), Perfluorocarbons (PFC's) and Sulphur hexafluoride (SF_6).

104 The Annex I countries must cut emissions to at least 5 per cent below the 1990 levels by 2008–2012. See X. WANG and G. WISER, "The Implementation and Compliance Regimes under the Climate Change Convention and its Kyoto Protocol", *RECIEL* (2002), p. 186.

105 Climate Change Secretariat, *Guide to the Climate Change Convention and its Kyoto Protocol* (2002), p. 24.

fuels such as coal, oil and gas (which have fuelled industrial development and continue to fuel the global economy) and that burning these fossil fuels emits greenhouse gases. These greenhouse gases are recognised as being responsible for climate change.

96. Therefore, there is a need for drastic measures in the energy and transport sector in order to achieve greenhouse gas emissions reduction and general enhancement of the environment. Hereto national governments should be encouraged to focus on reducing energy and carbon intensity, whereby the solution exists in energy efficiency improvements and a greater reliance on environmentally sound energy.[106] In December 2010 the UN climate change conference took place in Cancún. This conference adopted, contrary to the findings of the participating parties[107], a number of decisions with a very limited scope. The path towards a low-emissions future seems to be a less attractive policy goal on an international level than on an EU level.

97. Climate change increases the temperatures on sea and land, and alters precipitation quantity and patterns which result in the increase of the global average sea level and risks of coastal erosion. Moreover, it leads to an expected increase in the severity of weather-related natural disasters. Due to the changing water levels, temperatures and flow, the food supply, health, industry and transport and ecosystems integrity will be affected. Climate change will lead to significant social and economic impacts of which the severity varies by region, sector and section of society. Some regions, sectors and sections of society (such as the elderly, the disabled and low-income households) will most likely be confronted with more adverse affects than others.

98. In Europe, the most vulnerable regions are southern Europe, the Mediterranean Basin, outermost regions and the Arctic. Moreover, mountain areas, in particular the Alps, coastal, island and urban areas and densely populated floodplains might face particular problems due to climate change. Sectors which could be affected by climate change are for example agriculture, forest health and productivity, fisheries and aquaculture sectors, coasts and marine ecosystems, energy sector, infrastructure, tourism, human, animal and plant health, quality and availability of water resources, etc.[108]

---

106 *Ibid.*, p. 24–38.

107 For further details see http://unfccc.int/2860.php; European Union welcomes Cancún Agreement as important step towards global framework for climate action, Cancún 11 December 2010, IP/10/1699 and Press release UN Climate Change Conference in Cancún delivers balanced package of decisions, restores faith in multilateral process, UNFCCC secretariat.

108 White paper of 1 April 2009, *Adapting to climate change: Towards a European framework for action*, COM(2009)147, p. 3–6.

99. The European Union has always been very ambitious in climate change negotiations on an international level and has ratified the Kyoto Protocol together with its Member States.[109] This requires the EU to reduce its greenhouse gas emissions by 8% in the commitment period 2008 to 2012 compared to its 1990 levels.

100. A key driver in this context is the promotion of renewable energy sources. The EU has long recognised the need to promote renewable energy sources given that their exploitation not only contributes to climate change mitigation through the reductions of greenhouse gas emissions, but they also improve security and diversification of energy supply in the Union. Over the years, the focus in the EU renewable energy policy has shifted from the promotion of renewable energy through indicative targets for the electricity and transport sectors to legally binding targets supported by a comprehensive legislative framework.[110]

101. The Commission has highlighted in its communication of January 2011 that the review of Member States National Renewable Energy Action Plans shows that renewable energy will grow at a faster pace in the years up to 2020 than in the past. By 2020, on the basis of these plans, renewable energy should constitute 37% of Europe's electricity mix. Moreover, if all these production forecasts are fulfilled, the 20% target in 2020 will even be exceeded. The new approach of legally binding targets is starting to pay off. However, as acknowledged in the communication on energy infrastructure for 2020 and beyond, such an expansion of electricity from renewable sources has a number of implications, for example the need to accelerate the modernisation of the electricity grid, electricity systems should become more interconnected and flexible, new infrastructure should be developed and reinforcements are necessary, including the deployment of smart grid technologies, etc.[111]

102. The National Renewable Energy Action Plans contain encouraging signs, although the challenge is to ensure that these plans are transposed into concrete action on the ground in the Member States in the coming years.

103. One of the other key factors in achieving the sustainability and security of energy supply goals in the EU is energy efficiency. We can no longer afford to

---

109 Council Decision 2002/358/EC concerning the conclusion, on behalf of the European Community, of the Kyoto Protocol to the UNFCCC and the joint fulfillment thereunder, *OJ* L 130/1.

110 Communication from the Commission to the European Parliament and the Council of 31 January 2011, *Renewable Energy: Progressing towards the 2020 target*, COM(2011)31, 15 p. and see *Infra*.

111 *Ibid.*, p. 3–7.

waste energy. Energy efficiency needs to be mainstreamed in all relevant policy areas in order to change the current behavioural patterns and should cover the whole energy chain from production, via transmission and distribution, to final consumption.[112]

104. Currently, the quality of the National Energy Efficiency Actions Plans developed by the Member States on the basis of Directive 2006/32/EC is very weak and disappointing.[113] Thus, the potential of this central objective of the EU remains untapped by the Member States whereby the EU is currently far away from achieving the 20% objective set for energy efficiency by 2020.[114]

4. SECURITY OF ENERGY SUPPLY

105. During the First World War *"security of energy supply"* became a real policy issue. Even after the First World War it has been high on the agenda of any energy policy. Currently, guaranteeing both short- and long-term security of supply is central in the European energy policy in particular due to the Union's increasing dependency on imported oil and natural gas. In the years to come, scarcity of domestic resources will become even more pressing as a result of the geopolitical challenges which must be confronted in order to secure reliable volumes of energy imports into the Union.

106. The most recent update of the Energy Trends for 2030 foresees between today[115] and 2030 a slight growth of primary energy consumption according to the so-called baseline scenario, while growth is set to remain largely stable according to the reference scenario.[116] In these scenarios the share of coal and oil declines in the overall energy mix between today and 2030. Gas demand will remain largely stable until 2030. Oil demand is expected to decline in the EU-15

---

112 See *Supra.*

113 For a detailed analysis of the National Energy Efficiency Action Plans see the Commission staff working document of 23 June 2009, *Synthesis of the complete assessment of all 27 National Energy Efficiency Action Plans as required by the Directive 2006/32/EC on energy end-use efficiency and energy services, Moving forward together on saving energy*, SEC(2009)889, 137 p. and see the Study on the energy savings Potentials in EU Member States, Candidate Countries and EEA Countries Final Report for the European Commission Directorate-General Energy and Transport, EC Service Contract Number TREN/D1/239–2006/S07.66640, 313 p.

114 Communication from the Commission of 10 November 2010 to the European Parliament, the Council, the European Economic and Social Committee and the Committee of the Regions, *Energy 2020 A Strategy for competitive, sustainable and secure energy*, COM(2010)639, p. 3–7.

115 Today means August 2012.

116 It should be noted that these forecasts do not include energy efficiency policies to be implemented from 2010 onwards, an emission reduction target of 30% by 2020 or additional transport policies beyond CO_2 and cars emissions regulation.

Member States and in the new Member States it will grow constantly at 7.8% between 2010 and 2030. Under the PRIMES reference scenario the EU-27 gross electricity generation is projected to grow by at least 20% from about 3,362 TWh to 4,073 TWh in 2030 and to 4,192 TWh under the PRIMES baseline even without taking into account the possible effects of a strong development in electro-mobility. This growing demand of electricity and the increasing shares of generation from renewable sources, in addition to additional market integration, are the main challenges for the electricity infrastructure.[117]

107. Both in primary and final energy consumption the share of renewables is set to increase significantly. According to the reference scenario, the share of renewables in the gross electricity generation is expected to be around 33% in 2020 and 36% in 2030. In 2020, on the basis of the first 23 National Renewable Energy Action Plans (hereinafter NREAP)[118] there will be about 460 GW of renewable electricity installed capacity, against 244 GW at this moment. Of this total, about 63% would be related to variable energy sources such as wind (200 GW or 43%) and solar (90 GW or 20%). In the 23 Member States, renewables are projected to account for over 1,150 TWh of electricity generation of which approximately 50% will come from variable energy sources. The largest growth in wind capacities and generation will be in Germany, the United Kingdom, Spain, France, Italy and the Netherlands. Solar capacities and generation growth will be concentrated in Germany and Spain and to a lesser extent France and Italy.[119]

108. Nuclear is set to remain stable and amounts to 14% of the primary energy consumption. However, it remains to be seen whether the nuclear accident in Fukushima has had any impact on future investments in the EU. The import dependence of fossil fuels will continue to be high for oil, coal and gas. The gas dependency on imports will be growing further to reach about 73–79% of the total gas consumption by 2020 and about 81–89% by 2030. This growing dependency is mainly due to the depletion of indigenous resources. Compared to 2005, the additional import needs of gas ranges from 44 Mtoe to 148 Mtoe by 2020 and from 61 to 221 Mtoe by 2030.[120]

---

117 See *Supra*.

118 The 23 NREAP's are from the Member States: Austria, Bulgaria, Czech Republic, Cyprus, Germany, Denmark, Greece, Spain, Finland, France, Ireland, Italy, Latvia, Lithuania, Luxembourg, Malta, the Netherlands, Portugal, Romania, Sweden, Slovakia, Slovenia and the United Kingdom.

119 Communication from the Commission of 17 November 2010 to the European Parliament, the Council, the European Economic and Social Committee and the Committee of the Regions, *Energy infrastructure priorities for 2020 and beyond – A Blueprint for an integrated European energy network*, COM(2010)677, p. 19–25.

120 *Ibid.*

109. The foregoing makes clear that the biggest concern with respect to the increase in demand of energy is gas. A number of Member States appear to be overwhelmingly dependent on one single supplier. This is particularly the case for the 8 new eastern European Member States of which the dependence on Russian gas amounts to 77%.[121]

110. In early 2006, 2008, 2009 and 2010, the EU has witnessed gas supply interruptions. From 6–20 January 2009, gas supplies were interrupted from Russia to the EU via Ukraine which affected a majority of the Member States directly and indirectly. Since the breakup of the Soviet Union, tensions between Russia and Ukraine and Belarus have raised several times due to the continuing difficulties between these states on the details of a new gas transit and supply regime. Currently, 58% of the EU's gas consumption is imported, 42% of which comes from Russia, and approximately 80% of the EU's gas imports from Russia pass through Ukraine.[122]

111. It is possible that disruptions of the scale of those of January 2009 happen again. The EU is not only exposed to commercial disputes, but also to the risk of technical failure or accidents on major gas pipelines. The January 2009 gas crisis illustrates that the standard of preparation to deal with such circumstances varies significantly in the different Member States. Some are not even developed yet. It also illustrates the need for diversified resources.

112. Over the years, at different successive meetings of the European Council, echoed by the European Parliament, the need for a common approach has been emphasised. In line with this call, the European Commission has established enhanced cooperation dialogues with the major energy players on an international level. This has mainly been done in the form of Memoranda of Understanding or energy dialogues. Where possible, attempts have been made to underpin these energy cooperations with solid legal provisions.[123] Moreover, the coordination in international organisations such as the Internal Energy Agency,

121 Commission staff working document of 16 July 2009 accompanying document to the proposal for a regulation of the European Parliament and of the Council concerning measures to safeguard security of supply and repealing Directive 2004/67/EC, *The January 2009 Gas supply disruption to the EU: an assessment*, SEC(2009)977, p. 3.

122 Per day between 300–350 million cubic meters of gas pass through Ukraine towards the EU. Commission staff working document of 16 July 2009 accompanying document to the proposal for a regulation of the European Parliament and of the Council concerning measures to safeguard security of supply and repealing Directive 2004/67/EC, *The January 2009 Gas supply disruption to the EU: an assessment*, SEC(2009)977, p. 2–18.

123 Examples of bilateral agreements are the new EU-Russia agreement and the currently negotiated Free Trade Agreement and Partnership and Cooperation Agreement with Ukraine.

Energy Charter Treaty[124] and G8/G20 has improved; different efforts have been made to extend the Energy Community Treaty beyond southeast Europe[125]; furthermore, the EU took different initiatives to diversify its energy suppliers and routes.

113. Additionally, the transport sector in the EU, for example, is highly dependent on imports of oil with an increasing dependency on imports of diesel in order to balance it demands on the European fuel market. This dependency translates into exposure to global markets which is particularly the case for oil characterised by significant volatility. The EU has witnessed in 2008 oil prices reaching almost 150 dollars per barrel which subsequently collapsed due to the economic and financial crisis, before edging up in 2010 to more than 90 dollars per barrel. Recent events underlined that the main challenge concerns the future security of gas imports, however there have also been some oil supply disruptions, oil being more diverse in terms of supply, routes and infrastructure.

E. AN INSUFFICIENT, INCOHERENT AND FRAGMENTED EU ENERGY POLICY

114. Over the past years, as discussed above, the EU adopted a comprehensive EU energy policy based on mutually complementing pillars of competitiveness, sustainability and security of supply. Nevertheless, it is uncertain whether the existing strategy is sufficient to achieve all the EU energy policy aims.

---

124 The Energy Charter signed in December 1991 was the political foundation for the Charter process. This process has led to the adoption of the Energy Charter Treaty (hereinafter ECT) and the Energy Charter Protocol on Energy Efficiency and Related Environmental Aspects (hereinafter PEEREA) which were signed in December 1994 and entered into legal force in April 1998. From a security of supply point of view the impact of these Treaties is very limited. It mainly covers issues with respect to market governance and market access for both supply and transit. Currently, the ECT and the PEEREA have 53 signatories of which 46 states have ratified them. It should be noted that on 20 August 2009 the Russian Federation has officially informed the Depository that it did not intend to become a Contracting Party to the ECT and PEEREA. In accordance with Article 45(3(a)) of the ECT, such notification results in Russia's termination of its provisional application of the ECT and the PEEREA upon expiration of 60 calendar days of the date on which the notification is received by the Depository. Thus, Russian Federation has applied both the ECT and the PEEREA up to and including 18 October 2009 provisionally.

125 In May 2010, Moldova became a full member of the Energy Community Treaty, negotiations with Turkey are ongoing; Council Decision 2006/500/EC of 29 May 2006 on the conclusion by the European Community of the Energy Community Treaty, *OJ* L 198/15. The South East European Energy Community has been signed in 2005 (Council Decision 2005/905/EC of 17 October 2005 on the signing by the European Community of the Energy Community Treaty, *OJ* L 329/30) and is concluded for a period of 10 years from the date of its entry into force. This Treaty is an attempt to extend the European regulatory framework for the internal market to non-EU countries, *inter alia* the *acquis communautaire* for electricity, oil and gas.

115. One of the central objectives of the EU energy policy is energy efficiency and savings. Since the mid-1990s a number of initiatives were taken on an EU level with a view to realising the 20% energy efficiency and savings objective. However, recent Commission estimates highlight that the European Union is not on track to achieve the objective of saving 20% of the EU's energy consumption completely, compared to the projections for 2020. Although there is a break in the trend towards ever-increasing energy demand, the reduction of energy consumption in the European Union will be 9% and not the 20% target by 2020. The National Energy Efficiency Action Plans, adopted by the Member States since 2008, are disappointing and leave a great energy efficiency and saving potential untapped.[126]

116. The market for energy efficient products, building and services appears not to develop quickly enough due to a number of market and regulatory failures, *inter alia*, the fact that energy prices do not reflect all costs to society, asymmetric information on benefits and costs makes people reluctant, there are low numbers of trained professionals, harmful subsidies, a lack of comprehensive policy framework, a low uptake of new and innovative technologies.[127] In order to enable all the Member States to decouple their energy use further from economic growth, much more attention should be given on a European and national level to achieve the energy efficiency and saving potential.[128]

117. The process of opening the internal electricity and gas markets started more than ten years ago. This opening has the potential to deliver great benefits for households and companies in terms of growth and development. However, at this moment, a truly single EU internal market for electricity and gas is far from

126 Communication from the Commission of 10 November 2010 to the Council, the European Parliament, the European Economic and Social Committee and the Committee of the Regions, *Energy 2020 A Strategy for competitive, sustainable and secure energy*, COM(2010)639, p. 3 and Commission staff working document, *State of play in the EU energy policy*, Accompanying document to the Communication from the Commission to the European Parliament, the Council, the European Economic and Social Committee and the Committee of the regions, *Energy 2020 A Strategy for competitive, sustainable and secure energy*, SEC(2010)1346, p. 4–6.

127 Communication of the Commission to the European Parliament, the Council, the European Economic and Social Committee and the Committee of the Regions of 8 March 2011, *Energy Efficiency Plan 2011*, COM(2011)109, p. 2 and Commission staff working document, *Impact Assessment, Accompanying document to the Communication of the Commission to the European Parliament, the Council, the European Economic and Social Committee and the Committee of the Regions of 8 March 2011, Energy Efficiency Plan 2011*, SEC(2011)277, p. 6–17.

128 Communication from the Commission of 10 November 2010 to the Council, the European Parliament, the European Economic and Social Committee and the Committee of the Regions, *Energy 2020 A Strategy for competitive, sustainable and secure energy*, COM(2010)639, p. 6.

complete since too many barriers for competition and too many differences between Member States still exist.[129]

118. The 2009–2010 report of the Commission on progress in creating the internal gas and electricity market concludes that, on a wholesale level, more market opening and integration have been registered. However, there are still significant obstacles to open integrated and competitive markets in electricity and gas. A number of wholesale markets still suffer from a lack of liquidity and limited competition. Many consumers (citizens and companies) do not perceive that they are better off since the opening of the electricity and gas markets.[130] On a retail level, the integration in both markets remains insufficiently developed. With respect to price levels, the electricity and gas markets are on a retail level still characterised by disparities in the different Member States. The electricity and gas retail markets remain highly concentrated. Moreover, the switching rate for consumers on a household level is still very low.[131]

119. In many Member States, regulated energy prices which reduce competition remain in place.[132] A fluid transmission of energy within and between Member States is prevented due to insufficient interconnection capacity. Moreover, even if more interconnection capacity would become available, the absence of harmonised market rules leads to market segmentation and higher transaction costs which constitute a barrier for new market players.[133]

120. The internal energy market is not yet working as a single market. It still remains fragmented into national markets, often with incumbent companies having a *de facto* monopoly position. Moreover, the internal energy market has not yet achieved its potential for transparency, accessibility and choice for private and industrial customers.[134] The basic concepts of the internal energy market

---

129 *Ibid.*, p. 7 and Commission staff working document of 9 June 2011, *2009–2010 Report on the progress in creating the internal gas and electricity market*, p. 2.

130 Commission staff working document, *State of play in the EU energy policy, Accompanying document to the Communication from the Commission to the European Parliament, the Council, the European Economic and Social Committee and the Committee of the Regions, Energy 2020 A Strategy for competitive, sustainable and secure energy*, SEC(2010)1346, p. 8 and 12.

131 Commission staff working document of 9 June 2011, *2009–2010 Report on the progress in creating the internal gas and electricity market*, p. 14.

132 In 16 Member States regulated end user gas prices remained in place for household consumers and in 13 Member States for non-household consumers. In 19 Member States regulated end user electricity prices are used for household consumers and in 16 Member States for non-household consumers see Commission staff working document of 9 June 2011, *2009–2010 Report on the progress in creating the internal gas and electricity market*, p. 11.

133 Commission staff working document of 9 June 2011, *2009–2010 Report on the progress in creating the internal gas and electricity market*, p. 14.

134 Communication from the Commission of 10 November 2010 to the Council, the European Parliament, the European Economic and Social Committee and the Committee of the

have been embedded in terms of legal framework, institutional arrangements and the physical infrastructure. Nevertheless, different barriers to an open and fair competition remain in place as is evidenced by the energy sector inquiry.[135] Thus, meaningful competition currently does not exist in many Member States.[136] This prevents EU citizens and companies from receiving the full benefits of a liberalised energy market.[137]

121. The main aims of the internal market objective in the energy field are rather to secure access to competitive sources of fuel, to render the supply of energy as efficient as possible and to eliminate undue monopoly profits. Moreover, focusing on the internal market objective might even come at the expense of a lack of progress in the security of supply and sustainability objectives.

122. For many years, the European Commission has systematically prioritised one single objective, namely the creation of an internal energy market.[138] It has put this objective forward as the panacea allowing the EU to pursue its core objectives in the energy field: competitiveness, security of supply and sustainability.[139]

123. It was suggested that the creation of the internal market would contribute to the security of supply and that *"a power blackout like the one in California would not be possible in the internal market, which is reinforced by rules governing investment, competition and access to resources and transport networks which*

Regions, *Energy 2020 A Strategy for competitive, sustainable and secure energy*, COM(2010)639, p. 3 and 9.

135 See *Supra*.

136 Commission staff working document of 9 June 2011, *2009–2010 Report on the progress in creating the internal gas and electricity market*, 14 p.

137 Communication from the Commission of 10 January 2007, *Inquiry pursuant to Article 17 of Regulation (EC) No 1/2003 into the European gas and electricity sectors (Final Report)*, COM(2006)851, 15 p; Communication from the Commission of 10 January 2007 to the Council and the European Parliament, *Prospects for the internal gas and electricity market*, COM(2006)841, p. 2 and Communication from the Commission of 10 November 2010 to the Council, the European Parliament, the European Economic and Social Committee and the Committee of the Regions, *Energy 2020 A Strategy for competitive, sustainable and secure energy*, COM(2010)639, p. 2.

138 Commission working document of 2 May 1988, *The internal energy market*, COM(1988)238, 90 p.; Communication from the Commission to the Council and the European Parliament of 13 March 2001, *Completing the internal energy market*, COM(2001)125, p. 21–25; White Paper on an Energy Policy for the European Union, COM(95)682, 46 p.; Communication from the Commission to the Council and the European Parliament, *Report on the Green Paper Towards a European Strategy for the Security of Energy Supply*, COM(2002)321, 32 p. and Policy Proposal by Jacques Delors, *Towards a European Energy Community: A Policy proposal*, study by S. ANDOURA, L. HANCHER and M. VAN DER WOUDE, p. 27–71.

139 *Ibid.*

protect against this type of breakdown".[140] Although a long-standing electricity exchange was in place in order to deal with peaks in demand, the blackout in Italy in September 2003[141] left many parts of Italy without electricity for several hours. The final report on this blackout highlighted that in order to prevent similar incidents in the future, more reliability and a more secure exchange in electricity should be created and new market rules should be adopted.[142]

124. This incident demonstrates that the well-functioning of the internal market is far less effective as a tool to promote the security of supply objective. Instead of promoting the internal market, more attention should be given to the energy infrastructure which needs to be improved, developed and reinforced. Otherwise the internal market will not deliver on its promises.

125. Beside the assumption that the internal market fully guarantees the security of supply, it is uncertain whether an increased well-functioning of the internal market will lead to greater external security of supply, unless progress can be made on the external dimension[143], or lead to a demand and supply of energy which is more climate-friendly.

126. As far as the external security of supply is concerned, despite all the initiatives on a European level there is still no common approach towards partners, suppliers or transit countries. A unilateral approach on the part of the Member States to secure their energy supplies remains the rule. Bilateral deals with respect to energy supplies between separate Member States and third countries or external energy suppliers continue to prevail over a specific approach on a European level. Such separate bilateral approaches will continue to undermine directly the development of a comprehensive EU energy policy.[144]

127. The EU accounts for one fifth of the world's energy consumption, however it continues to have less weight on international energy markets than its economic weight suggests. Due to the fact that the EU is the world's largest

---

140 Communication from the Commission to the Council and the European Parliament, *Report on the Green Paper Towards a European Strategy for the Security of Energy Supply*, COM(2002)321, p. 8.

141 This blackout was initiated by a tree flashover in Switzerland which led to an overload in other electricity lines.

142 S. HAGHIGHI, *Energy Security – The External Legal Relations of the European Union with Major Oil- and Gas-Supplying Countries* (Oxford, Hart Publishing, 2007), p. 165 which refers to the Final Report on the Investigation Committee on the 28 September 2003 Blackout in Italy, *The System Security in the Light of System Failures in 2003*, p.13.

143 Policy Proposal by Jacques Delors, *Towards a European Energy Community: A Policy proposal*, study by S. ANDOURA, L. HANCHER and M. VAN DER WOUDE, p. 63.

144 *Ibid.*, p. 55.

importer of energy, it is likely to become more vulnerable to supply risks.[145] The tradition of diagnosing the energy security issues is now well-established, however concrete and coherent actions remain elusive. The policy framework for energy security issues is characterised by *ad hoc* arrangements such as the establishment of advisory bodies like the European Energy Supply Observatory.

128. Solutions to the gas disruptions in eastern Europe in January 2009 were hindered by a lack of reserve flow options, inadequate interconnection and storage facilities. The renewables potential in southern Europe and North Africa will not be developed without additional interconnections. Both off- and onshore projects in the North and Baltic Sea regions are hampered by insufficient grid connections. Today Europe is still lacking the grid infrastructure in order to allow renewables to develop and compete with other sources. Moreover, the current grid infrastructure will be unable to absorb the volumes of renewable power of the 2020 targets. The energy infrastructure in the European Union is outdated and poorly interconnected.

129. Without investments in smart, effective and competitive energy infrastructure in the coming years, the risk and cost of disruptions will become much higher. Moreover, due to the limited lifetime of the current generation capacity, some parts in the EU might lose more than a third of this capacity by 2020. This means that without replacing and expanding the existing capacities the internal energy supply could be undermined.[146]

130. The current measures on a European level, which promote a diversification of the economy away from oil, in particular in the transport sector, are not moving fast enough. Hence, the resistance of the EU economy to the volatility of the oil prices of the global market is unsatisfactory. This volatility is not only damaging the economic stability as a whole, but is also detrimental for energy investments and project planning mainly because gas prices currently remain linked to oil prices in most continental long-term contracts.[147] Oil supplies could

---

145 Stock taking document, *Towards a new Energy Strategy for Europe 2011–2020*, p. 7 and Communication from the Commission of 10 November 2010 to the European Parliament, the Council, the European Economic and Social Committee and the Committee of the Regions, *Energy 2020 A Strategy for competitive, sustainable and secure energy*, COM(2010)639, p. 2 and 3.

146 Communication from the Commission of 17 November 2010 to the European Parliament, the Council, the European Economic and Social Committee and the Committee of the Regions, *Energy infrastructure priorities for 2020 and beyond – A Blueprint for an integrated European energy network*, COM(2010)677, p. 4 and Communication from the Commission of 10 November 2010 to the Council, the European Parliament, the European Economic and Social Committee and the Committee of the Regions, *Energy 2020 A Strategy for competitive, sustainable and secure energy*, COM(2010)639, p. 3, 9–11.

147 Stock taking document, *Towards a new Energy Strategy for Europe 2011–2020*, p. 7.

become very tight before 2020: it is therefore important to increase the efforts to reduce oil demand, unfortunately, this is not happening at the moment.

131. On a European level, it seems to be underestimated that energy relates to a specific product and it uses and thus, requires specific rules. Compared to other products energy raises specific issues. For example its demand is inelastic, the construction of production and storage facilities requires large investments, the operation and construction of its networks (transmission and distribution) raises various issues from financial, technical and regulatory points of view. Energy is vital for households and industries, its rupture of supply could create panic and high costs, it is subject to a number of geopolitical problems, its consumption has environmental consequences, most of the energy sources are exhaustible, it is partly substitutable, etc.[148] So far, it has mainly been addressed only by the basic EU provisions on completing the internal market without taking fully into account these specificities of energy markets and the fact that in these markets state intervention will be an on-going reality.

132. Moreover, the well-known concept of packages (a first, second, third and maybe even a fourth one) in the EU energy policy confirms and reinforces the fragmented approach in this field. Their process inevitably brings about the postponing of resolutions of controversies to the next package.

133. Given the possible impacts of climate change, it is important that policy-makers on an International, European and national level understand these impacts and thus develop and implement policies addressing climate change which lead to mitigation and adaptation. Currently, in a piecemeal manner mitigation and adaptation are taking place. Nonetheless, a more strategic approach is necessary in order to ensure that adaptation and mitigation measures are taken effectively and timely which guarantee coherency across different sectors and levels of governance.[149]

134. On the basis of the foregoing, it can be concluded that the common EU energy policy needed for achieving the three central and ambitious goals remains insufficient in its results. Some progress has been made towards these goals, but Europe's energy systems are adapting too slowly nonetheless. The existing strategy for a common EU energy policy is currently unlikely to achieve

---

[148] Policy Proposal by Jacques Delors, *Towards a European Energy Community: A Policy proposal*, study by S. ANDOURA, L. HANCHER and M. VAN DER WOUDE, p. VII and F. DEHOUSSE, "Towards a Real New Energy Policy for the European Union? The 2007 Challenge", *Studia Diplomatica* (2007), p. 12–14.

[149] White paper of 1 April 2009, *Adapting to climate change: Towards a European framework for action*, COM(2009)147, p. 3–6.

all the 2020 targets and with respect to the longer-term challenges, it is incomplete.[150]

135. The current challenges of the EU energy policy seem to lack consistency in balancing the three central goals: competitiveness, security of supply and sustainability.[151] However, the need for consistency is not a novelty in the EU energy policy. For example, in order to achieve the security of supply objective, the first European Energy Efficiency measures confirmed nearly three decades ago, and even before climate change was on the political agenda of the EU and its Member States, that renewable energy sources should be developed in order to mitigate the supply risks of fossil fuel shortages.[152]

136. For many years the EU energy policy was essentially an internal market policy without an effective and concrete foreign policy. Both the security of supply and sustainability objectives have either been pursued as ancillary issues to the environmental policy of the Union or as legal or conceptual derogations to the internal market provisions.

137. Despite the intensive analytical activities on a European level in the form of policy papers it must be concluded that the current EU energy policy is inadequate for achieving long-term challenges in the energy field. The EU is unable to counter common threats collectively and/or to protect its own position at the international scene. Currently, the sustainability and, in particular, security of energy supply objectives are mainly pursued on a national level with the inherent risk of contradictions and divergences. Although, much has been achieved in the energy and climate change field the last decade, it can be concluded that this has been hampered by fragmentation. For the short-term results in the European energy field, this is not necessarily a negative situation because fragmentation accommodates flexibility. However, due to the fact that fragmentation has become institutionalised to such a degree in the current process, it could be a serious obstacle to the formation and implementation of a carbon-free or low-carbon economy by 2050.[153]

138. Unless obsolete power plants are replaced by new, competitive and environmentally-friendly alternatives and unless the energy infrastructure is

---

150 Communication from the Commission of 10 November 2010 to the Council, the European Parliament, the European Economic and Social Committee and the Committee of the Regions, *Energy 2020 A Strategy for competitive, sustainable and secure energy*, COM(2010)639, p. 3.

151 Policy Proposal by Jacques Delors, *Towards a European Energy Community: A Policy proposal*, study by S. ANDOURA, L. HANCHER and M. VAN DER WOUDE, p. 27–71.

152 *Ibid.*

153 Policy Proposal by Jacques Delors, *Towards a European Energy Community: A Policy proposal*, study by S. ANDOURA, L. HANCHER and M. VAN DER WOUDE, p. 62–71.

improved, developed and reinforced and energy is used more efficiently, the three central goals: competitiveness, security of supply and sustainability will not be achieved. Moreover, due to the huge need for investments in the energy sector, and to higher international energy prices and carbon pricing, the energy prices of households and industry will be affected. However, in order to maintain our quality of life, job creation and long-term sustainable growth, these energy prices must remain affordable. Given the scale of these energy challenges there is a strong need for more effective policies and action on an EU level.[154]

139. Thus, on the basis of the foregoing it can be concluded that the current Union powers and instruments in relation to the three objectives are not able to achieve the key objectives and challenges of a robust energy policy on a European level, namely competitiveness, security of supply and sustainability.[155]

F. CONCLUDING REMARKS

140. The aim of this chapter was to provide an overview of the various developments in the EU energy policy over the past years. This overview highlighted that the Union has been very active in adopting a large number of policy documents in the energy field which pursue three central goals: competitiveness, security of supply and sustainability.

141. Although there are currently three central objectives in the EU energy policy, this chapter has acknowledged that the internal energy market is the key objective, and it has been overemphasised for many years. This might even come at the expense of a lack of progress in the security of supply and sustainability objectives. Both the objectives of security of supply and sustainability have either been pursued as ancillary issues to the environmental policy of the Union or as legal or conceptual derogations to the internal market provisions.

142. During the last decade, much has been achieved in the energy and climate change field. However, it can be concluded that this has been hampered by fragmentation. Since this fragmentation has become institutionalised to such a degree in the current process, it could be a serious obstacle to the formation and implementation of a carbon-free or low-carbon economy by 2050. The current European energy policy framework, although it is very ambitious, will not

---

[154] Communication from the Commission of 10 November 2010 to the Council, the European Parliament, the European Economic and Social Committee and the Committee of the Regions, *Energy 2020 A Strategy for competitive, sustainable and secure energy*, COM(2010)639, p. 19–20.

[155] Policy Proposal by Jacques Delors, *Towards a European Energy Community: A Policy proposal*, study by S. ANDOURA, L. HANCHER and M. VAN DER WOUDE, p. 66.

achieve its three central goals. Thus, it is undeniable that urgent action is required.

143. Following this overview of policy initiatives on a European level in the EU energy policy, the next chapter discusses a number of concrete measures adopted on a European level translating these policy initiatives into the European energy framework of secondary legislation.

CHAPTER 3

KEY EU SECONDARY LEGISLATION IN THE ENERGY FIELD

A. INTRODUCTION

144. This chapter will assess in more detail the different measures of secondary legislation adopted on a European level in order to achieve the three central goals of competitiveness, security of supply and sustainability. It highlights that over the past years, the European legislator has been very active in adopting secondary legislation in the energy field. However, it remains uncertain whether these measures are effective.

145. Under the current legal framework several specific instruments on a European level have been adopted in order to achieve the full opening of the internal market for electricity and gas, to address security of supply crises leading to supply disruption, and to tackle the climate change challenge. The previous chapter has concluded that the current Union instruments, in relation to the three objectives, are not able to achieve the key objectives of a robust energy policy on a European level: competitiveness, security of supply and sustainability. Although the Union has been very active in adopting a large number of policy documents in the energy field, it does not seem to be able to establish, with its various instruments, an effective energy policy and law on a European level.

146. The purpose of this chapter is to provide for each of the three central goals an overview of the most relevant secondary legislation adopted over the past years. It concludes with some deficiencies currently suffered by the secondary legislation in the energy field.

B. INTERNAL MARKET FOR ELECTRICITY AND GAS

1. INTRODUCTION

147. Since 1996, the Commission has finally, after a lengthy and complex negotiation process, adopted the directives on the electricity and gas markets. The aim of these directives was to open the electricity and gas markets to competition, i.e. the liberalisation of the electricity and gas markets in the European Union as *de jure* or *de facto* national monopolies and/or oligopolies in the electricity and gas sector needed to be broken down.[156] Prior to the liberalisation process, the European electricity and gas markets were essentially characterised by vertically integrated national monopolies. These monopolies controlled in their national energy markets the generation, transmission, distribution and supply of energy. They have tried, and still do in some Member States, to retain their markets by ensuring that new entrants have limited opportunities to compete with them.[157]

148. A full opening of the electricity and gas markets should, according to the Commission, stimulate fair and competitive energy prices, as well as higher investments in the electricity and gas sectors. The next part will provide a short overview of the various measures adopted in order to achieve the fully liberalised single market for electricity and gas.

2. ELECTRICITY AND GAS DIRECTIVES AND REGULATIONS

(a) *Electricity and Gas Directives*

149. With the adoption of the First Electricity Directive[158] in late 1996 and the First Gas Directive[159] in 1998, the Commission has realised the foundations of

---

156 W. GELDHOF and F. VANDENDRIESSCHE, "European Electricity and Gas Market Liberalisation Background, Status, Developments" in B. DELVAUX, M. HUNT and K. TALUS (eds.), *EU Energy Law and Policy Issues* (Rixensart, Euroconfidentiel, 2008), p. 33–35.

157 K. TALUS, *Vertical Natural Gas Transportation capacity, upstream commodity contracts and EU Competition law*, Dissertation submitted to the faculty of law, University of Helsinki, 2010, p. 81 and 82. See also K. TALUS, *Vertical Natural Gas Transportation Capacity, Upstream Commodity Contracts and EU Competition Law* (The Netherlands, Kluwer Law International, 2010), 321 p.

158 Directive 96/92/EC of the European Parliament and the Council of 19 December 1996 concerning common rules for the internal market in electricity, *OJ* L 27/10 (hereinafter First Electricity Directive).

159 Directive 98/30/EC of the European Parliament and the Council of 22 June 1998 concerning common rules for the internal market in natural gas, *OJ* L 204/1 (hereinafter First Gas Directive).

the legislative alternatives for a progressive opening of the electricity and gas market[160], further supplemented with the adoption of the Second Electricity Directive[161], Second Gas Directive[162] and, since 2009, the adoption of the Third Electricity Directive[163] and Third Gas Directive.[164] Beside these directives, other milestones on the way to opening up both markets are directives covering price transparency[165] and the transit directives of electricity and gas through grids.[166]

150. On 26 June 2003, the Union adopted the Second Electricity Directive. This directive proposed a number of measures in order to open up the electricity market, to establish fair competition and to put in place a true single market in electricity within the EU. The objective was to achieve a secure, competitive and sustainable electricity market. This directive ensures that industrial clients and domestic customers have the freedom to choose their electricity supplier since 1 July 2004 and 1 July 2007 respectively. It has established common rules for the generation, transmission, distribution and supply of electricity. The Second Electricity Directive, moreover, laid down the rules with respect to the organisation and functioning of the internal market in electricity, authorisation procedures, access to this market and the various procedures and criteria for calls for tenders.[167]

---

160 H. BJØRNEBYE, *Investing in EU energy security – Exploring the regulatory approach to tomorrow's electricity production*, Dissertation submitted to the faculty of law, University of Oslo, 2009, p. 134. See also H. BJØRNEBYE, *Investing in EU Energy Security. Exploring the regulatory approach to tomorrow's electricity production* (The Netherlands, Kluwer Law International, 2010), 456 p.

161 Directive 2003/54/EC of the European Parliament and the Council of 26 June 2003 concerning common rules for the internal market in electricity and repealing Directive 96/92/EC, *OJ* L 176/37 (hereinafter Second Electricity Directive).

162 Directive 2003/55/EC of the European Parliament and the Council of 26 June 2003 concerning common rules for the internal market in natural gas and repealing Directive 98/30/EC, *OJ* L 176/57 (hereinafter Second Gas Directive).

163 Directive 2009/72/EC of the European Parliament and the Council of 13 July 2009 concerning common rules for the internal market in electricity and repealing Directive 2003/54/EC, *OJ* L 211/55 (hereinafter Third Electricity Directive).

164 Directive 2009/73/EC of the European Parliament and the Council of 13 July 2009 concerning common rules for the internal market in natural gas and repealing Directive 2003/55/EC, *OJ* L 211/94 (hereinafter Third Gas Directive).

165 Council Directive 90/377/EEC of 29 June 1990 concerning a Community procedure to improve the transparency of gas and electricity prices charged to industrial end-users, *OJ* L 185/16.

166 Council Directive 90/547/EEC of 29 October 1990 on the transit of electricity through transmission grids, *OJ* L 313/30 and Council Directive 91/296/EEC of 31 May 1991 on the transit of natural gas through grids, *OJ* L 147/37. Both directives have been repealed by the Second Electricity Directive and Second Gas Directive. For a brief discussion of these transit directives see M. M. ROGGENKAMP, A. RØNNE, C. REDGWELL and I. DEL GUAYO (eds.), *Energy law in Europe, National, EU and International Law and Institutions* (Oxford, Oxford University Press, 2001), p. 282–284.

167 For an analysis of these provisions see also E. D. CROSS, B. DELVAUX, L. HANCHER, P. J. SLOT, G. VAN CALSTER and W. VANDENBERGHE, "EU Energy Law (chapter 5)" in M. M.

151. However, as of 3 March 2011, the Second Electricity Directive has been replaced by the Third Electricity Directive due to the dysfunctions[168] in the internal market in electricity. The European Commission has redefined some of the rules applying to the electricity market in order to guarantee appropriate consumer protection and fair competition.

152. The Third Electricity Directive stipulates that Member States may impose public service obligations which cover issues of security of supply, regularity, quality of services, prices, energy efficiency and environmental protection on undertakings operating in the electricity sector. Member States must ensure that all customers have the right to choose their electricity supplier and change supplier within three weeks. Under this directive Member States must define the authorisation procedure for the construction of new generating capacity in their territory. This procedure must be conducted in accordance with objective, transparent and non-discriminatory criteria.

153. Without effective separation of electricity networks from activities of generation[169] and supply there is an inherent risk of discrimination in the operation of the network and in the incentives for vertically integrated undertakings to invest adequately in their networks. Since the provisions on legal and functional unbundling, as provided for in the Second Electricity Directive, have not led to effective unbundling of the transmission systems, Member State are required to unbundle these transmission systems in accordance with the new provisions of the Third Electricity Directive.[170]

154. Before an undertaking can officially be designated as a transmission system operator, it must be certified first. The designated transmission system operators are responsible for, *inter alia*, providing system users with the information to access their system, managing electricity flows on their system and contributing to security of supply.[171]

155. Member States must not only designate transmission system operators, but also distribution system operators or require from those that own or are

ROGGENKAMP, A. RØNNE, C. REDGWELL and I. DEL GUAYO (eds.), *Energy Law in Europe National, EU and International Regulation* (Oxford, Oxford University Press, 2007), p. 343–356.

168 See the energy sector inquiry briefly discussed above.

169 Producers of electricity active in Europe are for example GDFSUEZ, RWE, E.ON, EDF, ENEL, Endesa, British Energy, Centrica Energy, etc.

170 Preambule of the Third Electricity Directive.

171 Transmission operators active in Europe are for example Elia System Operator SA in Belgium; EnBW Transportnetze AG, TenneT TSO GmbH, Amprion GmbH, 50Hertz Transmission GmbH in Germany; Red Eléctrica de España S.A. in Spain; Réseau de Transport d'Electricité in France; Creos Luxembourg S.A. in Luxembourg; TenneT TSO B.V. in the Netherlands, etc.

responsible for distribution systems, to do so. These distribution system operators are, *inter alia*, responsible for ensuring transparency with respect to system users, providing system users with information, covering electricity losses and maintaining reserve electricity capacity. One of the novelties of the Third Electricity Directive is the option for Member States to put in place a closed distribution system to distribute electricity within a geographically confined industrial, commercial or shared services site.[172]

156. Under the Third Electricity Directive, Member States must organise a system of third party access to transmission and distribution systems. They must designate an independent regulatory authority on a national level which is responsible for matters such as fixing transmission or distribution tariffs, monitoring investment plans of the transmission system operators. With the exception of Article 11[173], Member States should transpose and apply, as from 3 March 2011, the provisions of the Third Electricity Directive, discussed above.

157. Beside the general provisions of the Third Electricity Directive, discussed above, this directive has introduced a number of important provisions dealing with the further unbundling of network operations from supply and/or generation activities[174], transparency and record keeping obligations, the protection of the consumer and the strengthening of national regulatory authorities.[175]

158. In the gas sector, the Union adopted the Second Gas Directive in order to achieve the full opening of the internal gas markets which should, *inter alia*, increase competitiveness, improve service quality, guarantee fair prices for customers and improve interconnection. On the basis of this directive, as the Third Electricity Directive, industrial clients and domestic customers have had the freedom to choose their gas supplier since 1 July 2004 and 1 July 2007 respectively. The provisions of this directive are similar to those of the Second Electricity Directive discussed above.[176]

---

172 M. HUNT, "Closed Distribution Systems: The End of Private Networks?" in B. DELVAUX, M. HUNT and K. TALUS (eds.), *EU Energy Law and Policy Issues* (Cambridge, Intersentia, 2011), p. 97–121.

173 Member States must apply these provisions as from 3 March 2013.

174 See *Infra* for a more detailed discussion.

175 For an analysis of these provisions see M. HUNT, "EU Energy Policy in 2007–2008" in N. HUNT and K. TALUS (eds.), *The EU Energy Directory* (Rixensart, Euroconfidentiel, 2008), p. 13–21, S. GOLDBERG and H. BJØRNEBYE, "The Third Energy Package of the European Union, Introduction and comment" in B. DELVAUX, M. HUNT and K. TALUS (eds.), *EU Energy Law and Policy Issues* (Cambridge, Intersentia, 2011), p. 9–31.

176 For an analysis of these provisions see also E. D. CROSS, B. DELVAUX, L. HANCHER, P. J. SLOT, G. VAN CALSTER and W. VANDENBERGHE, "EU Energy Law (chapter 5)" in M. M. ROGGENKAMP, A. RØNNE, C. REDGWELL and I. DEL GUAYO (eds.), *Energy Law in*

159. Since the internal market for gas suffered from a lack of transparency and other dysfunctions which impede its proper functioning, the European legislator adopted the Third Gas Directive on 13 July 2009. The purpose of this directive is to introduce common rules for the transmission, distribution, supply and storage of natural gas.[177] The Third Gas Directive also contains similar provisions, like the Third Electricity directive, relating to the public service obligations, unbundling requirements, distribution, third party access, etc. Member States had to transpose the provisions of the Third Gas Directive at the latest by 3 March 2011 with the exception of Article 11 of this directive which they should apply as from 3 March 2013. Moreover, as from 3 March 2011 the provisions of the Second Gas Directive are repealed.

(b) Electricity and Gas Regulations

160. On 26 June 2003, Regulation No 1228/2003 on conditions for access to the network for cross-border exchanges in electricity was adopted.[178] The purpose of this regulation is to stimulate cross-border exchanges in electricity. To this end, it establishes a compensation mechanism for transit flows of electricity and introduces harmonised principles on cross-border transmission charges. Under this regulation, transmission system operators will receive compensation for costs incurred as a result of hosting cross-border flows of electricity on their network. This compensation is paid by the operators of national transmission systems from which the cross-border flows originate and calculated on the basis of the costs of the infrastructure used for the cross-border flows.

161. The transmission system operators must install coordination and information exchange mechanisms to ensure security of the networks in the context of congestion management. Network congestion problems should be addressed with non-discriminatory solutions, i.e. methods which do not involve a selection between contracts of individual market participants. Market participants are obliged to inform the transmission system operators concerned whether they intend to use allocated capacity a reasonable time ahead of the relevant period. Any allocated capacity that will not be used is attributed to the market in an open, transparent and non-discriminatory manner.

162. This regulation has also established the possibility for the Commission to adopt guidelines. These guidelines specify in particular: details of methods to

Europe. National, EU and International Regulation (Oxford, Oxford University Press, 2007), p. 364–373.

177 It concerns mainly natural gas, liquefied natural gas or LNG, biogas and gas from biomass.

178 Regulation No 1228/2003 of the European Parliament and of the Council of 26 June 2003 on conditions for access to the network for cross-border exchanges in electricity, *OJ* L 176/1 (hereinafter Regulation No 1228/2003).

determine the quantity of cross-border flows hosted and the magnitudes of such flows, details of the treatment in the context of the inter-transmission system operator compensation mechanism of electricity flows originating or ending in countries outside the European Economic Area and harmonised, appropriate and efficient locational signals on a European level.[179]

163. However, as of 3 March 2011, Regulation No 1228/2003 is replaced by Regulation No 714/2009 on conditions for access to the network for cross-border exchanges in electricity.[180] Regulation No 714/2009 was part of the Third Package which was adopted in 2009 in order to overcome remaining barriers to competition in the energy markets. This regulation therefore aims at laying down rules for cross-border exchanges in electricity in order to improving competition and harmonisation in the internal market for electricity.

164. This regulation established a certification of transmission system operators. National regulatory authorities must send the Commission the notification of decisions concerning the certification of a transmission system operator. The Commission then has a period of 2 months to deliver its opinion to the national regulatory authority which then adopts the final decision concerning the certification.

165. This regulation establishes the European Network of Transmission System Operators for Electricity (hereinafter ENTSO-E). The ENTSO-E is responsible for managing the electricity transmission system and for allowing the trading and supplying of electricity across borders in the Union. It will be responsible for adopting: common network operation tools, a ten-year network development plan, recommendations relating to the coordination of technical cooperation between the Union and transmission system operators, an annual work programme, an annual report and annual summer and winter generation adequacy outlooks.

166. The costs related to the activities of the ENTSO-E must be borne by the transmission system operators. They must also establish regional cooperation within the ENTSO-E and publish a regional investment plan every 2 years, on which investments can be based. For the costs incurred as a result of hosting cross-border flows of electricity on their networks, the transmission system operators receive compensation. This compensation will be paid by the operators

179 For an analysis of these provisions see also E. D. CROSS, B. DELVAUX, L. HANCHER, P. J. SLOT, G. VAN CALSTER and W. VANDENBERGHE, "EU Energy Law (chapter 5)" in M. M. ROGGENKAMP, A. RØNNE, C. REDGWELL and I. DEL GUAYO (eds.), *Energy Law in Europe. National, EU and International Regulation* (Oxford, Oxford University Press, 2007), p. 356–364.

180 Regulation No 714/2009 of the European Parliament and of the Council of 13 July 2009 on conditions for access to the network for cross-border exchanges in electricity and repealing Regulation No 1228/2003, *OJ* L 211/15 (hereinafter Regulation No 714/2009).

of national transmission systems from which the cross-border flows originate. With a view to ensuring the security of networks in the context of congestion management, the transmission system operators must put in place an information exchange mechanism. Network congestion problems should be addressed with non-discriminatory market-based solutions which give economic signals to the market participants and transmission system operators. On the basis of Regulation No 714/2009, upon request, new interconnectors can be exempted, for a limited period of time, from the general provisions governing congestion management on the condition that: their installation increases competition in electricity supply, the level of risk necessitates the exemption, the interconnection must be owned by a natural or legal person, charges are levied on users of the interconnection and the exemption must not be to the detriment of competition in the internal market.

167. In order to ensure also a minimum level of harmonisation in the gas market, the European Parliament and the Council adopted Regulation No 1775/2005 on conditions for access to the natural gas transmission networks.[181] This Regulation 1775/2005 obliged gas transmission operators to offer their services to all users on a non-discriminatory basis which means that the contractual conditions must be identical. However, this does not mean that other transmission system operators are obliged to offer the same contractual conditions, apart from the minimum contractual requirements.

168. The system operator must make the technical network capacity available to users in its entirety, taking into account system integrity and efficient use of the network. This capacity allocation must be performed on the basis of non-discriminatory and transparent mechanisms. In the event of contractual congestion, i.e. when the level of demand for firm capacity exceeds the system's technical capacity, the system operator can offer the unused capacity of certain users to other users on a short-term basis. Under this regulation, the system operator is obliged to set fair and transparent technical balancing rules. The operator must also provide users with relevant information on the balancing status and if necessary adopt corrective measures.

169. The tariffs for network access must be set by the system operators in a transparent and non-discriminatory way, and reflect the actual costs borne by them. These tariffs not only take into account maintenance of system integrity, but also its improvement.[182]

---

181 Regulation 1775/2005 of the European Parliament and of the Council of 28 September 2005 on conditions for access to the natural gas transmission networks, *OJ* L 289/1.

182 For an analysis of these provisions see also E. D. CROSS, B. DELVAUX, L. HANCHER, P. J. SLOT, G. VAN CALSTER and W. VANDENBERGHE, "EU Energy Law (chapter 5)" in M. M. ROGGENKAMP, A. RØNNE, C. REDGWELL and I. DEL GUAYO (eds.), *Energy Law in*

170. Since a number of barriers to competition are still present within the internal gas market, the new Regulation No 715/2009 on conditions for access to the natural gas transmission networks and repealing Regulation No 1775/2005 was adopted on 13 July 2009 as part of the Third Package.[183] This Regulation No 715/2009 lays down rules for natural gas transmission networks, gas storage and liquefied natural gas (hereinafter LNG) facilities which entered into force as from 3 March 2011. It provides a framework for natural gas transmission and trade in order to counteract the barriers to competition within the gas market. This framework establishes rules concerning the access to infrastructures, services to be offered, allocation of capacity, transparency and balancing of the network.

171. Regulation No 715/2009 has established, as Regulation No 714/2009, an identical procedure for the certification of transmission system operators. Under this regulation, a European Network of Transmission System Operators for Gas (hereinafter ENTSO-G) is established with similar tasks as the ENTSO-E. Transmission operators, LNG and storage facilities are obliged to offer their services in a non-discriminatory way to all network users on a rolling basis in the long and short term. The infrastructure operators must implement and publish non-discriminatory and transparent congestion management procedures which facilitate cross-border exchanges in gas on a non-discriminatory basis. All market participants should have access to maximum network capacity as well as storage and LNG facilities.

3. ACER REGULATION

172. In order to address the lack of a regulatory framework for cross-border issues in the electricity and gas market and based on experiences with cooperation initiatives such as the European Regulators' Group for Electricity and Gas (ERGEG), the Electricity Regulatory Forum (Florence) and the Gas Regulatory Forum (Madrid), the Commission has decided that a more powerful cooperation between national regulators needed to be set up.

173. As a result, numerous structures were considered, along with their advantages and disadvantages, including the establishment of a network of

Europe. National, EU and International Regulation (Oxford, Oxford University Press, 2007), p. 373–375 and B. DELVAUX, "The gas transmission regulation 1775/2005 – Has a genius been born?" in M. M. ROGGENKAMP and U. HAMMER (eds.), *European Energy Law Report IV* (Antwerpen, Intersentia, 2007), p. 41–68.

183 Regulation No 715/2009 of the European Parliament and of the Council of 13 July 2009 on conditions for access to the natural gas transmission networks and repealing Regulation No 1775/2005, *OJ* L 211/36 (hereinafter Regulation No 715/2009).

national regulators, such as the network of Competition Authorities[184], and the establishment of a cooperation form, as is the case with the European banks. The Commission has concluded on the basis of this analysis that the most suitable structure for an institution of this nature would be the establishment of an Agency, i.e. the establishment of the *Agency for the Cooperation of Energy Regulators* (hereinafter Agency).

174. On 23 March 2009, the Czech Presidency and the responsible rapporteurs of the Committee on Industry, Research and Energy have reached a compromise on the Third Legislative Package. This compromise was the object of the European Parliament vote on 21 April 2009 (second reading) which was backed by a large majority in Parliament and contains the adoption of the new ACER Regulation[185], among other directives and regulations in the energy field.[186]

175. The purpose of the Agency, according to Article 1 of the ACER Regulation, is to assist the national regulatory authorities in exercising, on a Union level, the regulatory tasks performed in the Member States and if necessary to coordinate their action. This means that the Agency mainly has a supportive role to play on a European level. The name also makes clear that the Agency's purpose is to establish a European cooperation between the different national regulatory authorities on various cross-border issues. The Agency will be a Union body with legal personality. Its tasks will be performed by an administrative board, a Board of Regulators, a Director with staff and a Board of Appeal.

176. These bodies of the Agency are allowed to issue opinions and recommendations addressed to the TSOs, regulatory authorities, European Parliament, Council and Commission. In specific cases (technical issues, some regulatory issues with respect to cross-border infrastructure and exemptions as provided for in Article 17 (5) of the Electricity Regulation 714/2009) referred to in Article 7, 8 and 9 of the ACER Regulation, the Agency is authorised to adopt individual decisions. Furthermore, the Agency can submit non-binding framework guidelines, for the development of network codes relating to the areas identified in the Commission's priority list, to the Commission in accordance

---

184 Council Regulation No 1/2003 of 16 December 2002 on the implementation of the rules on competition laid down in Articles 81 and 82 of the Treaty, *OJ* L 1/1.

185 Regulation No 713/2009 of the European Parliament and of the Council of 13 July 2009 establishing an Agency for the Cooperation of Energy Regulators, *OJ* L 211/1 (hereinafter ACER Regulation).

186 Press release of 22 April 2009 of the European Parliament, "MEPs give green light to further liberalisation of EU electricity and gas markets" and MEMO/09/176 of 21 April 2009 of the European Commission, "Q&A on the European Parliament's vote on the Third Internal Energy Market Legislative Package."

with the Articles 6 of the Electricity Regulation 714/2009 and the Gas Regulation 715/2009.[187]

4. REGULATION ON DATA AND INFORMATION ON INVESTMENT PROJECTS IN ENERGY INFRASTRUCTURE

177. The European Commission additionally proposed measures to ensure greater transparency on the likely evolution of energy infrastructure in the main energy sectors. The overall objective of the Commission's proposal is to revise and to strengthen the existing system of Council Regulation No 736/96[188] further while easing the administrative burden. According to the Commission, the existing regulatory framework imposing reporting obligations on energy infrastructure on Member States is heterogeneous. Moreover, relevant and validated data is not sufficiently available and therefore should be updated.[189]

178. This resulted in the adoption of Council Regulation (EU, EURATOM) No 617/2010[190] on 24 June 2010 establishing a common framework for the notification to the Commission of data and information on investment projects in energy infrastructure in the oil, natural gas, electricity, including electricity from renewable sources, and bio-fuel sectors, and on investment projects related to the capture and storage of carbon dioxide produced by these sectors.[191] This regulation establishes that in the case of infrastructure which is planned or under construction, certain data should be notified to the Commission, *inter alia*, the volume of the capacities planned or under construction, the type and main characteristics of infrastructure or capacities planned or under construction, the probable year of commissioning, etc.

---

187 For further details see B. DELVAUX, "The Agency for the Cooperation of Energy Regulators: A New Beginning?" in B. DELVAUX, M. HUNT and K. TALUS (eds.), *EU Energy Law and Policy Issues* (Rixensart, Euroconfidentiel, 2010), p. 183–208 and L. HANCHER and A. DE HAUTECLOCQUE, "Manufacturing the EU Energy Markets: The Current Dynamics of Regulatory Practise", *Competition and Regulation in Network Industries* (2010), p. 307–334.

188 Council Regulation No 736/96 of 22 April 1996 on notifying the Commission of investment projects of interest to the Community in the petroleum, natural gas and electricity sectors, *OJ* L 102/1.

189 Proposal for a Council Regulation concerning the notification to the Commission of investment projects in energy infrastructure within the European Community and repealing Regulation (EC) No 736/96, COM(2009)361.

190 Council Regulation (EU, Euratom) No 617/2010 of 24 June 2010 concerning the notification to the Commission of investment projects in energy infrastructure within the European Union and repealing Regulation No 736/96, *OJ* L 180/7.

191 *Ibid.*, Article 2.

179. Under this regulation Member States must collect this data from the undertakings which make the investment concerned or by any equivalent means. Moreover, they must notify this data to the Commission in aggregated form by 31 July of the reporting year. The first reporting year is 2011 and from then onwards every two years. On the basis of this data, the Commission will prepare an analysis of investment projects and development perspectives for the European energy system.

C. SUSTAINABILITY AND CLIMATE CHANGE

1. INTRODUCTION

180. With respect to the energy-related measures to combat climate change and improving sustainability, a distinction can be made between measures relating to the supply of energy (generation, transmission and distribution) on the one hand and measures dealing with energy demand (consumption) on the other. The latter includes energy efficiency measures in the household, tertiary and industry sector. Beside the measures which can be considered as energy-related measures to combat climate change and improving sustainability, this part also discusses the European emission trading and energy taxation directives.

2. MEASURES DEALING WITH THE SUPPLY OF ENERGY

(a) Renewable Energy Directives

181. Since the European Commission adopted the Communication *"Energy for the Future: Renewable Sources of Energy"*[192] in 1997, the EU has recognised the need to promote renewable energy sources as a priority measure in various documents, given the important contribution that renewable energy can make towards improving security and diversification of energy supply. It also reduces the strain which is being placed by conventional energy on the environment.[193]

---

192 Communication of the Commission on Energy for the Future: Renewable Sources of Energy, White Paper for a Community Strategy and Action Plan of 26 November 1997, COM(97)599, p. 1–55.

193 See Council Resolution of 16 September 1986 concerning new Community energy policy objectives for 1995 and convergence of the policies of the Member States, *OJ* C 241/1, where the Council listed the promotion of renewable energy sources among its objectives and stated: "... *(f) to maintain the development of new and renewable energy sources, including conventional hydroelectricity, in particular by continuing with the effort made and by placing greater emphasis on arrangements for disseminating results and reproducing successful projects. The output from new and renewable energy sources in place of conventional fuels*

The European Commission has therefore set out a strategy and target to double the share of renewables in the gross energy consumption from 6% in 1997 up to 12% in 2010.[194]

182. One of the measures on a European level in this respect was the adoption of Directive 2001/77/EC on the promotion of electricity from renewable energy sources in the internal electricity market.[195] The objective of this directive was to create a framework which will facilitate the future development of renewable electricity ('RES-E') within the EU and its access to the internal electricity market.[196] Under this directive, Member States had the obligation to set national indicative targets for the encouragement of a greater consumption of electricity produced from renewable energy sources.

183. Beside Directive 2001/77/EC, since 1985 the European Commission and the Council have been fostering the development of renewable energies and in particular biofuels, more precisely through Directive 85/536/EEC relating to crude oil savings through the use of substitute fuel components.[197] The importance of biofuels to improve partly the security of energy supply was emphasised and this directive authorised the incorporation into petrol of up to 5% of ethanol by volume and up to 15% of ethyl-tertiary-butyl-ether (ETBE) by volume. In 1993[198] and 1997[199], the ALTENER programmes to promote the use of renewable energies in the European Community emphasised the need to increase the use of biofuels and also introduced the objective for biofuels of a 5% market share.[200]

---

should be substantially increased, thereby enabling them to make a significant contribution to the total energy balance."

194 European Commission Green Paper towards a European strategy for the security of the energy supply of 29 November 2000, COM (2000)769, p. 48.

195 Directive 2001/77/EC of the European Parliament and of the Council of 27 September 2001 on the Promotion of Electricity Produced from Renewable Energy Sources in the Internal Electricity Market, *OJ* L 283/33 (hereinafter Directive 2001/77/EC).

196 M. GUAL and P. DEL RIO, "The promotion of green electricity in Europe: present and future", *European Environment* (2004), p. 219–234.

197 Council Directive 85/536/EEC on crude-oil savings through the use of substitute fuel components in petrol, *OJ* L 34/20.

198 Council Decision 93/500/EEC of 13 September 1993 concerning the promotion of renewable energy sources in the Community (ALTENER programme), *OJ* L 235/41.

199 Council Decision 98/352/EC of 18 May 1998 concerning a multiannual programme for the promotion of renewable energy sources in the Community (ALTENER II programme), *OJ* L 159/53.

200 Communication from the Commission to the European Parliament, the Council, the Economic and Social Committee and the Committee of the Regions on alternative fuels for road transportation and on a set of measures to promote the use of biofuels, of 7 November 2001 COM(2001)547, 38 p.

184. In November 2001, the Commission submitted a communication and two draft directives concerning alternative fuels.[201] When considering these drafts the European Parliament adopted a series of amendments and therefore obliged the Commission to produce an amended proposal. The purpose of this proposed directive is to ensure and to promote a minimum percentage of biofuels for transport in the EU (starting with 2% in 2005) to be marketed and distributed. It is clear that biofuels will only be able to achieve market penetration if they are widely available and competitive. To realise this competitiveness and to promote biofuels, the proposal enables Member States to stimulate the technological development of biofuel production and of the companies involved in its production by using financial instruments for research, the environment and regional development.[202]

185. Finally, in May 2003, at the end of a long evolution, the Council and the European Parliament adopted Directive 2003/30/EC on the promotion of the use of biofuels or other fuels for transport which aims at promoting the use of biofuels by largely ensuring a minimum proportion of biofuels and other renewable fuels for transport.[203] This directive states in Article 3 that Member States: "*should ensure that a minimum proportion of biofuels and other renewable fuels is placed on their markets, and to that effect, shall set national indicative targets*". For these targets the directive sets reference values: 2% by the end of 2005 and 5.75% by the end of 2010. Furthermore, Member States are obliged to report to the Commission each year on the measures taken to promote biofuels and on the share of biofuels placed on the market in the previous year. Like those of Directive 2001/77/EC, the targets in the biofuels directive are of an indicative nature and are not binding, which makes their legal nature questionable.

186. In addition to the Third Package for the electricity and gas markets, the Commission proposed its climate and energy package on 23 January 2008. The European Parliament adopted this package on 17 December 2008 which led to the adoption of the Renewable Energy Directive 2009/28/EC.[204] This directive repealed Directives 2001/77/EC and 2003/30/EC, discussed above.

---

201 Proposal for a Directive of the European Parliament and of the Council on the promotion of the use of biofuels for transport, COM(2001)547, 50 p.

202 Articles 3 and 4 of the amended proposal for a Directive of the European Parliament and of the Council on the promotion of the use of biofuels for transport presented by the Commission pursuant to Art. 250 (2) of the EC-Treaty, COM(2002)508.

203 Directive 2003/30/EC of 8 May 2003 of the European Parliament and of the Council on the promotion of the use of biofuels or other renewable fuels for transport, *OJ* L 123/42. For a detailed analysis of the Directive promoting the use of biofuels see B. DELVAUX, "Promoting Biofuels in Energy Supply: the European Legal Framework", *EELR* (2004), p. 66 – 78.

204 Directive 2009/28/EC of the European Parliament and of the Council of 23 April 2009 on the promotion of the use of energy from renewable sources and amending and subsequently

187. The Renewable Energy Directive 2009/28/EC established a common framework according to which the Member States need to ensure that the share of renewable energy in the EU final consumption reaches in 2020 at least 20%. It therefore sets mandatory targets for the overall share of energy from renewable sources in energy consumption and for the share of energy from renewable sources in transport for each Member State. Moreover, the share of energy from renewable sources in the transport sector must amount to at least 10% of final energy consumption in the sector by 2020.

188. Under this Renewable Energy Directive 2009/28/EC, Member States are obliged to establish national action plans which set the share of energy from renewable sources consumed in transport, as well as in the production of electricity and heating, for 2020. Member State are allowed, on certain conditions, to exchange an amount of energy from renewable sources using a statistical transfer and to set up joint projects concerning the production of electricity and heating from renewable sources.[205]

189. Each Member State must be able to guarantee the origin of electricity, heating and cooling produced from renewable energy sources. The information mentioned in these guarantees of origin is normalised and should be recognised by all Member States. They could also be used to provide consumers with information on the composition of the different electricity sources. Member States must ensure furthermore that operators guarantee the transport and distribution, and provide for priority access, of electricity from renewable sources.

190. The Renewable Energy Directive moreover, takes into account energy from biofuels and bioliquids. To benefit from financial support biofuels and bioliquids must be qualified as sustainable in accordance with the criteria of the Renewable Energy Directive.[206]

(b) Directive on combined heat and power

191. Cogeneration, or combined heat and power (hereinafter CHP), is a technique allowing the production of heat and electricity in a single process. In

---

repealing Directives 2001/77/EC and 2003/30/EC, *OJ* L 140/16 (hereinafter Renewable Energy Directive 2009/28/EC).

205 See chapters 3, 4 and 5 of the Commission staff working document, Accompanying the document Communication from the Commission to the European Parliament, the Council, the European Economic and Social Committee and the Committee of the Regions, *Renewable Energy: a major player in the European energy market*, SWD(2012)164, p. 8–20.

206 See *Infra* and see the Communication from the Commission to the European Parliament, the Council, the European Economic and Social Committee and the Committee of the Regions, *Renewable Energy: a major player in the European energy market*, COM(2012)271, 14 p.

spite of the clear benefits of CHP[207] with regard to saving primary energy, avoiding network losses and reducing emissions, in particular greenhouse gases, the use of cogeneration as a measure to save energy remains underused in the EU. It was therefore necessary to establish a directive to facilitate the installation and operation of such plants, hence Directive 2004/8/EC was born.[208]

192. In the short term, this Directive 2004/8/EC should serve as a level playing field for existing cogeneration installations and for new plants, while in the medium to long term, the necessary framework for high efficiency cogeneration needs to ensure the reduction of CO_2 emissions. High efficiency cogeneration is in this Directive defined by the energy savings obtained by combined production instead of separate production of heat and electricity. Energy savings of more than 10% qualify as "high efficiency cogeneration" (Annex III).

193. Member States must ensure that all forms of electricity produced from high efficiency cogeneration can be covered by guarantees of origin, which are clearly distinguished from exchangeable certificates. To maximise the energy savings and to avoid energy savings being lost, great attention must be paid to the functioning conditions of cogeneration units. Moreover, Member States may operate different mechanisms of support for cogeneration on a national level, including investment aid, tax exemptions or reductions, green certificates and direct price support schemes.[209]

3. MEASURES DEALING WITH ENERGY DEMAND

194. Due to the on-going surge in oil prices and the vast threat of climate change, energy saving through energy efficiency must be seen as a new fundamental priority for the EU.[210] According to the European Commission, energy saving is allegedly the quickest and most (cost-)effective manner for

207 Electricity/heat cogeneration installations can achieve energy efficiency levels of around 90%. The process is more ecological, since during combustion natural gas releases less carbon dioxide (CO_2) and nitrogen oxide (NO_X) than oil or coal. The development of cogeneration could avoid the emission of 127 million tonnes of CO_2 in the EU in 2010 and 258 million tonnes in 2020.

208 Directive 2004/8/EC of the European Parliament and of the Council of 11 February 2004 on the promotion of cogeneration based on a useful heat demand in the internal energy market and amending Directive 92/42/EEC, *OJ* L 52/50 (hereinafter Directive 2004/8/EC).

209 See for further information on this topic, A. JOHNSTON, "Take-or-pay contracts for Renewables: An analysis of European Legal issues", in B. DELVAUX, M. HUNT and K. TALUS (eds.), *EU energy law and policy issues* (Rixensart, Euroconfidentiel, 2008), p. 269–289.

210 Studies indicate that an effective energy efficiency policy could allegedly save at least 20% of its 2005 energy consumption by 2020, equivalent to € 60 billion per year, or the combined energy consumption of Germany and Finland.

reducing greenhouse gas emissions and will therefore help Member States in meeting their Kyoto commitments.[211] Finally, energy efficiency is one of the key methods to cap EU energy demand and to promote security of energy supply for the European Union.

195. A spectrum of policy tools is at hand to increase the efficiency of energy use. These tools range from raising the energy price to more specific actions, such as detailed information (in the form of labels), efficiency regulations, or subsidies. The best policy includes elements that both "push" the market (i.e. mandatory efficiency requirements for manufacturers and real prices for consumers) and "pull" the market, encouraging consumers and/or manufacturers. In the following, the most important legislative measures on energy efficiency and on the energy-efficient use of certain particular appliances will be described and commented briefly.

(a) *Directive on energy end-use efficiency and energy services*

196. As there is relatively limited scope for further influence on energy supply and distribution conditions in the short to medium term, either by building new capacity or by improving transmission and distribution[212], the European Commission has proposed a plan for a Directive on energy end-use efficiency and energy services on the 10 December 2003, which would create a European market for energy services and remove existing market barriers for the efficient end-use of energy.[213]

197. On 5 April 2006 the Directive 2006/32/EC on the promotion of end-use efficiency and energy efficiency (hereinafter Directive 2006/32/EC)[214] was adopted as a framework directive which establishes energy end-use efficiency measures and management measures on energy demand. It applies to all categories of providers of energy efficiency improvement measures, but Member States may exclude small distributors, small distribution systems and small retail energy sales companies.

198. The initial proposal of the Commission aimed at an annual amount of energy saving equal to 1% of the amount of energy distributed and/or sold to

211 "Europe could save 20% of its energy by 2020", *Rapid Press Releases* 22 June 2005, following the adoption of the Green Paper on Energy Efficiency.
212 Green Paper, *Towards a European Strategy for Energy Supply*, COM(2000)769, 111 p.
213 Proposal for a Directive of the European Parliament and of the Council on energy end-use efficiency and energy services, COM(2003)739.
214 Directive 2006/32/EC on the promotion of end-use efficiency and energy efficiency (hereinafter Directive 2006/32/EC), *OJ* L 114/64 (hereinafter Directive 2006/32/EC).

final customers, and a 1.5% savings target for the public sector.[215] These – in principal indicative – targets were given a uniform six-year period for achievement, while the national target would be subdivided into two three-year targets of energy savings. The European Parliament, however, requested a binding 3% target for the period 2006–2009, 4% for 2009–2012 and 4.5% for 2012–2015. Moreover, the MEPs rejected requiring energy distributors or energy sales operators to offer their customers free audits to evaluate their energy saving needs. A later proposal by the European Council featured a national indicative savings target of 6% for the sixth year of application of the directive.[216] At the end of June 2005, the EU Council of Ministers and the European Parliament differed considerably in their views on the targets in the draft energy end-use efficiency directive: whereas Member States wish to set a 6% savings target for their energy consumption within eight years of the entry into force of the directive, the European Parliament supports a 15% savings proposal by 2015.

199. Finally, according to Article 4 of the Directive 2006/32/EC Member States are obliged to adopt and aim at achieving an overall national indicative energy savings target of 9% after the ninth year of application of this directive, i.e. by 2016 which can be reached by way of energy services and other energy efficiency improvements measures. Member States must also appoint one or more existing independent public sector authorities or agencies to ensure overall monitoring of the process set up to achieve these targets.

200. Member States must ensure that the public sector adopts measures to improve energy efficiency, inform the public and business of the measures adopted and promote the exchange of good practice. On the basis of Article 14(2) of Directive 2006/32/EC, Member States are obliged to submit National Energy Efficiency Action Plans (NEEAP) which describe their planned energy efficiency improvement measures to reach their national energy savings targets.[217, 218]

---

215 Proposal for a Directive of the European Parliament and of the Council on energy end-use efficiency and energy services, COM(2003)739.

216 Proposal for a Directive of the European Parliament and of the Council on energy end-use efficiency and energy services, 19 May 2005, Council of the European Union, interinstitutional file 2003/0300 (COD).

217 For a further analysis of the first NEEAP's see the Communication from the Commission to the Council and the European Parliament on a first assessment of national energy efficiency action plans as required by Directive 2006/32/EC on energy end-use efficiency and energy services – moving forward together on energy efficiency, COM(2008)11, 16 p.

218 W. VANDENDORPE, "Recent developments on Energy Efficiency on the European Union level", in B. DELVAUX, M. HUNT and K. TALUS (eds.), *EU energy law and policy issues* (Rixensart, Euroconfidentiel, 2008), p. 245–267.

(b) *Directive on the energy performance of buildings*

201. The residential and tertiary sector, of which the major part is buildings, represents approximately 40% of the final energy consumption in the European Community (and is expanding) and the potential energy saving accounts for more than 20%. Enough reasons for the EU to focus on energy efficiency legislation in this area, starting with key Directive 93/76/EC, which requires Member States to limit CO_2 emissions and which includes insulation measures and heating requirements.[219] The Action Plan to improve energy efficiency in the European Community[220], the SAVE Programme for the Promotion of Energy Efficiency[221], as well as the Green Paper on Security of Supply have emphasised the potential of this directive and called for continuation, expansion and more concreteness of the legal rules, the latter being realised through Directive 2002/91/EC.[222]

202. Directive 2002/91/EC has established a general methodology of calculation of the integrated energy performance of buildings (Article 3), the application of minimum requirements on the energy performance of new buildings and of large existing buildings subject to renovation (Articles 4 to 6), energy certification of buildings (Article 7), and a regular inspection of boilers and air-conditioning systems inside the buildings (Articles 8 and 9). The crucial role Directive 2002/91/EC can play in the public sector must also be emphasised, as public buildings displaying prominently and permanently a building certificate relating to energy efficiency benchmarks can give an excellent example to a large number of citizens (Article 7(3)).[223]

203. On 19 May 2010, a new Directive 2010/31/EU on the energy performance of buildings was adopted which repealed the Directive 2002/91/EC.[224] The aim of this directive is to promote the energy performance of buildings and building units. Under this directive Member States shall adopt, either on a national or regional level, a methodology for calculating the energy performance of

---

219 Council Directive 93/76/EEC of 13 September 1993 to limit carbon dioxide emissions by improving energy efficiency (SAVE), *OJ* L 237/28.

220 Committee and Committee of the Regions of 26 April 2000: *Action Plan to Improve Energy Efficiency in the European Community*, COM(2000)247, 23 p.

221 Council Decision no 647/2000/EC of the European Parliament and of the Council of 28 February 2000 adopting a multiannual programme for the promotion of energy efficiency (SAVE) (1998 to 2002), *OJ* L 79/6.

222 Directive 2002/91/EC of the European Parliament and of the Council of 16 December 2002 on the energy performance of buildings, *OJ* L 1/65.

223 W. VANDENDORPE, "Recent developments on Energy Efficiency on the European Union level", in B. DELVAUX, M. HUNT and K. TALUS (eds.), *EU energy law and policy issues* (Brussels, Euroconfidentiel, 2008), p. 245–267.

224 Directive 2010/31/EU of the European Parliament and of the Council of 19 May 2010 on the energy performance of buildings, *OJ* L 153/13.

buildings which takes into account a number of elements, specifically: the thermal characteristics of a building, heating insulation and hot water supply, the air-conditioning installation, the built-in lighting installation and indoor climate conditions. The positive influence of other aspects such as natural lighting, electricity produced by cogeneration, local solar exposure and district or block heating or cooling are also taken into account.

204. In compliance with the aforementioned calculation methodology, Member States must put in place minimum requirements for the energy performance of buildings in order to achieve cost-optimal levels. According to Directive 2010/31/EU, by 31 December 2020 all new buildings shall be nearly zero energy consumption buildings. The same criteria should be complied with by 31 December 2010 for new buildings occupied or owned by public authorities. Member States must also draw up a list of the existing and potential instruments used to promote improvements in the energy performance of buildings.

205. A system for the energy performance certification of buildings needs to be implemented by the Member States which shall include information on the energy performance of a building and recommendations for cost improvements. When a building or building unit is offered for sale or rent, the energy performance indicator of the energy performance certificate must be included into the advertisements in commercial media. In a building where a total floor area of over 500 m^2 is occupied by a public authority and frequently visited by the public, the energy performance certificate shall be displayed on a prominent place and be clearly visible. It should be noted that this threshold on 9 July 2015 shall be lowered to 250 m^2. Moreover, Member States are also obliged to adopt a system of regular inspections of heating and air-conditioning systems in buildings.

(c) Directive on Eco-design for energy-using appliances

206. Beside putting energy efficiency requirements to particular groups of products, it was considered necessary to focus also on energy-using products in general, as they account for a large proportion of the consumption of natural resources and energy in the EU. Action should be taken during the design phase of energy-using products, since it appears that the pollution caused during a product's life cycle is determined at that stage. According to Directive 2005/32/EC[225], manufacturers are now encouraged to produce products with an

---

[225] Directive 2005/32/EC of the European Parliament and of the Council of 6 July 2005 establishing a framework for the setting of ecodesign requirements for energy-using products and amending Council Directive 92/42/EEC and Directives 96/57/EC and 2000/55/EC of the European Parliament and of the Council, *OJ* L 191/29 (Directive 2005/32/EC).

environmentally-friendly design throughout their entire life-cycle, thus implementing integrated product policy and bearing in mind the goals and priorities of the Sixth Environmental Action Programme.

207. This Directive does not introduce directly binding requirements for specific products, but defines conditions and criteria for setting, through subsequent implementing measures, the requirements regarding environmentally relevant product characteristics (such as energy consumption) and allows them to be improved quickly and efficiently. Although these implementing measures are foreseen, priority should be given to alternative courses of action such as self-regulation by the industry where such action is likely to deliver the policy objectives faster or in a less costly manner than mandatory requirements. Legislative measures may still be needed where market forces fail to evolve in the right direction or at an acceptable speed.[226]

208. With the adoption of Directive 2009/125/EC on 21 October 2009 establishing a framework for the setting of ecodesign requirements for energy-using products[227], Directive 2005/32/EC was repealed. This new Directive 2009/125/EC established ecodesign requirements for energy-related products in the Union; however it does not apply to means of transport for persons or goods. These ecodesign requirements relate to different phases in the product life cycle which are raw material selection and use, manufacturing, packaging, transport and distribution, installation and maintenance, use and end-of-life. For each phase, according to Directive 2009/125/EC, certain aspects of the product must be assessed, *inter alia*, the predicted consumption of materials, energy and other resources, anticipated emissions to air, water or soil, anticipated pollution (noise, vibration, radiation, electromagnetic fields), etc.

209. Under Directive 2009/125/EC all products covered by implementing measures must bear CE marking and must undergo a conformity assessment concerning all of the ecodesign requirements before they are placed on the market. Products which bear the Union eco-label are presumed to comply with the ecodesign requirements of the applicable implementing measures. This directive also requires that manufacturers provide consumers with information on the role they can play in the sustainable use of the given product, as well as the ecological profile of the product and the advantages of ecodesign.

---

226 W. VANDENDORPE, "Recent developments on Energy Efficiency on the European Union level", in B. DELVAUX, M. HUNT and K. TALUS (eds.), *EU energy law and policy issues* (Rixensart, Euroconfidentiel, 2008), p. 245–267.

227 Directive 2009/125/EC of the European Parliament and of the Council of 21 October 2009 establishing a framework for the setting of ecodesign requirements for energy-using products, *OJ* L 285/10 (hereinafter Directive 2009/125/EC).

(d) *Directives on efficiency in energy using products*

i) Household appliances

210. The labelling of household appliances was the first step in applying the policy of energy efficiency to specific products.[228] The provision of accurate, relevant and comparable information on the specific energy consumption of household appliances may influence the public's choice in favour of those appliances which consume less energy, thus indirectly encouraging the efficient use of these appliances. However, although the fact that this measure will play a more important role as the energy prices will continue to increase, one cannot lose sight of the reality that it remains to be seen whether this information will give sufficient stimulus to potential buyers.

211. To this end, and to avoid the creation of barriers to intra-Union trade, a uniform label for all appliances of the same type became obligatory with the establishment of Directive 92/75/EC on the labelling of household appliances.[229]

212. Suppliers have the duty to provide a fiche and a label which details the consumption of electric energy, other forms of energy and other important resources, while it is the task of the dealer to attach this product information onto the household appliances. The energy label thus indicates the energy efficiency class of the appliance. The provision of this standardised information on labels must be insured for all the appliances concerned through the rejection of voluntary schemes. As a continuation of this directive, various daughter directives for specific product groups have been drawn up, *inter alia*, for household electric refrigerators, freezers and their combinations[230], washing machines[231], dishwashers[232], ovens, tumble dryers, household lamps, and so on.

---

[228] V. BRUGGEMAN and B. DELVAUX, "EU Energy Policy and Legislation under Pressure since the UNFCC and the Kyoto Protocol?" in M. PEETERS and K. DEKETELAERE (eds.), *EU Climate Change policy – The Challenge of New Regulatory Initiatives* (London, Edward Elgar Publishing, 2006), p. 223–239.

[229] Council Directive 92/75/EEC of 22 September 1992 on the indication of labelling and standard product information of the consumption of energy and other resources by household appliances, *OJ* L 29/16.

[230] Commission Directive 2003/66/EC of 3 July 2003 amending Directive 94/2/EC implementing Council Directive 92/75/EEC with regard to energy labelling of household electric refrigerators, freezers and their combinations, *OJ* L 170/10. As of 2004, consumers can easily identify those refrigerators and freezers which consume the least energy through the labels A++ and A+.

[231] Commission Directive 95/12/EC of 23 May 1995 implementing Council Directive 92/75/EEC with regard to energy labelling of household washing machines, *OJ* L 136/35, as amended by Directive 96/89/EC, *OJ* L 388/85.

[232] Commission Directive 1999/9 of 26 February 1999 amending Directive 97/17/EC implementing Council Directive 92/75/EEC with regard to energy labelling of household dishwashers, *OJ* L 56/46.

213. Since Directive 92/75/EC only applies to household appliances, the Commission adopted Directive 2010/30/EU on the indication by labelling and standard products information of the consumption of energy and other resources by energy-related products.[233] Member States must apply the provisions of the new Directive 2010/30/EU as from 20 July 2011. It applies to products which are likely to have a direct or indirect impact on the consumption of energy and on other potential resources during use. However, it does not apply to second-hand products, any means of transport for persons or goods and product rating plates.

214. By virtue of Directive 2010/30/EU, suppliers are obliged to place on the market products that have a label containing information on the product's consumption of electric energy or other forms of energy. In the event that the final consumer does not see the product, in particular when purchasing by mail order, by catalogue or through the internet, the consumer must have access to product information through delegated acts[234] which specify the way in which the label or the fiche is displayed or provided to the end-user.

By virtue of Directive 2010/30/EU, the Commission has the power to adopt delegated acts for a period of 5 years from 19 June 2010. This period will be renewed automatically unless the European Parliament or the Council revoke this right. Both the European Parliament and the Council also have the power to object to a delegated act.

ii) Office appliances

215. A large proportion of electricity consumption in the tertiary sector can be attributed to office equipment (such as personal computers, monitors, fax machines, scanners, etc.). In this area, and following the Union's climate change commitments, energy efficiency has a significant role to play. A very important voluntary labelling system referring to the qualified energy-efficient product types has been elaborated, known as the "Energy Star". This energy label has been established in the USA and was adopted some years later in December 2000 by the European Union[235] through the signature of an agreement with the US government on the cooperation and coordination of energy-efficient labelling

[233] Directive 2010/30/EU of the European Parliament and of the Council of 19 May 2010 on the indication by labelling and standard products information of the consumption of energy and other resources by energy-related products, *OJ* L 153/1 (hereinafter Directive 2010/30/EC).

[234] Such a delegated act shall indicate in particular: a discription of the product, measurement standards and methods, details of technical documentation, the design and content of the label, etc.

[235] A Commission Decision of 11 March 2003 has established a European Community Energy Star Board (ECESB) to manage and coordinate the introduction of the Energy Star in Europe: Commission Decision 2003/168/EC of 11 March 2003 establishing the European Community Energy Star Board, *OJ* L 67/22 and Commission Decision 2003/367/EC of 15 May 2003

programs for office equipment, which was a logical step as the Energy Star was already the *de facto* standard for office equipment manufacturers. This Agreement has been incorporated in the Council Decision of 14 May 2001.[236]

216. Early 2006, the Council entrusted the Commission with a mandate to negotiate with the USA on a new Energy Star Agreement for office equipment. This agreement was finalised in June 2006 and incorporated in the Council Decision of 18 December 2006.[237, 238]

217. The labelling programme will act as a world-wide efficiency standard for office information and communication technology equipment. Its logo will enable consumers to identify energy-efficient office equipment which should therefore result in the maximisation of electricity savings and environmental benefits while stimulating the supply of and demand for energy-efficient appliances. Hereto, the most effective measure for reducing electrical consumption of office equipment is to reduce the standby consumption and to support switching off when not needed. In October 2003, a Working Plan on the development of the Energy Star programme was delivered, announcing the revision of all Energy Star office equipment over the next years.[239] In this programme, Member States are encouraged to incorporate the Energy Star requirements in policies for 'Green Public Procurement', considering energy efficiency requirements at the time of public purchasing.

---

establishing the Rules of Procedure of the European Community Energy Star Board, *OJ* L 125/9.

236 Council Decision 2001/469/EC of 14 May 2001 concerning the conclusion on behalf of the European Community of the Agreement between the Government of the United States of America and the European Community on the coordination of energy-efficient labelling programs for office equipment, *OJ* L 172/1. The detailed arrangements for implementing this Agreement on a Community level have been put forward in Regulation No 2422/2001 of the European Parliament and of the Council of 6 November 2001 on a Community energy-efficient labelling programme for office equipment, *OJ* L 332/1.

237 Agreement between the Government of the United States of America and the European Community on the coordination of energy-efficiency labelling programs for office equipment, *OJ* L 381/26.

238 Council Decision 2006/1005/EC of 18 December 2006 concerning the conclusion on behalf of the European Community of the Agreement between the Government of the United States of America and the European Community on the coordination of energy-efficient labelling programs for office equipment, *OJ* L 381/24. The detailed arrangements for implementing this Agreement on a Community level have been put forward in Regulation No 106/2008 of the European Parliament and of the Council of 15 January 2008 on a Community energy-efficient labelling programme for office equipment, *OJ* L 39/1.

239 Energy Star – European Community – Energy Efficiency labelling programme for office equipment – Working Plan, 9 October 2003.

iii) Tyre labelling

218. On 25 November 2009, Regulation No 1222/2009 on the labelling of tyres with respect to fuel efficiency and other essential parameters was adopted.[240] The aim of this regulation is to provide end-users of vehicles with clear and relevant information about the quality of the tyre, and to guide them towards choosing a product which is more fuel efficient, has better wet braking and is less noisy. Since tyres are responsible for 20 to 30% of vehicles' fuel consumption the purchase of good quality tyres allows fuel consumption and CO_2 emissions to be reduced considerably. Regulation No 1222/2009 therefore harmonises information concerning the energy performance of tyres, as well as on wet braking and external rolling noise.

219. Under this regulation tyre suppliers are obliged to monitor the labelling of C1, C2 and C3 tyres delivered to distributors and end-users. This labelling includes a sticker which must indicate: the fuel efficiency class (letter A to G), the wet grip class (letter A to G)[241] and the external rolling noise measures value (in decibels). Tyre suppliers must declare this information in all technical promotion literature such as on their website and in their catalogues. Distributors of tyres must ensure that the labels are clearly visible on the tyres presented or stored at the point of sale. If tyres cannot be seen by end-users, the regulation obliges distributors to provide this information to them. If end-users of vehicles can choose their tyres, car suppliers and distributors must give them the performance of each option of tyre in terms of fuel efficiency, wet braking and rolling noise. This regulation entered into force on 11 October 2010.

4. EMISSION TRADING DIRECTIVE

220. One of the key instruments on a European level to achieve the emission reduction goals in a cost-effective way is the EU-wide emission trading regime, under which operators must cover their emissions of carbon dioxide by the allowances they have been granted by the Member States or through trading in those allowances. Directive 2003/87/EC of 13 October 2003[242] provides the

[240] Regulation No 1222/2009 of the European Parliament and of the Council of 25 November 2009 on the labelling of tyres with respect to fuel efficiency and other essential parameters was adopted, *OJ* L 342/46 (hereinafter Regulation No 1222/2009).

[241] At this stage Regulation 1222/2009 only provides classification for wet grip performance for C1 tyres, but intends to introduce a classification for C2 tyres and eventually C3 tyres once tests are available.

[242] European Parliament and Council Directive 2003/87/EC establishing a scheme for greenhouse gas emission allowance trading within the Community and amending Council Directive 96/61/EC *OJ* L 275/32 (hereinafter Directive 2003/87/EC). For a more detailed analysis on the greenhouse gas emissions allowance trading scheme, see J. DELBEKE (ed.),

regulatory framework for the implementation of a compulsory greenhouse gas allowance trading scheme in all the EU Member States as from 1 January 2005.[243]

221. The 1st phase of the European emission trading scheme (2005 to 2007) has established a free trade of emission allowances across Europe which developed into the world's largest single carbon market, accounting for 67% in terms of volume and 81% in terms of value of the global market. However, due to the growing pains (lack of verified emission data, excessive allocation of allowances in some Member States and some sectors, etc.) it should be noted that the environmental outcome of this 1st phase was very limited. The 2nd phase of the European emission trading scheme (2008 to 2012) has gathered experience from the 1st phase and led to the approval of National Allocation Plans (hereinafter NAPs) of Member States which should in principle result in a real emission reduction of at least 6.5% compared to the 2005 verified emissions.

222. The experience of the 1st phase gave strong reasons to believe that the overall functioning of the European emission trading scheme could be improved on a number of aspects. Beside the need to improve the emission trading scheme, the European Spring Council of 8 and 9 March 2007 has also made a firm commitment of at least a 20% reduction of greenhouse gas emissions by 2020, irrespective of any international agreement. These elements and the outcome of various meetings of the European Climate Change Program Working group on Emission Trading, led to the review of the European emission trading Directive 2003/87/EC.[244] The proposed amendments to this Directive 2003/87/EC were guided by three objectives:

(a) fully exploiting the potential of the European emission trading scheme contributing to the EU's overall greenhouse gas reduction commitments in an economically efficient manner;

EU Energy Law – EU Environmental Law – The EU Greenhouse Gas Emission Trading Scheme (Leuven, Claeys & Casteels, 2006), p. 1–436; D. POCKLINGTON, "European Emission Trading – the Business Perspective", *EELR* (2002), p. 209–218; M. PEETERS, "Emissions Trading a New Dimension to European Environmental law: The Political Agreement of the European Council on Greenhouse Gas Allowance Trading", *EELR* (2003), p. 83–92; B. OLE GRAM MORTENSEN, "The EU Emission Trading Directive", *EELR* (2004), p. 275–284 and G. H. KELLY, "An Evaluation of the European Union's Emissions Trading Scheme in Practice", *EELR* (2006), p. 175–185.

243 For further details reference can be made to http://ec.europa.eu/clima/news/index_en.htm; M. PEETERS and K. DEKETELAERE (eds.), *EU Climate Change policy – The Challenge of New Regulatory Initiatives* (London, Edward Elgar Publishing, 2006), 334 p. and J. DE CENDRA DE LARRAGAN, "EU Climate change mitigation policy: is it adequate, is it fair?", in B. DELVAUX, M. HUNT and K. TALUS (eds.), *EU energy law and policy issues* (Rixensart, Euroconfidentiel, 2008), p. 193–222.

244 See *Supra*.

(b) refining and improving the European emission trading scheme in the lights of experience gathered; and
(c) contributing to transforming Europe into a low greenhouse-gas-emitting economy and creating the right incentives for forward looking low-carbon investment decisions by reinforcing a clear, undistorted and long-term carbon price signal.[245]

223. These amendments led to the adoption of Directive 2009/29/EC[246] which provides that as from 2013, new sectors and gases will be included into the European emission trading scheme.[247]

224. Besides this scheme, in order to achieve the 20% reduction of greenhouse gas emissions by 2020 in a cost-effective manner, additional policies and measures must be adopted which limit the emissions from sources that are not covered under the European emission trading scheme. Hence, there is need for a European framework with national commitments covering the remaining emissions in sectors such as buildings, transport, waste, agriculture and industrial plants that fall under the threshold for inclusion in the European emission trading scheme. Such a European framework has been agreed upon in the Decision No 406/2009/EC. Under this decision each Member State should, by 2020, limit its greenhouse gas emissions at least by the percentage set for that Member State in Annex II to the decision in relation to its emissions in 2005.[248]

225. To obtain their emission reduction target, Member States are free to determine their measures and have access to CDM credits covering almost one third of their emission reduction effort. Furthermore, some of the emission reductions will also be based upon EU measures such as stricter CO_2-emission standards for cars and fuel, rules promoting energy efficiency, etc.[249]

---

245 Proposal for a Directive of the European Parliament and of the Council amending Directive 2003/87/EC so as to improve and extend the greenhouse gas emission allowance trading system of the Community, COM(2008)16, p. 2 and 3.

246 Directive 2009/29/EC of the European Parliament and of the Council of 23 April 2009 amending Directive 2003/87/EC so as to improve and extend the greenhouse gas emission allowance trading scheme of the Community, *OJ* L 140/63.

247 For more details see http://ec.europa.eu/clima/policies/brief/eu/index_en.htm.

248 Decision No 406/2009/EC of the European Parliament and of the Council of 23 April 2009 on the effort of Member States to reduce their greenhouse gas emissions to meet the Community's greenhouse gas emission reduction commitments up to 2020, *OJ* L 140/136.

249 Communication from the Commission to the European Parliament, the Council, the European Economic and Social Committee and the Committee of the Regions – *2020 by 2020 – Europe's climate change opportunity*, COM(2008)30, p. 7 and Proposal for a Decision of the European Parliament and of the Council on the effort of Member States to reduce their greenhouse gas emissions to meet the Community's greenhouse gas emission reduction commitments up to 2020, COM(2008)17.

5. UNION FRAMEWORK FOR THE TAXATION OF ENERGY PRODUCTS AND ELECTRICITY

226. With the adoption of Council Directive 2003/96/EC of 27 October 2003 restructuring the Community framework for the taxation of energy products and electricity[250], the Union established a general framework for the taxation of energy products and electricity. The aim of Directive 2003/96/EC is to improve the functioning of the internal market by reducing distortions in competition between mineral oils and other energy products. In the interest of protecting the environment, it allows Member States to grant tax advantages to businesses that take specific measures to reduce their emissions.

227. Directive 2003/96/EC sets minimum rates of taxation for motor fuel, motor fuel for industrial or commercial use, heating and electricity. It specifies that industrial and commercial uses are: agricultural, horticultural or piscicultural works, and forestry, stationary motors, plant and machinery used in construction, civil engineering and public works and vehicles intended for use off the public roadway. The levels of taxation applied by the Member States cannot be lower than the minimum rates set in the directive.

It allows the Member States to differentiate between commercial and non-commercial use of gas-oil used as propellant, provided that the minimum rates of the directive are observed and the rate for commercial gas-oil as propellant does not fall below the national level of taxation in force on 1 January 2003, notwithstanding any derogations for this use provided for in the directive.

228. Moreover, differentiated tax rates are allowed by Directive 2003/96/EC in certain cases, provided that they comply with the minimum levels of taxation prescribed by the directive and are compatible with EU law. These cases are the following: when differentiated rates are directly linked to product quality, when differentiated rates depend on quantitative consumption levels for electricity and energy products used for heating purposes, for the following uses such as local public passenger transport (including taxis), waste collection, armed forces, public administration, disabled people and ambulances, and between business and non-business use, for energy products and electricity.

Directive 2003/96/EC also establishes that certain energy products and electricity are exempt from taxation such as energy products used to produce electricity and electricity used to maintain the ability to produce electricity. However, for reasons of environmental policy, Member States are allowed to subject these products to taxation, energy products supplied for use as fuel for

[250] Council Directive 2003/96/EC of 27 October 2003 restructuring the Community framework for the taxation of energy products and electricity, *OJ* L 283/51, last amended by Directives 2004/74/EC and 2004/75/EC, *OJ* L 157/87 and 100.

the purpose of air navigation for flying other than for private pleasure and energy products supplied for use as fuel for the purposes of navigation within EU waters, including fishing, other than private pleasure crafts, and electricity produced on board of a craft.

229. By virtue of Directive 2003/96/EC, Member States are also allowed to refund, fully or in part, taxes paid by businesses that have invested in the rationalisation of their energy use. In the case of energy intensive businesses this refund can be as much as 100% and up to 50% for other businesses. With respect to Directive 2003/96/EC the Commission has emphasised that it intends to come forward with a proposal to revise the Energy Taxation Directive.[251] On 13 April 2011, the Commission presented its new Energy Taxation Directive. This proposal reforms the outdated rules on the taxation of energy products. It restructures the way energy products are taxed in order to remove the existing imbalances and take into account their CO_2 emissions and energy content.[252]

D. SECURITY OF ENERGY SUPPLY

1. INTRODUCTION

230. Over time, the security of energy supply has become an increasingly topical issue which is high on the political agenda. Hence the increasing pressure to consider security of energy supply as a central element when the European legislator adopts new legislation. This has led to a primarily positive and necessary development. However, since this development was realised through a political process it has also created certain negative consequences.

231. On a European level, security of energy supply related provisions are included in a number of legislative instruments with the purpose of increasing the security of energy supply internally. Beside these internal measures, the Union adopted a number of provisions in order to develop the external dimension of security of energy supply. The next part will provide an overview of the most relevant directives and regulations which deal specifically with the security of energy supply aspects.[253]

---

251 Communication from the Commission to the European Parliament and the Council of 31 January 2011, *Renewable Energy: Progressing towards the 2020 target*, COM(2011)31, p.14.

252 IP/11/468 of 13 April 2011, *Energy taxation: Commission promotes energy efficiency and more environmental friendly products* and Proposal for a Council Directive amending Directive 2003/96/EC restructuring the Community framework for the taxation of energy products and electricity, COM(2011)169/3.

253 K. TALUS, "Energy Security in European Energy Law" in M. HUNT and K. TALUS (eds.), *The EU Energy Directory* (Rixensart, Euroconfidentiel, 2008), p. 86 and 100.

2. SECURITY OF ENERGY SUPPLY DIRECTIVES AND REGULATIONS

(a) *Directives on minimum stocks of crude oil and/or petroleum*

232. Due to the importance of oil in the EU's energy mix, the strong external dependence of the EU for supply of petroleum products and the unstable geopolitical environment where the balance between supply and demand is generally uneasy, the Union's dependency on imports of petroleum products has become of increasing concern. A supply crisis caused by suppliers from third countries would most likely have a serious impact on the economic activities of the EU. In order to ensure the security of its oil supply, the EU obliges Member States, in accordance with Directive 2006/67/EC[254], to guarantee minimum stocks of petroleum products which can be used in the event of a supply crisis.

233. Directive 2006/67/EC obliges Member States to build up and constantly maintain minimum stocks of petroleum products equal to at least 90 days of the average daily internal consumption during the previous calendar year. The daily internal consumption is calculated on the basis of motor spirit and aviation fuel, gas oil, diesel oil, kerosene and jet fuel of the kerosene type, as well as fuel oils. The stockholding arrangements under Directive 2006/67/EC must ensure that the stocks are available to and accessible by Member States in order to allow them to react immediately in the event of a supply crisis. In the event of a supply crisis, a coordinated operation is put in place in which the Commission organises a consultation between the Member States, either on its own initiative or at the request of one of the Member States. In principle, Member States are not allowed to make withdrawals from the stocks that would bring them below compulsory minimum level before a consultation took place, except in a particularly urgent situation.

234. In 2009, with the adoption of Council Directive 2009/119/EC[255], the EU has revised its oil stockholding system in order to enhance the system, bring it into line with the existing rules of the International Energy Agency and to optimise the administrative obligations in Member States. The new Directive 2009/119/EC lays down rules which aim at: making oil supply in the Union more secure through reliable and transparent mechanisms based on solidarity amongst Member States, maintaining minimum stocks of crude oil and/or

[254] Council Directive 2006/67/EC of 24 July 2006 imposing an obligation on Member States to maintain minimum stocks of crude oil and/or petroleum products, *OJ* L 217/8 (hereinafter Directive 2006/67/EC).

[255] Council Directive 2009/119/EC of 14 September 2009 imposing an obligation on Member States to maintain minimum stocks of crude oil and/or petroleum products, *OJ* L 265/9.

petroleum products and putting in place emergency procedures in the event of a shortage.

235. Directive 2009/119/EC has established that Member States must maintain a total level of oil stocks corresponding, at the very least, to 90 days of average daily net imports or 61 days of average daily inland consumption, whichever quantity is the greatest. The average daily net imports should be calculated on the basis of the method explained in Annex I of the directive, whilst the procedure for calculating average daily inland consumption is given in Annex II. The procedure for calculating stock levels is laid down in Annex III. Member States are obliged to ensure that stocks are available and physically accessible. They are responsible for putting in place arrangements for the identification, accounting and control of these oil stocks. Each Member State is allowed to set up a Central Stockholding Entity (hereinafter CSE) in the Union in the form of a non-profit making body or service.

236. Each Member State must under Directive 2009/119/EC maintain a minimum level, defined in terms of number of days of consumption, of specific stocks. These specific stocks will be owned by the Member State concerned or the CSE set by it. In accordance with the provisions of the directive, Member States must be able to release all or part of their emergency stocks and specific stocks if required. Members States must comply with the provisions of Directive 2009/119/EC by 31 December 2012 and as from that date, this directive replaces Directive 2006/67/EC.

(b) Security of electricity supply and infrastructure directive and security of natural gas supply directive

237. Blackouts in both the EU and the US have highlighted the need to define clear operational standards, correct maintenance and development of the transmission networks. The European legislator therefore adopted Directive 2005/89/EC concerning measures to safeguard security of electricity supply and infrastructure investment.[256] Directive 2005/89/EC established measures to safeguard security of electricity supply in order to ensure the proper functioning of the EU internal market for electricity, an adequate level of interconnection between Member States, an adequate level of generation capacity and a balance between supply and demand.

[256] Directive 2005/89/EC of the European Parliament and of the Council of 18 January 2006 concerning measures to safeguard security of electricity supply and infrastructure investment, *OJ* L 33/22 (hereinafter Directive 2005/89/EC).

238. With respect to the operational network security, Directive 2005/89/EC obliges transmission system operators to set minimum rules and obligations to ensure continuous operation of the transmission and, where appropriate, the distribution network under foreseeable circumstances. Member States can decide to submit these rules and obligations for approval to competent authorities and, where appropriate, can decide that these rules and obligations also be respected by the transmission network operators. Network operators must also set and meet quality of supply and network security performance objectives.

239. Member States must have a clear policy in place to maintain the balance between electricity supply and demand. They must, in particular, encourage the establishment of wholesale markets, require network operators to ensure that an appropriate level of generation reserve capacity is maintained, facilitate the development of new generation capacity, or encourage energy conservation and technology for demand management in real time.[257]

240. For gas, the European legislator adopted Council Directive 2004/67/EC concerning measures to safeguard security of natural gas supply in 2004.[258] The objective of Council Directive 2004/67/EC is to establish ways to safeguard the security of gas supply and to enhance the functioning of the internal energy market. The added value of this directive for the gas sector is limited and its wording is very general.[259] However, this does not mean that Council Directive 2004/67/EC should be considered redundant. On the contrary, it addresses an aspect of security of energy supply which was previously not dealt with in an adequate manner: the systemic aspect of energy security. The provisions of the directive establishing enhanced cooperation in the form of reporting and monitoring, and the work of the gas coordination group are in this respect a positive contribution in order to enhance the security of supply.[260]

241. Following the proposals made in the Strategic Energy Review and in response to the Russian-Ukrainian gas crisis in January 2009, causing an unprecedented disruption of gas supplies to the Union via Ukraine[261], the

---

257 For a detailed analysis of the Directive 2005/89/EC see H. BJØRNEBYE, "The Security of Supply Directive – The Right Aims but the Wrong Means" in B. DELVAUX, M. HUNT and K. TALUS (eds.), *EU Energy Law and Policy Issues* (Rixensart, Euroconfidentiel, 2008), p. 173–188.

258 Council Directive 2004/67 of 26 April 2004 concerning measures to safeguard security of natural gas supply in 2004, *OJ* L 127/92.

259 K. TALUS, "Energy Security in European Energy Law" in M. HUNT and K. TALUS (eds.), *The EU Energy Directory* (Rixensart, Euroconfidentiel, 2008), p. 86.

260 *Ibid.*

261 For more details see the Commission staff working document of 16 July 2009 accompanying document to the Proposal for a Regulation of the European Parliament and of the Council

European Commission adopted a draft Regulation in 2009 repealing the existing Directive 2004/67/EC concerning measures to safeguard security of natural gas supply.[262, 263]

242. Directive 2004/67/EC was, according to the Commission, no longer sufficient for the new realities of a growing import dependence of the EU and increased supply and transit risks in third countries.[264] Therefore, the European legislator adopted Regulation No 994/2010 concerning measures to safeguard security of gas supply on 20 October 2010 which entered into force on 2 December 2010 and which repealed Directive 2004/67/EC.[265]

243. Regulation No 994/2010 has defined more clearly the roles of the gas industry, the Member States and the Union in order to deal with a supply disruption in the short term, but also with respect to the necessary infrastructure in the longer term. It has strengthened the existing system of Directive 2004/67/EC and ensures that all Member States and their gas market participants take effective action well in advance to prevent and mitigate the consequences of potential disruptions to gas supplies. It establishes a mechanism for Member States to work together, in a spirit of solidarity, when they deal with any major gas disruptions that might arise.

244. The regulation has established common infrastructure and supply standards. According to the infrastructure standard, Member States must ensure at the latest by 3 December 2014 that in the event of disruption of the single largest infrastructure, they are able to satisfy total gas demand during a day of exceptionally high gas demand. According to the supply standard, for protected customers which are all households, natural gas undertakings are obliged to secure supplies to protected customers under severe conditions: in the event of a

---

concerning measures to safeguard security of gas supply and repealing Directive 2004/67/EC, COM(2009)977 and S. PIRANI, J. STERN and K. YAFIMAVA, *The Russo-Ukrainian Gas Dispute of January 2009: A comprehensive assessment* (Oxford, Oxford University Studies, 2009), 66 p.

262 Council Directive 2004/67/EC of 26 April 2004 concerning measures to safeguard security of natural gas supply, *OJ* L 127/92.

263 Also on 3 February 2009, the European Parliament adopted a Resolution on the Second Strategic Energy Review (2008/2239(INI).

264 Proposal of 16 July 2009 for a Regulation of the European Parliament and of the Council concerning measures to safeguard security of gas supply and repealing Directive 2004/67/EC, COM(2009)363, p. 2–4 and see also the Commission staff working document of 16 July 2009 accompanying document to the Proposal for a Regulation of the European Parliament and of the Council concerning measures to safeguard security of gas supply and repealing Directive 2004/67/EC, COM(2009)978.

265 Regulation No 994/2010 of the European Parliament and of the Council of 20 October 2010 concerning measures to safeguard security of gas supply and repealing Council Directive 2004/67/EC, *OJ* L 295/1.

7-day temperature peak and for at least 30 days of high demand, as well as in the case of an infrastructure disruption in normal winter conditions.

245. A competent authority must, by 3 December 2011, make a full assessment of the risks affecting the security of supply. This assessment must take into account the supply and infrastructure standards, all relevant national and regional circumstances, various scenarios of exceptionally high gas demand and supply disruption and the interaction and correlation of risks with other EU Member States. No later than 3 December 2012, the competent authority must, on the basis of the results of the risk assessment, adopt a preventive action plan containing the measures required to remove or mitigate the risk identified, and an emergency plan containing the measures required to remove or mitigate the impact of a gas supply disruption. Every 2 years these plans must be updated.

246. Regulation No 994/2010 has defined three main crisis levels: early warning, alert level and emergency level. The Commission can declare an EU or regional emergency at the request of a competent authority. The regulation has established, as the repealed Directive 2004/67/EC had, a gas coordination group to facilitate the coordination of measures concerning security of gas supply. This group will be composed of representatives of the EU Member States, in particular of their competent authorities, the Agency, the ENTSO-G and representative bodies of the industries concerned and those of relevant customers. During emergencies, the natural gas undertakings concerned are obliged to provide certain information on a daily basis to the competent authority. In the event of an EU or regional emergency, the Commission is entitled to request that the competent authority provides at least information on the measures planned to be taken and already implemented to mitigate the emergency.

3. EXTERNAL SECURITY OF ENERGY SUPPLY

247. Due to the increasing import dependence of the Union energy security has become increasingly intertwined with its foreign relations and the stability of the key producing countries. The main threats of the external aspects of energy supply include reserve depletion, insufficient import infrastructure, insecurity of transit routes, insufficient investment in gas fields, etc. External measures on a European level with respect to energy security can be divided into two categories, on the one hand funding, and international agreements on the other.[266] The next part will discuss in more detail the latter category with respect to security of energy supply.

[266] K. TALUS, "Energy Security in European Energy Law" in M. HUNT and K. TALUS (eds.), *The EU Energy Directory* (Rixensart, Euroconfidentiel, 2008), p. 109.

(a) *Energy Charter Treaty*

248. The process towards the adoption of the Energy Charter Treaty dates back to the early 1990s. At the Dublin European Council in June 1990, the Prime Minister of the Netherlands suggested to establish a framework for international cooperation in the energy sector with the Eastern European and former Soviet Union countries, with the aim of developing the energy potential of Central and Eastern European Countries and of ensuring security of energy supply for the EU. The Commission was invited by the Council to look into the best way of establishing such cooperation and in 1991 the Commission proposed the Energy Charter Treaty. The purpose of this Treaty is to promote East-West industrial cooperation through legal guarantees concerning investments, transit and trade.[267]

249. On 17 December 1994 the Energy Charter Treaty and the Energy Charter Protocol on energy efficiency and related environmental aspects were signed in Lisbon by all signatories to the 1991 Charter except for the United States of America and Canada. The objective of this Treaty, according to its Article 2, is *"to establish a legal framework to promote long-term cooperation in the energy sector based on the principles enshrined in the European Energy Charter"*. The main provisions of the Treaty concern the protection of investment, trade in energy materials and products, transit and dispute settlement. Despite the fact that this Treaty has been ratified by various countries, Russia among other oil and gas producing countries has not ratified the Treaty. Russia argues that the Treaty is from a producer's perspective too consumer-friendly and does not contain the necessary balance between the consumer and producer countries.[268] Thus, the scope of the Treaty is rather limited.

(b) *Energy Community Treaty*

250. Another initiative established in order to increase the Union's international relations in the energy area is the adoption of the Treaty establishing the Energy Community for South East Europe or the so-called Energy Community Treaty. This Treaty has created an internal market in electricity and gas between the EU and 7 European states in the Balkans.[269] It

---

[267] Council and Commission Decision 98/191/EC, ECSC, EURATOM of 23 September 1997 on the conclusion, by the European Communities, of the Energy Charter Treaty and the Energy Charter Protocol on energy efficiency and related environmental aspects, *OJ* L 69/1.

[268] K. TALUS, "Energy Security in European Energy Law" in M. HUNT and K. TALUS (eds.), *The EU Energy Directory* (Rixensart, Euroconfidentiel, 2008), p. 109.

[269] The 7 European states in the Balkan are Albania, Bosnia and Herzegovina, Croatia, the former Yugoslav Republic of Macedonia, Montenegro, Serbia and the United Nations Interim Administration Mission in Kosovo pursuant to United Nations Security Council Resolution 1244.

applies to the territory of the contracting parties and has entered into force on 1 July 2006 for a term of ten years.

251. The objectives of the Energy Community Treaty are to (a) create a stable legal framework capable of attracting investments in order to ensure a stable and continuous energy supply, (b) create a single regulatory space for trade in network energy, enhance security of supply, (c) enhance security of energy supply and develop cross-border relations, (d) improve energy efficiency and the environmental situation related to network energy, (e) develop renewable energy sources and (f) develop network energy market competition. This Treaty is a key instrument of the Union's strategy in South East Europe, but it is also an instrument of pre-accession for countries in that region.[270]

252. The main part of the Energy Community's activities of the signatory parties is the adoption of the *acquis communautaire* within their national legislation. With respect to the notion of *acquis,* the Treaty adopted a clear definition of its content. On the basis of this definition only the Union's energy, environment, competition and renewable energies *acquis*, as well as compliance with certain general Union standards relating to technical systems, must be taken into account. The Treaty, in addition, establishes a mechanism for operation of regional energy markets which covers the territory of the parties to the Energy Community Treaty and the some EU Member States (Austria, Bulgaria, Greece, Hungary, Italy, Romania and Slovenia). This mechanism established a framework of measures relating to long-distance transportation of network energy, security of supply, harmonisation, etc.

253. The Energy Community Treaty has also created an energy market without internal frontiers between the parties. In this energy market, customs duties and quantitative energy import and export restrictions, and any measures having equivalent effect are prohibited between the parties unless exceptional circumstances apply. The Treaty also contains provisions on relations with third countries and mutual assistance in the event of disturbance.

254. The institutional framework of the Energy Community Treaty is to a large extent similar to that on a European level. The most important institution is the Ministerial Council which provides the general policy guidelines, takes measures to meet the objectives of the Treaty and adopts procedural acts. Another institution is the Permanent High Level Group which prepares the work of the

---

[270] M. HUNT and R. KAROVA, "The Energy Acquis Under the Energy Community Treaty and the Integration of South East European Electricity Markets: An Uneasy Relationship?" in B. DELVAUX, M. HUNT and K. TALUS (eds.), *EU Energy Law and Policy Issues* (Rixensart, Euroconfidentiel, 2010), p. 54.

Ministerial Council and consists of one representative of each party of the Treaty. The third institution is the Regulatory Board which advises the other institutions and issues recommendations in the event of cross-border disputes. The Energy Community is also advised by two fora composed of representatives of all interested parties. Finally, there is the permanent Secretariat which has its seat in Vienna and provides, amongst other things, administrative support to all the other institutions of the Energy Community.[271]

255. In addition to the two treaties discussed above, the European Union established a number of bilateral measures with third countries such as the EU-Ukraine cooperation, Black Sea Synergy, Africa-EU Energy Partnership, Energy Facility for ACP countries, etc.[272]

E. PROBLEMS FACED WITH EU SECONDARY LEGISLATION IN THE ENERGY FIELD

256. It is essential that the current legal framework adopted on a European level, in order to liberalise the energy market, is properly transposed into national legislation. However, the implementation of this European legal framework is very disappointing. In June 2009, the European Commission has launched more than 40 infringement procedures on the second internal energy market package.

257. The key violations identified lack of transparency, insufficient coordination efforts by transmission system operators (hereinafter TSOs) to make maximum interconnection capacity available, absence of regional cooperation, lack of enforcement action by the competent authorities in Member States and the lack of adequate dispute settlement procedures.[273, 274]

---

271 For more details See M. HUNT and R. KAROVA, "The Energy Acquis Under the Energy Community Treaty and the Integration of South East European Electricity Markets: An Uneasy Relationship?" in B. DELVAUX, M. HUNT and K. TALUS (eds.), *EU Energy Law and Policy Issues* (Rixensart, Euroconfidentiel, 2010), p. 51–86.

272 K. TALUS, "Energy Security in European Energy Law" in M. HUNT and K. TALUS (eds.), *The EU Energy Directory* (Rixensart, Euroconfidentiel, 2008), p. 109.

273 Communication from the Commission of 11 March 2010 to the Council and the European Parliament, *Report on progress in creating the internal gas and electricity market*, COM(2010)84, p. 2.

274 One of those Member States which had not properly transposed the European legal framework is Belgium. Thus in October 2009, the European Commission has initiated further infringement proceedings. Article 2 of the Belgian Law of 10 March 2009 introduces exemptions for natural gas transit contracts from the general rules on access to the network for the transit of natural gas and provides for negotiated tariffs which apply for a period fixed by contract. Moreover, the law makes it possible to set a fair profit margin for transit which is clearly higher than that applicable to other transmission activities and makes a distinction

258. In June 2010, the European Commission sent 35 separate requests to 20 Member States to implement and apply in full, the different aspects of EU legislation creating a single market for gas and electricity. These 20 Member States had two months to respond to the requests which took the form of reasoned opinions under the EU infringement procedures. In the absence of satisfactory responses from Member States with the reasoned opinions, the European Commission may decide to refer them to the Court of Justice.

The main violations identified by the European Commission concern: lack of information provided by electricity and gas TSOs which thereby obstruct effective access for supply companies, lack of coordination and cooperation across borders by electricity TSOs which is necessary in order to allocate better any network capacity on cross-border interconnections, inadequate efforts by gas TSOs to offer the possibility to network users to enter upstream gas markets, lack of effective enforcement action by the competent authorities in Member States in the event of violations of the EU regulations, including the absence of effective penalties on a national level, etc.[275]

259. The adoption of different legislative provisions is one thing, applying and respecting these provisions is another. As we have seen above, different Member States are still in breach with the implementation of the second package of secondary energy market legislation. This lack of implementation and enforcement of secondary legislation by the Member States undermines the credibility of the common EU energy policy. Moreover, given the fact that it takes several years before the Court of Justice reaches a final judgment, in the meantime Member States often continue to ignore their legal obligations.

---

between existing and future installations. According to the European Commission, the Belgian Act of 10 March 2009 discriminates between network users carrying on similar activities, a practice which is incompatible with Union law. The European Commission has emphasised that non-discriminatory third-party access to the natural gas transmission network has become a central aspect of the opening up of the markets to competition. Since the markets were fully opened up to competition on 1 July 2007, Union law no longer makes any distinction between transmission of gas involving transit of natural gas and the transmission of gas intended for national customers. Hence, the European Commission has concluded that Belgium, with the adoption of Article 2 of the Belgian Act of 10 March 2009 introducing exemptions for natural gas transit contracts, infringes Union law. See IP/09/1490 of 8 October 2009, The Commission opens infringement proceedings against Belgium concerning its gas transit system.

275 Communication from the Commission of 10 January 2007 to the Council and the European Parliament, *Prospects for the internal gas and electricity market*, COM(2006)841, p. 6 and 7; Communication from the Commission of 10 November 2010 to the Council, the European Parliament, the European Economic and Social Committee and the Committee of the Regions, *Energy 2020 A Strategy for competitive, sustainable and secure energy*, COM(2010)639, p. 2 and IP/10/836 of 24 June 2010, Commission requests 20 Member States to implement and apply Single Market rules without delay.

260. As highlighted by the Commission in its sector inquiry, there are different distortions of competition in both the electricity and gas markets which market players could raise. However, enforcement by market players remains the exception in the energy market. The Commission is today still the main enforcer in the electricity and gas markets.[276]

261. The traditional reliance on framework directives in the energy field ensuring minimal harmonisation and allowing much room for Member States to adopt divergent and heterogeneous implementing rules and regulations has resulted in a patchwork structure for the energy sector. In various Member States the divergent national provisions continue to frustrate the completion and accessibility of the internal markets for electricity and gas. Also the possibility for Member States to derogate from secondary legislation and the fact that they can sometimes tag additional measures on the back of European directives (so-called gold-plating), undermines the consistency of the EU energy policy.

262. Many recently adopted energy provisions are of a declaratory or facilitative nature and thus leave it to the Member States or the market players to take particular action of initiative, if required. It has also become the order of the day that *ad hoc* expert bodies adopt non-binding guidelines and declarations which are constantly fine-tuning the measures in the energy field.

263. There is also an apparent failure to learn from experience gained in enforcing one set of objectives to another.[277] For example, binding targets for renewable energy sources have been ignored for a decade; however in 2009 they were finally accepted with the adoption of the Renewable Energy Directive 2009/28/EC.[278] A similar process can be expected for investments in energy infrastructure.

264. Not only energy companies, but also all the European consumers cannot yet reap the fruits of a true internal market in energy.[279] Consumers should not

276 Policy Proposal by Jacques Delors, *Towards a European Energy Community: A Policy proposal*, study by S. ANDOURA, L. HANCHER and M. VAN DER WOUDE, p. 32 and 67.

277 Various reports of the past years and in particular the Renewable Energy Roadmap, have highlighted the slow progress Member States were making in achieving the 12% share of renewable energy in gross inland consumption by 2010. It was even unlikely that the EU as a whole would exceed 10% by 2010. See Communication from the Commission to the Council and the European Parliament, *Renewable Energy Road Map Renewable energies in the 21st century: building a more sustainble future*, COM(2006)848, 20 p.

278 See in particular Article 3 and Annex I.

279 Communication from the Commission of 10 November 2010 to the Council, the European Parliament, the European Economic and Social Committee and the Committee of the Regions, *Energy 2020 A Strategy for competitive, sustainable and secure energy*, COM(2010)639, p. 3.

only have a legal right on paper to choose their electricity and/or gas supplier freely between any EU energy company, but it should be a real, effective and easily exercised right. Nonetheless, a wide range of EU legislative and policy initiatives currently contribute directly or indirectly to improve consumer welfare in the energy sector.

265. The Union toolbox is inadequate and incomplete. For example, the SET Plan towards a low-carbon future is lacking significant (financial) incentives or legal obligations for Member States to make incentives available[280], the move to greater energy efficiency is too slowly, there is still no common approach towards external energy suppliers and transit countries, Europe's energy markets are still characterised by fragmentation, there are still numerous barriers to open and fair competition in the electricity and gas market, the functioning of the retail markets needs to be improved, etc.[281]

266. In addition, the current decision-making process is not efficient. A very effective tool to enhance the rational use of energy and energy sustainability could be the adoption of tax measures, in particular a CO_2 tax on a European level. However, due to the fact that the TFEU requires unanimity for the adoption of measures of a fiscal nature in the decision-making process, it will be very difficult to implement such kind of measures. With this unanimity requirement, only one Member State could in practice impede the adoption of tax measures on a European level.

267. Besides the inefficient decision-making process, the success of a sustainable, competitive and secure European legal framework in the energy field depends on the willingness of the different actors directly and indirectly involved in the drafting and adoption of secondary legislation. Some Member States are still reluctant to transfer more and more competences in the energy sector to EU institutions. Consequently, the result of the adoption process of secondary legislation is often a political compromise. A recent example in the energy field is the debate on a European level among Member State governments, the European Parliament and the Commission with respect to the ownership unbundling provisions of transmission networks in the energy sector.[282] Such political compromises undoubtedly have an impact on the quality of adopted regulations, directives, decisions, etc. in the energy field.

---

280 Policy Proposal by Jacques Delors, *Towards a European Energy Community: A Policy proposal*, study by S. ANDOURA, L. HANCHER and M. VAN DER WOUDE, p. 65–71.

281 Communication from the Commission of 10 November 2010 to the Council, the European Parliament, the European Economic and Social Committee and the Committee of the Regions, *Energy 2020 A Strategy for competitive, sustainable and secure energy*, COM(2010)639, 20 p.

282 See *Infra*.

268. Energy markets are complex markets which often raise new issues which require decisions to be taken sooner rather than later. However, the current decision-making process of the institutional machinery on a European level is by far sufficiently flexible to adopt at any given moment in time the necessary measures.[283]

269. In conclusion, given the current institutional setting on a European level it is uncertain whether the Union is capable of achieving its three central goals by 2020. Even with the entry into force of the Lisbon Treaty, it cannot be denied that the decision-making process on a European level for the adoption of secondary legislation is far from efficient.

F. CONCLUSION

270. This chapter has provided an overview of the most relevant secondary legislation currently adopted in order to achieve the three central goals: competitiveness, security of supply and sustainability of the European energy policy and law. It was argued that, at the moment, the Union has a relatively well-developed set of measures to achieve these three central goals of its energy policy. Moreover, the on-going attempts on a European level to improve its secondary legislation cannot be denied.

271. However, it is uncertain whether the Union is capable of achieving its three central goals by 2020. The EU is currently a champion in setting ambitious targets for 2020 and beyond 2030, but the requisite legal instruments are seldom aligned to deliver these targets.[284] Secondary legislation in the energy field is confronted so far with a lack of implementation and enforcement by the Member States. This undermines the credibility of a common EU energy policy.

272. The traditional reliance on framework directives in the energy field gives Member States too much room and often results in inconsistencies in the energy field. Moreover, the current decision-making process for the adoption of secondary legislation in the energy field is far from efficient. Even with the entry into force of the Lisbon Treaty, it cannot be denied that the decision-making process on a European level for the adoption of secondary legislation is far from efficient. Moreover, since the adoption of legislation in the energy field often the result is of a political comprise, it will certainly not improve its quality.

---

283 *Ibid.*, p. 82.

284 Policy Proposal by Jacques Delors, *Towards a European Energy Community: A Policy proposal*, study by S. ANDOURA, L. HANCHER and M. VAN DER WOUDE, p. 27–60.

273. Since energy is indispensable for the daily functioning of society, it is undeniable that urgent action is required. In order to determine who (EU and/or Member States) is competent to act in the energy field and to which extent, the next chapter will discuss in detail the relevant primary provisions of EU law.

274. The current energy and climate change challenges can hardly be conceived without cooperation and therefore preference should be given to action on a European level. Such action will likely lead to faster and better results than national action in meeting the energy and climate change challenges. Moreover, national action could lead to diverging energy policies which increases the risk of inconsistencies.[285]

275. However, action on a European and national level in the energy field must be compatible with the provisions of the Union Treaties. The next chapter will therefore identify, on the basis of the principles of attribution, subsidiarity and proportionality and the provisions of the Union Treaties, to which extent, before the adoption of the Lisbon Treaty, the Union was competent to adopt measures in the energy field. It also discusses some of the most relevant provisions of primary law which could limit the adoption of measures in the energy field both on a European and national level.

---

[285] *Ibid.*, p. 78 and 79.

CHAPTER 4

PRINCIPLES OF EU LAW AND LEGAL BASES IN THE TFEU ALLOWING UNION ACTION IN THE FIELD OF ENERGY PRIOR TO THE LISBON TREATY

A. INTRODUCTION

276. The previous chapters have discussed the objectives and key EU secondary legislation in the energy field. The overall objective of this book is to investigate whether the current legal framework on a European level for energy law is sufficient to address the three objectives. Thus, this chapter analyses the provisions of primary law which allowed Union action in the energy field prior to the adoption of the Lisbon Treaty. Moreover, chapter 4 discusses some of the relevant general TFEU provisions and their application in the energy sector both on a European and national level. This analysis is based upon the existing legal doctrine and case law of mainly the Court of Justice.

277. It should be highlighted that although chapter 4 will analyse the primary Treaty provision prior to the adoption of the Lisbon Treaty, it will apply the new numbering of the provision of the TFEU in order to facilitate its reading.

278. Until recently, before the adoption of the Lisbon Treaty, the Union did not have explicit competence in the energy area although energy-related Union legislation has been adopted since the late 1960s. The lack of any specific articles, chapter or title on energy in the EC Treaty has not prevented the Union from enacting various measures in the energy sphere,[286] based on non-energy provisions and principles (attribution, subsidiarity and proportionality) of the TFEU. Thus, in the energy sphere, the Union intervened, on the one hand, by enforcing the general TFEU provisions and, on the other hand, by adopting secondary legislation on the basis of its competences in the internal market, tax,

[286] E. D. CROSS, B. DELVAUX, L. HANCHER, P. J. SLOT, G. VAN CALSTER and W. VANDENBERGHE, "EU Energy Law (chapter 5)" in M. M. ROGGENKAMP, A. RØNNE, C. REDGWELL and I. DEL GUAYO (ed.) *Energy Law in Europe. National, EU and International Regulation*, 2nd ed. (Oxford, Oxford University Press, 2007), p. 227.

infrastructure and environmental measures. The analysis of these provisions is important because the adoption of the new energy title in the Lisbon Treaty does not necessarily mean that all measures in the energy field need to be based on this new title.[287]

279. The purpose of this chapter is to provide firstly an historical overview of the Treaty context with respect to energy. This part is followed by an analysis of the principles of attribution, subsidiarity and proportionality which govern the existence and exercise of Union powers and their application to the energy field. The next part of this chapter is a detailed discussion of the most relevant general provisions of the TFEU and their impact on the adoption of legislation in the energy field. This part is followed by an analysis of the different legal bases for the adoption of measures in the energy field on a European Level.[288]

B. ENERGY AND THE TREATY CONTEXT: HISTORICAL OVERVIEW

1. INTRODUCTION

280. Since the post-World War reconstruction, Europe has recognised the necessity of a coordinated action in the energy sector. The primary concern to both the Communities and its Member States in that period was the issue of energy supply.[289]

281. Energy was considered as *"a production factor of fundamental and universal importance to economic activity, which cannot be ignored or downgraded in the construction of an economic union"*.[290] This is reflected in the

287 See *Infra*.

288 Please note that in order to simplify the readability of the different provisions of the Treaties, the new Treaty numbers and wording of the Treaty provisions have been used in this chapter.

289 J. SCHWARZE, "European Energy Policy in Community Law" in E. J. MESTMÄCKER (ed.), *Natural Gas in the Internal Market – A Review of Energy Policy* (London, Graham and Trotman, 1993), p. 153 and 154. See furthermore, E. F. SCHUMACHER, "The Struggle for a European Energy Policy", *Journal of Common Market Studies* (1964), p. 199 *et seq.*; P. O. LAPIE, "La Politique énergétique commune", *Revue Marché Commun* (1964), p. 134–141; W. MÜLLER-MICHAELIS, *Die Integration der westeuropäischen Energiemärkte* (Göttingen, Vandenhoeck u. Ruprecht, 1968); S. ANADUCCI, "Towards a Comprehensive Community Energy Policy: Some Legal Considerations", *Diritto comunitario e degli scambi internazionali* (1976), p. 247–281; M. BYWATER, "La Politique énergétique commune", *Revue Marché Commun* (1977), p. 409 *et seq.*; H. MICHAELIS, *Energiemarkt und Energiepolitik in einer Europäischen Union* (Frankfurt am Main, A. Metzner, 1976), 96 p. and N. GREEN, "The legal basis of a Community Energy Policy", *European Law Review* (1983), p. 52 *et seq.*

290 T. DAINTITH and S. F. WILLIAMS, *The legal integration of energy markets* (Berlin, W. De Gruyter, 1987), p. 5.

European Treaties of the 1950s since two of the three treaties are specifically directed at the energy sector. This part seeks to provide an overview of the main developments in the Union Treaties that have arisen prior to the adoption of the Lisbon Treaty.

2. TREATY ESTABLISHING THE EUROPEAN COAL AND STEEL COMMUNITY

282. The 1951 Treaty of Paris has established the Treaty establishing the European Coal and Steel Community (hereinafter ECSC Treaty) with the aim of transferring the control of coal and steel industries from national authorities to the High Authority. The adoption of the ECSC Treaty in the 1950s was based upon the scarcity of coal and the high demand for it. In those years most of the energy consumed was coal. Therefore six Member States[291] have tried to establish a common European market for coal with free circulation and undistorted competition. Fifty years after entering into force, the ECSC Treaty expired on 23 July 2002 as planned.[292]

283. Article 3 of the ECSC Treaty contained the first expression of the concept of security of supply in Union law since it mentioned that a regular supply of coal and steel to the common market was one of the main objectives of the Union. Not only security of supply was listed as one of the main objectives, but also sustainability by promoting *"a policy of using natural resources rationally and avoiding their unconsidered exhaustion"*.[293]

284. According to Article 4 of the ECSC Treaty, the objective of the common European market for coal, with free circulation and undistorted competition, should be attained by abolishing and prohibiting within the Union (a) import and export duties, or charges with an equivalent effect, and quantitative restrictions on the movement of coal; (b) measures or practices discriminating among producers, buyers or consumers; (c) subsidies or state assistance, or special charges imposed by the state and (d) restrictive practices tending towards the division of markets or the exploitation of the consumer. However, the competitive pattern for the coal market has to a certain extent been limited by other objectives in the ECSC Treaty with respect to employment, cost of living and supply.

---

291 France, Germany, Italy, Belgium, Luxembourg and the Netherlands.

292 Article 97 of the ECSC Treaty.

293 Policy Proposal by Jacques Delors, *Towards a European Energy Community: A Policy proposal*, study by S. ANDOURA, L. HANCHER and M. VAN DER WOUDE, p. 7 and 8.

285. The complex system of regulation laid down in the ECSC Treaty granted rigid competences to intervene and to influence developments in the coal market effectively. It allowed the High Authority[294] to intervene in matters of the pricing, competition and commercial policy, crisis management and matters of financial and social concern.[295] Moreover, these competences were supported by powers of information, control and sanctions. Given these competences the High Authority had truly supranational powers.[296] However, during the 1950s up to 1960s, the general shortage of coal turned into a surplus which led to a radical decline of the coal industry. Consequently, the comprehensive regulatory powers of the High Authority remained dead letter.[297]

3. TREATY ESTABLISHING THE EUROPEAN ATOMIC ENERGY COMMUNITY

286. Another instrument aiming at integrating the energy sector on a European level was the Treaty establishing the European Atomic Energy Community (hereinafter EURATOM Treaty), signed in Rome on 25 March 1957 for an unlimited period, establishing the European Atomic Energy Community.[298] At that moment the use of nuclear energy was still in an exploratory stage. This can be derived from the Treaty itself which was essentially research-driven.[299]

287. Moreover, it aimed at reducing the energy import dependency and to counter the dominance of the former USSR and the USA in the nuclear field at that moment. During the uncertainties in the supply of oil caused by the Suez crisis in 1956[300], nuclear energy was furthermore advocated as a safe source of energy for the European countries. Another impetus to cooperate in a Union structure was the very considerable investment that would be required to fund

---

294 The High Authority was essentially financed by levies imposed on coal and steel production.

295 See Articles 53–54, 60–64, 65–70, 72, 57–59 of the ECSC Treaty.

296 Policy Proposal by Jacques Delors, *Towards a European Energy Community: A Policy proposal*, study by S. ANDOURA, L. HANCHER and M. VAN DER WOUDE, p. 7 and 8.

297 U. IMMENGA, "The Development of European Energy Policy: From the ECSC Treaty to the Internal market" in E. J. MESTMÄCKER (ed.), *Natural Gas in the Internal Market – A Review of Energy Policy* (London, Graham and Trotman, 1993), p. 47–49 and J. SCHWARZE, "European Energy Policy in Community Law" in E. J. MESTMÄCKER (ed.), *Natural Gas in the Internal Market – A Review of Energy Policy* (London, Graham and Trotman, 1993), p. 153 and 154.

298 The six founding states were France, Germany, Italy, Belgium, Luxembourg and the Netherlands.

299 Policy Proposal by Jacques Delors, *Towards a European Energy Community: A Policy proposal*, study by S. ANDOURA, L. HANCHER and M. VAN DER WOUDE, p. 9.

300 See OEEC, *Europe's Need for Oil: Implications and Lessons of Suez Crisis* (Paris, OEEC, 1958), 106 p.

the development of the nuclear fuel cycle. However, due to the lack of consensus in this industry between the Member States and the fact that the nuclear industry is well-established and needs little financial support from the EU since it receives enough from national bodies, the latter impetus to cooperate has minor significance.[301]

288. The overall aim of the EURATOM Treaty was to promote nuclear energy for civil purposes and to regulate the emerging nuclear industry. This can be derived from Article 1 of the EURATOM Treaty which states that: *"It shall be the task of the Community to contribute to the raising of the standard of living in the Member States and to the development of relations with the other countries by creating the conditions necessary for the speedy establishment and growth of nuclear industries"*. In order to perform this task, the EURATOM Treaty contains a number of provisions regulating the supply of nuclear material, prices, establishment of safety standards, investments, property questions, external relations, research and free trade, in order to establish a common market for the substances and the equipment used in the nuclear energy sector.[302]

289. The main body of the EURATOM Treaty is the Supply Agency. This Agency has the right of option to procure materials produced in the territories of Member States. Thus, every producer of nuclear materials in the Union must offer its output first to the Agency. When the Agency exercises this right it will acquire property of these nuclear materials except for special fissile materials the property of which is vested in the Union.[303] The Agency also has the exclusive right to conclude contracts with respect to the supply of nuclear materials coming from inside or outside the Union. In practice the need for such a central intervention by the Agency appeared not to be necessary since market forces were able to match supply and demand.[304]

---

301 P. CAMERON, *Competition in Energy Markets, Law and Regulation in the European Union* (Oxford, Oxford University Press, 2007), p. 44–46 and 245 and 246.

302 See the Articles 4, 12, 24–27, 30–39, 40–52, 52–76, 67–69, 86, 92–100 of the EURATOM Treaty.

303 See Article 86 of the EURATOM Treaty.

304 Policy Proposal by Jacques Delors, Towards a European Energy Community: A Policy proposal, study by S. ANDOURA, L. HANCHER and M. VAN DER WOUDE, p. 10. For further details see EAEC Council: The Statutes of the Euratom Supply Agency, *OJ* L 27/534; EAEC Supply Agency: Rules of the Supply Agency of the European Atomic Energy Community determining the manner in which demand is to be balanced against the supply of ores, source materials and special fissile materials, *OJ* L 32/777 and Regulation of the Supply Agency of the European Atomic Energy Community amending the rules of the Supply Agency of 5 May 1960 determining the manner in which demand is to be balanced against the supply of ores, source materials and special fissile materials, *OJ* L 193/37. See C. TRUE, "Legislative competences of Euratom and the European Community in the energy sector: the Nuclear Package of the Commission", *European Law Review* (2003), p. 664–685; P. BOWDEN, "Nuclear Regulation: a brave new world?", *EELR* (1995), p. 334–336; I. CENEVSKA, "The European Parliament and the European Atomic Energy Community: a legitimacy crisis?",

290. In this area, the EURATOM Treaty remains currently the primary source of competence for the Union. Given the new trends in energy law and policy on a European level, one may wonder whether the provisions of the EURATOM Treaty are still relevant and whether they should not be adjusted. Should the EURATOM Treaty not contain provisions with respect to the decommissioning of nuclear power plants or the management of high-level radioactive waste in order to avoid market distortions, but also from an environmental and sustainable point of view? Another element which could be taken into account in this context is the issue of security of supply. Should the EURATOM Treaty not contain any provisions in this respect? In my view there is undoubtedly a need to evaluate the provisions of the EURATOM Treaty and to adopt them in coherence with the new developments in energy law and policy in the EU.[305]

4. TREATY ESTABLISHING THE EUROPEAN ECONOMIC COMMUNITY

291. At the same time as the signing of the EURATOM Treaty, the Treaty of Rome has founded the European Economic Community designed to create an internal market. The EEC Treaty contains no provisions on energy policy.[306] Hence the energy policy had to be adopted in Union law on the basis of the more general provisions of the EEC Treaty such as the rules on fundamental freedoms and the rules on the internal market. Since the EEC Treaty does not set a regulatory framework for an energy policy as a whole, but rather contains some indications of how energy should be dealt with, the adoption of a general energy policy on a European level was to some extent a patchwork.[307]

292. Due to the fragmentary nature of the EEC Treaty with regard to energy policy the main integration in this field needs to be provided through secondary legislation. This process was even more complex because of the immense differences in Members States' attitudes towards energy markets (dirigisme *versus* free market philosophy), but also because of the national interests in the

---

European Law Review (2010), p. 415–424 and P. D. CAMERON, "The revival of nuclear power: an analysis of the legal implications", *Journal of Environmental Law* (2007), p. 71–87.

305 Similarly, P. CAMERON, *Competition in Energy Markets, Law and Regulation in the European Union* (Oxford, Oxford University Press, 2007), p. 249 and 250.

306 E. D. CROSS, L. HANCHER, and P. J. SLOT, "EC Energy Law" in M. M. ROGGENKAMP, A. RØNNE, C. REDGWELL and I. DEL GUAYO (eds.), *Energy Law in Europe. National, EU and International Law and Institutions* (Oxford, Oxford University Press, 2001), p. 214–215 and P. CAMERON, *Competition in Energy Markets, Law and Regulation in the European Union* (Oxford, Oxford University Press, 2007), p. 46.

307 J. SCHWARZE, "European Energy Policy in Community Law" in E. J. MESTMÄCKER (ed.), *Natural Gas in the Internal Market – A Review of Energy Policy* (London, Graham and Trotman, 1993), p. 155.

energy sector.[308] Energy legislation has been confronted for a long time with the paradox that whenever the Commission proposed a coherent set of measures, the Council refused to take a decision on principles. By contrast, if the Commission suggested a specific measure, the Council refused to decide because of the absence of a broader context.[309] From the adoption of the three Treaties until 1973 no effective energy policy was developed. Nevertheless, during this period the energy policy on a European and national level was one of the most discussed topics.[310]

(a) Adoption of the Single European Act

293. With the entry into force of the Single European Act[311] on 1 July 1987, the integration process was growing.[312] The provisions of the SEA were the result of various concurrent efforts to revive the European integration process from a political and economic point of view. The SEA introduced a number of important modifications to the EEC Treaty the main goal of which was a closer economic cooperation. This can be found in Article 26 of the TFEU which establishes the objective of creating an internal market by the end of 1992. This article defines an internal market as *"an area without internal frontiers in which the free movement of goods, persons, services and capital is ensured in accordance with the provisions of this Treaty"*.

294. In order to achieve the internal market, the SEA has amended the legislative procedures to allow for more decisions on the completion of the internal market to be taken by a qualified majority instead of unanimity. The qualified majority vote was adopted in order to facilitate the adoption of legislation. For example, many of the measures to complete the internal market by the end of 1992 required a qualified majority vote and consequently denied the ability of a single Member State or a small minority of Member States to block this kind of measures.[313] A new Court of First Instance (now General Court) was established and in certain fields the powers of the Union institutions

308 T. DAINTITH and S. F. WILLIAMS, *The legal integration of energy markets* (Berlin, W. De Gruyter, 1987), p. 136 en 137.

309 J. SCHWARZE, "European Energy Policy in Community Law" in E. J. MESTMÄCKER (ed.), *Natural Gas in the Internal Market – A Review of Energy Policy* (London, Graham and Trotman, 1993), p. 155 and R. DE BAUW, "La politique énergétique" in D. LASOK and P. SOLDATOS (ed.), *Les Communautés Européennes en fonctionnement* (Bruxelles, Bruylant 1981), p. 309–332, quoted from T. DAINTITH and S. F. WILLIAMS, *The legal integration of energy markets* (Berlin, W. De Gruyter, 1987), p. 137.

310 *Ibid.*

311 Single European Act (hereinafter SEA), signed on 17 February 1986, *OJ* L 169/1.

312 P. CAMERON, *Competition in Energy Markets, Law and Regulation in the European Union*, (Oxford, Oxford University Press, 2007), p. 51–54.

313 S. WEATHERILL and P. BEAUMONT, *EU law – The essential guide to legal workings of the European Union* (London, Penguin Books, 1993), p. 9.

were codified. The SEA also strengthened the powers of the European Parliament and increased the Parliamentary involvement by allowing its cooperation in the adoption of Union legislation.

295. With respect to energy law and policy, the SEA and the Commission's White Paper on the completion of the internal market[314], despite the central importance of the energy sector, did not expressly refer to an energy policy. The SEA itself did not introduce a provision in the EC Treaty covering only energy law and policy. This was because the Member States expressed that under the SEA the Union should not assume additional powers in the energy field.[315]

296. The SEA also introduced new fields of activity such as economic and social cohesion, research and technological development. Since these provisions to some extent confirmed initiatives which had already been embarked on, the SEA has put these developments on a formal footing.[316]

297. Another relevant provision was the introduction of the new environmental title into the EEC Treaty. This title included Article 130r (now Article 191 of the TFEU and ex Article 174 of the EC Treaty), which refers to the *"prudent and rational utilisation of natural resources"* and Article 130s (now Article 192 of the TFEU and ex Article 175 of the EC Treaty) that refers to the adoption of measures *"significantly affecting a Member State's choice between different energy sources and the general structure of its energy supply"*.

298. With respect to Article 130r, the Intergovernmental Conference noted in a declaration adopted on 9 September 1985 and annexed to the SEA[317], that the Union's activities in environmental matters may not interfere with national policies on the exploitation of energy resources.[318] In the view of some authors the legal effect of this declaration is doubtful.[319] However, the continuing political sensitivity of the Member States towards energy policy matters has been confirmed by the mere existence of this declaration. The reluctance of the Member States, during the thirty years since the EEC Treaty came into force, to

314 Commission of the European Communities, *Completing the Internal Market: White Paper from the Commission to the European Council* of 14 June 1985, COM(1985)310.

315 L. HANCHER, "A single European Energy Market – rhetoric or reality?", *Energy Law Journal* (1990), p. 217–219.

316 S. WEATHERILL and P. BEAUMONT, *EU law – The essential guide to legal workings of the European Union* (London, Penguin Books, 1993), p. 10.

317 This declaration can be found on the following link: http://europa.eu/abc/treaties/archives/en/entr14b.htm.

318 P. CAMERON, *Competition in Energy Markets, Law and Regulation in the European Union*, (Oxford, Oxford University Press, 2007), p. 53.

319 A.G. TOTH, "The Legal status of the Declarations Annexed to the Single European Act", *Common Market Law Review* (1986), p. 811 and 812.

transfer any sovereignty to the Union institutions in the energy field has undoubtedly led to the continuing failure on the part of the Commission to adopt meaningful measures in order to tackle the energy questions.[320]

299. Following the adoption of the SEA, the European Community Council adopted in its resolution of September 1986 its energy objectives for 1995 which recognised the need for *"greater integration, free from barriers to trade, of the internal energy market with a view to improving security of supply, reducing costs, and improving economic competiveness"*. This resolution furthermore stipulated that the Commission must ensure a proper balance between the energy and environmental objectives *"taking into account the desire to limit distortions of competition in the energy markets by a more coordinated approach to environmental affairs in the Community"*.[321]

(b) Adoption of the European Union Treaty

300. After the adoption of the SEA, the Heads of Government and State of the Member States met in December 1991 in Maastricht and have agreed to adopt further changes to the Union structure and have devised the European Union as a broader structure. The new European Union Treaty (hereinafter EU Treaty) was signed on 7 February 1992 and entered into force on 1 November 1993. It contains various amendments to the Treaty of Rome and in doing so it altered its name from the EEC Treaty to the European Community Treaty (hereinafter EC Treaty). The EU Treaty also established the so-called pillars of activity to be pursued by the Union which were the Common Foreign and Security Policy and Justice and Home Affairs. It has also added several Union competences such as culture, public health, consumer protection, industry, development cooperation and has strengthened the provisions with respect to economic and social cohesion.[322]

301. The EU Treaty has also made some changes to the EEC Treaty which are relevant to energy law and policy. It has introduced (a) Article 3(1)(u) which lists measures in the *"sphere of energy, civil protection and tourism"* as one of the Union's common policies[323]; (b) the introduction of the notion of subsidiarity as a principle of general application in Article 3b (now Article 5 of the TEU and ex

320 L. HANCHER, "A single European Energy Market – rhetoric or reality?", *Energy Law Journal* (1990), p. 219 and 220.

321 Council Resolution of 16 September 1986 concerning new Community energy policy objectives for 1995 and convergence of the policies of the Member States, *OJ* C 241/1.

322 S. WEATHERILL and P. BEAUMONT, *EU law – The essential guide to legal workings of the European Union* (London, Penguin Books, 1993), p. 10–17.

323 U. EHRICKE and D. HACKLÄNDER, "European Energy Policy on the Basis of the New Provisions in the Treaty of Lisbon" in A. BAUSCH and B. SCHWENKER (eds.), *Handbook Utility Management* (Berlin Heidelberg, Springer-Verlag, 2009), p. 742.

Article 5 of the EC Treaty)[324] and (c) Article 129b (now Article 170 of the TFEU and ex Article 154 of the EC Treaty) on Trans-European Networks, which provides for a financing vehicle to correct the legacy of network; the energy transportation networks were not designed with a view to achieve a single European energy market.

302. It also contained a Declaration on Civil Protection, Energy and Tourism (hereafter: "Declaration of the EU Treaty") stating that Union action on the basis of the provisions of the EC Treaty was required in these fields. Besides, the Conference declared that, on the basis of a Commission report to be submitted by 1996, it had to be considered whether new Treaty titles were necessary in these fields'.[325]

303. In accordance with the Declaration of the EU Treaty, the Commission duly submitted its report in April 1996 and recommended that an in-depth consolidation of the ECSC, EURATOM and EC Treaties be carried out in order to bring the different legal bases together in a single chapter and bring coherence to the energy policy of the Union. Alternatively to this approach, the report suggested that a chapter on energy be inserted into the EC Treaty spelling out the energy policy objectives of the Union and providing the relevant instruments and procedures.[326] Nevertheless, there has been disagreement whether the insertion of a special chapter on energy in the EC Treaty could remedy the hindered progress towards a common energy policy.[327]

304. However, the increased focus on the energy sector in the 1990s resulted in a further adoption of European energy law and policy. The 1993 consultations by the Commission on new energy policy guidelines led in December 1995 to the adoption of the White Paper on *"An Energy Policy for the European Union"*.[328]

---

324 See *Infra.*

325 E. D. CROSS, L. HANCHER, and P. J. SLOT, "EC Energy Law" in M. M. ROGGENKAMP, A. RØNNE, C. REDGWELL and I. DEL GUAYO (eds.), *Energy Law in Europe. National, EU and International Law and Institutions* (Oxford, Oxford University Press, 2001), p. 217–218 and P. CAMERON, *Competition in Energy Markets, Law and Regulation in the European Union* (Oxford, Oxford University Press, 2007), p. 54–56.

326 Report from the Commission to the Council on Civil Protection, Tourism and Energy, SEC(1996)496, 15 p.

327 G. RASHBROOKE, "Clarification or Complication? The New Energy Title in the Draft Constitution for Europe", *Journal of Energy & Natural Resources Law* (2004), p. 375 and 377; E. D. CROSS, L. HANCHER, and P. J. SLOT, EC Energy Law" in M. M. ROGGENKAMP, A. RØNNE, C. REDGWELL and I. DEL GUAYO (eds.), *Energy Law in Europe. National, EU and International Law and Institutions* (Oxford, Oxford University Press, 2001), p. 219–220 and P. CAMERON, *Competition in Energy Markets, Law and Regulation in the European Union* (Oxford, Oxford University Press, 2007), p. 51–53.

328 White Paper of the European Commission of 13 December 1995, *An energy policy for the European Union*, COM(1995)628.

This White Paper was welcomed by the Council *"as a useful basis for further work in view of the fact that it endorses the necessity of reconciling the objectives of competitiveness, security of supply and protection of the environment taking into account the principle of subsidiarity and economic and social cohesion"*.[329] Nonetheless, various deficiencies remained in the framework of the European energy law and policy, mainly due to the lack of political support.

305. Until recently with the adoption of the Lisbon Treaty, the treaty rules applicable or relevant to the energy sector have not been amended significantly. From time to time there have been inter-institutional disputes over the correct legal basis for Union measures which led to annulment by the Court of Justice in some cases.[330]

(c) Treaty framework after the adoption of the Treaty of Amsterdam

306. The Treaty framework was further amended by the Treaty of Amsterdam[331] which was signed in October 1997 and entered into force on 1 May 1999. This Treaty amended the EU Treaty, the Treaties establishing the three European Communities and certain related acts. Within the first pillar, the European Community, the Amsterdam Treaty has strengthened the position of the European Parliament, modified the co-decision procedure in order to improve the efficiency of its operation, moved twenty-four fields from unanimity towards qualified majority vote and extended the list of Union competences (such as employment, the competence to take appropriate action to combat discrimination based on sex, racial or ethnic origin, religion or belief, disability, age or sexual orientation). The Amsterdam Treaty also modified the other two pillars, concerning cooperation in foreign and security policy and justice and home affairs.[332]

307. The amendments by this Treaty with relevance to energy law and policy were the adoption of a Protocol relating to the application of the principle of

---

329 Council Resolution of 8 July 1996 on the White Paper An energy policy for the European Union, *OJ* C 224/1.

330 See Cases C-413/04 and C-414/04, *European Parliament v Council of the European Union*, [2006] ECR I-11221 and Case C-281/01, *Commission of the European Communities v Council of the European Union*, [2002] ECR I-12049 and the Case C-411/06, *Commission of the European Communities v European Parliament and Council of the European Union*, [2007] ECR I-6199.

331 Treaty of Amsterdam amending the Treaty on European Union, the Treaties establishing the European Communities and Certain related acts, *OJ* C 340/1. Note that one of the most immediately striking features of the Amsterdam Treaty was the renumbering of the EC Treaty and the switch from letters to numbers in the EU Treaty.

332 S. WEATHERILL and P. BEAUMONT, *EU law – The essential guide to legal workings of the European Union*, (London, Penguin Books, 1993), p. 17–23.

subsidiarity, the insertion of an article relating to the "services of general economic interest", the further clarification of the constitutional status of environmental protection and the inclusion of the integration principle in Article 6 of the EC Treaty (now Article 11 of the TFEU) which states that *"environmental protection requirements must be integrated into the definition and implementation of the Community policies and activities referred to in Article 3, in particular with a view to promoting sustainable development"*.[333]

(d) Developments with the adoption of the Treaty of Nice

308. In February 2001 the Member States have signed the Treaty of Nice[334] which entered into force on 1 February 2003. This Treaty made some amendments to the Treaties framework and in particular concerning the impact of the enlargement of the EU.[335] In preparation of the "big bang" expansion of the Union, the Treaty of Nice has primarily included adjustments to the composition and operation of the European Parliament, the Council and the Commission.

309. This Treaty did not adopt any changes with relevance to energy law and policy. Nonetheless, during the discussions on the Treaty of Nice text, the Portuguese presidency in 2000 has questioned whether in areas such as energy, external competence and the establishment of decentralised agencies, the repeated use of Article 308 of the EC Treaty justified the adoption of a specific legal basis in the EC Treaty which required a qualified majority instead of a unanimous vote. However, this proposal of the Portuguese presidency was dropped.[336]

(e) Constitutional Treaty of Europe

310. The Convention on the future of Europe started its work on 28 February 2002 and applied a three-stage procedure: listening, examining and proposing. The final output of the Convention exceeded all initial expectations since on 13 June 2003 the Praesidium of the Convention obtained at its

---

333 J. H. JANS, *European Environmental Law* (Groningen, Europa Law Publishing, 2000), p. 8–9 and 17–23.

334 Treaty of Nice amending the Treaty on European Union, the Treaties establishing the European Communities and certain related acts.

335 E. D. CROSS, L. HANCHER, and P. J. SLOT, "EC Energy Law" in M. M. ROGGENKAMP, A. RØNNE, C. REDGWELL and I. DEL GUAYO (eds.), *Energy Law in Europe. National, EU and International Law and Institutions* (Oxford, Oxford University Press, 2001), p. 218.

336 P. CAMERON, *Competition in Energy Markets, Law and Regulation in the European Union*, (Oxford, Oxford University Press, 2007), p. 58 and Conference of the Representatives of the Governments of the Member States of 22 February 2000, CONFER 4711/00, p. 1–4.

plenary session a broad consensus on a Draft Treaty establishing a Constitution for Europe.[337]

311. On 29 October 2004 the Constitutional Treaty of Europe was signed in Rome but following the two negative votes on the text in France and the Netherlands in 2005, the ratification process was stalled.[338] Nevertheless, it's worth mentioning that in Article I-14 of the Constitutional Treaty energy was listed as one of the eleven areas of shared competence along with internal market, environment, trans-European networks, transport and consumer protection. How the energy competence had to be shared was specified in Article III-256 of the Constitutional Treaty which provided that:

> *"1. In establishing an internal market and with regard for the need to preserve and improve the environment, Union policy on energy shall aim to:*
> *(a) ensure the functioning of the energy market,*
> *(b) ensure security of energy supply in the Union, and*
> *(c) promote energy efficiency and saving and the developments of new and renewable forms of Energy".*
> *2. The measures necessary to achieve the objectives in paragraph 1 shall be enacted in European laws or framework laws. Such laws shall be adopted after consultation of the Committee of the Regions and the Economic and Social Committee.*
> *3. Such laws or framework laws shall not affect a Member State's choice between different energy sources and the general structure of its energy supply, without prejudice to Article III-234 (2) (c)".*

312. According to this provision, the Union policy on energy shall ensure the functioning of the European energy market, the security of energy supply, the promotion of energy efficiency and the development of renewable energy sources. The European laws or framework laws to achieve these objectives had to be adopted jointly by the European Parliament and the Council of Ministers acting on a proposal from the Commission, in accordance with the ordinary legislative procedure described in Article III-396 of the Constitutional Treaty.[339]

---

337 K. LENAERTS, and P. VAN NUFFEL, in BRAY, R. (ed.), *Constitutional Law of the European Union*, (London, Thomson Sweet & Maxwell, 2005), p. 67–69 and N. FOSTER, *Foster on EU Law* (Oxford, Oxford University Press, 2009), p. 42 and 43.

338 *Ibid.*, p. 11–15 and see also Eurelectric, *Legal Note on the Energy Chapter of the Constitution of the European Union*, Legal affairs working group (February 2005), 16 p.

339 See also Article I-34 of the Constitutional Treaty. The adoption of these measures had to be done on the basis of the new system of qualified majority defined in Article I-25 of the Constitutional Treaty. According to this Article a qualified majority had to be achieved if a decision is supported by 55% of Member States, including at least fifteen of them, representing at the same time at least 65% of the Union's population. It should be noted that the provision according to which a qualified majority must be supported by 55% of the Member States, "*including at least fifteen of them*" is a transitional provision since in a Union of 25 Member States, 15 Member States represent 60% of the total, however, in a Union of more than 25 Member States, this provision will lose importance. From the moment the Union enlarges to

313. Article III-256, paragraph 3 of the Constitutional Treaty furthermore stipulated that such laws or framework laws to obtain the objectives mentioned in paragraph 1 must not effect a Member State's choice between different energy sources and the general structure of its energy supply without prejudice to Article III-234 (2) (c) of the Constitutional Treaty. This appears to mean that measures which affect a Member State's choice or the general structure of its energy supply must be adopted on the basis of unanimity. This unanimity rule, according to Article III-171 of the Constitutional Treaty, will also apply for energy taxation measures. This new provision is a compromise which ensured the Member States' sovereignty over natural resources and energy taxation issues and a shared Union competence to obtain the objectives in the energy field which are mentioned in Article III-256 of the Constitutional Treaty.[340]

314. With respect to Article III-256 of the Constitutional Treaty, Declaration nr. 22 annexed to the final Act to the Constitution Treaty stated that: "*the Conference believes that Article III-256 does not affect the right of the Member States to take the necessary measures to ensure their energy supply under the conditions provided for in Article III-131*". The wording of this declaration seems to indicate that in the event of an energy crisis, the Union institutions could not usurp the competences of the national governments. Furthermore, paragraph 3 of Article III-256 of the Constitutional Treaty does not seem to allow the Union institutions to oblige Member States to produce sufficient volumes from their own energy resources for the benefit of the other Member States or the Union, not even when this is in the interest of security of supply of the Union.[341]

315. However, this provision will not prevent the Union institutions to adopt precautionary measures (for example a measure which requires the Member States to ensure oil stocks for emergency purposes) in order to secure the energy supply in the Union on the basis of paragraph 1 of Article III-256 of the Constitutional Treaty. Even with this kind of precautionary measures which for example provide for oil stocks, it seems that a Member state, on the basis of the Declaration 23 and paragraph 3 of Article III-256 of the Constitutional Treaty, remains competent to decide whether these oil stocks can be used or not. It remains uncertain, whether with this energy provision in the Constitutional framework, if adopted, laws or framework laws in the energy field would more easily obtain a qualified majority of Member States and the necessary political support. Some commentators argue that, despite the inclusion of such a separate

---

26 Member States, 55% of the total number of Member States will, mathematically, comprise at least 15 of them. See http://europa.eu/scadplus/constitution/doublemajority_en.htm.

340 See *Infra*.

341 L. HANCHER, "The new EC Constitution and the European Energy market" in M. M. ROGGENKAMP and U. HAMMER (eds.) *European Energy Law Report II* (Antwerpen, Intersentia, 2005), p. 6 and 7.

energy section in the Constitutional Treaty, it will not resolve all legal problems in relation to the common energy policy of the EU.[342]

316. The adoption of Article III-256 of the Constitutional Treaty would have given formal recognition to a situation which already existed in practice, with the adoption of a broad range of energy and energy-related measures on a European level.[343]

317. Following the failed referenda on the Constitutional Treaty of Europe, a "period of reflection" on the future of Europe was launched to reconnect the citizens with the European project and to decide the fate of the Constitutional Treaty of Europe. At the European Council in June 2007, the EU leaders managed to overcome the institutional impasse and agreed on the outlines of a new Treaty which replaces the draft Constitutional Treaty of Europe.[344]

5. CONCLUSION

318. This part has highlighted that over the past years the Union Treaties, prior to the adoption of the Lisbon Treaty, have been amended several times. These amendments have conferred more powers to the Union in multiple policy areas, but with respect to energy, as discussed above, they were limited.

319. The lack of any specific articles, chapter or title on energy in the EC Treaty has not prevented the Union from enacting various measures in the energy sphere,[345] based on non-energy provisions of the TFEU. However, this legislative action of the Union has been limited through the application of the principles of attribution of competences, subsidiarity and proportionality. The next part will discuss these principles and their application towards the energy field in more detail.

---

342 *Ibid.*, p. 7 and 8 and, G. RASHBROOKE, "Clarification or Complication? The New Energy Title in the Draft Constitution for Europe", *Journal of Energy & Natural Resources Law* (2004), p. 373–387.

343 P. CAMERON, *Competition in Energy Markets, Law and Regulation in the European Union* (Oxford, Oxford University Press, 2007), p. 58 and 59.

344 Www.consilium.europa.eu/ueDocs/cms_Data/docs/pressData/en/ec/94932.pdf.

345 E. D. CROSS, B. DELVAUX, L. HANCHER, P. J. SLOT, G. VAN CALSTER and W. VANDENBERGHE, "EU Energy Law (chapter 5)" in M. M. ROGGENKAMP, A. RØNNE, C. REDGWELL and I. DEL GUAYO (eds.), *Energy Law in Europe. National, EU and International Regulation* (Oxford, Oxford University Press, 2007), p. 227.

C. DIVISION OF COMPETENCES BETWEEN THE UNION AND ITS MEMBER STATES IN THE ENERGY FIELD

1. INTRODUCTION

320. In this part, the principles of attribution of competences, subsidiarity and proportionality which govern the existence and exercise of Union powers are discussed. It has always been recognised that the Union only has those powers within the limited areas in which it has been given powers. It is the legal basis which in the first place determines the division of competences between the Union and its Member States i.e. the vertical division of competences. This vertical division and the attribution of more competences to the Union have given rise to a continuous debate on the most appropriate division of powers between the Union and its Member States.[346]

321. The vertical division generally means the division of legislative, jurisdictional and executive competences between the Union and its Member States, whereas a horizontal division means the division of competences within the Union between the relevant Union institutions: Council, European Commission, European Parliament and the Court of Justice.[347] Notwithstanding the fact that vertical and horizontal division of competences partly overlap, since the Union institutions act and use or are accused of misusing the Union competences, the author considers that this distinction facilitates the systematisation of the division of competences between the Union and the Member States.

322. However, it is not possible to define this vertical division exhaustively on the basis of the requirement for an appropriate Treaty basis. In the Nice Declaration of December 2000 on the future of the Union, the delimitation of competences was one of the items listed upon request of the government of Germany which criticised this delimitation. According to the German government, the lack of clarity in the delimitation of these competences makes it difficult for citizens of the Union to understand who does what on a European level. It risks allowing a creeping increase of the powers of the institutions on a European level. Thus, the Intergovernmental Conference agreed to include in the Nice Declaration of December 2000 the question *"[h]ow to establish and monitor*

346 K. LENAERTS, and P. VAN NUFFEL, in BRAY, R. (ed.), *Constitutional Law of the European Union* (London, Thomson Sweet & Maxwell, 2005), p. 87 and 88.

347 For a discussion on the vertical division of competences see A. VON BOGDANDY and J. BAST, "The European Union's vertical order of competences: the current law and proposals for its reform", *Common Market Law Review* (2002), p. 227–268.

a more precise delimitation of powers between the European Union and the Member States, reflecting the principle of subsidiarity".

323. The idea of establishing a precise delimitation of powers between the Union and its Member States has been discussed in the Convention, however it was quickly abandoned. It was argued that such delimitation of powers was too difficult to establish and, given the characteristics of the EU, inappropriate. Therefore, the Convention focused on the definition of the different categories of competences and the development of a mechanism which ensures a stricter implementation by the European institutions of the principles of subsidiarity and proportionality.[348]

324. The next part will not only provide a clear picture of the meaning of the principle of attribution of powers it will also apply this principle to the energy field and draw some conclusions.

2. THE PRINCIPLE OF ATTRIBUTION OF POWERS

(a) Meaning of the principle of attribution of powers on a European level

325. A prominent feature of the EU legal order is that the Union only has those competences, which the Member States have attributed to it. Lawyers acquainted with EU law commonly speak of the principle of attribution of powers, conferral of competences or enumerated competences.

326. The principle of attributed powers was already included in the initial Treaty of Rome and can now be deduced from Article 13 of the TEU identifying the main institutions of the Union. It has always included a sub-paragraph stating: *"Each institution shall act within the limits of the powers conferred on it in the Treaties, and in conformity with the procedures, conditions and objectives set out in them. The institutions shall practice mutual sincere cooperation"*. This paragraph of Article 13 of the TEU is known as the principle of attribution (or conferral) of powers. It is noteworthy that, according to Article 263 of the TFEU, lack of competence is one of the grounds for declaring a Union act void by the Court of Justice.

327. With the introduction of Article 5 into the EC Treaty (now paragraph 2 of Article 5 of the TEU) by the Maastricht Treaty, paragraph 2 of Article 13 of the

348 J.-C. PIRIS, *The Lisbon Treaty: A Legal and Political Analysis* (Cambridge, Cambridge University Press, 2010), p. 11 and 76.

TFEU is paralleled with a twin principle[349], which provides that: *"[u]nder the principle of conferral, the Union shall act only within the limits of the competences conferred upon it by the Member States in the Treaties to attain the objectives set out therein. Competences not conferred upon the Union in the Treaties remain with the Member States"*. The European Union also made clear that this principle of attribution of powers not only applies to the institutions, but also to the Union as such.[350] It is pivotal to the relationship between the Union and its Member States.[351]

328. According to Court of Justice case law, Union competences are also considered to be specific.[352] Specificity means that competences not conferred upon the Union remain with the Member States.

329. In its Opinion 2/94 the Court of Justice ruled that the first paragraph of Article 5 of the EC Treaty (now Article 5(2) of the TEU) means that the Union *"has only those powers which have been conferred on it"*.[353] This opinion makes clear that the Union is not a self-authenticating order. It does not have what German scholars call *Kompetenz Kompetenz*. This means that the Union order is unable to generate new powers within itself over and above the powers which have been conferred by the constitution-making authority. The competences of the Union are derived from the Member States and are subject to the limits imposed by the Treaty. This contrasts with the position of the Member States because their powers are derived from their own sovereignty.[354]

330. Furthermore according to the Opinion 2/94 of the Court of Justice the principle of attribution must be applied in both the internal and international actions of the Union.[355] It is the task of the Court of Justice, *inter alia*, to ensure

---

349 J.-C. PIRIS, *The Lisbon Treaty A Legal and Political Analysis* (Cambridge, Cambridge University Press, 2010), p. 82.

350 A. ARNULL, A. DASHWOOD, M. DOUGAN, M. ROSS, E. SPAVENTA and D. WYATT, *European Union Law* (London, Sweet & Maxwell, 2006), p.85.

351 A. DASHWOOD, "The relationship between the Member States and the European Community/European Union" in A. MCDONELL (ed.), *A review of forty years of Community Law* (The Hague, Kluwer Law International, 2002), p. 39.

352 Opinion 2/94, *Accession by the Communities to the Convention for the Protection of Human Rights and Fundamental Freedoms* [1996] ECR 1759, paragraph 25.

353 Opinion 2/94, *Accession by the Communities to the Convention for the Protection of Human Rights and Fundamental Freedoms* [1996] ECR 1759, paragraph 23.

354 A. ARNULL, A. DASHWOOD, M. DOUGAN, M. ROSS, E. SPAVENTA and D. WYATT, *European Union Law* (London, Sweet & Maxwell, 2006), p. 85 and A. DASHWOOD, "The relationship between the Member States and the European Community/European Union" in A. MCDONELL (ed.), *A review of forty years of Community Law* (The Hague, Kluwer Law International, 2002), p. 39 and A. DAHWOOD, "States in the European Union", *European Law Review* (1998), p. 201–216.

355 Opinion 2/94, *Accession by the Communities to the Convention for the Protection of Human Rights and Fundamental Freedoms* [1996] ECR 1759, paragraph 24.

that the Union respects the limits of its powers. In that regard, the Court emphasised in its case law that the choice of legal basis for a measure may not simply depend on an institution's conviction as to the object pursued.[356] Instead, it held that *"in the context of the organisation of the powers of the Community, the choice of a legal basis for a measure must be based on objective factors which are amenable to judicial review. Those factors include in particular the aim and content of the measure"*.[357]

331. Obviously, the principle of conferral of competences does not function independently. All Union actions must be based upon a power-conferring provision or a legal basis of the Treaty. Such a provision defines the action the Union is authorised to take in a given area of policy or the *ratione materiae* of the Union's competence. It specifies the decision-making procedure for enacting measures and the legislative instruments that may be used. According to case law of the Court of Justice, failure to refer to a precise provision *"is indispensable where, in its absence, the parties concerned and the Court of Justice are left uncertain as to the precise legal basis"*.[358] It follows from the foregoing that provisions which do not contain any power-attributing provision, such as Article 3 of the TEU, which identify the objectives of the Union, and Articles 3, 4, 5, 6, 8 and 119 of the TFEU which identify the activities of the Union, cannot be used as a legal basis for Union action.[359] Thereby the principle of conferral of competences seems to be well-established both in the TFEU, TEU and the Court of Justice's case law.

(b) *Energy and the principle of attribution of powers*

332. In accordance with the principle of attribution of powers, the Union legislator needs to select the appropriate legal basis according to the content and main object of the act in question.[360] The legal basis can be explicit such as

---

356 Case C-300/89, *Commission of the European Communities v Council of the European Communities* [1991] ECR I-2867, paragraph 10.

357 Case C-269/97, *Commission of the European Communities v Council of the European Union* [2000] ECR I-2257, paragraph 43 and Case C-233/94, *Federal Republic of Germany v European Parliament and Council of the European Union* [1997] ECR I-2405, paragraph 10 and Opinion of Advocate General Fennelly in Case C-376/98, *Federal Republic of Germany v European Parliament and Council of the European Union* and Case C-74/99, *The Queen v Secretary of State for Health and Others, ex parte Imperial Tobacco Ltd and Others*, paragraph 61.

358 Case 45/86, *Commission of the European Communities v Council of the European Communities* [1987] ECR 1493, paragraph 9.

359 K. LENAERTS, and P. VAN NUFFEL, in BRAY, R. (ed.), *Constitutional Law of the European Union*, (London: Thomson Sweet & Maxwell, 2005), p. 86 and 87 and A. ARNULL, A. DASHWOOD, M. DOUGAN, M. ROSS, E. SPAVENTA and D. WYATT, *European Union Law* (London, Sweet & Maxwell, 2006), p. 85.

360 E.g. Case C-436/03, *European Parliament v Council of the European Union* [2006] ECR I-3733, paragraph 35.

Article 192 of the TFEU concerning the environmental policy, or more general such as Article 114 of the TFEU concerning the establishment and functioning of the internal market, or even more general, the so-called loophole Article 352 of the TFEU. The latter enables Union action where it proves necessary to attain, in the course of the operation of the common market, one of the objectives of the Union and the Treaty does not provide the necessary powers.[361]

333. It is clear that the principle of attribution of powers requires that all Union action must be founded upon a legal basis laid down in the Union Treaties. However, for many years, before the adoption of the Lisbon Treaty, the Union Treaties contained no legal basis with respect to energy. Hence, the measures in the energy field in the Union had to be adopted on the basis of the more general provisions of the Treaties such as the rules on fundamental freedoms, the rules on internal market and the so-called loophole Article 352 of the TFEU.

334. With the adoption of the Single Act, the completion of the internal market became one of the key objectives of the EU. This Act introduced Article 26(2) of the TFEU which defines internal market as *"an area without internal frontiers in which the free movement of goods, persons, services and capital is ensured in accordance with the provisions of the Treaties"*. This definition includes energy.[362] Moreover, given the importance of energy in all economic activity in the Union, the completion of the internal market is inconceivable without an integrated internal market for energy.[363]

335. The past years national energy markets have been liberalised and harmonised as part of the establishment of the internal market.[364] The main provision in the TFEU concerning the establishment and functioning of the internal market is Article 114. This article has served for many years as the legal basis for the adoption of measures in the energy field.[365] Consequently, the measures in the energy field comply with the principle of attribution.

---

361 Opinion of Advocate General Fennelly in Case C-376/98, *Federal Republic of Germany v European Parliament and Council of the European Union*, paragraph 62 and Opinion of Advocate General Jacobs in Case C-350/92, *Kingdom of Spain v Council of the European Union*, paragraph 26.

362 See *Infra*.

363 Commission of the European Communities, Proposal for a Council Directive concerning common rules for the internal market in electricity and Proposal for a Council Directive concerning common rules for the internal market in natural gas, COM(91)548, p. 2.

364 K. LENAERTS, and P. VAN NUFFEL, in BRAY, R. (ed.), *Constitutional Law of the European Union* (London, Thomson Sweet & Maxwell, 2005), p. 323.

365 Examples of energy measures based on Article 114 of the TFEU are Directive 96/92/EEC of 19 December 1996 concerning common rules for the internal market in electricity, *OJ* L 27/20; Directive 2003/54/EC of 26 June 2003 concerning the common rules for the internal market in electricity and repealing Directive 96/92/EC, *OJ* L 176/37; Regulation No 1228/2003 of 26 June 2003 on conditions for access to the network for cross-border exchanges in

336. The competence attributed to the Union by Article 114 of TFEU is of a horizontal nature since it is in advance not limited by reference to a particular subject-matter defined *ratione materiae*. Instead, on the basis of this provision the Union is authorised to adopt approximation measures which have the establishment and functioning of the internal market as their object. This means that the scope of the Union's competence is defined by reference to a criterion of a functional nature which extends laterally to all measures designed to ensure attaining the single market.[366] It leads to Union legislation that touches upon the most diverse areas of national law (e.g. energy) provided that the adoption of this legislation is relevant for the establishment and functioning of the internal market.[367]

337. Beside Article 114 of the TFEU which is a broad legislative provision, the Treaty contains an even broader provision, namely Article 352 of the TFEU. When the Union legislator in certain areas did not possess a specific legislative authority, it often had recourse to Article 352 of the TFEU. This provision has been used to legitimate legislation in areas such as regional, environmental and development cooperation policies, before these areas were dealt with through later amendments to the Treaty.[368]

338. Also for the adoption of measures in the energy field, before the Treaties conferred on the Union any specific powers in the environmental field, the European legislator took recourse to Article 352 of the TFEU.[369] This allowed the Union legislator to comply with the principle of attribution when it adopted measures in the energy field.

electricity, *OJ* L 20/24; Directive 98/30/EC of 22 June 1998 concerning the common rules for the internal market in natural gas, *OJ* L 204/1; Directive 2003/55/EC of 26 June 2003 concerning the common rules for the internal market in natural gas and repealing Directive 98/30/EC, *OJ* L 176/57; Regulation No 1775/2005 of 28 September 2005 on conditions for access to the natural gas transmission networks, *OJ* L 289/1; Directive 2005/89/EC of 18 January 2006 concerning measures to safeguard security of electricity supply and infrastructure investment, *OJ* L 33/22; Council Directive 92/75/EEC of 22 September 1992 on the indication by labelling and standard product information of the consumption of energy and other resources by household appliances and its amendments and implementing measures, *OJ* L 297/16.

366 Opinion of Advocate General Tesauro in Case C-300/89, *Commission of the European Communities v Council of the European Communities*, paragraph 10.

367 Opinion of Advocate General Fennelly in Case C-376/98, *Federal Republic of Germany v European Parliament and Council of the European Union*, paragraph 62 and Opinion of Advocate General Jacobs in Case C-350/92, *Kingdom of Spain v Council of the European Union*, paragraph 26.

368 P. CRAIG and G. DE BÚRCA, *EU Law Text, Cases and Materials* (Oxford, Oxford University Press, 2008), p. 93.

369 An example of an energy measures based on Article 352 of the TFEU is Council Directive 93/76/EEC of 13 September 1993 to limit carbon dioxide emissions by improving energy efficiency, *OJ* L237/28.

339. As Articles 114 and 352 of the TFEU are relatively ambiguous and open-ended in their scope of application, these legal provisions gave rise to some difficulties in applying the principle of conferral of competences. The Union institutions interpreted these provisions of the Treaty in a relatively generous way which led to the phenomenon commonly known as "competence creep" or an extension of EU competences.[370]

340. It can be concluded in the light of the foregoing that although the Treaties did not confer any specific powers to the European legislator to adopt measures in the energy field, the recourse to a more general legal basis in the TFEU such as Article 114 of the TFEU concerning establishment and functioning of the internal market, or even more general, the so-called loophole Article 352 of the TFEU for the adoption of energy measures, was sufficient in order to comply with the principle of attribution. The next part will discuss the second principle which governs the existence and exercise of Union powers, the principle of subsidiarity.

3. THE PRINCIPLE OF SUBSIDIARITY

(a) Introduction

341. The subsidiarity principle was included in Article 5 of the EC Treaty by the Maastricht Treaty (now Article 5 of the TEU). As with the principle of conferral of powers, the subsidiarity principle was a reaction from the Member States meant to restrict the Union's creeping competences. This reaction contained a binding commitment of Member States acting on a Union level, whereby the action of the Union legislator would be restricted to what was strictly necessary. To make this commitment credible, the Member States converted the concept of subsidiarity into a legal principle inserted into the EC Treaty that can be reviewed by the Court of Justice.

342. The idea behind the subsidiarity principle is that public powers should be located where they can be exercised most effectively. The function of the principle in the constitutional order of the Union is to guide the exercise of non-exclusive competences, and to limit the recourse to these competences, by the Union institutions.[371]

---

[370] A. ARNULL, A. DASHWOOD, M. DOUGAN, M. ROSS, E. SPAVENTA and D. WYATT, *European Union Law* (London, Sweet & Maxwell, 2006), p. 81 and 86.

[371] A. DASHWOOD, "Relationship between Member States and EC/EU", in A. MCDONNELL (ed.), *A review of forty years of Community law – Legal developments in the European Communities and the European Union* (The Hague, Kluwer Law International, 2005), p. 48 and 49; N. EMILIOU, "Subsidiarity: An effective Barrier against "The Enterprises of

343. The next part will define the meaning of the principle of subsidiarity and apply this principle to the energy field.

(b) *Meaning of the principle of subsidiarity on a European level*

i) Principle of subsidiarity: not without reason

344. The discontent with the European integration process since the inclusion of the majority principle into the Union's institutional structure has been clearly pronounced by the United Kingdom Prime Minister Margaret Thatcher in 1988 during the opening ceremony at the College of Europe in Bruges. Margaret Thatcher stated that: *"[c]loser co-operation requires neither centralization of power in Brussels nor that decisions shall be taken by a bureaucratic apparatus which is not accountable before the electorate [...]. We have not successfully reduced the State to its just terms in order that the State re-establishes its old borders at European level, constituting a super-European State that exerts a new power from Brussels"*.[372]

345. This was not the only sign of Member States' discontent. The Danish referendum in 1992 resulted in a vote against the ratification of the adopted Maastricht Treaty. This referendum was followed by a referendum in France which resulted in a very narrow majority in favour of the Maastricht Treaty.

346. Another sign of the crisis suffered by the European integration process is the *Maastricht* judgment of the German Constitutional Court.[373] The German Constitutional Court held that the ratification of the Maastricht Treaty was permissible. However, it emphasised that Union law only applies in Germany to the extent permitted by the German Constitution. The growing discomfort in the public opinion with the European integration can be interpreted as a shared perception among the Member States of pure and simple defencelessness in the

---

Ambition", *European Law Review*, p. 383–391; J. PETERSON, "Subsidiarity: A definition to Suit Any Vision?", *Parliamentary Affairs* (1994), p. 118 *et seq*; C. HENKEL, "The allocation of Powers in the European Union: A closer look at the principle of subsidiarity", *Berkeley Journal of International Law* (2002), p. 359; N. MACCORNICK, "Democracy and Subsidiarity in the European Commonwealth", in N. MACCORNICK (ed.), *Questioning Sovereignty* (Oxford, Oxford University Press, 1999), p. 152–155.

372 A. ESTELLA, *The EU Principle of Subsidiarity and its Critique* (Oxford, Oxford University Press, 2002), p. 6 and 82.

373 *Brunner v European Union Treaty*, German Constitutional Court (Bundesverfassungsgericht) BverfGE 89, 155 (October 12,1993), translated in English at *Common Market Law Review* (1994) 57 and 33 *I.L.M.* (1994) p. 388.

process of the Union intervention. Due to this perception, the Member States pushed for the introduction of the principle of subsidiarity into the Treaties.[374]

347. In the Union context, the principle of subsidiarity was not unknown. The first references to this principle can be found in the Tindemans Report on the European Union[375] and in the MacDougall Report on fiscal federalism.[376] Furthermore, the draft Treaty on the European Union of the European Parliament made in Article 12, 2° implicit reference to the concept of subsidiarity. In 1987 the principle of subsidiarity was introduced into the Union legal order by the SEA in Article 130r, 4°. This provision stipulated that the Union should take action to the extent to which environmental objectives *"can be attained better at Community level than at the level of the individual Member States"*.[377]

348. The principle of subsidiarity was included by the Maastricht reform in Article 3b, 2° of the EEC Treaty (now Article 5 of the TEU). The final outcome was a compromise between two camps on a European level. One set of players (Spain, Italy and France, among others) favoured a mere mentioning of the principle in the Preambles of the Treaties. This political camp succeeded in limiting the scope of the principle to Union concurrent competences and to preserve the "better attainment" clause in the last sentence of Article 3b, 2° of the EEC Treaty (now Article 5 of the TEU). The other camp (Commission, European Parliament, UK and Germany, among others) defended the inclusion of the principle in the operative part of the EEC Treaty. This camp which defended the legal profile of the principle was able to claim victory because the principle of subsidiarity became reviewable before the Court of Justice.[378]

349. In the early days after the insertion of the principle of subsidiarity into the EC Treaty, it was debated whether this principle was justifiable.[379] However,

---

374 A. ESTELLA, *The EU Principle of Subsidiarity and its Critique* (Oxford, Oxford University Press, 2002), p. 83 and 84; G. DE BURCA and B. DE WITTE, "The delimitation of powers between the EU and its Member States" in A. ARNULL and D. WINCOTT (eds.), *Accountability and Legitimacy in the European Union* (Oxford, Oxford University Press, 2002), p. 201–222 and J. H. H. WEILER, "A Constitution for Europe", *Journal of Common Market Studies* (2002), p. 563 and 571–573.

375 Report on the European Union of 7 January 1976 by Mr. Leo Tindemans, Prime Minister of Belgium, to the European Council. Bulletin of the European Communities, Supplement 1/76.

376 Report of 1977 of the Study Group on the Role of the Public Finance in European Integration.

377 See for the history and development D. Z. CASS, "The word that saves Maastricht? The principle of subsidiarity and the division of powers within the European Community, *Common Market Law Review* (1992), p. 1110–1128 and N. W. BARBER, "The limited Modesty of Subsidiarity", *European Law Journal* (2005), p. 309–313.

378 A. ESTELLA, *The EU Principle of Subsidiarity and its Critique* (Oxford, Oxford University Press, 2002), p. 85–87 and A. ARNULL, A. DASHWOOD, M. DOUGAN, M. ROSS, E. SPAVENTA and D. WYATT, *European Union Law* (London, Sweet & Maxwell, 2006), p. 97.

379 A. G. TOTH, "Is subsidiarity justifiable?", *European Law Review* (1994), p. 268 and A. G. TOTH, "The principle of subsidiarity in the Maastricht Treaty", *Common Market Law Review*

there is nothing in the text of Article 5 of the TEU which suggests the contrary, nor that the Court of Justice does not have the power to interpret this provision.[380] Nowadays on the basis of case law of the Court of Justice, the justifiability of the principle of subsidiarity can be considered as a *fait accompli*.[381, 382]

ii) Present form and meaning of the principle of subsidiarity

350. The principle of subsidiarity in its present form is defined in the third paragraph of Article 5 of the TEU which provides that: "*[u]nder the principle of subsidiarity, in areas which do not fall within its exclusive competence, the Union shall act only if and in so far as the objectives of the proposed action cannot be sufficiently achieved by the Member States, either at central level or at regional and local level, but can rather, by reason of the scale or effects of the proposed action, be better achieved at Union level.*

The institutions of the Union shall apply the principle of subsidiarity as laid down in the Protocol on the application of the principles of subsidiarity and proportionality. National Parliaments ensure compliance with the principle of subsidiarity in accordance with the procedure set out in that Protocol". The purpose of this principle is not determined when the Union legislator has competence since this question is already answered, as discussed above, by the principle of attribution of powers provided for by the first and second paragraph of Article 5 of the TEU.

351. Not only Article 5(3) of the TEU, which is the most obvious reference to the concept of subsidiarity, holds the concept of subsidiarity, but also Article 1(2)

(1992), p. 1079–1105. This author analysed three conditions of justifiability and concluded that: "*in the absence of a distinct procedure specifically designed to bring subsidiarity issues before the Court of Justice, subsidiarity is justifiable before the Court only up to a certain point and subject to the restrictions arising from the limited nature of the Court's jurisdiction in cases involving judicial review. Within those limits, however, the Court can play an important part in clarifying many confusing aspects of this new concept.*"

380 Lord Mackenzie-Stuart, "Subsidiarity – A Busted Flush?", in D. CURTIN and D. O'KEEFE (eds.), *Constitutional Adjudication in European and National Law: essays for T. F. O'Higgins* (London, Butterworths, 1992), p. 19.

381 A. ARNULL, *The European Union and its Court of Justice* (Oxford, Oxford University Press, 2006), 699 p.

382 A. ARNULL, A. DASHWOOD, M. DOUGAN, M. ROSS, E. SPAVENTA and D. WYATT, *European Union Law* (London, Sweet & Maxwell, 2006), p. 101.

of the TEU by stating that: *"This Treaty marks a new stage in the process of creating an ever closer union among the peoples of Europe, in which decisions are taken as openly as possible and as closely as possible to the citizen"*. Compared to Article 5(3) of the TEU this provision gives a rather different interpretation to the concept of subsidiarity because it is more a political statement than a precise legal definition.[383]

352. The principle of subsidiarity is designed to determine whether Union intervention, in accordance with a number of criteria of the third paragraph of Article 5 of the TEU, is appropriate or not. Subsidiarity does not indicate when the Union legislator has competence or has not but, if it is assumed that the Union legislator has competence, when he should act.

353. The principle seems to solve a question of opportunity and not of competence.[384] This means that before the Union legislator intervenes, it must meet the test of the principle of attribution of powers and the test of the principle of subsidiarity. If both tests are not passed by the Union legislator, the adopted acts are deemed unconstitutional and should therefore be annulled by the Court of Justice. The principle of subsidiarity already applies at the stage when the Union legislator reflects whether or not to adopt a piece of Union law.[385]

iii) Subsidiarity test

354. It is clear from Article 5(3) of the TEU that subsidiarity only needs to be considered in relation to areas which do not fall within the exclusive competence of the Union. As regards the term "exclusive competence", legal doctrine had broad and narrow views.[386] Since the areas which fall within the exclusive competences of the Union are excluded from the potential application of the subsidiarity, the principle of subsidiarity regulates the exercise of the Union competences and it applies, to any situation in which the Union takes action unless, as noted above, the Union competences are exclusive. It should be noted

383 S. DOUGLAS-SCOTT, *Constitutional Law of the European Union* (Harlow, Pearson education Limited, 2002) p. 177 and G. BERMANN, "Taking Subsidiarity Seriously", *Columbia Law Review* (1994), p. 332.

384 A. ESTELLA, *The EU Principle of Subsidiarity and its Critique* (Oxford, Oxford University Press, 2002), p. 97.

385 A. DASHWOOD, "States in the European Union", *European Law Review* (1998), p. 211 and A. VON BOGDANDY and J. BAST, "The European Union's Vertical Order of Competences: The Current Law and Proposals for its Reform", *Common Market Law Review* (2002), p. 261.

386 P. CRAIG and G. DE BÚRCA, *EU law text, cases and materials* (Oxford, Oxford University Press, 2003) p. 132 and 133; T. C. HARTLEY, *The Foundations of Community Law An*

that for those actions/provisions in the energy field on a European level which fall within one of the exclusive competences the subsidiarity must not be applied.

355. This means that for those fields of Union law characterised by shared or supporting competence, the test of subsidiarity applies. This test has a dual aspect in requiring, on the one hand, the impossibility to obtain certain objectives by action on a Member States level and, on the other hand, the superior efficacy of action on a Union level (hereinafter dual test). Therefore, in order to justify action on a Union level it must be shown that both conditions of the test are fulfilled.[387]

356. In order to examine whether these conditions are fulfilled, the Amsterdam Protocol has provided guidelines to be followed by the Union institutions.[388]

Introduction to the Constitutional and administrative law of the European Community (Oxford, Oxford University Press, 2007), p. 113 and 114. According to A. G. TOTH all the areas of power granted to the Community under the EEC Treaty as originally concluded are exclusive see A. G. TOTH, "The principle of Subsidiarity in the Maastricht Treaty", *Common Market Law Review* (1992), p. 1079 and 1080; According to J. STEINER on the other hand, only the areas in which the Community has exclusive competence for the purpose of the Article 5, 3b are those in which it has already legislated see J. STEINER, "Subsidiarity under the Maastricht Treaty" in D. O'KEEFFE and P. M. TWOMY (eds.), *Legal Issues of the Maastricht Treaty* (London, Chancery Press, 1994), p. 49 and 55–58 and N. EMILIOU, "Subsidiarity: Panacea or Fig Leaf?" in D. O'KEEFFE and P. M. TWOMY (eds.), *Legal Issues of the Maastricht Treaty* (London, Chancery Press, 1994), p. 65–83; T. SCHILLING, "A new dimension of Subsidiarity: Subsidiarity as a Rule and a Principle", *Yearbook of European Law* (1994) p. 217; V. CONSTANTINESCO, "Who's Afraid of Subsidiarity?", *Yearbook of European Law* (1992), p. 38; N. EMILIOU, "Subsidiarity: An Effective Barrier Against 'the Enterprises of Ambition'?", *European Law Review* (1992), p. 388. Before the adoption of the Lisbon Treaty, the Commission argued that an area falls within the exclusive competence of the Union if the Treaties impose an obligation to act on the Union because it is regarded as having the sole responsibility for the performance of a particular task. On the basis of this criterion, the Commission has identified the following areas as exclusive competences: the common commercial policy; the removal of the barriers to the free movement of goods, persons, services and capital; the general rules on competition; the conservation of fisheries resources; the organisation of agricultural markets and the essential elements of transport policy. See T.C. HARTLEY, *The Foundations of Community Law An Introduction to the Constitutional and administrative law of the European Community*, (Oxford, Oxford University Press, 2007), p. 114 and Communication from the Commission to the Council and the European Parliament, *The principle of subsidiarity*, SEC(92)1990, p. 5 and 7.

387 A. ARNULL, A. DASHWOOD, M. DOUGAN, M. ROSS, E. SPAVENTA and D. WYATT, *European Union Law* (London, Sweet & Maxwell, 2006), p. 98–100.

388 *This Protocol states that: "(a) the issue under consideration has transnational aspects which cannot be satisfactorily regulated by action by Member States; (b) actions by Member States alone or lack of Community action would conflict with the requirements of the Treaty (such as the need to correct distortion of competition or avoid disguised restrictions on trade or strengthen economic and social cohesion) or would otherwise significantly damage Member States' interests and (c) action at Community level would produce clear benefits by reason of its scale or effects compared with action at the level of the Member States"* see Treaty of

These guidelines however, do not add that much to the general wording of Article 5(3) of the TEU.[389]

357. The Amsterdam Protocol in its third provision also emphasises that the principle of subsidiarity is a dynamic concept which allows for *"Community action within the limits of its powers to be expanded where circumstances so require, and conversely, to be restricted or discontinued where it is no longer justified"*. This means that certain events (e.g. discovery of dioxins) could for instance result in a need to adopt a new programme of food safety measures on a Union level. We have to ask ourselves continuously whether or not the proposed action can be tackled sufficiently well by the Member States or whether the issue should instead be addressed on a Union level. On the contrary, if proposals or legislation no longer meet the subsidiarity test, these proposals should be withdrawn and the adopted legislation should be repealed.[390]

358. Even before the Amsterdam Protocol was adopted, concrete substance was given to the subsidiarity principle, and more generally, the principles of Article 5 by the Edinburgh Conclusions of 11 and 12 December 1992.[391] The political energy expended on the negotiations of the Edinburgh text emphasised the importance at least some Member States attached to these principles in order to help secure the ratification of the Treaty of the European Union. This Edinburgh text has been supplemented by the Interinstitutional Agreement on procedures for implementing the principle of subsidiarity of 25 October 1993 between the European Parliament, the Council and the Commission.[392] The

---

Amsterdam amending the Treaty on European Union, the Treaties establishing the European Communities and certain related acts – Protocol annexed to the Treaty of the European Community – Protocol on the application of the principles of subsidiarity and proportionality, *OJ* C 340/105 (hereinafter Amsterdam Protocol) and see provision 5 of the Amsterdam Protocol. These criteria can be viewed as two criteria since the first and the third can be combined to form a single criterion. Therefore the two main criteria in order to justify action on a Community level are the "transnational dimension" and the "market distortion."

389 A. ESTELLA, *The EU Principle of Subsidiarity and its Critique* (Oxford, Oxford University Press, 2002), p. 107.

390 A. ARNULL, A. DASHWOOD, M. DOUGAN, M. ROSS, E. SPAVENTA and D. WYATT, *European Union Law* (London, Sweet & Maxwell, 2006), p. 97 and Commission of the European Communities of 3 November 1999, *Commission report to the European Council – Better lawmaking 1999*, COM(1999) 562, p. 2.

391 European Council in Edinburgh on 11 and 12 December 1992, see www.europarl.europa.eu/summits/edinburgh/default_en.htm.

392 Resolution on democracy, transparency and subsidiarity and the Interinstitutional Agreement on procedures for implementing the principle of subsidiarity; the regulations and general conditions governing the performance of the Ombudsman' s duties; the arrangements for the proceedings of the Conciliation Committee under Article 189 b EC, *OJ* C 329/132.

main elements of both political texts have been incorporated into the Amsterdam Protocol.[393]

iv) Conclusion

359. Subsidiarity is, as discussed, foremost a political principle since it was adopted as a response to the wide and expanding scope of Union law-making competence and the exercise of this competence. Nonetheless, it became a fundamental principle in the Union legal order which needs to be interpreted in the light of its text, its objectives and the principle of effectiveness. One could argue that from a textual point of view it is possible to interpret the dual test of subsidiarity in such a way that it minimises or even eliminates its potential to inhibit Union action. This means that proposals on Union-wide rules covering any subject-matter at all within the Union competences would pass the dual test if Member States cannot sufficiently achieve the objectives of the proposal, while the effects of the proposed action can be better achieved on a Union level. However, such an approach would deprive the useful effect of the subsidiarity principle.

360. The rationale of the tests is therefore to determine whether the effects of the proposed action justify the adoption of Union-wide rules instead of leaving the policy choices over the subject-matter in issue to the Member States. The objectives of the Union action in question, referred to in the subsidiarity test, are objectives which can only be achieved by Union-wide action. Consequently, in areas of non-exclusive competence, the Union must act, however only if by Union-wide action, and not by the Member States' action, the objectives of the action in question can be achieved. Nonetheless, the application of the subsidiarity principle is not necessarily straightforward because rational individuals may conclude differently upon the same action in question due to for example mixed objectives (some objectives may require Union-wide action and others may not).[394]

361. In order to facilitate the analysis of the foregoing issues in a transparent and systematic way, first the Edinburgh Conclusions[395], the Inter-institutional Agreement on subsidiarity of 25 October 1993[396] and the Amsterdam Protocol

---

393 A. DASHWOOD, "The relationship between the Member States and the European Community/European Union" in A. MCDONELL (ed.), *A review of forty years of Community Law* (The Hague, Kluwer Law International, 2002), p. 49 and 50.

394 A. ARNULL, A. DASHWOOD, M. DOUGAN, M. ROSS, E. SPAVENTA and D. WYATT, *European Union Law* (London, Sweet & Maxwell, 2006), p. 103–108.

395 European Council in Edinburgh on 11 and 12 December 1992, see www.europarl.europa.eu/summits/edinburgh/default_en.htm.

396 Resolution on democracy, transparency and subsidiarity and the Interinstitutional Agreement on procedures for implementing the principle of subsidiarity; the regulations and

introduced some procedural requirements which apply to the subsidiarity appraisal during the legislative process. However, even with such procedural requirements and guidelines on a European level, the application of the subsidiarity principle often remains unclear. This is mainly because one has to rely on general terms which involve values and subjective elements of policy choices.[397]

362. Besides, the effectiveness of the principle as a check on Union intervention is questionable because the aforementioned criteria and guidelines are in itself incapable of offering a clear-cut answer to whether the Union should act or not. In most areas where there seem to be clear reasons in favour of national, regional or local intervention, it could be argued that due to the development of the single market and the close relationship between these areas, Union intervention is necessary.[398] The Protocols certainly were a step forward, however, except for some innovations, they merely incorporated what already existed in soft law acts. Consequently, it seems that the attempts to define *ex ante* criteria of a general and abstract nature in order to limit Union intervention stands little chance of success.[399]

363. Instead of providing a method balancing the interests between the Union and the Member States, the principle of subsidiarity simply asks who should be doing the implementing work. Thus, it seems to protect the right of Member States to be co-opted by the Union to do its work. However, it does not protect their right to do their own work.[400]

---

general conditions governing the performance of the Ombudsman' s duties; the arrangements for the proceedings of the Conciliation Committee under Article 189b EC, *OJ* C 329/132.

397 S. DOUGLAS-SCOTT, *Constitutional Law of the European Union* (Harlow, Pearson education Limited, 2002), p. 180.

398 A. ESTELLA, *The EU Principle of Subsidiarity and its Critique* (Oxford, Oxford University Press, 2002), p. 131 and 132.

399 *Ibid.*, p. 113 and 114.

400 G. DAVIES, "Subsidiarity: The wrong idea, in the wrong place, at the wrong moment", *Common market Review* (2006), p. 67–72. The author clarifies with two hypothetical legislative proposals the shortcomings of the subsidiarity principle. On the one hand, the adoption of a common contract code replacing the contract law of the Member States and on the other hand, the adoption of a common high school curriculum with a common syllabus and the same methods of streaming and division of pupils. The existence of different high school systems and national contract laws both create obstacles to the free movement in the Union. According to the author most politicians will argue that the competence for civil law and education should remain with the Member States. Such a broad use of competences by the Union legislator has led in the first place to the dissatisfaction on a European level and to the call for a more powerful role of the subsidiarity principle. However, this principle does not seem to give a convincing reason why these measures should not be taken on a Union level. It is clear that common contract laws and educational systems would be good for some Union goals but, in many ways, bad for various things that the Member States are legitimately concerned with and wish to protect. Consequently, it should be analysed whether the given Union measures in these areas are proportionate or disproportionate.

(c) *Energy and the principle of subsidiarity*

364. When the Union finds a sufficient legal basis in the Treaty and thus has the power to act, the next question is whether it should act. This question of the need for Union action, as discussed above, involves the assessment with respect to the principle of subsidiarity *sensu stricto*.[401] Each legislative measure on a European level must pass the subsidiarity test. Currently, all legislation provides in this respect a justification. However, energy measures often have in their proposals and preamble a somewhat standard reference and there is little explicit discussion of the principle to be found.[402]

365. See for instance the reference to the subsidiarity principle in recital 62 of the preamble of the Third Electricity Directive which states that *"[s]ince the objective of this Directive, namely the creation of a fully operational internal electricity market, cannot be sufficiently achieved by the Member States and can therefore be better achieved at Community level, the Community may adopt measures, in accordance with the principle of subsidiarity as set out in Article 5 of the Treaty"*. The proposal of the Third Electricity Directive did not even justify how it satisfies the subsidiarity principle.[403]

With respect to the Renewable Directive 2009/28/EC, the Commission states in its proposal that: *"[...t]he objectives of the Proposal cannot be sufficiently achieved by the Member States for the following reasons:*

It is clear from the experience with the promotion of renewable energy sources in the European Union that real progress only began to be made when the European Union adopted legislative instruments containing targets to be reached by a given deadline. This is true for Directive 2001/77/EC on the promotion of electricity from renewable energy sources and for Directive 2003/30/EC on the promotion of the use of biofuels. No such legal framework exists to promote the penetration of renewable energy sources in the heating and cooling sector. The development of renewable energy in this sector is nearly stagnant.

[...]

Leaving action to the Member States would put the achievement of this share at risk and would not realize an equitable distribution of the efforts needed to arrive at the 20% overall share. In addition, leaving action completely to the

401 K. LENAERTS, "The Principle of Subsidiarity and the Environment in the European Union: Keeping the Balance of Federalism", *Fordham International Law Journal* (1993), p. 875.

402 D. SHAW, V. NADIN and K. SEATON, "The application of subsidiarity in the making of European environmental law, *European Environment* (2000), p. 91 and J. H. JANS and H. H. B. VEDDER, *European Environmental Law* (Groningen, Europa Law Publishing, 2008), p. 11.

403 Proposal for a Directive of the European Parliament and of the Council amending Directive 2003/54/EC concerning common rules for the internal market in electricity of 19 September 2007, COM(2007)528.

Member States would also create investor uncertainty as to the objectives to be reached and the pathway toward these objectives.

[...] A Community approach to promoting renewables by these means is proportionate, because the level of ambition of the target requires coordinated action which addresses the sectors where most progress can be made.

Community action in the field of biofuel sustainability is justified, because it avoids the development of multiple national schemes which might impede trade to and within the Community.

In the Proposal, Member States retain wide discretion to favour the development of the renewable energy sector in the way that suits their national potential and circumstances best, including the option of achieving their targets by supporting the development of renewable energy in other Member States.

The Proposal therefore complies with the subsidiarity principle".

366. If we apply the guidelines of the Amsterdam Protocol, discussed above, to the energy field, the following can be noted. Firstly, the issue under consideration should have transnational aspects which cannot be regulated satisfactorily by Member States. As far as energy is concerned, energy markets cross national boundaries more and more. The establishment of an internal energy market within the EU is of a transnational nature. Moreover, due to the territorial limitations of national powers, unilateral action by Member States will clearly be less effective than concerted action on a Union level for example to restrict the effects of CO_2 emissions, to facilitate cross-border energy trade, to secure the external supply of energy, to minimise the risk of ionising radiation or shipments of radioactive waste, etc.

367. Secondly, according to the Amsterdam Protocol, the actions by Member States alone or lack of Union action would conflict with the requirements of the Treaty (such as the need to correct distortion of competition or avoid disguised restrictions on trade or strengthen economic and social cohesion) or would otherwise significantly damage Member States' interests. With respect to energy for instance, this could be the case when Member States establish national mandatory energy labelling schemes. Such schemes would create barriers to trade and free movement of goods within the Union, and generate undue costs for the industry. Moreover, without action on an EU level with respect to the liberalisation process, market participants were exposed to conflicting and uncoordinated regimes.

368. The third guideline of the Amsterdam Protocol with respect to the application of the subsidiarity principle requires that action on a Union level produces clear benefits by reason of its scale or effects compared with action on the level of the Member States. In the energy field, a minimum level of harmonisation throughout the Union can only be achieved effectively with the adoption of measures on a European level (e.g. 20% share of renewable energy sources in final energy consumption by 2020, sustainability criteria for biofuels and bioliquids, unbundling provisions, tasks of the national regulators in the electricity and gas markets, etc.).

369. The foregoing demonstrates that in the energy field as well as in the environmental field, in the light of the discussed guidelines above, arguments can be found which allow proposed measures to pass the subsidiarity test.[404] It also shows that the subsidiarity test is a very elastic test. Nonetheless, in order to determine whether a given energy measure complies with the principle of subsidiarity the European legislator must always analyse this on a case-by-case basis. If the assessment of the need for Union action is positive, then a final test is required. This is the proportionality principle which considers whether the nature and intensity of the Union action of the given measure is justified.[405] The next part will analyse the principle of proportionality and its application with respect to energy.

4. THE PRINCIPLE OF PROPORTIONALITY

(a) Introduction

370. The third principle in Union law relevant for the existence and exercise of Union powers is the principle of proportionality. Before the proportionality principle was inserted into the third paragraph of Article 5 of the EC Treaty, it had mainly been used for two purposes: to control the extent to which Union measures are allowed to override the interests of particular individuals and to limit the exceptions Member States had been given to protect public interests through derogations from certain fundamental provisions of the Union framework.[406] This principle of proportionality and its impact on the energy field will be discussed in more detail in the next part.

---

404 J. H. JANS and H. H. B. VEDDER, *European Environmental Law* (Groningen, Europa Law Publishing, 2008), p. 12 en 13.

405 K. LENAERTS, "The Principle of Subsidiarity and the Environment in the European Union: Keeping the Balance of Federalism", *Fordham International Law Journal* (1993), p. 875 and D. SHAW, V. NADIN and K. SEATON, "The application of subsidiarity in the making of European environmental law", *European Environment* (2000), p. 89.

406 A. ARNULL, A. DASHWOOD, M. DOUGAN, M. ROSS, E. SPAVENTA and D. WYATT, *European Union Law* (London, Sweet & Maxwell, 2006), p. 110 and 240–244. See also G. DE BÚRCA, "The Principle of Proportionality and its Application in EC law", *Yearbook of European law* (Oxford, Clarendon,1993), p. 105–150; U. BERNITZ and J. NERGELIUS, *General Principles of European Community Law* (The Hague, Kluwer Law International, 2000), 244 p.

(b) *Meaning of the principle of proportionality on a European level*

371. The treaties do not contain an explicit set of principles against which the legality of Union action or Member State action which falls within the sphere covered by Union law, can be tested. Consequently, the Court of Justice and the Court of First Instance had to fashion these principles, among which the principle of proportionality.[407]

372. The first reference to this principle in the Union can be found in the *Fédéchar* case[408] of 1955 and the most often cited definition of proportionality in case law of the Court of Justice is the one given in the *Fromançais* case which states that: *"[i]n order to establish whether a provision of Community law is consonant with the principle of proportionality it is necessary to establish, in the first place, whether the means it employs to achieve its aim correspond to the importance of the aim and, in the second place, whether they are necessary for its achievement"*.[409] This definition has been further developed in the jurisprudence of the Court of Justice into a definition which contains, like the German definition of proportionality[410], three subsidiary principles (suitability, necessity and proportionality *sensu stricto*).[411]

373. In the *Fedesa* case the Court of Justice has used this definition and stated that: *"[...] the principle of proportionality is one of the general principles of Community law. By virtue of that principle, the lawfulness of the prohibition of an economic activity is subject to the condition that the prohibitory measures are appropriate and necessary in order to achieve the objectives legitimately pursued by the legislation in question; when there is a choice between several appropriate*

---

407 P. CRAIG, *EU Administrative Law* (Oxford, Oxford University Press, 2006), p. 655 and 656.

408 Case C-55/8, *Fédération Charbonnière de Belgique v High Authority of the European Coal and Steel Community* [1954–1956] ECR 245.

409 Case C-66/82, *Fromançais SA v Fonds d'orientation et de régularisation des marchés agricoles (FORMA)* [1983] ECR 395, paragraph 8; Case C-118/89, *Lingenfelser v Federal Republic of Germany* [1990] ECR I-2637 and Case C-155/89, *Belgian State v Philipp Brothers SA.* [1990] ECR 3265, paragraph 34.

410 P. CRAIG, *EU Administrative Law* (Oxford, Oxford University Press, 2006), p. 655 and 656 which refers to J. SCHWARZE, *European Administrative Law* (London, Sweet & Maxwell, 2006), p. 685–687. According to this author the concept of proportionality initially appeared in the context of policing. It was used as a ground for challenging measures on the basis that they were excessive or unnecessary in relation to the pursued objective. The current German formulation involves three factors which will be examined by the given court: (a) was the measure suitable for the attainment of the desired objective; (b) was the measure necessary and (c) was the measure disproportionate.

411 *Ibid.* and G. DE BÚRCA, "The Principle of Proportionality and its Application in EC law", *Yearbook of European law* (Oxford, Clarendon,1993), p. 113 and N. EMILIOU, *The Principle of Proportionality in European Law A Comparative Study* (London, Kluwer Law International, 1996), p. 24 and D. CHALMERS, G. DAVIES and G. MONTI, *European Union Law* (Cambridge, Cambridge University Press, 2010), p. 367.

measures recourse must be had to the least onerous, and the disadvantages caused must not be disproportionate to the aims pursued".[412]

374. In this case and other cases, the Court of Justice has acknowledged the proportionality principle as a *general principle* of EU law. The word *general* in *general principle* seems to mean that the principle under discussion is inherent in a series of infinite applications of the law.[413] This particular feature means that a general principle of law, as opposed to a regular principle of law, is applicable to a greater variety of cases. The word *general* furthermore refers to the fact that the principle is also part of other systems of law. With respect to the principle of proportionality, this principle can for example be found in German and French administrative law and German constitutional law.[414] A *principle of law* is a rule which is presumed to form the basis of the very institution of law and which represents an ideal of reason and/or justice.[415]

375. It is clear from the Court's case law that different definitions are often used in the context of different types of cases. Besides, the origin of the proportionality principle in Union law has never been the subject of discussion and has been deemed as self-evident with no need for further explanation and clarification.[416] However, there has been some uncertainty in legal doctrine as to whether the third element of the three subsidiary principles, referred to as proportionality *sensu stricto* in the modern German formulation of the concept of proportionality, is also part of the Union test.[417]

---

412 Case C-331/88, *The Queen v Minister of Agriculture, Fisheries and Food and Secretary of State for Health, ex parte: Fedesa and others* [1990] ECR 4023, paragraph 13 and Joined cases C-133/93, C-300/93 and C-362/93, *Antonio Crispoltoni v Fattoria Autonoma Tabacchi and Giuseppe Natale and Antonio Pontillo v Donatab Srl* [1994] ECR 4863, paragraph 10.

413 T.-I. HARBO, "The function of the Proportionality Principle in EU Law", *European Law Journal* (2010), p. 159 which refers to N. EMILIOU, *The Principle of Proportionality in European Law A Comparative Study* (London, Kluwer Law International, 1996), p. 115.

414 T.-I. HARBO, "The function of the Proportionality Principle in EU Law", *European Law Journal* (2010), p. 159.

415 *Ibid.* which refers to F. GENY, *Méthode d'interprétation et sources en droit privé positif* (Paris, Librairie générale de droit et de jurisprudence, 1954) and N. EMILIOU, *The Principle of Proportionality in European Law A Comparative Study* (London, Kluwer Law International, 1996), p. 115.

416 N. EMILIOU, *The Principle of Proportionality in European Law: A Comparative Study* (London, Kluwer Law International, 1996), p. 134 and 135.

417 P. CRAIG, *EU Administrative Law* (Oxford, Oxford University Press, 2006), p. 657 which refers to W. VAN GERVEN, "The Effect of Proportionality on the Actions of Member States of the European Community: National Viewpoints from Continental Europe", in E. ELLIS (ed.), *The Principle of Proportionality* (Oxford, Hart Publishing, 1999), p. 37 and 38 and T.-I. HARBO, "The function of the Proportionality Principle in EU Law", *European Law Journal* (2010), p. 165.

376. This discussion on whether the Union courts apply a two- or three-tier proportionality test is relevant since it would change the nature of the test considerably. If a measure under the proportionality test complies with the suitability and necessity requirements it would be regarded as legal even if the burden imposed on an individual might be considered as disproportionate to the desired objective. Under the suitability test, the question is asked whether the chosen measure is suitable or appropriate in order to achieve the given aim. Under the necessity test, the Union courts assess whether the chosen measure is necessary to achieve the given aim. This implies that the chosen measure should be the one which is the least restrictive on the given norm.[418]

377. In the Union courts, the normal judicial formulation of proportionality appears to be based on the two-part proportionality test which comprises the suitability and necessity of the given measure.[419] There are clear traces of some guideline according to which the Union courts assess these two tests.[420]

378. The Union courts also seem to apply a three-part proportionality test which includes the *sensu stricto* proportionality when the applicant, contesting the legality of a given measure, raises arguments that place the matter before the Union courts.[421] This means that the onus falls upon the applicant.[422] Compared to the suitability and necessity tests, the exact content of the *sensu stricto* proportionality test by the Union Courts is less clear.[423]

379. Some case law has indeed confirmed that the Union courts are willing to consider the *sensu stricto* proportionality test. However, it does not tell how they undertake this exercise. The depth of the exercise by the Courts in this respect differs considerably. On the basis of the foregoing, Union Courts are willing to consider the *sensu stricto* proportionality test if the applicant presents arguments

---

418 T.-I. HARBO, "The function of the Proportionality Principle in EU Law", *European Law Journal* (2010), p. 165.

419 P. CRAIG, *EU Administrative Law* (Oxford, Oxford University Press, 2006), p. 657 and 670.

420 T.-I. HARBO, "The function of the Proportionality Principle in EU Law", *European Law Journal* (2010), p. 165.

421 Case C-331/88, *The Queen v Minister of Agriculture, Fisheries and Food and Secretary of State for Health, ex parte: Fedesa and others* [1990] ECR 4023; Joined cases C-133/93, C-300/93 and C-362/93, *Antonio Crispoltoni v Fattoria Autonoma Tabacchi and Giuseppe Natale and Antonio Pontillo v Donatab Srl* [1994] ECR 4863; Case C-426/93, *Federal Republic of Germany v Council of the European Union* [1995] ECR I-3723; Case 183/95, *Affish BV v Rijksdienst voor de keuring van Vee en Vlees* [1997] ECR I-4315; Case T-13/99, *Pfizer Animal Health SA v Council of the European Union* [2002] ECR II-3305; Joined cases T-125/96 and T-152/96, *Boehringer Ingelheim Vetmedica GmbH and C.H. Boehringer Sohn v Council of the European Union and Commission of the European Communities* [1999] ECR II-3427.

422 P. CRAIG, *EU Administrative Law* (Oxford, Oxford University Press, 2006), p. 670–672.

423 T.-I. HARBO, "The function of the Proportionality Principle in EU Law", *European Law Journal* (2010), p. 165.

directed specifically to it. However, in practice such applicant appears to face an uphill battle in order to convince the Union courts that the given measure should be struck down for being disproportionate *sensu stricto*. The arguments of the applicant will be balanced against the objective sought by the Union. In the cases where this objective involves the protection of public health, the balance is unlikely to tilt in favour of the applicant.[424]

380. The proportionality principle was included by the Maastricht reform in Article 3b, 3° of the EEC Treaty and is currently defined in Article 5(4) of the TEU which provides that: *"[u]nder the principle of proportionality, the content and form of Union action shall not exceed what is necessary to achieve the objectives of the Treaties. The institutions of the Union shall apply the principle of proportionality as laid down in the Protocol on the application of the principles of subsidiarity and proportionality"*.[425] This provision has put the proportionality principle on a Treaty footing and provides for a proportionality test which answers the question: *"What should be the intensity or nature of the Union action?"*[426]

381. The logic behind the principle, as discussed above, remains the same. However, it operates on the constitutional plane now with respect to the exercise and existence of the competences of the Union legislator. It will guide the Union legislator towards the most appropriate form of action on a Union level which will intrude into the Member States' powers, but to the smallest practicable extent.[427]

382. In order to establish whether a provision of Union law complies with the principle of proportionality, it must be ascertained, according to the case law of the Court of Justice, whether the means it employs are suitable to achieve the desired objective and whether they do not go beyond what is necessary to achieve it.[428] It serves to control the scope of Union legislation, by avoiding excessively

---

424 P. CRAIG, *EU Administrative Law* (Oxford, Oxford University Press, 2006), p. 670–672.

425 It should be highlighted that the wording of the principle of proportionality has been amended with the adoption of the Lisbon Treaty. It now refers to the criteria of content and form which were previously mentioned in Protocol 7 to the EC Treaty.

426 A. ARNULL, A. DASHWOOD, M. DOUGAN, M. ROSS, E. SPAVENTA and D. WYATT, *European Union Law* (London, Sweet & Maxwell, 2006), p. 111 and Joined Cases 279/84, 280/84, 285/84 and 286/84, *Walter Rau Lebensmittelwerke and others v Commission of the European Communities* [1987] ECR 1069, paragraph 34 and Case 426/93, *Federal Republic of Germany v Council of the European Union* [1995] ECR 3723.

427 *Ibid.*

428 Joined Cases 279/84, 280/84, 285/84 and 286/84, *Walter Rau Lebensmittelwerke and Others v Commission*, [1987] ECR 1069 and Case 426/93, *Federal Republic of Germany v Council of the European Union* [1995] ECR I-3723.

detailed provisions, and to determine the least intrusive means for the implementation of the given Union legislation.[429]

383. To evaluate the proportionality of a provision of Union law, reference should be made to the guidelines of the Edinburgh Conclusions of 11 and 12 December 1992, the Amsterdam Protocol.[430] To conclude, the proportionality principle is not only a judicial doctrine which needs to be applied by the Court of Justice when it evaluates the legality of Union action, but also a legislative provision which the Union legislators must apply in the exercise of their decision-making process.

(c) Energy and the principle of proportionality

384. As discussed above, the principle of proportionality ensures a proportionate relationship between the means and the ends in legislative measures. In other words it means *"do not use a sledgehammer to crack a nut if a nutcracker will do"*.[431] The same applies when the European legislator adopts measures in the energy field. It must determine the nature and intensity of its action in the energy field in the light of the different aspects of the principle of proportionality. This means that each energy measure on a European level must comply with the three-tier proportionality test, namely the suitability, necessity and *sensu stricto* proportionality requirements.

385. Energy measures have in their preamble with respect to the proportionality principle, like the subsidiarity principle, very often a somewhat standard reference. See for instance the reference to the proportionality principle in recital 60 of the preamble of the Third Gas Directive which states that: *"[i]n accordance with the principle of proportionality, […], this Directive does not go beyond what is necessary in order to achieve that objective"*.[432]

429 N. EMILIOU, *The Principle of Proportionality in European Law A Comparative Study* (London, Kluwer Law International, 1996), p. 142.

430 See in particular provisions §and 7 which state that: "*the form of Community action shall be as simple as possible, consistent with satisfactory achievement of the objective of the measure and the need for effective enforcement. The Community shall legislate only to the extent necessary. Other things being equal, directives should be preferred to regulations and framework directives to detailed measures. […] While respecting Community law, care should be taken to respect well-established national arrangements and the organisation and working of Member States' legal systems.*"

431 D. SHAW, V. NADIN and K. SEATON, "The application of subsidiarity in the making of European environmental law, *European Environment* (2000), p. 89.

432 Other directives and regulations which in similar words refer to the proportionality principle are for example Regulation No 715/2009, Third Electricity Directive, Regulation No 994/2010.

386. Regardless of this standard reference, it must be assessed whether each energy measure complies with the three-tier proportionality test. This means that with respect to the *sensu stricto* proportionality requirement the European legislator must determine whether a given energy measure imposes any disadvantages which are excessive compared to the perceived advantages. In other words, it must be analysed whether, due to the adoption of the energy measures concerned, any huge burdens are imposed on Member States which are disproportionate to the benefits gained.[433] For example, if an energy measure of the Union would impose not only a legally binding target for renewable energy sources on Member States, but also determine for each Member State which kind of renewable energy technology (solar, wind, etc.) must be used in order to achieve the binding target. Such a measure has certain advantages, although, in addition to other legal issues, it imposes on a Member State an inflexible regime which is disproportionate to the benefits it will gain.

387. The second requirement which needs to be assessed is whether, under the necessity test, the given energy measures on a Union level is necessary to achieve the given aim. As discussed above, this means that the chosen energy measure should be the one which is the least restrictive on the given norm. For instance it was necessary for the European legislator to adopt directives with the purpose of opening up the electricity and gas markets. Any other measure less than directives, i.e. non-binding measures on a Union level, would have been insufficient to achieve this aim.

388. The final requirement with respect to the principle of proportionality is the suitability test under which the question is asked whether the chosen measure is suitable or appropriate to achieve the given aim. This test analyses the nature of the action and evaluates whether the proposal is both an appropriate and effective mechanism for achieving its aims. In this context, preference should be given to directives as the most "proportional" normative instrument on a European level since it leaves national authorities the choice of form and methods.[434] In particular, the first electricity and gas directives left as much scope for national decision as possible. This approach changed in the energy directives and the scope for Member States became more limited. This scope might become even more limited if the European legislator would legislate more often through regulations. Unfortunately, in the Third Package, this approach

433 D. SHAW, V. NADIN and K. SEATON, "The application of subsidiarity in the making of European environmental law, *European Environment* (2000), p. 93.

434 K. LENAERTS, "The Principle of Subsidiarity and the Environment in the European Union: Keeping the Balance of Federalism", *Fordham International Law Journal* (1993), p. 885 and D. SHAW, V. NADIN and K. SEATON, "The application of subsidiarity in the making of European environmental law, *European Environment* (2000), p. 94.

was not chosen for all measures, although it would have been possible, in the light of the foregoing.

389. This part is followed by a discussion in great detail of the different provisions of the TFEU which could serve as a legal basis allowing Union action in the energy field before the adoption of the Lisbon Treaty.

5. CONCLUSION

390. This part discussed the principles of attribution, subsidiarity and proportionality which govern the existence and exercise of the Union powers. Union action in the energy field is limited through the application of these principles. Before the Union legislator can adopt measures in the energy field the foregoing demonstrated that on the basis of the principle of attribution an appropriate legal basis is required. Since the Treaties, prior to the adoption of the Lisbon Treaty, did not confer any specific powers in the energy field, the European legislator took its recourse to more general legal bases such as Article 114 of the TFEU or Article 352 of TFEU in order to comply with the principle of attribution.

391. The second principle analysed in this part is the subsidiarity principle. This principle is designed to determine whether Union intervention is appropriate or not. Legislative measures adopted on a European level must pass the so-called subsidiarity test. In order facilitate this test a number of procedural requirements and guidelines were adopted. Nonetheless, the application of the subsidiarity principle often remains unclear. With respect to the energy field, this part stated that, in the light of the discussed guidelines, arguments could be found which allow energy measures to pass the subsidiarity test. Moreover, it argued that this test is very elastic.

392. The last principle analysed in this context was the proportionality principle which ensures a proportionate relationship between the means and the ends of legislative measures. Very often, like the subsidiarity principle, energy measures contain in their preamble a somewhat standard reference which confirms the accordance with the proportionality principle. Regardless of this standard reference, this part has argued that each energy measure must pass the three-tier proportionality test and a case-by-case analysis is therefore required.

393. On the basis of the analysis above, it can be concluded that the principles of attribution, subsidiarity and proportionality do not seem to constitute a barrier for the adoption of measures in the energy field. Their main objective is to ensure that action in the energy sector on a European level falls within the competences of the Union. Any shortcomings in the current European energy legal framework should not be attributed to the existence and application of these principles.

D. THE PROVISIONS OF THE TFEU ALLOWING UNION ACTION IN THE FIELD OF ENERGY

1. INTRODUCTION

394. The TFEU contains a number of provisions applicable to the energy sector. The most important ones are the rules on the free movement of goods set out in Articles 34–37 of the TFEU, the rules on the right of establishment and the free movement of services and capital set out in Articles 49–66 of the TFEU, the rules on competition set out in Articles 101–102 and 106 of the TFEU and the rules on State aid set out in 107–108 of the TFEU.

Nevertheless, these provisions on the free movement and competition are subject to exceptions in the EU Treaty which led to some modifications when applied to the energy sector.[435]

395. The Union also intervened in the energy markets by secondary legislation primarily on the basis of Article 114 of the TFEU which empowers the Union and sets out the legal procedures for the adoption of measures in achieving the internal market. Beside Article 114 of the TFEU, the European legislator had recourse to Title XX of the TFEU, comprising Articles 191–193, and establishes the Union's competence in the environmental field, for the adoption of measures in the energy sphere. These were not the only provisions; other examples are Article 113 of the TFEU on taxation, and Articles 170–172 of the TFEU on the promotion of trans-European Networks (TENs) in the areas of transport, telecommunications and energy infrastructure.

396. The purpose of this part of chapter 4 is to discuss these provisions and their impact on the development of energy legislation in the Union.

---

[435] G. RASHBROOKE, "Clarification or Complication? The New Energy Title in the Draft Constitution for Europe", *Journal of Energy & Natural Resources Law* (2004), p. 375–376; E. D. CROSS, L. HANCHER, and P. J. SLOT, "EC Energy Law (chapter 5)" in M. M. ROGGENKAMP, A. RØNNE, C. REDGWELL and I. DEL GUAYO (eds.), *Energy Law in Europe. National, EU and International Law and Institutions* (Oxford, Oxford University Press, 2001), p. 222 and see for a more detailed analysis of these provisions E. D. CROSS, B. DELVAUX, L. HANCHER, P. J. SLOT, G. VAN CALSTER and W. VANDENBERGHE, "EU Energy Law (chapter 5)" in M. M. ROGGENKAMP, A. RØNNE, C. REGWELL and I. DEL GUATO (eds.), *Energy Law in Europe National, EU and International Regulation* (Oxford, Oxford University Press, 2007), p. 234–303.

2. THE GENERAL TFEU PROVISIONS AND THEIR APPLICATION IN THE ENERGY SECTOR

(a) Introduction

397. The creation of competitive electricity and gas markets in Europe began slowly and has accelerated in recent years. On a European level the process of opening the electricity and gas markets started seriously in the early 1990s when the Commission adopted the directive on the transparency of gas and electricity prices[436], the transit of electricity and gas directives[437], and formally proposed its first electricity and gas directives.[438]

398. During the period before the adoption of these proposals, most utilities in the energy sector and other sectors were run by national and very often state-owned monopolies. Member States granted some of these monopolies in the energy sector, either *de jure* or *de facto,* the exclusive right to sell, import and export electricity and/or gas and to construct infrastructure in their particular area. In the late 1980s and early 1990s the Commission began to challenge, first in the telecommunications and then the electricity and gas sectors, the existence of such monopolies and their exclusive rights because they made the existence of a European internal market for these goods impossible.[439] As far as the European electricity and gas sector are concerned, it should be noted that the Commission was very reluctant to challenge the existence of these monopolies and other national measures. Consequently, Member States have effectively been allowed to organise their electricity and gas market as they preferred.[440]

399. As it became clear that in the early 1990s there still was no realistic chance of getting the agreement of all or a qualified majority of Member States to adopt

---

436 Council Directive 90/377/EEC of 29 June 1990 concerning a Community procedure to improve the transparency of gas and electricity prices charged to industrial end-users, *OJ* L 185/16. This directive has been repealed by Directive 2008/92/EC of the European Parliament and of the Council of 22 October 2008 concerning a Community procedure to improve the transparency of gas and electricity prices charged to industrial end-users, *OJ* L 298/9.

437 Council Directive 90/547/EEC of 29 October 1990 on the transit of electricity through transmission grids, *OJ* L 313/30 and Council Directive 91/296/EEC of 31 May 1991 on the transit of natural gas through grids, *OJ* L 147/37.

438 Proposal for a Council Directive concerning common rules for the internal market in electricity, COM(1991)548 and Proposal for a Council Directive concerning common rules for the internal market in natural gas, COM(1991)548.

439 C. W. JONES and W. WEBSTER, *EU ENERGY LAW – Volume I The Internal Energy Market* (Leuven, Claeys & Casteels, 2006), p. 1.

440 L. HANCHER, "Delimitation of Energy Law Jurisdiction: the EU and its Member States: From Organisational to Regulatory Conflicts", *Journal of Energy & Natural Resources Law* (1998), p. 44 and 45 and A. JOHNSTON, "Maintaining the Balance of Power: Liberalisation, Reciprocity and Electricity in the European Community", *Journal of Energy and Natural Resources Law* (1999), p. 121–125.

legislative measures leading towards a liberalisation process on a Union level of the energy sector or other sectors, the Commission started to use the Treaty provisions relating to competition law, beginning with the telecom sector. With these provisions the Commission tried to force Member States to abandon the monopolies in these sectors. It argued that the granted rights to these monopolies were contrary to the Treaty requirements on the free movement of goods and establishment of a single European market which could not be justified on the grounds of public service obligations, pursuant to Article 106(2) of the TFEU. On the basis of this reasoning, the Commission took a number of decisions in the telecom and energy sectors.

400. With respect to the energy sector, the Commission started infringement proceedings in 1991, on the basis of Article 258 of the TFEU, against some Member States regarding electricity and gas monopolies on the grounds that their exclusive import and export rights were infringements of Articles 36, 40 and 43 of the TFEU. The Court of Justice before which these matters were brought, decided in essence that the Commission was allowed to take action under the TFEU to prohibit such exclusive rights, but required it to meet a high burden of proof to demonstrate that the existence of such monopoly rights were not essential to achieve legitimate public service obligations.[441, 442]

401. These judgments are of a fundamental importance as they confirmed that the Commission has the power to abolish monopoly rights under certain circumstances. This brought the Member States to the negotiating table as they believed it was better to agree upon the text of a directive requiring the progressive opening of the electricity and gas markets on the basis of Article 114 of the TFEU, rather than leave it to the Commission to take initiatives under Article 106(2) of the TFEU.[443]

402. Before discussing the legal bases which led to the adoption of secondary legislation, the next two parts analyse first the limits of some of the most relevant general TFEU provisions applicable to the energy sector.

---

441 C. W. JONES and W. WEBSTER, *EU ENERGY LAW – Volume I The Internal Energy Market* (Leuven, Claeys & Casteels, 2006), p. 2 and 3.

442 Case C-158/94, *Commission of the European Communities v Italian Republic* [1997] ECR I-5789 and Case C-159/94, *Commission of the European Communities v French Republic* [1997] ECR I-5815.

443 C. W. JONES and W. WEBSTER, *EU ENERGY LAW – Volume I The Internal Energy Market* (Leuven, Claeys & Casteels, 2006), p. 3.

(b) *Electricity and gas markets and the limits of the general TFEU provisions*

403. One of the main legal problems in applying the provisions of the TFEU on free movement and undistorted competition to the electricity and gas markets is that the TFEU leaves considerable scope to the Member States to determine how they organise these markets.

404. This is partially the result of Article 106(1) of the TFEU which recognises that Member States may continue to grant special and exclusive rights as long as these rights are not in breach with the provisions of Article 18 of the TFEU, which prohibits any discrimination on the basis of nationality, and Articles 101 and 109 of the TFEU which lays down the rules on competition.[444] Another provision of the TFEU which has led to a difficult application of the provisions of the TFEU on free movement and undistorted competition to the electricity and gas markets is Article 345 of the TFEU. This Article states that: "*the treaties shall in no way prejudice the rules in Member States governing the system of property ownership*".

405. It should be highlighted that Articles 106(1) and 345 of the TFEU are not the only reasons why the TFEU leaves considerable scope to the Member States to determine how they organise these electricity and gas markets. Another reason is the division of competences between the Member States and the Union in the energy field which can be derived in particular form the Articles 191–193 of the TFEU, establishing the Union's competence in the environmental field.[445]

406. Before analysing these provisions of TFEU, the purpose of this part is to have a clear understanding of the meaning and application of the Articles 106(1), (2) and 345 of the TFEU since they could hamper the application of some of the TFEU provisions, and secondary legislation, to the electricity and gas markets. The analysis below of the legal doctrine and case law of the Court of Justice, has led to a better understanding of both Articles and some conclusions when they are applied in the energy field.

---

444 L. HANCHER, "Delimitation of Energy Law Jurisdiction: the EU and its Member States: From Organisational to Regulatory Conflicts", *Journal of Energy & Natural Resources Law* (1998), p. 45 and 46.

445 See *Infra*.

i) The application of Article 106 of the TFEU

1. Article 106(1) of the TFEU

407. Article 106(1) of the TFEU declares that: *"[i]n the case of public undertakings and undertakings to which Member States grant special or exclusive rights, Member States shall neither enact nor maintain in force any measure contrary to the rules contained in the Treaties, in particular to those rules provided for in Article 18 and Articles 101 to 109".*

408. This article is normally applied in conjunction with another article of the treaty because its purpose is to limit the ways in which State measures protect undertakings which might hinder the operation of the TFEU. It does not only deal with competition, but with all the provisions of the TFEU, although it only refers in particular to the prohibition on discrimination on the grounds of nationality (Article 18 of the TFEU) and the competition rules (Article 102 of the TFEU). Apart from these rules, the most likely concerned rules are those on free movement and State monopolies.

(a) Case law of the Court of Justice interpreting Article 106(1) of the TFEU

409. The prohibition of Article 106(1) of the TFEU is addressed to Member States and not to undertakings.[446] This provision prohibits Member States to adopt or maintain measures in force in relation to public undertakings to which they have granted special or exclusive rights contrary to the provisions of the TFEU. It is designed to prevent Member States from depriving the effectiveness of the Treaty rules through measures that they enact in respect of public undertakings or through measures which enable private undertakings to escape the constraints of the competition provisions.[447]

410. It is not always clear which measure from a Member State violates the provisions of Article 106(1) of the TFEU. Situations where the State's arrangements inherently infringe a rule of TFEU in itself, as in the *Merci*

[446] Case C-41/90, *Höfner v Macroton* [1991] ECR I-1979, paragraph 16 and Case C-320/91, *Criminal proceedings against Paul Corbeau* [1994] ECR I-1783, paragraphs 10–12.

[447] A. JONES and B. SUFRIN, *EC Competition Law – Text, cases and materials* (Oxford, Oxford University Press, 2008), p. 624 and 625 and J. FAULL and A. NIKPAY (eds.), *The EC Law of Competition* (Oxford, Oxford University Press, 2007), p. 596 and 597. For further reading see G. MARENCO, "Public Sector and Community Law", *Common Market Law Review* (1983), p. 495–527; D. EDWARD and M. HOSKINS, "Article 90: Deregulation and EC Law Reflections arising from the XVI FIDE Conference", *Common Market Law Review* (1995), p. 157–186 and D. G. GOYDER, *EC Competition Law* (Oxford, Oxford University Press, 2003), p. 482–486.

Convenzionali case[448], are clearly a violation of Article 106(1) of the TFEU. In this case the Italian laws reserved the loading and unloading of ships at an Italian port to certain dock-work companies whose workers had to be of Italian nationality. This was clearly an infringement of Article 45 of the TFEU on the free movement of workers and Article 18 of the TFEU which prohibits discrimination on the grounds of nationality. In other cases the Court of Justice has ruled that the measures taken by the Member States result in violations of the Treaty. If in some cases the measures taken by the Member States result in a violation of one of the competition rules, it is suggested that the very fact of granting monopoly rights may in itself be a violation.[449]

411. The Court of Justice rejected in its case law on Article 106 of the TFEU the idea that Member States have an absolute sovereignty to determine whether or not a sector is subject to the Treaty rules on competition. In the *France v Commission* case[450] the Court of Justice held that the conferral and exercise of exclusive rights could be controlled by the Court.[451] The Court furthermore stressed in this case that Article 106 of the TFEU presupposes the existence of undertakings which have certain special or exclusive rights. However, it does not follow that all the special or exclusive rights are necessarily compatible with the Treaty provisions. This seems to indicate that the creation of such monopolies needs to be in balance with the principle of free competition. However, it is not that clear at which precise point the balance is to be struck.[452] Some guidance can be derived from the case law of the Court of Justice.[453]

412. Nonetheless, the rulings of the Court of Justice do not seem to suggest that the Member States have lost all power to establish monopolies. In Case 155/73[454] the Court of Justice held that Article 106(1) of the TFEU ensured that the grant of exclusive rights in respect of television broadcasts to RAI under Italian law was not of itself incompatible with Union law.[455]

---

448 Case C-179/90, *Merci Convenzionali Porto di Genova v Siderurgica Gabrielli SpA* [1991] ECR I-5889.

449 A. JONES and B. SUFRIN, *EC Competition Law – Text, cases and materials* (Oxford, Oxford University Press, 2008), p. 629.

450 Case C-202/88, *France v Commission* [1991] ECR I-1223.

451 *Ibid.*

452 D. EDWARD and M. HOSKINS, "Article 90: Deregulation and EC Law. Reflections arising from the XVI Fide conference", *Common Market Law Review* (1995), p. 160.

453 A. JONES and B. SUFRIN, *EC Competition Law – Text, cases and materials* (Oxford, Oxford University Press, 2008), p. 630.

454 Case C-155/73, *Giuseppe Sacchi* [1974] ECR 409.

455 *Ibid.*, paragraph 14.

413. Also in the *Höfner v Macroton* case[456] the Court of Justice reiterated in paragraph 29 that *"the simple fact of creating a dominant position of that kind by granting an exclusive right within the meaning of Article 90(1) is not as such incompatible with Article 86 of the Treaty (see Case 311/84 CBEM, above, paragraph 17). A Member State is in breach of the prohibition contained in those two provisions only if the undertaking in question, merely by exercising the exclusive right granted to it, cannot avoid abusing its dominant position"*. Consequently, the simple fact of creating a dominant position by granting exclusive rights is as such not incompatible with the Treaty provisions. Nonetheless, according to the Court a Member State will breach the Treaty provisions if the undertaking in question, merely by exercising the exclusive right granted to it, cannot avoid abusing its dominant position.[457, 458]

414. In the *Elliniki Radiophonia Tileorassi Anonimi Etaira (hereinafter ERT) v Dimotiki Etairia Pliroforissis* case[459] the Court of Justice took a slightly different approach since it ruled that the manner in which the monopoly (*in casu* a television monopoly) is organised or exercised, may breach the Treaty rules and in particular those relating to the free movement of goods, the freedom to provide services and the rules on competition.[460]

415. In this case the accumulation of exclusive rights did not result in an unavoidable infringement of Article 102 of the TFEU, but led to a situation where the monopolist was allowed to breach Article 102 of the TFEU because it would inevitably discriminate in favour of retransmitting its own programmes rather than the programmes of any other undertaking. The rights conferred on the undertaking in the *ERT* case, but also the *RTT* case[461] have been found in

---

456 Case C-41/90, *Höfner v Macroton* [1991] ECR I-1979.

457 *Ibid.*

458 A. JONES and B. SUFRIN, *EC Competition Law – Text, cases and materials* (Oxford, Oxford University Press, 2008), p. 630–632.

459 Case C-260/89, *Elliniki Radiophonia Tileorassi Anonimi Etaira v Dimotiki Etairia Pliroforissis* [1991] ECR I-2925.

460 *Ibid.*, paragraph 11. In paragraph 37 and 38 of the *ERT* case the Court of Justice held that *"[...] it should be observed that Article 90(1) of the Treaty prohibits the granting of an exclusive right to retransmit television broadcasts to an undertaking which has an exclusive right to transmit broadcasts, where those rights are liable to create a situation in which that undertaking is led to infringe Article 86 of the Treaty by virtue of a discriminatory broadcasting policy which favours its own programmes. The reply to the national court must therefore be that Article 90(1) of the Treaty prohibits the granting of an exclusive right to transmit and an exclusive right to retransmit television broadcasts to a single undertaking, where those rights are liable to create a situation in which that undertaking is led to infringe Article 86 by virtue of a discriminatory broadcasting policy which favours its own programmes, unless the application of Article 86 obstructs the performance of the particular tasks entrusted to it"*. A. JONES and B. SUFRIN, *EC Competition Law – Text, cases and materials* (Oxford, Oxford University Press, 2008), p. 633 and 634.

461 Case C-18/88, *RTT v GB-INNO-BM SA* [1991] ECR I-5941.

violation of Article 106 of the TFEU because they created a conflict of interest. This situation is different from the one in the *Höfner v Macroton* case[462] where the infringement was unavoidable due to the given monopoly rights by the State to a body manifestly unable to deal with the demand for services.[463]

416. In its *Merci Convenzionali* case[464], the Court of Justice cited its judgments in the *Höfner v Macroton* and *ERT* cases and held that the simple fact of creating a dominant position by granting exclusive rights is not incompatible with Article 106 of the TFEU, but the granting of rights which induce or lead an undertaking to commit abuses, is incompatible. In the *Corbeau* case[465] the Court of Justice seemed to go even further by stating that Article 106(1) of the TFEU "*[...] must be read in conjunction with Article 90(2) which provides that undertakings entrusted with the operation of services of general economic interest are to be subject to the rules on competition in so far as the application of such rules does not obstruct the performance, in law or in fact, of the particular tasks assigned to them.*

That latter provision thus permits the Member States to confer on undertakings to which they entrust the operation of services of general economic interest, exclusive rights which may hinder the application of the rules of the Treaty on competition in so far as restrictions on competition, or even the exclusion of all competition, by other economic operators are necessary to ensure the performance of the particular tasks assigned to the undertakings possessed of the exclusive rights".[466]

417. In this case the Court of Justice appeared to suggest, in order to ensure the *effet utile* or effectiveness of Articles 106(1) and 102 of the TFEU, that the very granting of special or exclusive rights by a Member State which confer a dominant position on an undertaking is unacceptable unless the rights at issue can be justified under Article 106(2) of the TFEU.[467] In effect, the Court of Justice seems to reverse the burden of proof in this judgment. Until the *Corbeau* case, exclusive rights were *prima facie* legal unless they led the given undertakings to engage in abusive behaviour or they constituted an extension of a dominant position.[468] With the judgment in the *Corbeau* case, exclusive rights

---

462 Case C-41/90, *Höfner v Macroton* [1991] ECR I-1979.

463 A. JONES and B. SUFRIN, *EC Competition Law – Text, cases and materials* (Oxford, Oxford University Press, 2008), p. 634 and 638.

464 Case C-179/90, *Merci Convenzionali Porto di Genova v Siderurgica Gabrielli SpA* [1991] ECR I-5889.

465 Case C-320/91, *Criminal proceedings against Paul Corbeau* [1994] ECR I-1783.

466 *Ibid.*, paragraphs 13 and 14.

467 A. JONES and B. SUFRIN, *EC Competition Law – Text, cases and materials* (Oxford, Oxford University Press, 2008), p. 640 which referred in footnote 91 to L. HANCHER, "Case note on Corbeau", *Common Market Law Review* (1994), p. 111.

468 See *Supra.*

were presumed not legal, but *prima facie* illegal unless they can be objectively justified or fulfil the criteria of Article 106(2) of the TFEU.[469, 470]

418. Contrary to its ruling in the *Corbeau* case, the Court of Justice in subsequent cases such as the *La Crespelle* case[471] in which it was faced with a similar situation as in the *Corbeau* case abandoned its automatic abuse doctrine. The Court ruled that the creation of a dominant position by granting an exclusive right is not as such incompatible with Article 106 of the TFEU. The application of Article 106 of the TFEU could be justified on the fact that the monopolist had abused its dominant position by demanding excessive prices. However, this could not lead to an action against a Member State on the basis of Articles 106(1) and 102 of the TFEU. Such an action could only be justified if a causal link can be demonstrated between the State measure and the abuse.[472] The *La Crespelle* case was not the last view of the Court of Justice on this issue. In subsequent cases the Court of Justice relied again on the automatic abuse doctrine of the *Corbeau* case.[473]

419. Conversely, in other cases the Court of Justice continued to hold the principle that the creation of a dominant position through the grant of exclusive rights within Article 106(1) of the TFEU is not itself contrary to the Treaty provisions, unless the rights create a situation whereby the undertaking, merely by exercising its rights, is led to commit an abuse or if these rights make an abuse unavoidable. If this is the undertaking's position, it must be verified whether the justification under Article 106(2) of the TFEU applies.[474] In some cases the Court did not even proceed to see whether the undertaking was put in such a position,

---

469 See *Infra*.

470 J. FAULL and A. NIKPAY (eds.), *The EC Law of Competition* (Oxford, Oxford University Press, 2008), p. 613 and 614 and L. HANCHER, "Casenote on Corbeau", *Common Market Law Review* (1994), p. 113.

471 Case C-323/93, *Société Civile Agricole du Centre d'Insémination de la Crespelle v Coopérative d'Elevage et d'Insémination Artificielle du Département de la Mayenne* [1997] ECR I-5077.

472 J. FAULL and A. NIKPAY (eds.), *The EC Law of Competition* (Oxford, Oxford University Press, 2008), p. 614.

473 Case C-203/96, *Chemische Afvalstoffen Dusseldorp BV and Others v Minister van Volkshuisvesting, Ruimtelijke Ordening en Milieubeheer* [1998] ECR I-4075, paragraphs 53–68 and Joined cases C-147 and 148/97, *Deutsche Post AG v Gesellschaft für Zahlungssysteme mbH GZS) (C-147/97) and Citicorp Kartenservice GmbH (C-148/97)* [2000] ECR I-825, paragraphs 39–41.

474 See Case C-18/93, *Corsica Ferries Italia Srl v Corpo dei Piloti del Porto di Genova* [1994] ECR I-1783; Case C-55/96, *Job Centre Coop. arl* [1997] ECR-I-7119; Case C-163/96, *Criminal proceedings against Silvano Raso and Others* [1998] ECR-I-533 and Case C-266/96, *Corsica Ferries France SA v Gruppo Antichi Ormeggiatori del porto di Genova Coop. arl, Gruppo Ormeggiatori del Golfo di La Spezia Coop. arl and Ministero dei Trasporti e della Navigazione* [1998] ECR-I-3949.

but instead analysed whether the justification under Article 106(2) of the TFEU could be applied.[475, 476]

420. Consequently, the competition law provisions in the TFEU and in particular Article 102 of the TFEU prohibit an abuse of dominant position and not the establishment of a monopoly. This provision means that there is no rule against the existence of a dominant position created by national acts, but rather a control of the consequences on the basis that the exercise of economic strength could be abusive.[477] The Court of Justice ruled on various occasions that certain forms of monopoly rights were in breach with Articles 102 and 106 of the TFEU since these rights create a situation whereby the undertaking, merely by exercising its rights, is led to commit an abuse or if these rights make an abuse unavoidable.

(b) Conclusion

421. It seems not to be possible to categorise neatly which types of measures infringe Article 106(1) of the TFEU. Even if this would be possible the categories would not be closed. Nonetheless, on basis of the case law discussed above, one could identify some general themes.

422. Firstly, Article 106(1) of the TFEU seems to be infringed if a monopoly is set up in such a way that it is incapable of meeting the demand or enabled to behave inefficiently in the reserved sector.[478] Under this head the abuses at issue are a violation of Article 102(b) of the TFEU. This provision considers the limiting of production and markets or technical development to the prejudice of consumers to be an abuse which is incompatible with the internal market in so far as it may affect trade between Member States.

423. Secondly, in a number of judgments the Court of Justice ruled that the exclusive rights conferred on an undertaking were found to be contrary to

---

475 See Case C-67/96, *Albany International BV v Stichting Bedrijfspensioenfonds Textielindustrie* [1991] ECR I-5751; Case C-209/98, *Entreprenørforeningens Affalds/Miljøsektion (FFAD) v Københavns Kommune* [2000] ECR I-3743; Case C-147–8/97, *Deutsche Post AG v Gesellschaft für Zahlungssysteme mbH GZS) (C-147/97) and Citicorp Kartenservice GmbH (C-148/97)* [2000] ECR I-825; Case C-475/99, *Firma Ambulanz Glöckner v Landkreis Südwestpfalz* [2001] ECR I-8089 and L. GYSELEN, "Case note on Albany, Brentjens' and Drijvende Bokken" *CML Rev* (2000), p. 444–448.

476 A. JONES and B. SUFRIN, *EC Competition Law – Text, cases and materials* (Oxford, Oxford University Press, 2008), p. 642–650.

477 S. WEATHERILL and P. BEAUMONT, *EU law –The essential guide to legal workings of the European Union* (London, Penguin Books, 1993), p. 1003 and 1004.

478 See e.g. Case C-41/90, *Höfner v Macroton* [1991] ECR I-1979; Case C-179/90, *Merci Convenzionali Porto di Genova v Siderurgica Gabrielli SpA* [1991] ECR I-5889 and Case C-55/96, *Job Centre coop. arl.* [1997] ECR I-7119.

Article 106(1) of the TFEU since they created a conflict of interest.[479] Thirdly, some undertakings to which exclusive rights have been conferred may be enabled to commit pricing abuses which was held to be contrary to Article 106(1) of the TFEU.[480] Fourthly, the Court of Justice concluded in some cases that if undertakings with exclusive rights were enabled to extend their monopoly into neighbouring markets, this was incompatible with the provisions of Article 106(1) of the TFEU.[481] Fifthly, situations where an undertaking could refuse to supply are an infringement of Article 106(1) of the TFEU.[482, 483]

2. Article 106(2) of the TFEU

424. Under Article 106(2) of the TFEU *"Undertakings entrusted with the operation of services of general economic interest or having the character of a revenue-producing monopoly shall be subject to the rules contained in the Treaties, in particular to the rules on competition, in so far as the application of such rules does not obstruct the performance, in law or in fact, of the particular tasks assigned to them. The development of trade must not be affected to such an extent as would be contrary to the interests of the Union".*[484]

425. This provision provides for a limited derogation from the Treaty rules in order to deal with the activities of undertakings entrusted by the State with certain tasks.[485] To two sorts of undertakings, in particular the undertakings to

479 See e.g. Case C-18/88, *RTT v GB-INNO-BM SA* [1991] ECR I-5941; Case C-260/89, *Elliniki Radiophonia Tileorassi Anonimi Etaira v Dimotiki Etairia Pliroforissis* [1991] ECR I-2925 and Case C-163/96, *Criminal proceedings against Silvano Raso and Others* [1998] ECR-I-533.

480 See e.g. Case C-147–8/97, *Deutsche Post AG v Gesellschaft für Zahlungssysteme mbH GZS (C-147/97) and Citicorp Kartenservice GmbH (C-148/97)* [2000] ECR I-825; Case 266/96, *Corsica Ferries France SA v Gruppo Antichi Ormeggiatori del porto di Genova Coop. arl, Gruppo Ormeggiatori del Golfo di La Spezia Coop. arl and Ministero dei Trasporti e della Navigazione* [1998] ECR I-3949 and C-163–99, *Portuguese Republic v Commission of the European Communities* [2001] ECR I-2613.

481 See e.g. Case C-18/88, *RTT v GB-INNO-BM SA* [1991] ECR I-5941 and Case C-475/99, *Firma Ambulanz Glöckner v Landkreis Südwestpfalz* [2001] ECR I-8089.

482 See e.g. Case C-147–8/97, *Deutsche Post AG v Gesellschaft für Zahlungssysteme mbH GZS) (C-147/97) and Citicorp Kartenservice GmbH (C-148/97)* [2000] ECR I-825.

483 A. JONES and B. SUFRIN, *EC Competition Law – Text, cases and materials* (Oxford, Oxford University Press, 2008), p. 650–652 and J. FAULL and A. NIKPAY (eds.), *The EC Law of Competition* (Oxford, Oxford University Press, 2008), p. 614 and 615.

484 See *Infra*.

485 This provision was introduced at the insistence of France which wanted to secure that the new competition rules on a European level would not affect the areas which fell under the State's responsibility. France feared that undertakings entrusted with tasks of services of general economic interest would be unable to perform their tasks due to the application of the competition law provisions. Thus, Article 106 of the TFEU (ex Article 86 of the EC Treaty) can be considered as a genuine compromise which on the one hand, avoids distortions of competition from the public sector. On the other hand, it protects the services of a general

which services of general economic interest have been entrusted and the revenue-producing monopolies, the Treaty rules shall apply only in so far as they do not obstruct the performance of tasks of these undertakings. However, this exception is subject to the condition that it should not affect trade to an extent contrary to the interest of the Union.[486]

426. The second category of undertakings referred to above has the nature of a revenue-producing monopoly which, through the exploitation of their exclusive rights, raises revenue for the national exchequer.[487]

(a) Services of general economic interest

427. The most important category of undertakings in Article 106(2) of the TFEU is that of the entrusted undertakings. Whether the given undertaking is private or public is immaterial. However, the undertaking should be entrusted with a specific task to provide services of general economic interest by "an act of the public authority".[488] This means that the given Member State must have

---

economic interest from the competition law provisions. See L. N. NISTOR, *Public Services and the European Union A Study of the Impact of Community Law on Health Care, Health Insurance and Education Services*, Proefschrift ter verkrijging van het doctoraat in de Rechtsgeleerdheid aan de Rijksunversiteit Groningen, 2009, p. 219.

486 A. ARNULL, A. DASHWOOD, M. DOUGAN, M. ROSS, E. SPAVENTA and D. WYATT, *Wyatt and Dashwoods's European Union Law* (London, Sweet & Maxwell, 2006), p. 1137. These authors stated that this *provisio* of the second sentence of Article 106(1) of the TFEU (ex Article 86(1) of the EC Treaty) has received little discrete interpretation mainly due to the fact that for many years attempts to rely on Article 106(1) of the TFEU (ex Article 86(1) of the EC Treaty) were consistently rejected on other grounds. They furthermore referred to the Opinion of Advocate General Cosmans in the Joined Opinion in Cases C-157/94 *Commission* v *Netherlands*, C-158/94 *Commission* v *Italy*, Case C-159/94 *Commission* v *French Republic* and Case C-160/94 *Commission* v *Spain*, paragraph 126 and the Opinion of Advocate General Léger in the Case C-438/02, *Krister Hanner*, paragraph 143 which stated that: "*[...] the last condition in Article 86(2) EC requires that '[t]he development of trade must not be affected to such an extent as would be contrary to the interests of the Community'. Even though the Court has not yet ruled on the meaning of that requirement, certain Advocates General have already adopted a position on the issue. In their view, effect on the development of intra-Community trade within the meaning of Article 86(2) EC, unlike the classic definition of the concept of 'measures having an effect equivalent to a quantitative restriction', calls for proof that the measure in issue has in fact had a substantial effect on intra-Community trade. That assessment does seem to me to be supported by the wording of Article 86(2) EC.*"

487 A. ARNULL, A. DASHWOOD, M. DOUGAN, M. ROSS, E. SPAVENTA and D. WYATT, *Wyatt and Dashwoods's European Union Law* (London, Sweet & Maxwell, 2006) p. 1130–1133.

488 Case C-127/73, *Belgische Radio en Televisie v SV SABAM and NV Fonior* [1974] ECR 313, paragraph 20. Case C-159/94 *Commission* v *France* [1997] ECR I-5815 in its paragraphs 65 and 66 has made it clear that tasks to provide services of general economic interest could also be entrusted through a concession governed by public law.

taken legal steps in order to secure the provision of services of general economic interest by the given undertaking.[489]

428. With respect to the concept of *"services of general economic interest"* Member States appear to have a broad margin of discretion to determine which services could be classified as services of general economic interest.[490] Article 14 of the TFEU also supports the discretion of Member States to define the services of general economic interest and reads as follows[491]: *"Without prejudice to Article 4 of the Treaty on European Union or to Articles 93, 106 and 107 of this Treaty, and given the place occupied by services of general economic interest in the shared values of the Union as well as their role in promoting social and territorial cohesion, the Union and the Member States, each within their respective powers and within the scope of application of the Treaties, shall take care that such services operate on the basis of principles and conditions, particularly economic and financial conditions, which enable them to fulfil their missions. The European Parliament and the Council, acting by means of regulations in accordance with the ordinary legislative procedure, shall establish these principles and set these conditions without prejudice to the competence of Member States, in compliance with the Treaties, to provide, to commission and to fund such services"*.[492]

---

489 A. ARNULL, A. DASHWOOD, M. DOUGAN, M. ROSS, E. SPAVENTA and D. WYATT, *Wyatt and Dashwood's European Union Law* (London, Sweet & Maxwell, 2006) p. 1131 which refers in this respect to Case C-159/94, *Commission of the European Communities v French Republic* [1997] ECR I-5815, paragraph 66 and emphasises that such an act should not be in any particular form.

490 See also U. NEERGAARD, "Chapter 1 Services of General Economic Interest: The Nature of the Beast" in M. KRAJEWSKI, U. NEERGARD and J. VAN DE GRONDON (eds.), *The Changing Legal Framework for Services of General Interest in Europe Between Competition and Solidarity* (The Hague, T.M.C. Asser Press, 2009), p. 17–50; M. ROSS, "A healthy approach to services of general economic interest? The BUPA judgment of the General Court", *European Law Review* (2009), p. 127–140; W. SAUTER, "Services of general economic interest and universal service in EU law", *European Law Review* (2008), p. 167–193 and C. BRIGHT, "Article 90, economic policy and the duties of Member States", *E.C.L.R.* (1993), p. 263–272.

491 H. BJØRNEBYE, *Investing in EU energy security – Exploring the regulatory approach to tomorrow's electricity production*, Dissertation submitted to the faculty of law, University of Oslo, 2009, p. 107. See also H. BJØRNEBYE, *Investing in EU Energy Security. Exploring the regulatory approach to tomorrow's electricity production* (The Netherlands, Kluwer Law International, 2010), 456 p.

492 As a side note, the scope of ex Article 16 of the EC Treaty and Article 14 of the TFEU is similar. However, the new Article 14 of the TFEU adds that the European Parliament and the Council shall establish these principles and set these conditions to provide, to commission and to fund such services. This provision seems to allow that the Union can adopt a legislative framework on a European level for services of general economic interest. The reference on two occasions to the powers and rights of the Member States in this provision seems to indicate that a difficult balance needs to be achieved when this provision will be applied on a European level. Since Article 14 of the TFEU is included under the heading II of the TFEU called *"Provisions having general application"* this could imply that it should be interpreted as another horizontal norm. Furthermore under Article 14 of the TFEU, the services of a general economic interest are still considered as one of the shared values of the Union. This highlights

429. The Commission's Communication on services of general interests in Europe also acknowledges the wide margin of discretion by stating that "*Member States are primarily responsible for defining what they regard as services of general economic interest on the basis of the specific features of the activities. This definition can only be subject to control for manifest error. They may grant special or exclusive rights that are necessary to the undertakings entrusted with their operation, regulate their activities and, where appropriate, fund them. In areas that are not specifically covered by Community regulation Member States enjoy a wide margin for shaping their policies, which can only be subject to control for manifest error. Whether a service is to be regarded as a service of general interest and how it should be operated are issues that are first and foremost decided locally. The role of the Commission is to ensure that the means employed are compatible with Community law. However, in every case, for the exception provided for by Article 86(2) to apply, the public service mission needs to be clearly defined and must be explicitly entrusted through an act of public authority (including contracts) (5). This obligation is necessary to ensure legal certainty as well as transparency vis-à-vis the citizens and is indispensable for the Commission to carry out its proportionality assessment*".[493]

430. As emphasised in this Communication, the Commission shall monitor whether the choices made by the Member States are not made with manifest error. A wide range of activities such as waste management functions[494], airline

---

the importance of the concept services of general economic interest which has been confirmed by the Protocol 26 on services of general interest. See M. ROSS, "Article 16 E.C. and services of general interest: from derogation to obligation?", *European Law Review* (2000), p. 22–38 and L. HANCHER and S. JANSSEN, "Shared Competences and Multi-Faceted Concepts – European Legal Framework for Security of Supply", in B. BARTON, C. REDGWELL, A. RØNNE and D. ZILLMAN (eds.), *Energy Security – Managing Risk in a Dynamic Legal and Regulatory Environment* (Oxford, Oxford University Press, 2004), p. 91. See also, since the adoption of the Lisbon Treaty, the new Protocol 26 on services of general interest. For further details see S. WERNICKE, "Chapter 3 Taking Stock: The EU Institutions and Services of General Economic Interest" in M. KRAJEWSKI, U. NEERGARD and J. VAN DE GRONDON (eds.), *The Changing Legal Framework for Services of General Interest in Europe Between Competition and Solidarity* (The Hague: T.M.C. Asser Press, 2009), p. 69–72 and J.-C. PIRIS, *The Lisbon Treaty A legal and Political Analysis* (Cambridge: Cambridge University Press, 2010), p. 313–316. For further reading related to the debate on a European level to adopt an EC general framework of instruments for services of general economic interest see S. RODRIGUES, "Towards a General EC Framework Instrument Related to SGEI? Political Considerations and Legal Constraints" in M. KRAJEWSKI, U. NEERGARD; J. VAN DE GRONDON (eds.), *The Changing Legal Framework for Services of General Interest in Europe Between Competition and Solidarity* (The Hague: T.M.C. Asser Press, 2009), p. 255–266 and D. SOLE, "Services of General Economic Interest in the EU: A 50-Year-Battle between Liberalization, Deregulation and Subsidiarity", *Eur. J.L. Reform* (2003), p. 335–408.

493 Communication from the Commission of 2001 – *Services of general interest in Europe*, OJ C 17/4, paragraph 22.

494 Case C-209/98, *Entreprenørforeningens Affalds/Miljøsektion (FFAD) v Københavns Kommune* [2000] 3743.

carriers[495], mooring services[496], responsibility for radio and television broadcasts[497], ensuring the navigability of waterways[498], etc., beyond the various obvious examples such as energy and telecom utilities, water companies and postal services, appear to qualify as *services of general economic interest.*[499]

(b) Addressee of Article 106(2) of the TFEU

431. The provisions of Article 106(2) of the TFEU are, unlike Article 106(1) of the TFEU, addressed to undertakings and not to Member States. This however, does not mean that Member States cannot rely on them. Article 106(2) of the TFEU is particularly important in relation to the application of the provisions of Article 102 of the TFEU since it could allow a derogation or exemption from this Article.[500] If an undertaking would like to benefit from Article 102 of the TFEU it must prove that compliance with the Treaty provisions would obstruct the performance of its assigned tasks.

432. In the Albany International case the Court of Justice clarified its position in this respect and stated that "*it is not necessary, in order for the conditions for the application of Article 90(2) of the Treaty to be fulfilled, that the financial balance or economic viability of the undertaking entrusted with the operation of a service of general economic interest should be threatened. It is sufficient that, in the absence of the rights at issue, it would not be possible for the undertaking to perform the particular tasks entrusted to it, defined by reference to the obligations and constraints to which it is subject (Commission v Netherlands, cited above, paragraph 52) or that maintenance of those rights is necessary to enable the holder of them to perform tasks of general economic interest which have been assigned to it under economically acceptable conditions (Case C-320/91 Corbeau [1993] ECR*

495 Case C-66/86, *Ahmed Saeed Flugreisen and Silver Line Reisebüro GmbH v Zentrale zur Bekämpfung unlauteren Wettbewerbs e.V.* [1989] ECR 803.

496 Case C-266/96, *Corsica Ferries France SA v Gruppo Antichi Ormeggiatori del porto di Genova Coop. arl, Gruppo Ormeggiatori del Golfo di La Spezia Coop. arl and Ministero dei Trasporti e della Navigazione* [1998] ECR I-3949.

497 Case C-155/73, *Giuseppe Sacchi* [1947] ECR 409.

498 Case C-10/71, *Ministère public luxembourgeois v Madeleine Muller, Veuve J.P. Hein and others* [1971] ECR 723.

499 Joined cases 96–102, 104, 105, 108 and 110/82, *NV IAZ International Belgium and others v Commission of the European Communities* [1983] ECR 3369; Case C-19/93 P, NV, *Centraal Overijsselse Nutsbedrijven NV and Regionaal Energiebedrijf Salland NV v Commission of the European Communities* [1995] ECR I-3319; Case C-320/91, *Criminal proceedings against Paul Corbeau* [1994] ECR I-1783; etc.

500 A. JONES and B. SUFRIN, *EC Competition Law – Text, cases and materials* (Oxford, Oxford University Press, 2008), p. 652–667.

I-2533, paragraphs 14 to 16, and Commission v Netherlands, cited above, paragraph 53)".[501]

433. This position appears to be standard. However, in some cases the Court of Justice applied other formulations for example in the *Dusseldorp* case[502] in which the Court of Justice stated that *"even if the task conferred on that undertaking could constitute a task of general economic interest, however, it is for the Netherlands Government, as the Advocate General points out at paragraph 108 of his Opinion, to show to the satisfaction of the national court that that objective cannot be achieved equally well by other means. Article 90(2) of the Treaty can thus apply only if it is shown that, without the contested measure, the undertaking in question would be unable to carry out the task assigned to it".*[503]

434. In order to invoke Article 106(2) of the TFEU in this case the Court of Justice required from the Dutch Government to proof that the entrusted objective could not be achieved equally well by other means. In conclusion, the analysis of case law highlights that the Court of Justice has clearly expanded the scope of Article 106(2) of the TFEU.[504]

3. Article 106 of the TFEU and the energy sector

435. After the discussion on the application of the Union's rules to undertakings which fall under the application of Article 106 of the TFEU, the next part of this book will analyse the cases of the Court of Justice in which this provision has been used in cases related to the energy sector.

436. This case law highlights that the Commission's attempts to demonstrate breaches of the primary Union provisions in the energy field were severely circumscribed. It has obliged the Commission to propose a legislative solution

---

501 Case C-67/96, *Albany International BV v Stichting Bedrijfspensioenfonds Textielindustrie* [1999] ECR 5751, paragraph 107.

502 Case C-203/96, *Chemische Afvalstoffen Dusseldorp BV and Others v Minister van Volkshuisvesting, Ruimtelijke Ordening en Milieubeheer* [1998] ECR I-4075.

503 *Ibid.*, paragraph 67.

504 A. ARNULL, A. DASHWOOD, M. DOUGAN, M. ROSS, E. SPAVENTA and D. WYATT, *Wyatt and Dashwoods's European Union Law* (London, Sweet & Maxwell, 2006) p. 1136 and H. BJØRNEBYE, *Investing in EU energy security – Exploring the regulatory approach to tomorrow's electricity production*, Dissertation submitted to the faculty of law, University of Oslo, 2009, p. 104–109. See also H. BJØRNEBYE, *Investing in EU Energy Security. Exploring the regulatory approach to tomorrow's electricity production* (The Netherlands, Kluwer Law International, 2010), 456 p.

through the adoption of appropriate legislative alternatives in the energy field, instead of seeking a judicial one.[505]

437. With the adoption of the First Electricity Directive late 1996 and the First Gas Directive in 1998, the Commission has realised its first legislative alternatives. The purpose of this part is to analyse whether with the adoption of secondary legislation in the energy field, the scope for Member States to organise their electricity and gas markets in a monopolistic way has been reduced.

(a) Case law of the Court of Justice

438. The first energy case in which the Court of Justice has ruled that the task of ensuring the supply of electricity could amount to a service of general economic interest was the *Almelo* case.[506] The *Almelo* case concerned a dispute between the Municipality of Almelo and other local distributors of electric energy and Energiebedrijf IJsselmij NV, an undertaking engaged in the regional distribution of electric energy.

At that moment in the Netherlands, electric energy was produced by four undertakings holding shares in a jointly owned undertaking, NV Samenwerkende Elektriciteits-Produktiebedrijven, which was intended to serve as a vehicle for cooperation between generators. The distribution of electricity was organised on a regional and local level. The regional distribution undertakings supplied, within the territory granted to them, the local distribution undertakings owned by the municipalities and, in some cases, the end-users. The local distribution undertakings handled the supply to customers within the municipalities.

439. The generation and distribution undertakings were owned, directly or indirectly, by the provinces and municipalities. A Royal Order of 1918 granted the Energiebedrijf IJsselmij NV a non-exclusive concession allowing it to distribute electricity within the territory covered by that concession. Energiebedrijf IJsselmij NV supplied electricity to local distributors, in particular the Municipality of Almelo and the other plaintiffs in the main proceedings, and also made direct supplies to consumers in rural areas.

440. Between 1985 and 1988, local distributors were prohibited from importing electricity due to an exclusive purchasing clause which obliged municipalities to obtain electric energy for supply in its territory exclusively from the NV Electriciteits-Maatschappij Ijsselcentrale (the former name of Energiebedrijf

505 L. HANCHER, "Delimitation of Energy Law Jurisdiction: the EU and its Member States: From Organisational to Regulatory Conflicts", *Journal of Energy & Natural Resources Law* (1998), p. 51.

506 Case C-393/92, *Municipality of Almelo and others v NV Energiebedrijf Ijsselmij* [1994] ECR I 1477.

IJsselmij NV) and to use that power only for their own consumption or to supply third parties in the territory of the given municipality. Besides, on the basis of the cooperation agreement of 22 March 1986 between the electricity generation companies and NV Samenwerkende Elektriciteits-Produktiebedrijven also a prohibition against imports of electric power existed between the generators and the regional distributors.

441. Finally, Energiebedrijf IJsselmij NV, as from 1 January 1985, charged local distributors an equalisation supplement which is a surcharge intended to offset the difference between the higher cost incurred by it in distributing electricity to consumers in rural areas and the lower cost incurred by the local distributors in distributing electricity to consumers in urban areas. Against this background, the local distributors challenged the import ban of the cooperation agreement; the exclusive purchasing obligation and the power of the Energiebedrijf IJsselmij NV to determine prices unilaterally and to impose the equalisation supplement.

442. With respect to Article 106(2) of the TFEU, the Court of Justice in the *Almelo* case was asked to answer the question whether the Energiebedrijf IJsselmij NV had been "*entrusted with the operation of services of general interest*" and noted that" *it should be borne in mind that it has been given the task, through the grant of a non-exclusive concession governed by public law, of ensuring the supply of electricity in part of the national territory*". It continued by stating that undertakings such as Energiebedrijf IJsselmij NV must "*ensure that throughout the territory in respect of which the concession is granted, all consumers, whether local distributors or end-users, receive uninterrupted supplies of electricity in sufficient quantities to meet demand at any given time, at uniform tariff rates and on terms which may not vary save in accordance with objective criteria applicable to all customers*".

443. Thus, the Court of Justice ruled that the restrictions on competition from other economic operators were allowed in so far as they are necessary in order to enable Energiebedrijf IJsselmij NV to perform its entrusted task of general economic interest. According to the Court of Justice, the national court must apply the proportionality test in order to determine whether the exclusive purchasing clause prohibiting local distributors from importing electricity, is necessary so that Energiebedrijf IJsselmij NV could perform its task of general interest.[507]

444. In the *Commission v. the Netherlands* case[508] the Commission brought an action before the Court of Justice because the Netherlands conferred on the

507 *Ibid.*, paragraphs 47–49.

508 Case C-157/94, *Commission of the European Communities v Kingdom of the Netherlands* [1997] ECR I-5699.

basis of Article 34 of the Elektriciteitswet of 16 November 1989[509] (hereinafter EW) to the NV Samenwerkende Elektriciteitsproduktiebedrijven (hereinafter SEP) an exclusive right to import electricity for public supply. Article 37(1) of the EW furthermore stated that the supply to another person of electricity imported by any person other than SEP is prohibited. The Commission challenged these provisions of the EW because due to the fact that SEP enjoyed a national import monopoly, producers in other Member States were prevented from selling their production to customers in the Netherlands other than the monopoly holder. Moreover, potential customers in the Netherlands were also prevented from freely choosing their sources of supply of electricity from other Member States.

445. In the Commission's view, the exclusive import rights of SEP were therefore liable to restrict trade between Member States and thus measures having an effect equivalent to quantitative restrictions on imports which are contrary to the provisions of 34 of the TFEU. They further constituted in the Commission's view a discrimination within the meaning of Article 37 of the TFEU – the provision which requires Member States to adjust their existing state trading monopolies and prohibits them from conferring any new import rights upon them – not only as regards exporters established in other Member States, but also as regards users established in the Member State concerned.

446. The Court of Justice ruled that the exclusive import rights of SEP are discriminatory and thus contrary to Article 37 of the TFEU. Since all electricity must be channelled through SEP, free trade is prevented. The Court of Justice did not consider whether SEP's exclusive import rights are contrary to 34 of the TFEU because they were found contrary to Article 37 of the TFEU.[510]

447. The Court of Justice continued to consider whether SEP's exclusive import rights could be justified under Article 106(2) of the TFEU. With respect to Article 106(2) of the TFEU the Commission argued that this provision did not apply to Member State measures, but only to the activities of the firms in question. This approach was rejected by the Court of Justice. It held that the provisions of Article 106(1) and (2) of the TFEU should be interpreted so as to prevent Member States from avoiding their obligations under Articles 34, 35 and 37 of the TFEU by conferring certain exclusive rights on undertakings when the exercises of such rights by the Member States itself would be contrary to those Articles.

---

509 Law laying down rules for the generation, import, transmission and sale of electricity.

510 L. HANCHER, "Delimitation of Energy Law Jurisdiction: the EU and its Member States: From Organisational to Regulatory Conflicts", *Journal of Energy & Natural Resources Law* (1998), p. 48.

448. The Commission furthermore proposed a restrictive interpretation of the scope of Article 106(2) of the TFEU and argued that SEP should only be exempted from the Treaty rules if this was necessary for it to continue to fulfil its public service duties on acceptable conditions and in a state of financial equilibrium. The Court of Justice rejected this approach of the Commission and stated that: *"[t]he Member States' interest being so defined, they cannot be precluded, when defining the services of general economic interest which they entrust to certain undertakings, from taking account of objectives pertaining to their national policy or from endeavouring to attain them by means of obligations and constraints which they impose on such undertakings.*

[...] contrary to the Commission's contention, it is not necessary, in order for the conditions for the application of Article 90(2) of the Treaty to be fulfilled, that the financial balance or economic viability of the undertaking entrusted with the operation of a service of general economic interest should be threatened. It is sufficient that, in the absence of the rights at issue, it would not be possible for the undertaking to perform the particular tasks entrusted to it, defined by reference to the obligations and constraints to which it is subject".[511]

449. The Court of Justice found that Article 106(2) of the TFEU did not prevent Member States themselves from defining the nature of the public service duties, in accordance with their own internal policy goals, which they entrusted to certain undertakings as well as from specifying the various legal and other means by which these duties should be performed.

450. The Court of Justice furthermore emphasised that Article 106(2) of the TFEU can be applied whenever the undertaking in question would be prevented from fulfilling its public duties. Consequently, contrary to the Commission's view, it is not necessary that the financial viability of the undertaking in question is threatened. Then, the Court of Justice turned to the issue whether the Member State or the Commission must prove whether the undertaking in question could not perform its public service duties. In the present case, the Court held that this is a Member State's task in which the Dutch government succeeded, by proving that the relaxation of SEP's import monopoly would prevent SEP to meet its public service duties to provide a secure and reliable supply of electricity at the lowest cost possible.

451. Finally, the Court of Justice examined whether the application of the exemption in Article 106(2) of the TFEU would adversely affect the development of trade in electricity to such an extent that it would be contrary to the interests of the Union.

---

[511] See paragraphs 40 and 52 of the Case C-157/94, *Commission of the European Communities v Kingdom of the Netherlands* [1997] ECR I-5699.

On the one hand, the Dutch government explained that trade did take place, the imports of electricity by SEP in recent years have accounted for about 15% of total domestic demand for electricity, and that the available capacity of cross-frontier lines was thus fully taken up. On the other hand, the Commission did not provide any explanation to demonstrate that due to SEP's exclusive import rights, the development of intra-Union trade in electricity has been and continues to be affected to an extent contrary to the interest of the Union. Consequently, the Court of Justice dismissed the Commission's case and allowed that the Netherlands conferred to SEP an exclusive right to import electricity for public supply.[512]

452. In the *Commission v Italy* case[513], the Court of Justice was confronted with a similar situation as the above discussed *Commission v the Netherlands* case. In this case, the Italian government had conferred to the *Ente Nazionale per l'Energia Elettrica* (hereinafter ENEL) on the basis of the Law No 1643 of 6 December 1962 and the Decree No 342 of 18 March 1965, the exclusive right to import and export electricity within Italy. The Court of Justice ruled that these exclusive import and export of ENEL were contrary to Article 37 of the TFEU.

453. The Italian government argued that electricity displays much greater similarity to the category of services than to that of goods and therefore does not fall within the scope of Articles 34 to 37 of the TFEU *ratione materiae*. This view was rejected by the Court of Justice which referred to its judgment in the *Almelo and others v Energiebedrijf IJsselmij* case[514] and the *Costa v ENEL* case.[515] The Court of Justice then examined whether the exclusive rights at issue conferred to ENEL, could be justified under 106(2) of the TFEU.

454. The Italian government pointed out that the tasks entrusted to ENEL consisted in particular of "*ensuring at minimum management cost the availability of electrical energy of a quantity and at a price appropriate to the requirements of balanced economic development of the country*".[516] According to the Italian government the elimination of those exclusive rights would inevitably obstruct the performance of the particular tasks assigned to ENEL. It furthermore argued that if those exclusive rights were removed, most industrial consumers

[512] L. HANCHER, "Delimitation of Energy Law Jurisdiction: the EU and its Member States: From Organisational to Regulatory Conflicts", *Journal of Energy & Natural Resources Law* (1998), p. 48 and 49.

[513] Case C-158/94, *Commission of the European Communities v Italian Republic* [1997] ECR I-5789.

[514] Case C-393/92, *Municipality of Almelo and others v NV Energiebedrijf Ijsselmij* [1997] ECR I-1477, paragraph 28.

[515] Case C-6/64, *Flaminio Costa v E.N.E.L.* [1964] ECR 585.

[516] See paragraph 46 of the Case C-158/94, *Commission of the European Communities v Italian Republic*, [1997] ECR I-5789.

established in northern Italian regions near the frontiers would choose to resort to foreign suppliers. This would deprive ENEL of the means of equalising the costs of distributing electricity and cause an increase in the average price of electricity. Thereby, ENEL's ability to perform its task of delivering low cost electricity throughout the country was threatened.

The Court of Justice ruled that "*it must nevertheless be pointed out that by thus enumerating, in general terms, certain means as alternatives to the rights at issue, the Commission has neither taken account of the particular features of the national system of electricity supply (in particular those imposed by geography) to which the Italian Government drew attention nor specifically considered whether those means would enable ENEL to perform the tasks of general economic interest entrusted to it under economically acceptable conditions*".[517]

455. From the foregoing, it is clear that the Commission failed to respond to the arguments of the Italian government, in particular the arguments relating to the geographical peculiarities at issue. Consequently, the Court of Justice followed the same reasoning as in its judgment *Commission v the Netherlands* and dismissed the Commission's application against Italy.[518]

456. A similar pattern was followed by the Court of Justice in the *Commission v France* case[519], however the Court not only dealt with electricity, but also gas. The Commission challenged in this case the exclusive rights conferred on *Électricité de France* (hereinafter EDF) to import and export electricity and on *Gaz de France* (hereinafter GDF) to import and export gas. As in the previous cases, the Court of Justice ruled that the exclusive import and export rights at issue are contrary to Article 37 of the TFEU. The Court of Justice then turned to the issue whether Article 106(2) of the TFEU could be applied, contrary to the similar judgments, the Court of Justice was confronted with some difficulties.

457. Firstly, the various duties imposed upon EDF and GDF were not inserted into the Law No 46–628 of 8 April 1946 on the nationalisation of electricity and gas, but for both electricity and gas in concession agreements with the French State. Secondly, the French government argued that the duties imposed upon EDF and GDF were far-reaching since they not only required to supply demand at competitive prices, but also to contribute actively to the implementation of national environment and regional policies. Thirdly, with respect to pricing policy the concessions did not contain any direct requirements as such.

---

517 *Ibid.*, paragraph 53.

518 L. HANCHER, "Delimitation of Energy Law Jurisdiction: the EU and its Member States: From Organisational to Regulatory Conflicts", *Journal of Energy & Natural Resources Law* (1998), p. 49 and 50.

519 Case C-159/94, *Commission of the European Communities v French Republic* [1997] ECR I-5815.

458. Contrary to the view of the Commission, the Court of Justice held that since the concessions for EDF and GDF were public acts, Article 106(2) of the TFEU can be applied.[520] However, since no reference can be found in these concessions to duties relating to the implementation of national environment and regional policies, the performance of these duties did not fall within the scope of the exemption.[521] With respect to the pricing policy, the Court of Justice ruled that neither EDF nor GDF had a duty to ensure that prices were as low as possible for the consumers. However, according to the Court of Justice it was apparent from the concessions that both EDF and GDF were under a duty to supply and should maintain an equal treatment between different customers.[522] Consequently, the Court of Justice concluded that in order to perform the albeit more limited public service duties, the exclusive rights conferred upon EDF and GDF were necessary.[523]

459. In the *Commission v Spain* case the Court of Justice concluded that the Spanish government on the basis of the Law No 49/84 of 26 December 1984 on the unified operation of the national electricity system and the Royal Decree No 91/85 of 23 January 1985 did not confer upon *Red Eléctrica de España SA* (hereinafter RED) an exclusive right to either import or export electricity. The Spanish legal provisions merely gave RED a duty to ensure the unified operation of the national electricity system over the high-voltage networks. The Commission failed to distinguish between RED's rights to coordinate, the activities by which the various electricity-generating and distribution undertakings contribute to the country's electricity supply, from the supposed concentration of those rights in the hands of a single operator. The Court of Justice therefore concluded that the Spanish legal provisions were not contrary to Article 37 of the TFEU.

460. It can be concluded on the basis of the foregoing, that these judgments of the Court of Justice seemed to send to the Commission the signal that the Court was unwilling to deny competence to Member States in these sensitive areas as long as the Commission itself had not acted by adopting appropriate legislative alternatives. Consequently, in order to ensure that Member States would opt for a less restrictive approach to guarantee its chosen public service goals, the Commission is obliged to propose a legislative solution and not a judicial one.

With the adoption of the First Electricity Directive in late 1996 and the First Gas Directive in 1998, the Commission has realised its first legislative

---

520 *Ibid.* paragraph 65 and 66.

521 *Ibid.* paragraph 69.

522 *Ibid.* paragraph 106 and 107.

523 L. HANCHER, "Delimitation of Energy Law Jurisdiction: the EU and its Member States: From Organisational to Regulatory Conflicts", *Journal of Energy & Natural Resources Law* (1998), p. 50.

alternatives, further supplemented with the adoption of the Second Electricity Directive, Second Gas Directive and, since 2009, the adoption of the Third Electricity Directive and Third Gas Directive.[524]

(b) Situation after the adoption of the secondary legislation in the energy field

461. However, it is necessary to consider whether these legislative alternatives are viable alternatives. Will they reduce the scope for Member States to organise their electricity and gas markets along monopolistic lines?[525] The electricity and gas directives seem to provide expressly that Member States will continue to define the extent of public service provisions and allow them to determine how these services are to be delivered.

462. This can be derived from Article 3(2) of the Third Electricity Directive which provides that: "*[h]aving full regard to the relevant provisions of the Treaty, in particular Article 86 thereof, Member States may impose on undertakings operating in the electricity sector, in the general economic interest, public service obligations which may relate to security, including security of supply, regularity, quality and price of supplies and environmental protection, including energy efficiency, energy from renewable sources and climate protection [...]*" and Article 3(14) of the Third Electricity Directive which states that "*Member States may decide not to apply the provisions of Articles 7, 8, 32 and/or 34 insofar as their application would obstruct the performance, in law or in fact, of the obligations imposed on electricity undertakings in the general economic interest and insofar as the development of trade would not be affected to such an extent as would be contrary to the interests of the Community. The interests of the Community include, inter alia, competition with regard to eligible customers in accordance with this Directive and Article 86 of the Treaty*".

463. A similar provision can be found in Articles 3(2) and 3(10) of the Third Gas Directive. However, compared to the provisions in the Third Electricity Directive, the scope of Article 3(10) of the Third Gas Directive seems to be more limited since Member States can only decide not to apply the authorisation procedure with respect to distribution.

464. The 44th recital of the Third Gas Directive expressly states that the Third Gas Directive lays down only minimum common standards as regards public service obligations which take into account the objectives of common protection,

524 See *Supra*.

525 L. HANCHER, "Delimitation of Energy Law Jurisdiction: the EU and its Member States: From Organisational to Regulatory Conflicts", *Journal of Energy & Natural Resources Law* (1998), p. 57.

security of supply, environmental protection and equivalent levels of competition in all Member States. It furthermore emphasises that the requirements of these public service obligations can be interpreted on a national basis, taking into account national circumstances and subject to the respect of Union law.[526] The Third Electricity Directive contains in its 46th recital similar provisions.[527]

465. As a side note, reference should be made to the uncertainty on whether the reasons/circumstances listed in Article 3(2) of the Third Electricity Directive[528] for Member States to impose public service obligations on undertakings in the general economic interest, are exhaustive or not.

466. It has been argued in legal doctrine that this list is exhaustive.[529] Thus, Member States were excluded to impose public service obligations for other reasons/circumstances of general interest.[530] However, Advocate General Ruiz-Jarabo Colomer stated in his opinion on the *Federutility and others* case[531] that: *"Article 3(2) of Directive 2003/55 [...] permits Member States to impose certain burdens on gas undertakings provided that these are essential to protect a 'general economic interest' whose precise scope remains undefined"*.[532]

In footnote 30, Advocate General Ruiz-Jarabo Colomer argued that: "[t]*he directive indicates only that public service obligations are created in the general economic interest and that they 'may relate to security, including security of supply, regularity, quality and price of supplies, and environmental protection, including energy efficiency and climate protection'. There are those who regard this list as a closed category of reasons of 'general economic interest' capable of*

526 See also the 47th recital of the Third Gas Directive.

527 See also the 50th recital of the Third Electricity Directive.

528 Article 3, §2 of the Third Electricity Directive states that: *"Having full regard to the relevant provisions of the Treaty, in particular Article 86 thereof, Member States may impose on undertakings operating in the electricity sector, in the general economic interest, public service obligations which may relate to security, including security of supply, regularity, quality and price of supplies and environmental protection, including energy efficiency, energy from renewable sources and climate protection"*. Similar provision can be found in Article 3, §2 of the Third Gas Directive.

529 A. DOHERTY, "Retail market development and public service objectives" in C. JONES (ed.) *EU Energy law, The international energy market* (Leuven, Claeys & Casteels, 2010) pp. 393–441.

530 T. DERUYTTER, W. GELDHOF and F. VANDENDRIESSCHE, "Public Service Obligations in the Electricity and Gas Markets" in B. DELVAUX, M. HUNT and K. TALUS (eds.), *EU Energy Policy and Law* (Cambridge, Intersentia, 2011), p. 63–95.

531 Case C-265/08, *Federutility, Assogas, Libarna Gas SpA, Collino Commercio SpA, Sadori Gas Srl, Egea Commerciale Srl, E.On Vendita Srl, Sorgenia SpA v Autorità per l'energia elettrica e il gas*, not yet reported.

532 Opinion of Advocate General Ruiz-Jarabo Colomer in the Case C-265/08, *Federutility, Assogas, Libarna Gas SpA, Collino Commercio SpA, Sadori Gas Srl, Egea Commerciale Srl, E. On Vendita Srl, Sorgenia SpA v Autorità per l'energia elettrica e il gas*, paragraph 45. Emphasis added.

constituting grounds for state involvement (Geldhof, W., and Vanden Driessche, F., 'Chapter 1: European Electricity and Gas Market Liberalisation. Background, Status, Developments', in Delvaux, B., Hunt, M. and Talus, K. (eds.) EU Energy Law and Policy Issues, Euroconfidentiel, p. 48, citing, in turn, Jones, C.W., op. cit. p. 230). In my opinion, however, the provision describes the types of action in which the aforementioned public service obligations should result, rather than a list of circumstances which justify them, although the two aspects are closely linked. It follows from Article 3(2) of the directive that Member States can create public service obligations relating to gas prices, but simply stating that is not enough. In each case it is necessary to examine whether there exists a task of general economic interest".

467. Contrary to the legal doctrine, Advocate General Ruiz-Jarabo Colomer argued that Article 3(2) of the Second Gas Directive describes the types of action in which the public service obligations result. Thus, this Article does not contain a list of circumstances (such as security of supply, energy efficiency, climate protection) which justify the adoption of public service obligations. Moreover, on the basis of this Article Member States are allowed to adopt public service obligations, but in each case it must be determined whether there exists a task of general economic interest. This is in the author's opinion a good approach.

Although, the Advocate General Ruiz-Jarabo Colomer did not settle the issue of whether the list of circumstances is exhaustive or indicative, I would argue that given the wording of Article 3(2) of the Second Gas Directive which states *"[...] in the general economic interest, public service obligations which may relate to security, including security of supply, [...]"*[533], this is an indicative list. If the European legislator preferred and exhaustive list, it could not mentioned the word "may" in this Article.

468. Moreover, an exhaustive list, as discussed above, would be contrary to Article 106(2) of the TFEU which appears to give Member States a broad margin of discretion to determine which services could be classified as services of general economic interest. However, it must be highlighted that the list of circumstances can be interpreted very broadly. Consequently, a very large number of public service obligations could fall within the scope of the categories listed in Article 3(2) of the Second Gas Directive.[534] If this is the correct approach it would mean that in this context Article 106(2) of the TFEU has become irrelevant.

---

533 Emphasis added.

534 T. DERUYTTER, W. GELDHOF and F. VANDENDRIESSCHE, "Public Service Obligations in the Electricity and Gas Markets" in B. DELVAUX, M. HUNT and K. TALUS (eds.), *EU Energy Policy and Law* (Cambridge, Intersentia, 2011), p. 63–95.

(c) Regulated gas prices an acceptable public service obligation – Federutility and others case

469. In this context, reference should be made to the judgment of the Court of Justice in its *Federutility and others* case.[535] This case concerned the Decision No 79/07 adopted on 29 March 2007 by the *Autorità per l'energia elettrica e il gas* (hereinafter the Italian gas regulator) revising the economic terms for the supply of gas for the period between 1 January 2005 and 31 March 2007 and established the criteria for future revisions of these terms. In particular, paragraph 1.3.1. of the decision provided that the calculation formulae approved for the revision of the variable wholesale cost, would apply until 30 June 2008. Moreover, paragraph 1.3.2. allowed the Italian gas regulator to determine whether the conditions were met for this power to be extended until 30 June 2009.

470. This Decision No 79/07 and other subsequent decisions of the *Autorità* (Decision Nos 80/07, 158/07, 242/07 and 346/07 of the Italian gas regulator) were challenged before the fourth Chamber of the *Tribunale Administrativo Regionale della Lombardia* by different undertakings and associations of undertakings active in the natural gas market of Italy.[536]

The main argument of these applicants concerned a supposed breach of the Second Gas Directive, pursuant to which the gas sector was to have been entirely liberalised as from 1 July 2007. They argued that in order to ensure that all customers can freely choose their supplier, the gas prices from that date on must be determined exclusively by supply and demand without interference from the authorities. Thus, they claimed that after the second quarter of 2007 the reference prices of the Decision No 79/07 were infringing Union law. In this context, the fourth Chamber of the *Tribunale Administrativo Regionale della Lombardia* decided to refer for a preliminary ruling two questions to the Court of Justice.

471. In essence, the fourth Chamber of the *Tribunale Administrativo Regionale della Lombardia* asked whether Articles 3(2) and (3) and 23 of the Second Gas Directive preclude national legislation such as that at issue in the main proceedings which allows the determination of the level of the price of the supply of natural gas by the definition of reference prices after 1 July 2007.

472. Given this question, the Court examined whether State intervention was justified in the general economic interest. It argued that the Second Gas Directive

535 Case C-265/08, *Federutility, Assogas, Libarna Gas SpA, Collino Commercio SpA, Sadori Gas Srl, Egea Commerciale Srl, E.On Vendita Srl, Sorgenia SpA v Autorità per l'energia elettrica e il gas*, not yet reported.

536 These undertakings and associations of undertakings are Federutility, Assogas, Libarna Gas SpA., Collino Commercio SpA, Sadori Gas SpA., Egea Commerciale, E.On Vendita srl and Sorgenia SpA.

provides no definition of that expression, but the reference in Article 3(2) of the Second Gas Directive, both to this condition and to Article 106 of the TFEU, implies that this condition should be interpreted in the light of Article 106 of the TFEU. The Court of Justice referred to its case law and stated that: *"that provision is designed to reconcile the Member States' interest in using certain undertakings as an instrument of economic or social policy with the Community's interest in ensuring compliance with the rules on competition and preservation of the unity of the common market (see, to that effect, Case C-67/96 Albany [1999] ECR I-5751, paragraph 103 and case-law cited).*

In that context, Member States are entitled, while complying with the law of the Union, to define the scope and the organisation of their services in the general economic interest. In particular, they may take account of objectives pertaining to their national policy (see, to that effect, Albany, paragraph 104). [...]".[537]

473. The Court Justice ruled that *"Directive 2003/55 allows Member States to assess whether, in the general economic interest, after 1 July 2007, it is necessary to impose on undertakings operating in the gas sector public service obligations in order, in particular, to ensure that the price of the supply of natural gas to final consumers is maintained at a reasonable level having regard to the reconciliation which Member States must make, taking account of the situation in the natural gas sector, between the objective of liberalisation and that of the necessary protection of final consumers pursued, as mentioned in paragraphs 18 and 20 of this judgment, by the Union legislature".*[538]

474. Thus, on the basis of the Second Gas Directive, Member States have certain discretion in adopting public service obligations. However, the Court of Justice made clear that Member States do not have absolute freedom in this respect. Member States must prove that the adopted measure pursues a general economic interest; complies with the principle of proportionality and that the public service obligation is clearly defined, transparent, non discriminatory and verifiable, and that it guarantees equal access for EU companies to consumers.[539]

The Court of Justice in this case did not explicitly answer the questions of the fourth Chamber of the *Tribunale Administrativo Regionale della Lombardia.* However, it draws up some guidelines for the Italian court in order to assess whether certain public service obligations could be allowed under the gas and electricity directives.[540]

---

537 Case C-265/08, *Federutility, Assogas, Libarna Gas SpA, Collino Commercio SpA, Sadori Gas Srl, Egea Commerciale Srl, E.On Vendita Srl, Sorgenia SpA v Autorità per l'energia elettrica e il gas*, not yet reported, paragraphs 26–31.

538 *Ibid.*, paragraph 32.

539 *Ibid.*, paragraphs 32, 33 and 44.

540 See also T. DERUYTTER, W. GELDHOF and F. VANDENDRIESSCHE, "Public Service Obligations in the Electricity and Gas Markets" in B. DELVAUX, M. HUNT and K. TALUS

475. Consequently, the provisions of the Third Electricity and Gas Directives allow Member States to decide to reserve rights and privileges on a case-by-case basis. One of the final important questions in this respect is the burden of proof. The Court of Justice in its case law mentioned above, pointed out that it was for the Member States to rely on the exception to make its case in which it was sufficient for them to establish that the undertaking concerned would be prevented or hindered in carrying out these tasks.

476. Then the burden of proof to establish that the same objectives could be achieved by less restrictive measures is shifted to the Commission.[541] It remains questionable whether the same process would apply if a Member State on the basis of the Third Electricity Directive chooses to derogate from, for example, the third party access regime of Article 32 of the Third Electricity Directive. The burden of proof would be on the Member State which needs to demonstrate that the derogation is necessary in order to guarantee the performance of the public service obligations in question.

4. Conclusion

477. It can be concluded, as pointed out in the judgments above, that the Court of Justice, before the adoption of the secondary legislation, was sending to the Commission the signal that it was unwilling to deny competence to Member States in the energy field, by justifying a number of national measures on the basis of Article 106(2) of the TFEU, as long as the Commission itself had not adopted appropriate measures. With the adoption of the electricity and gas directives, the Commission has realised its legislative alternatives in the energy field.

478. The discussion above highlights that even with the adoption of the electricity and gas directives, Member States seem to remain substantial discretion with respect to the adoption of public service obligations in the energy field and which services could be classified as services of general economic interest.

Court of Justice argued in its *Federutility and other* case, discussed above, on the basis of Article 106(2) of the TFEU and the wording of the Second Gas Directive that Member States are in the general economic interest allowed to impose public service obligations on undertakings operating in the gas sector. However, it also argued that Member States still need to prove that the adopted

(eds.) *EU Energy Policy and Law* (Cambridge, Intersentia, 2011), p. 63–95.

541 L. HANCHER, "Delimitation of Energy Law Jurisdiction: the EU and its Member States: From Organisational to Regulatory Conflicts", *Journal of Energy & Natural Resources Law* (1998), Vol 16 No 1, p. 57–59.

measure is in the general economic interest and complies with the principle of proportionality. Public service obligation must furthermore be clearly defined, transparent, non-discriminatory and verifiable, and guarantee equal access for EU companies to consumers.

479. This ruling of the Court of Justice resulted in some guidelines to assess whether certain public service obligations could be allowed under the electricity and gas directive. However, if the national court on the basis of these guidelines comes to the conclusion that reference prices are allowed it could be argued that in order to achieve the creation of an EU internal market in electricity and gas, one of the three objectives of the European energy policy[542], we should side-line Article 106(2) of the TFEU and rephrase the provisions of secondary legislation.

ii) The application of Article 345 of the TFEU

480. As Article 106 of the TFEU, Article 345 of the TFEU is one of the provisions which could impede the application of the free movement and undistorted competition provisions of the TFEU in the electricity and gas markets. The purpose of this part is to analyse to which extent Article 345 of the TFEU could hamper the application of some of the TFEU provisions in the energy field.

481. According to Article 345 of the TFEU, the Treaties[543] shall in no way prejudice the rules in Member States governing the system of property ownership. Anyone on the basis of this wording of Article 345 of the TFEU could draw the conclusion that property law is exclusively for the Member States and exempted from interference by the Union, as the latter has no power to legislate in this area.[544] However, contrary to the presumption of the wording of Article 345 of the TFEU this provision has not prevented the European legislature to adopt measures and legislative proposals in this area[545], nor does it seem to prevent the Court of Justice to pass judgments which have far-reaching consequences in the area of property law.[546] The wording of Article 345 of the

---

542 See *Supra*.

543 This refers to the Treaty on the European Union (hereinafter TEU) and the TFEU.

544 B. AKKERMANS and E. RAMAEKERS, "Article 345 TFEU (ex Article 295 EC), Its Meaning and Interpretations", *European Law Journal* (2010), p. 292 and 293.

545 For example Directive 2002/47EC of the European Parliament and the Council of 6 June 2002 on financial collateral arrangements, *OJ* L 168/43; Green paper of the European Commission on successions and wills of 1 March 2005, COM(2005)65, 11p.; White paper of the European Commission on the integration of EU Mortgage Credit Markets of 18 December 2007, COM(2007)807, 11p.

546 Case C-182/83, *Robert Fearon & Company Limited v Irish Land Commission* [1984] 3677, paragraph 7; Case C-302/97, *Klaus Konle v Republik Österreich* [1999] ECR I-3099, paragraph

TFEU is very broad, which makes its exact meaning, scope and rationale behind it not difficult to determine, nor does it clarify its exact meaning and effect on the rest of the Treaty provisions or on national law.[547]

It is argued in some legal doctrine that since there is no explicit competence of the Union to intervene in the Member State's property mechanisms, one may conclude that a negative provision as Article 345 of the TFEU seems to be unnecessary.[548] This view with respect to Article 345 of the TFEU is not correct. If the Treaty lacked this provision, the Member State's property mechanism would not have a similar legal certainty.

1. INTERPRETING ARTICLE 345 OF THE TFEU

482. A possible interpretation of Article 345 of the TFEU is that this provision in the Treaty *"limits, but not prevents, the application of the TFEU as a whole to the way in which rules of a Member State deal with the right of ownership of undertakings"*.[549] Moreover, this Article does not appear to provide powers to the Union or the Member States. Another author states that if the Union *"has powers in a given area, they must not be employed to interfere with the rules in Member States governing the system of property ownership"*.[550] Furthermore, according to this author, the interpretation of Article 345 of the TFEU within its overall framework clearly leaves the power to the Union to legislate with the effect of regulating the use of property since this does not immediately affect ownership. This author emphasises that if the system of ownership is touched, only the Member States will be competent to regulate this.[551]

483. Also the Court of Justice interpreted in its case law Article 345 of the TFEU. The first case in this context was the *Fearon* case[552] in which the Supreme

38; Case C-367/98, *Commission of the European Communities v Portuguese Republic* [2002] ECR I-4731, paragraph 48 and Joined Cases C-515/99 and C-527/99 to C-540/99 and Joined Cases C-519/99 to C-524/99 and C-526/99), *Hans Reisch and Others (joined cases C-515/99 and C-527/99 to C-540/99) v Bürgermeister der Landeshauptstadt Salzburg and Grundverkehrsbeauftragter des Landes Salzburg and Anton Lassacher and Others (joined cases C-519/99 to C-524/99 and C-526/99) v Grundverkehrsbeauftragter des Landes Salzburg and Grundverkehrslandeskommission des Landes Salzburg* [2002] ECR I-2157.

547 B. AKKERMANS and E. RAMAEKERS, "Article 345 TFEU (ex Article 295 EC), Its Meaning and Interpretations", *European Law Journal* (2010), p. 293 which referred to W. DEVROE, "Privatization and Community Law: Neutrality versus Policy", *Common Market Law Review* (1997), p. 267–306.

548 F. BECKER, "Market Regulation and the 'Right to Property' in the European Economic Constitution" in *Yearbook of European Law 2007* (Oxford, Clarendon, 2007), p. 292.

549 B. AKKERMANS and E. RAMAEKERS, "Article 345 TFEU (ex Article 295 EC), Its Meaning and Interpretations", *European Law Journal* (2010), p. 292 and 293.

550 F. BECKER, "Market Regulation and the 'Right to Property' in the European Economic Constitution" in *Yearbook of European Law 2007* (Oxford, Clarendon, 2007), p. 292.

551 *Ibid.*, p. 292.

552 Case C-182/83, *Robert Fearon & Company Limited v Irish Land Commission* [1984] 3677.

Court of Ireland brought a preliminary question to the Court of Justice on the conformity of the Irish Land Act 1933 as amended by section 35 of the Irish Land Act of 1965 with the freedom of establishment. The Court of Justice seems to rule that the meaning of Article 345 of the TFEU *"is not to exclude the application of the Treaty provisions to questions of state or private ownership at all, but is rather to emphasise how according to the Treaty these powers might belong to the Member States, but not as far as the exercise of those powers is concerned".*[553] In view of the European economic integration it must be emphasised that a certain degree of unification of property law is unavoidable. This undoubtedly results in a complex relationship between the provisions of the TFEU and Article 345 of the TFEU.[554] However with respect to this relationship, the Court of Justice ruled in its *Fearon* case and the *Commission v Portugal* case, the *Commission v France* case and *Commission v Belgium* case[555] that Article 345 of the TFEU does not allow Member States to exempt their systems of property ownership from the fundamental Treaty provisions.[556]

484. Beside the Court of Justice, the European Free Trade Association Court (hereinafter EFTA Court) interpreted in its judgment E-2/06 of 26th June 2007 Article 125 of the Agreement on the European Economic Area (hereinafter EEA).

Since 2001 there was a dispute between the European Free Trade Association (hereinafter EFTA) Surveillance Authority and Norway as to whether under the Norwegian Industrial Concession Act of 1917, the acquisition of waterfalls by entities other than the Norwegian State is subject to concession. These concessions are granted for maximum 60 years. When the concession expires,

553 B. AKKERMANS and E. RAMAEKERS, "Article 345 TFEU (ex Article 295 EC), Its Meaning and Interpretations", *European Law Journal* (2010), p. 309.

554 K. TALUS and A. JOHNSTON, "Comment on Pielow, Brunekreeft and Ehlers on ownership unbundling", *Journal of World Energy Law & Business* (2009), p. 151 and F. BECKER, "Market Regulation and the 'Right to Property' in the European Economic Constitution" in *Yearbook of European Law 2007* (Oxford, Clarendon, 2007), p. 292.

555 Case C-367/98, *Commission of the European Communities v Portuguese Republic* [2002] ECR 4731, paragraph 48; Case C-483/99, *Commission of the European Communities v French Republic* [2002] ECR I-4781, paragraph 44; Case C-503/99, *Commission of the European Communities v Kingdom of Belgium* [2002] ECR I-4809, paragraph 44; Joined Cases C-92/92 and C-326/92, *Phil Collins v Imtrat Handelsgesellschaft mbH and Patricia Im- und Export Verwaltungsgesellschaft mbH and Leif Emanuel Kraul v EMI Electrola GmbH* [1993] I-5145, paragraphs 18–19 and 23 and Case C-309/26, *Daniele Annibaldi v Sindaco del Comune di Guidonia and Presidente Regione Lazio* [1997] ECR I-7493, paragraphs 17 and 23. Other examples of case law in this respect are Case C-300/01, *Doris Salzmann* [2003] ECR I-4899, paragraph 39 and Case C-452/01, *Margarethe Ospelt and Schlössle Weissenberg Familienstiftung* [2003] ECR I-9743.

556 The Court of Justice has, apart from the cases mentioned above, adopted several cases with respect to intellectual property law. For further analysis of these cases see B. AKKERMANS and E. RAMAEKERS, "Article 345 TFEU (ex Article 295 EC), Its Meaning and Interpretations", *European Law Journal* (2010), p. 310 and 311.

the concessionaire must transfer to the Norwegian State the waterfall concerned and all the installations belonging to the power plant. Under Norwegian law this obligation is referred to as "reversion" or "hjemfall". However, this provision does not apply to public undertakings[557]; they could be granted concessions for an unlimited period of time. Since the Norwegian government was not willing to amend the Industrial Concession Act of 1917, the EFTA Surveillance Authority brought the dispute before the EFTA Court.[558]

The EFTA Surveillance Authority did not dispute the concession regime or the reversion. However, the EFTA Surveillance Authority claimed that the distinction between private and public undertakings under the concession regime infringed the fundamental principles, the right of establishment and free movement of capital of the EEA.

485. The main argument for Norway's contention was that on the basis of Article 125 of the EEA the contested provisions fell outside the scope of the EEA. Norway argued that although the wording of Article 125 of the EEA corresponds to the wording of Article 345 of the TFEU, these provisions have a different scope of application given the fundamental differences between the TFEU and EEA. The EFTA Court stated that in the case there are no specific circumstances at hand which could warrant an interpretation of Article 125 of the EEA different from the interpretation of Article 345 of the TFEU. Moreover, according to the EFTA Court, case law of the Court of Justice on Article 345 of the TFEU is of relevance when interpreting Article 125 of the EEA.

486. With reference to case law of the Court of Justice, the EFTA Court stated that: "*[i]t follows from the case law of the ECJ on Article 295 EC that Article 125 EEA is to be interpreted to the effect that, although the system of property ownership is a matter for each EEA State to decide, the said provision does not have the effect of exempting measures establishing such a system from the fundamental rules of the EEA Agreement, including the rules on free movement of capital and freedom of establishment [...]*".[559]

487. With respect to the interpretation of Article 125 of the EEA, the EFTA Court: "*holds that Article 125 EEA is to be interpreted to the effect that an EEA State's right to decide whether hydropower resources and related installations are*

---

557 In accordance with the provisions of this Act, public undertakings are all the undertakings if the Norwegian State and/or Norwegian municipalities directly or through the ownership of other undertakings, individually or jointly own at least 2/3 of the shares.

558 F. ARNESEN, "Chapter XVI The EFTA Court Decision on Norwegian Hydropower and Norway's response to that decision" in M. M. ROGGENKAMP and U. HAMMER (eds.), *European Energy Law Report VI* (Antwerp, Intersentia, 2009), p. 287–289.

559 Case E-2/06 of 26 June 2007, *EFTA Surveillance Authority v The Kingdom of Norway*, paragraph 62.

in private or public ownership is, as such, not affected by the EEA Agreement. The corollary of this is that Norway may legitimately pursue the objective of establishing a system of public ownership over these properties, provided that the objective is pursued in a non-discriminatory and proportionate manner.

However, for the Norwegian concession regime, with reversion to the State as an essential part of it, to qualify as aiming at establishing a "system of property ownership" within the meaning of Article 125 EEA, that regime must aim at attaining a situation where, as a matter of principle, the assets at issue are owned by public entities. A regime which merely brings or keeps this class of assets predominantly under public ownership, while at the same time leaving it to the discretion of the relevant authorities whether private ownership of the assets should be re-established, cannot be said to aim at establishing a system of property ownership under Article 125 EEA. Rather, such a regime aims at achieving a certain level of public control of the relevant sector of the economy – by means of securing public ownership of most of the relevant assets".[560]

488. The foregoing means that Article 125 of the EEA does not hinder Norway in pursuing the objective of establishing a "system of property ownership" with respect to certain categories of assets. In order to comply with the ruling of the EFTA Court, Norway has amended its Industrial Licensing Act of 1917 in the sense that only the Norwegian State, Norwegian municipalities and public undertakings can acquire waterfalls.[561] This judgment also proved that states are not allowed to let private investors invest in undertakings while at the same reserving special rights for the shares held by public authorities.

489. It can be concluded that, given the response of the Norwegian State to the EFTA Court decision, the EFTA Surveillance Authority efforts resulted, instead of removing restrictions on establishment and free movement of capital, in a more restrictive regime with no movement.[562]

2. Article 345 of the TFEU and the Energy Field, in Particular the Case of Ownership Unbundling

490. With respect to the energy sector, Article 345 of the TFEU has recently been placed in the forefront due to the Commission's proposals in the Third Energy Package requiring the full ownership unbundling of transmission

560 *Ibid.*, paragraphs 72 and 73.

561 F. ARNESEN, "Chapter XVI The EFTA Court Decision on Norwegian Hydropower and Norway's response to that decision" in M. M. ROGGENKAMP and U. HAMMER (eds.), *European Energy Law Report VI* (Antwerp, Intersentia, 2009), p. 291–293.

562 *Ibid.*, p. 293 and 294.

networks in the energy sector.[563] These proposals have provoked a heated debate on a European level among Member State governments and the European Parliament which finally led to a political compromise in the Third Electricity Directive and the Third Gas Directive. One key problem in the energy sector which has been highlighted for many years in the liberalisation process of the electricity and gas sector is the degree of vertical integration by incumbent companies.

491. In some Member States these vertically integrated incumbents appear to hold a natural monopoly for assets like transmission networks and to use this infrastructure in such a way that it forecloses competition in the electricity and gas market.[564] Vertically integrated companies have a myriad of possibilities at their disposal to discriminate against competitors, combined with a clear economic incentive to discriminate, which led to the awareness on a European level that without an effectively separated network company from its competitive activities such as generation, supply and sales, effective competition would simply not be realised. One of the remedies, beside regulated third-party access to the network and the imposition of fines, is to force Member States to unbundle the assets of network operators.[565]

(a) Unbundling process in the electricity and gas markets

492. In this respect, the European energy policy has cranked up the unbundling of these natural monopoly assets in the electricity and gas market slowly. The First Electricity Directive provided for accounting and management unbundling. With respect to gas, the First Gas Directives provided only for accounting unbundling and did not require the Member States to nominate an identifiable transmission system operator.

---

563 Proposal of the European Commission of 19 September 2007 for a Directive of the European Parliament and of the Council amending Directive 2003/54/EC concerning common rules for the internal market in electricity, COM(2007)528, p. 5–7 and Proposal of the European Commission of 19 September 2007 for a Directive of the European Parliament and of the Council amending Directive 2003/55/EC concerning common rules for the internal market in natural gas, COM(2007)529, p. 5–7.

564 A. JOHNSTON, "Ownership Unbundling: Prolegomenon to a Legal Analysis" in M. BULTERMAN, L. HANCHER, A. MCDONNELL and H. SEVENSTER (eds.), *Views of European Law from the Mountain – Liber Amicorum Piet Jan Slot* (The Netherlands, Kluwer Law International, 2009), p. 277–279.

565 C. W. JONES and W. WEBSTER, *EU Energy Law – Volume I The Internal Energy Markets* (Leuven, Claeys & Casteels, 2006), p. 69–74 and for further reading with respect to unbundling see B. MALMENDIER and J. SCHENDEL, "Unbundling Germany's Energy Networks", *Journal of Energy & Natural Resources Law* (1998), p. 362–383 and N. KROES, "Improving Competition in European Energy Markets through Effective Unbundling", *Fordham International Law Journal* (2008), p. 1387–1441.

493. After the implementation of these directives, it became immediately clear that these forms of unbundling were insufficient both on a transmission and distribution level. Consequently, the Second Electricity and Gas Directives required accounting, management and legal unbundling for both the transmission and distribution system operators which were part of a vertically integrated undertaking.[566] However, the Commission found in its sector inquiry[567] that even these forms of unbundling of the Second Electricity and Gas Directives were inadequate, in those Member States where a joint ownership of the network and supply business existed, to ensure a competitive market. Often new entrants lack effective access to networks, operators of networks are suspected of favouring their own affiliation. Besides, operational and investment decisions are taken on the basis of the supply interests of the integrated company and not in the interest of customer welfare.

494. Thus, to remedy this situation the Commission has proposed in its Third Package ownership unbundling of the transmission network assets. This means in practice that under the Third Electricity and Gas Directives that Member States from 3 March 2012 must ensure that the same person or persons are:

(a) not able to have direct or indirect control of a generation or supply undertaking and direct or indirect control or any right over, or hold any interest in, a transmission system operator or a transmission system (and vice versa);
(b) not entitled to appoint members of the supervisory, administrative board or bodies legally representing the undertaking of a transmission system operator or a transmission system and directly or indirectly exercise control or exercise any right over a generation or supply undertaking;
(c) not entitled to be members of the supervisory, administrative board or bodies legally representing the undertaking of both a generation or supply undertaking and a transmission system operator or a transmission system.[568]

495. Under the Third Electricity and Gas Directives the ownership unbundling requirements are not required for distribution system operators. The European legislature argued that the unbundling of the distribution system operators

566 *Ibid.*

567 DG Competition Report on Energy Sector Inquiry of 10 January 2007, SEC(2007)1724 and http://ec.europa.eu/competition/sectors/energy/inquiry/index.html.

568 See Articles 9, (1) of the Third Electricity Directive and Article 9, (1) of the Third Gas Directive. It should be noted that both directives also defined rights as (a) the power to exercise voting rights; (b) the power to appoint members of the supervisory board, the administrative board or bodies legally representing the undertaking; or (c) the holding of a majority share.

became compulsory as of 1 July 2007 in accordance with the Second Electricity and Gas Directives and its effects on the internal market still needs to be evaluated. Moreover, according to the European legislator the rules on functional and legal unbundling of the basis of the Second Electricity and Gas Directives can lead to effective unbundling provided that they are more clearly defined, properly implemented and closely monitored on a national level.[569]

496. While the Commission has a clear preference for ownership unbundling the proposals of the Third Package also allowed for the so-called Independent System Operator (hereinafter ISO). This alternative option allows Member States to derogate from the ownership unbundling and should be as effective as ownership unbundling. The ISO model, however, must provide the same guarantees as the ownership unbundling regarding the independence of action of the network in question and the same level of incentives on the network to invest in new infrastructure which may benefit competitors. It enables vertically integrated companies[570] to retain ownership of their network assets, but requires that the transmission network itself is managed by an ISO. The ISO is an undertaking or entity having wide-ranging powers to operate all the functions of a network operator and to decide on necessary investments, subject to regulatory approval. The ISO must be entirely separated from the vertically integrated company.[571]

497. Since both France and Germany opposed to ownership unbundling and the ISO model was an insufficient alternative option, a third option as a political

569 Recitals 26 of the Third Electricity Directive and 25 of the Third Gas Directive.

570 The term *"vertically integrated undertaking"* is defined in Article 1(21) of the Third Electricity Directive which states that *"vertically integrated undertaking"* means an electricity undertaking or a group of electricity undertakings where the same person or the same persons are entitled, directly or indirectly, to exercise control, and where the undertaking or group of undertakings perform at least one of the functions of transmission or distribution, and at least one of the functions of generation or supply of electricity. For gas a similar definition can be found in Article 1(20) of the Third Gas Directive which states that *"vertically integrated undertaking"* means a natural gas undertaking or a group of natural gas undertakings where the same person or the same persons are entitled, directly or indirectly, to exercise control, and where the undertaking or group of undertakings perform at least one of the functions of transmission, distribution, LNG or storage, and at least one of the functions of production or supply of natural gas. For a further discussion of this term see the Commission Staff Working Paper of 22 January 2010 – Interpretative note on Directive 2009/72/EC concerning common rules for the internal market in electricity and Directive 2009/73/EC concerning common rules for the internal market in natural gas – The unbundling Regime, p. 11.

571 Proposal of the European Commission of 19 September 2007 for a Directive of the European Parliament and of the Council amending Directive 2003/54/EC concerning common rules for the internal market in electricity, COM(2007)528, p. 5–7 and Proposal of the European Commission of 19 September 2007 for a Directive of the European Parliament and of the Council amending Directive 2003/55/EC concerning common rules for the internal market in natural gas, COM(2007)529, p. 5–7.

compromise was adopted in the Third Package, the so-called "Third Way" or Independent Transmission Operator (hereinafter ITO). Within the ITO model, the TSO remains within the integrated company and, like the ISO option, the integrated companies retain ownership of their network assets.

498. However, to ensure effective unbundling numerous details have been laid down in the Third Electricity and Gas Directives. Within this model, the ITO must have effective decision-making rights with respect to the assets necessary to operate, maintain and develop the transmission system independent of any other part of the vertically integrated undertaking. In order to guarantee the independence of the ITO from the vertically integrated undertaking both the Third Electricity and Gas Directive imposed various regulatory conditions on the ITO such as setting-up the Supervisory Body, to establish and implement a compliance programme, submit annually to the regulatory authority a ten-year network development plan, etc. Thus, since the adoption of the Third Electricity and Gas Directives, Member States are free to opt for one of the three models (ownership unbundling, ISO or ITO) which are on equal footing in the Directives for those companies which are vertically integrated companies at the moment the Third Electricity and Gas Directives entered into force on 3 September 2009.[572]

(b) Ownership unbundling and Article 345 of the TFEU

499. With respect to ownership unbundling, one of the legal obstacles[573] is whether the European directives requiring ownership unbundling are precluded by Article 345 of the TFEU. In legal doctrine opinions differ on this. Some authors argue that as long as individual decisions by the Union to expropriate do not concern a "*system of ownership*", which means that the ownership structure within an entire industry sector is addressed, these decisions are not precluded by Article 345 of the TFEU. Conversely, if individual decisions are only

---

572 Commission Staff Working Paper of 22 January 2010 – Interpretative note on Directive 2009/72/EC concerning common rules for the internal market in electricity and Directive 2009/73/EC concerning common rules for the internal market in natural gas – The unbundling Regime, 29 p.

573 Other legal issues with respect to ownership unbundling are whether the right to property laid down in the European Convention for the protection of Human Rights and Fundamental Freedoms precludes the application of ownership unbundling. It seems that if the current owners of the transmission assets receive some compensation for their inability to own these assets any longer, the proportionality requirement imposed by this provision appears to be satisfied. Other legal obstacles could be national constitutional constraints, the free movement provisions, proportionality requirement, etc. For further details see A. JOHNSTON, "Ownership Unbundling: Prolegomenon to a Legal Analysis" in M. BULTERMAN, L. HANCHER, A. MCDONNELL and H. SEVENSTER (eds.), *Views of European Law from the Mountain – Liber Amicorum Piet Jan Slot* (The Netherlands, Kluwer Law International, 2009), p. 277–290.

individual elements of an overarching plan this would amount to a system of ownership which is precluded by Article 345 of the TFEU. Moreover, the European Commission will not be able to circumvent the Union's lack of competence for legislative expropriation by adopting a string of individual decisions. Thus, according to this author the Union lacks due to Article 345 of the TFEU, the competence to adopt the model of ownership unbundling.[574]

500. On the contrary, it is argued in other legal doctrine that, as we discussed above, that the Union has to some extent competences to deal with property right issues. They emphasise that since the ownership unbundling model will lead to more efficient competition, which is a matter of public interest, Article 345 of the TFEU is no legal obstacle for the Commission to provide this model in the Third Package.

501. Furthermore, the fact that the Commission adopts the ownership unbundling model does not exclude the freedom for Member States to choose the property ownership regime (e.g. public, private or mixture of both) under which they wish to place the transmission activities.[575] Other authors also argue that the underlying objective of Article 345 of the TFEU is to guarantee that if Member States wish to retain certain industries within state ownership, they may continue to do so. This means that the Commission, even if it wishes to do so, could not challenge the fact that in some Member States electricity or gas companies remain in state ownership. However, this does not prevent the Commission from adopting the ownership unbundling. Consequently, a state-owned company would be split into two different and independent companies, but they remain under state control.[576]

3. Conclusion

502. On the basis of the analysis above it can be concluded that, contrary to some legal doctrine, Article 345 of the TFEU does not prevent the Commission

574 F. BECKER, "Market Regulation and the 'Right to Property' in the European Economic Constitution" in *Yearbook of European Law 2007* (Oxford, Clarendon, 2007), p. 292 and 293; Z. ZAFIROVA, "Unbundling the network: the case for ownership unbundling?", *International Energy Law & Taxation Review* (2007), p. 29–36.

575 A. JOHNSTON, "Ownership Unbundling: Prolegomenon to a Legal Analysis" in M. BULTERMAN, L. HANCHER, A. MCDONNELL and H. SEVENSTER (eds.), *Views of European Law from the Mountain – Liber Amicorum Piet Jan Slot* (The Netherlands, Kluwer Law International, 2009), p. 280 and 281 and K. TALUS and A. JOHNSTON, "Comment on Pielow, Brunekreeft and Ehlers on ownership unbundling", *Journal of World Energy Law & Business* (2009), p. 151 and F. BECKER, "Market Regulation and the 'Right to Property' in the European Economic Constitution" in *Yearbook of European Law 2007* (Oxford, Clarendon, 2007), p. 292.

576 C. W. JONES and W. WEBSTER, *EU Energy Law – Volume I The Internal Energy Markets* (Leuven, Claeys & Casteels, 2006), p. 74 in footnote 87.

from adopting the ownership unbundling model in the energy sector. Moreover, this model does not exclude the freedom of Member States to choose the property ownership regime under which they wish to place the transmission activities.[577]

503. The EFTA Court case has highlighted that Article 125 of the EEA does not hinder Norway in pursuing the objective of establishing a "system of property ownership" with respect to certain categories of assets. Moreover, it demonstrates that states are not allowed to let private investors invest in undertakings while at the same reserving special rights for the shares held by public authorities.

504. On the basis of the foregoing it could be argued that Article 345 of the TFEU does not seem to hamper the application of some of the TFEU provisions and secondary legislation to the electricity and gas markets.

(c) The application of the rules on the free movement of goods

505. The freedom of movement of goods, services, workers and capital are the four fundamental freedoms of the Union which should guarantee, or are the means towards, the creation of a common market. The Court of Justice has emphasised in its case law that Member States must exercise their powers in a way which is consistent with the fundamental Treaty principles. It has furthermore set the boundaries within which the political institutions of the European Union must work in their elaboration of new Union measures, including secondary legislation.[578]

506. This part addresses the Treaty rules on free movement of goods which are also general TFEU provisions. Since it is settled case law that the Treaty rules on the free movement of goods are applicable to the energy sector, this part will

---

577 See also M. HUNT, "Ownership Unbundling: the Main Legal Issues in a controversial Debate" in B. DELVAUX, M. HUNT and K. TALUS (eds.), *EU Energy Law and Policy Issues* (Rixensart, Euroconfidentiel, 2008), p. 57–90.

578 E. D. CROSS, B. DELVAUX, L. HANCHER, P. J. SLOT, G. VAN CALSTER and W. VANDENBERGHE, "EU Energy Law (chapter 5)" in M. M. ROGGENKAMP, A. RØNNE, C. REDGWELL and I. DEL GUAYO (eds.), *Energy Law in Europe. National, EU and International Regulation*, 2nd ed. (Oxford, Oxford University Press, 2007), p. 234–235. With respect to secondary legislation see for example Case C-543/08, *European Commission v Portuguese Republic*, not yet published, paragraph 89 in which the Court of Justice states that: *"[e]ven were it accepted that, pursuant to provisions of European Union secondary legislation, a Member State has an obligation to guarantee the supply of energy within its territory, as is claimed by the Portuguese Republic, compliance with such an obligation cannot be relied on to justify any measure which is contrary in principle to a fundamental* freedom."

provide an overview of these provisions and their application to the energy sector.

These fundamental provisions of the TFEU have set boundaries which also apply to the electricity and gas markets and must be taken into account by the Member States and the European legislator when they adopt measures in these markets. The purpose of this part is to have a clear understanding of these boundaries and to see to which extent they apply to the energy sector.

i) Fiscal barriers to trade or measures having an effect equivalent to customs duties

507. The provisions in the TFEU on the free movement of goods designed to prohibit quantitative restrictions between Member States and which are not fiscal in nature are Article 34 to 37 of the TFEU. The Court of Justice in its case law insisted on the need to maintain a distinction between the treatment of fiscal and non-fiscal barriers to trade under Union law.

508. According to settled case law, the prohibition of quantitative restrictions in Article 34 of the TFEU does not extend to the obstacles to trade covered by other specific provisions and obstacles of a fiscal nature or having an effect equivalent to customs duties, covered by Articles 28, 30 and 110 of the TFEU.[579] The next part will discuss the obstacles of a fiscal nature.

1. Meaning of Article 110 of the TFEU

509. The Treaty provisions relating to the abolition of customs duties and charges with equivalent effect between the Member States (Articles 28 and 30 of the TFEU) and those relating to discriminatory taxation (Article 110 of the TFEU) have been applied by the Court of Justice in various rulings of the Court. Article 110 of the TFEU declares that: "*no Member State shall impose, directly or indirectly, on the products of other Member States any internal taxation of any kind in excess of that imposed directly or indirectly on similar domestic products.*

Furthermore, no Member State shall impose on the products of other Member States any internal taxation of such a nature as to afford indirect protection to other products".

579 Case C-315/05, *Maciej Brzeziński v Dyrektor Izby Celnej w Warszawie* [2007] ECR I-513, paragraph 50; Case C-228/98, *Charalampos Dounias v Ypourgio Oikonomikon* [2000] ECR I-577, paragraph 39 and Case C-383/01, *De Danske Bilimportører v Skatteministeriet, Told- og Skattestyrelsen* [2003] ECR I-6065, paragraph 32. See also S. WEATHERILL and P. BEAUMONT, *EU law –The essential guide to legal workings of the European Union* (London, Penguin Books, 1993), p. 500.

510. The aim of Article 110 of the TFEU is to ensure free movement of goods between the Member States in normal conditions of competition by the elimination of all forms of protection which may result from the application of internal taxation that discriminate against products from other Member States.[580] Thus, it applies the basic Treaty rule of Article 18 of the TFEU against discrimination on grounds of nationality to the area of national taxation. Nonetheless, it does not ensure equalisation of tax treatment between Member States as this remains different state by state.[581]

511. According to the Court of Justice, the first paragraph of Article 110 of the TFEU is infringed where the tax charged on the imported product and that on the similar domestic product are calculated in a different manner on the basis of different criteria which lead, if only in certain cases, to higher taxation being imposed on the imported product. This means that under this paragraph of Article 110 of the TFEU an excise duty must not affect products originating from other Member States more onerously than similar domestic products.[582]

512. One of the key questions with respect to this provision is the meaning of the concept "similar". In the beginning, the Court of Justice applied a formal test in which it examined whether the products in question came within the same fiscal, customs or statistical classification. This formal test was followed by a broader test in which the Court of Justice combined a factual comparison of the products with an economic analysis of their use. The broader test has been applied in the *Commission v Denmark* case[583] in which the Court of Justice ruled

---

580 Case C-393/04 and C41/05, *Air Liquide Industries Belgium SA v Ville de Seraing and Province de Liège* [2006] ECR I- 5293, paragraph 55; Case C-168/04, *Commission of the European Communities v French Republic* [1980] ECR 347, paragraph 4 and Case 169/78, *Commission of the European Communities v Italian Republic* [1980] ECR 385, paragraph 4 and Case 171/78, *Commission of the European Communities v Kingdom of Denmark* [1980] ECR 447, paragraph 4.

581 S. WEATHERILL and P. BEAUMONT, *EU law –The essential guide to legal workings of the European Union* (London, Penguin Books, 1993), p. 466–499 and K. J. M. MORTELMANS, "Chapter VIII The Functioning of the Internal Market: The Freedoms" in P. J. G. KAPTEYN, A. M. MCDONNELL, K. J. M. MORTELMANS and C. W. A. TIMMERMANS (eds.), *The Law of the European Union and the European Communities*, fourth edition (The Netherlands, Kluwer Law International, 2008), p. 609.

582 Case C-375/95, *Commission of the European Communities v Hellenic Republic* [1997] ECR I-5981, paragraphs 20 and 29; Case C-213/96, *Outokumpu Oy* [1998] ECR I-1777, paragraph 34; Case C-393/98, *Ministério Público and António Gomes Valente v Fazenda Pública* [2001] ECR I-1327, paragraph 21; Case C-387/01, *Harald Weigel and Ingrid Weigel v Finanzlandesdirektion für Vorarlberg* [2004] ECR I-4981, paragraph 67 and Case C-313/05, *Maciej Brzeziński v Dyrektor Izby Celnej w Warszawie* [2007] ECR I-513, paragraph 29.

583 Case C-106/84, *Commission of the European Communities v Kingdom of Denmark* [1986] ECR 833. This approach has also been applied by the Court of Justice in the Case C-302/00, *Commission of the European Communities v French Republic* [2002] ECR I-2055, paragraphs 24–29.

that *"in order to determine whether products are similar it is necessary first to consider certain objective characteristics of both categories of beverages, such as their origin, the method of manufacture and their organoleptic properties, in particular taste and alcohol content, and secondly to consider whether or not both categories of beverages are capable of meeting the same need from the point of view of consumers".*[584]

513. This means that the similarity of products in the context of paragraph 1 of Article 110 of the TFEU will be determined on their similar and comparable use.[585] If similarity has been established and paragraph 1 of Article 110 of the TFEU applies, the tax rate must be the same for the domestic and similar imported product. If the Court of Justice comes to the conclusion that goods for the purposes of paragraph 1 of Article 110 of the TFEU are deemed not similar, it will examine whether they are under paragraph 2 of Article 110 of the TFEU in competition (even partial, indirect or potential) with each other. Contrary to paragraph 1 of Article 110 of the TFEU, paragraph 2 of Article 110 of the TFEU applies a more general criterion, due to the difficulty of making a sufficiently precise comparison between the products in question, namely the indirect protection afforded by a domestic tax system.[586]

2. Meaning of Articles 28 and 30 of the TFEU

514. Article 28 of the TFEU states in its first paragraph that *"the Union shall comprise a customs union which shall cover all trade in goods and which shall*

---

584 Case C-106/84, *Commission of the European Communities v Kingdom of Denmark* [1986] ECR 833, paragraph 12.

585 C. BARNARD, *The substantive Law of the EU The Four Freedoms* (Oxford, Oxford University Press, 2007), p. 49 and 50 and A. ARNULL, A. DASHWOOD, M. DOUGAN, M. ROSS, E. SPAVENTA and D. WYATT, *Wyatt and Dashwoods's European Union Law* (London, Sweet & Maxwell, 2006) p. 556.

586 For further details see A. ARNULL, A. DASHWOOD, M. DOUGAN, M. ROSS, E. SPAVENTA and D. WYATT, *Wyatt and Dashwoods's European Union Law* (London, Sweet & Maxwell, 2006) p. 547–560 and K. J. M. MORTELMANS, "Chapter VIII The Functioning of the Internal Market: The Freedoms" in P. J. G. KAPTEYN, A. M. MCDONNELL, K. J. M. MORTELMANS and C. W. A. TIMMERMANS (eds.), *The Law of the European Union and the European Communities*, fourth edition (The Netherlands, Kluwer Law International, 2008), p. 609–616; B. J. M. TERRA and P. J. WATTEL, *European Tax Law* (Alphen aan den Rijn, Kluwer Law International, 2008), p. 862; D. WEBER, *Tax avoidance and the EC treaty freedom: a study on the limitations under European law to the prevention of tax avoidance* (The Hague, Kluwer Law International, 2005), p. 290; T. LYONS, *EC customs law* (Oxford, Oxford University Press, 2008), p. 552; L. W. GORMLEY, *EU law of free movement of goods and customs union* (Oxford, Oxford University Press, 2009), p. 637 and case law of the Court of Justice among which Case C-156/04, *Commission of the European Communities v Hellenic Republic* [2007] ECR I-4129 paragraph 86; Case C-319/81, *Commission of the European Communities v Italian Republic* [1983] ECR 601; Case C-169/78, *Commission of the European Communities v Italian Republic* [1980] ECR 385; Case C-168/78, *Commission of the European Communities v French Republic* [1980] ECR 347; etc.

involve the prohibition between Member States of customs duties on imports and exports and of all charges having equivalent effect, and the adoption of a common customs tariff in their relations with third countries".[587] This paragraph makes clear that the customs of the Union have both internal and external aspects. The internal aspects of this provision require within the Union on goods the free circulation, the elimination of customs duties and charges having equivalent effect. The external aspect of Article 28 of the TFEU involves the adoption of the common customs tariff.[588]

515. The elimination of customs duties on intra-Union trade is achieved by virtue of Article 30 of the TFEU which states that "*customs duties on imports and exports and charges having equivalent effect shall be prohibited between Member States. This prohibition shall also apply to customs duties of a fiscal nature*".[589] Thus, Member States are prohibited to collect charges (import and export duties, but also transit charges) where goods cross its borders. The prohibition of Article 30 of the TFEU, which is directly applicable[590], has been interpreted strictly by the Court of Justice in its rulings.

516. The Court has also widened the scope of this provision through its interpretation of the prohibited notion of "*charges having equivalent effect*" to a customs duty.[591] This notion has been used by the Court of Justice to secure the

---

587 See also P. OLIVER, "Some Further Reflections on the Scope of Articles 28–30 (ex 30–36) EC", *Common Market Law Review* (1999), p. 783–806.

588 S. WEATHERILL and P. BEAUMONT, *EU law –The essential guide to legal workings of the European Union* (London, Penguin Books, 1993), p. 454–465. See also K. LENAERTS, and P. VAN NUFFEL, in BRAY, R. (ed.), *Constitutional Law of the European Union* (London, Thomson Sweet & Maxwell, 2005), p. 149–162 and K. J. M. MORTELMANS, "Chapter VIII The Functioning of the Internal Market: The Freedoms" in P. J. G. KAPTEYN, A. M. MCDONNELL, K. J. M. MORTELMANS and C. W. A. TIMMERMANS (eds.), *The Law of the European Union and the European Communities*, fourth edition (The Netherlands, Kluwer Law International, 2008), p. 601.

589 C. BARNARD, *The substantive Law of the EU The Four Freedoms* (Oxford, Oxford University Press, 2007), p. 35–39.

590 Case C-26/62, *NV Algemene Transport- en Expeditie Onderneming van Gend & Loos v Netherlands Inland Revenue Administration* [1963] ECR 1.

591 Examples of such charges having equivalent effect to a customs duty are: a tax on the export of articles of artistic, historic, archaeological or ethnographic value (Case 7/68, *Commission of the European Communities v Italian Republic* [1968] ECR 423); a duty for administrative services and a statistical levy on the importation of meat (Case 43/71, *Politi s.a.s. v Ministry for Finance of the Italian Republic* [1971] ECR 1039); a charge for a mandatory quality inspection of products when they cross a frontier resulting in the issue of a certificate and the affixing of a national export stamp (Case C 63/74, *W. Cadsky SpA v Istituto nazionale per il Commercio Estero* [1975] ECR 281); storage charges on imported products deposited in public warehouses prior to the completion of customs formalities (Case 132/82, *Commission of the European Communities v Kingdom of Belgium* [1983] ECR 1649); charges for health inspections of imported products required for the protection of public health (Case 158/82, *Commission of the European Communities v Kingdom of Denmark* [1983] ECR 3573, for

abolition of charges imposed at frontiers even where those charges would not be classified as customs duties in the normal sense.[592] It has interpreted the concept of *"charges having equivalent effect"* very broadly in the following terms: *"Any pecuniary charge, however small and whatever its designation and mode of application, which is imposed unilaterally on domestic or foreign goods by reason of the fact that they cross a frontier, and which is not a customs duty in the strict sense, constitutes a charge having equivalent effect within the meaning of Articles 9 and 13 of the EC Treaty, even if it is not imposed for the benefit of the State, is not discriminatory or protective in effect or if the product on which the charge is imposed is not in competition with any domestic product"*.[593] The emphasis of this definition of the Court of Justice lies on the effect the charge has as an obstacle to the free movement of goods.[594]

517. Article 30 of the TFEU and Article 110 of the TFEU have elements in common since both contemplate control of fiscal charges imposed by Member States on goods that have the capacity to restrict interstate trade. Both provisions contribute to the process of market integration on a European level. The Court of Justice has long held that Articles 30 and 110 of the TFEU are mutually exclusive.[595] The same charge cannot belong to both categories. Article 30 of the TFEU looks to the state's frontier controls and prohibits certain charges on imported goods. Whereas the trust of Article 110 of the TFEU is directed to the internal situation of a Member States by demanding that systems of internal taxation shall apply without discrimination on the basis of the origin of goods.

---

further examples see L. W. GORMLEY, *EU law of free movement of goods and customs union* (Oxford, Oxford University Press, 2009), p. 387–390.

592 S. WEATHERILL and P. BEAUMONT, *EU law –The essential guide to legal workings of the European Union* (London, Penguin Books, 1993), p. 458–459.

593 Joined cases 2 and 3/69, *Sociaal Fonds voor de Diamantarbeiders v S.A. Ch. Brachfeld & Sons and Chougol Diamond Co* [1969] ECR 211, paragraphs 15/18; Case 24/68, *Commission of the European Communities v Italian Republic* [1969] ECR 193, paragraph 9; Case 46/76, *W. J. G. Bauhuis v The Netherlands State* [1977] ECR 5, paragraph 10; Case C-130/93, *Lamaire NV v Nationale Dienst voor Afzet van Land- en Tuinbouwprodukten* [1994] ECR I-3215, paragraph 13 and Case C-90/94, *Haahr Petroleum Ltd v Åbenrå Havn, Ålborg Havn, Horsens Havn, Kastrup Havn NKE A/S, Næstved Havn, Odense Havn, Struer Havn and Vejle Havn, and Trafikministeriet* [1997] ECR I-4085, paragraph 20.

594 K. J. M. MORTELMANS, "Chapter VIII The Functioning of the Internal Market: The Freedoms" in P. J. G. KAPTEYN, A. M. MCDONNELL, K. J. M. MORTELMANS and C. W. A. TIMMERMANS (eds.), *The Law of the European Union and the European Communities*, fourth edition (The Netherlands, Kluwer Law International, 2008), p. 604 and L. W. GORMLEY, *EU law of free movement of goods and customs union* (Oxford, Oxford University Press, 2009), p. 384.

595 L. W. GORMLEY, *EU law of free movement of goods and customs union* (Oxford, Oxford University Press, 2009), p. 390 and Case C-57/65, *Alfons Lütticke GmbH v Hauptzollamt Sarrelouis* [1966] ECR 205; Case C-90/94, *Haahr Petroleum Ltd v Åbenrå Havn, Ålborg Havn, Horsens Havn, Kastrup Havn NKE A/S, Næstved Havn, Odense Havn, Struer Havn and Vejle Havn, and Trafikministeriet* [1997] ECR I-4085 and Case C-234/99, *Niels Nygård v Svineafgiftsfonden, and Ministeriet for Fødevarer, Landbrug og Fiskeri* [2002] ECR I-3657.

518. This provision allows for the imposition of taxes on imported goods on the condition that similar domestic products are subject to the same taxes. It is clear that it would be fruitless to outlaw customs duties if Member States could then prejudice at the frontier the opportunities for imported goods by subjecting them to a discriminatory taxation.[596] Nonetheless, the Court of Justice has ruled in its case law that a charge which appears to be a non-discriminatory internal taxation, could qualify as a charge having effect equivalent to customs duty if the revenue is intended exclusively to support activities which specifically benefit the taxed domestic product.[597]

519. Between both provisions there is no room for overlap. Under Article 30 of the TFEU, a fiscal charge is either considered to be a customs duty or a charge having equivalent effect, or else it forms part of a general system of internal taxation under Article 110 of the TFEU. It is necessary that these provisions do not overlap, since the consequences of their application are different.

In accordance with Article 110 of the TFEU, a charge is lawful unless it contains element that discriminate according to nationality, whereas a charge caught by Article 30 of the TFEU is unlawful. Nonetheless, the demarcation between a charge forming part of a general system of internal taxation and a customs duty is not always clear-cut.[598]

3. Fiscal barriers and energy

520. These provisions have been applied to the energy sector on various occasions by the Court of Justice. For example in the *Outokumpu Oy* case[599] in

---

596 K. J. M. MORTELMANS, "Chapter VIII The Functioning of the Internal Market: The Freedoms" in P. J. G. KAPTEYN, A. M. MCDONNELL, K. J. M. MORTELMANS and C. W. A. TIMMERMANS (eds.), *The Law of the European Union and the European Communities*, fourth edition (The Netherlands, Kluwer Law International, 2008), p. 604.

597 L. W. GORMLEY, *EU law of free movement of goods and customs union* (Oxford, Oxford University Press, 2009), p. 391 and Case 77/72, *Carmine Capolongo v Azienda Agricole Maya* [1973] ECR 611; Case 105/76, *Interzuccheri S.p.A. v Societa Rezzano e Cavassa* [1977] ECR 1029; Joined cases C-78/90, C-79/90, C-80/90, C-81/90, C-82/90 and C-83/90, *Compagnie Commerciale de l'Ouest and others v Receveur Principal des Douanes de La Pallice Port* [1992] ECR 1874, paragraphs 23 and 24; Case C-234/99, *Niels Nygård v Svineafgiftsfonden, and Ministeriet for Fødevarer, Landbrug og Fiskeri* [2002] ECR I-3657, paragraph 19 and Case C-517/04, *Visserijbedrijf D.J. Koornstra & Zn. Vof v Productschap Vis* [2006] ECR I-5015, paragraphs 15–24. For further analysis see L. W. GORMLEY, *EU law of free movement of goods and customs union* (Oxford, Oxford University Press, 2009), p. 390–392.

598 For a further analysis on this aspect see S. WEATHERILL and P. BEAUMONT, *EU law –The essential guide to legal workings of the European Union* (London, Penguin Books, 1993), p. 454–465 and K. J. M. MORTELMANS, "Chapter VIII The Functioning of the Internal Market: The Freedoms" in P. J. G. KAPTEYN, A. M. MCDONNELL, K. J. M. MORTELMANS and C. W. A. TIMMERMANS (eds.), *The Law of the European Union and the European Communities*, fourth edition (The Netherlands, Kluwer Law International, 2008), p. 608.

599 Case C-213/96, *Outokumpu Oy* [1998] ECR I-1777.

which a Finnish electricity producer brought an action before a national judge against a Finnish law (No 1473/94 of 29 December 1994 on excise duty on certain sources of energy) that established a levying as excise duties determined on the basis of the method of generation of electricity generated in Finland. At the same time, on imported electricity the excise duty charged was a flat rate, regardless of the method of generation of the electricity. Furthermore, according to this Finnish Law the excise duty chargeable on imported electricity is higher than the lowest excise duty chargeable on electricity generated in Finland, but lower than the highest excise duty chargeable on electricity generated in Finland.

Thus, the tax burden on imported electricity was in certain cases higher than that imposed on domestic generated electricity. The purpose of this Finnish law was to promote clean energy by imposing a lower duty on electricity generated, by using environmentally-friendly raw materials or, in an environmentally-friendly manner.[600]

521. The Court of Justice first stated that it has: "*consistently held (see, inter alia, Case C-90/94 Haahr Petroleum v Åbenrå Havn and Others [1997] ECR I-4085, paragraph 19) that provisions relating to charges having equivalent effect and those relating to discriminatory internal taxation cannot be applied together, so that under the system established by the Treaty the same charge cannot belong to both categories at the same time.*

The Court has also consistently held (see, inter alia, Haahr Petroleum, paragraph 20) that any pecuniary charge, whatever its designation and mode of application, which is imposed unilaterally on goods by reason of the fact that they cross a frontier, and which is not a customs duty in the strict sense, constitutes a charge having equivalent effect within the meaning of Articles 9, 12, 13 and 16 of the EC Treaty. However, such a charge may not be so characterised if it forms part of a general system of internal dues applying systematically to categories of products according to objective criteria applied without regard to the origin of the products, in which case it falls within the scope of Article 95 of the Treaty. [...]"[601]

522. Consequently, the Court of Justice ruled that the excise duty of the kind at issue constituted an internal taxation within the meaning of Article 110 of the TFEU, rather than a charge having equivalent effect to a customs duty within the meaning of Article 30 of the TFEU. The Finnish government claimed that the distinction between domestic *versus* imported electricity could not be applied to imported electricity due to a practical reason since once the imported electricity

---

600 *Ibid.*, paragraphs 1–18 and E. D. CROSS, B. DELVAUX, L. HANCHER, P. J. SLOT, G. VAN CALSTER and W. VANDENBERGHE, "EU Energy Law (chapter 5)" in M. M. ROGGENKAMP, A. RØNNE, C. REDGWELL and I. DEL GUAYO (eds.), *Energy Law in Europe. National, EU and International Regulation* (Oxford, Oxford University Press, 2007), p. 324.

601 Case C-213/96, *Outokumpu Oy* [1998] ECR I-1777, paragraphs 19 and 20.

entered the network it was impossible to determine the raw materials used or its method of generation.

523. The Court of Justice rejected this argument by stating that practical difficulties cannot justify the application of internal taxation which discriminated against products from other Member States. It furthermore emphasised that the Finnish legislation at issue does not even stipulate that an importer has the opportunity of demonstrating that the electricity imported by him has been generated by a particular method in order to qualify for the rate applicable to electricity of domestic origin generated by the same method. The Court therefore ruled that Article 110 of the TFEU precludes a duty under a national tax system that essentially imposes a flat rate of tax on imported electricity, but allows lower rates in certain instances for electricity of domestic origin.[602]

524. Thus, this case demonstrated that discriminatory taxation exists if the tax on imported products (such as electricity) is calculated differently on the basis of different criteria than that on similar domestic products. Even, if this leads to a higher taxation only in certain cases. Moreover, in the *Outokumpu Oy* case[603] it was irrelevant that imported electricity was taxed less heavily than domestic electricity. More important was the fact that imported electricity on occasions was taxed more heavily than domestic electricity. Consequently, in order to establish whether discriminatory taxation exists, the tax burden on imported products must be compared with the lowest tax burden on similar domestic products. Imported products cannot be taxed more heavily than domestic products.[604]

525. In the *Air Liquide* case[605], the Court of Justice illustrated the scope and limits of Articles 30 of the TFEU and 110 of the TFEU. This case concerned the adoption by the municipal council of the *ville de Seraing* of a regulation which introduced a tax on motive force. Under this regulation, industrial, commercial, financial and agricultural undertakings established in the territory of that

602 E. D. CROSS, B. DELVAUX, L. HANCHER, P. J. SLOT, G. VAN CALSTER and W. VANDENBERGHE, "EU Energy Law (chapter 5)" in M. M. ROGGENKAMP, A. RØNNE, C. REDGWELL and I. DEL GUAYO (eds.), *Energy Law in Europe. National, EU and International Regulation* (Oxford, Oxford University Press, 2007), p. 324.

603 Case C-213/96, *Outokumpu Oy* [1998] ECR I-1777.

604 T. LYONS, *EC customs law* (Oxford, Oxford University Press, 2008), p. 38 and 39 and see also Case C-152/89, *Commission of the European Communities v Grand Duchy of Luxembourg* [1991] ECR 3141, paragraphs 20–22; Case C-393/98, *Ministério Público and António Gomes Valente v Fazenda Pública* [2001] ECR I-1327 paragraphs 21 and 37 and Case C-228/98, *Charalampos Dounias v Ypourgio Oikonomikon* [2000] ECR I-577, paragraph 41.

605 Joined Cases C-393/04 and C-41/05, *Air Liquide Industries Belgium SA v Ville de Seraing and Province de Liège* [2006] ECR I- 05293.

municipality were required to pay an annual tax on the motors used in the establishments or plants linked to them, whatever the substance or the energy source used to operate them. The tax rate is proportionate to the power of the motor used. In certain situations, this regulation provides for exemptions. Article 3(9) in particular stipulates that motors used in natural gas compression stations in order to drive the compressors which create pressure in the supply pipes are exempt from the tax on motive force. A similar regulation was adopted by the provincial council of Liège.[606]

526. The applications of *Air Liquide* to have the demands to pay the tax on motive force withdrawn were rejected. *Air Liquide* brought an action before the Court of First Instance in Liège to withdraw the said demands and to have the tax paid prior to the 1999 tax year reimbursed. It claimed, in the context of that action, that the tax on motive force was discriminatory, in particular in view of the fact that undertakings transporting natural gas benefited from an exemption, and that this tax was incompatible with the TFEU. In the course of these proceedings, the Court of First Instance in Liège referred some questions to the Court of Justice for a preliminary ruling. Thus, the Court of Justice ruled that "*it is the functioning of these motors and their consumption of energy which constitute the chargeable event, whatever the substance or the energy source used to operate them. Since it is not levied because a frontier is crossed, the said tax is thus not connected with the import or export of goods*".[607]

527. Therefore, the tax at issue on motive force levied in particular on motors used for transporting industrial gas through very high pressure pipes, does not constitute a charge having equivalent effect within the meaning of Article 30 of the TFEU. The Court of Justice furthermore reasoned that the tax on motive force at issue is not specifically imposed on exported or imported products or in such a way as to differentiate between them, given that it applies to economic activities carried out by industrial, commercial, financial or agricultural undertakings and not to products as such.

Although, the tax exemption under the regulations granted to motors used for transporting natural gas could be considered favourable, the Court of Justice ruled that "*the fact that the exemption granted concerns motors used for transporting natural gas is not sufficient to establish that the tax is discriminatory, since, as is clear from the oral argument before the Court, it does not appear that natural gas is a product analogous to industrial gas or a product that competes with it*".[608] The Court of Justice consequently, held that the tax at issue on motive

606 *Ibid.*, paragraphs 6–19.
607 *Ibid.*, paragraph 53.
608 *Ibid.*, paragraphs 55–60.

force did not constitute discriminatory internal taxation for the purposes of Article 110 of the TFEU.[609]

528. In the *Commission v Italy* case[610] the Commission brought an action against the Italian Republic. The Commission claimed that the Italian Republic, by establishing and retaining an environmental tax on gas pipelines containing methane gas laid down by Article 6 of the Sicilian law, has failed to fulfil its obligations under Articles 28, 30, 31 and 207 of the TFEU. As well as Articles 4 and 9 of the Cooperation Agreement between the European Economic Union and the People's Democratic Republic of Algeria signed in Algiers on 26 April 1976.

529. The Court of Justice ruled that the tax in the given case "*on goods imported from a non-member country, in this case the People's Democratic Republic of Algeria, is contrary to both Articles 23 EC and 133 EC and Article 9 of the Cooperation Agreement. The Court further notes that, in so far as the Algerian gas which is taxed under the Sicilian Law is imported into Italy and subsequently exported to other Member States, the disputed tax is likely to affect intra-Community trade, contrary to Article 25 EC.*

[...] regarding the Italian Government's argument to the effect that the Commission's action is unfounded because the disputed tax was introduced with the sole aim of protecting the environment, in the light of, inter alia, the requirements of the precautionary principle, the Court notes that charges having equivalent effect to customs duties are prohibited irrespective of the purpose for which they were introduced and the destination of the revenue from them (see Case C-72/03 Carbonati Apuani [2004] ECR I-8027, paragraph 31)".[611]

530. The Court of Justice consequently ruled that the Sicilian tax was a charge levied on goods imported from a non-member country for distribution and consumption in Italy or in transit to another Member States prohibited by the Articles 28 and 30 of the TFEU and contrary to Article 207 of the TFEU as well as Article 9 of the Cooperation Agreement between the European Economic Community and Algeria.[612]

---

609 E. D. CROSS, B. DELVAUX, L. HANCHER, P. J. SLOT, G. VAN CALSTER and W. VANDENBERGHE, "EU Energy Law (chapter 5)" in M. M. ROGGENKAMP, A. RØNNE, C. REDGWELL and I. DEL GUAYO (eds.), *Energy Law in Europe. National, EU and International Regulation*, 2nd ed. (Oxford, Oxford University Press, 2007), p. 324 and 325.

610 Case C-173/05, *Commission of the European Communities v Italian Republic* [2007] ECR I-4917.

611 *Ibid.*, paragraphs 40–42.

612 L. W. GORMLEY, *EU law of free movement of goods and customs union* (Oxford, Oxford University Press, 2009), p. 388.

4. CONCLUSION

531. In the foregoing, it is argued that a fiscal charge should either be considered under Article 30 of the TFEU as a customs duty or a charge having equivalent effect thereto, or else, under Article 110 of the TFEU, it forms part of a general system of internal taxation. Such a demarcation is important since the consequences of the application of the two Treaty provisions are different. However, such demarcation is not always clearly visible.

532. It can be concluded on the basis of the analysis of case law of the Court of Justice above, that the use, generation, transport, consumption, etc. of energy products in a number of Member States is subject to a tax, customs duties or charges having equivalent effect.

533. Depending on their contents, they are not prohibited by the Articles 28, 30 and 110 of the TFEU. Consequently, in order determine whether such a charge on energy products is compatible with the provisions of Articles 28, 30 and 110 of the TFEU, a case-by-case analysis will be necessary in the light of the existing case law of the Court of Justice. The next part will discuss the non-fiscal barriers to trade.

ii) Non-fiscal barriers to trade

1. MEANING OF ARTICLES 34 AND 35 OF THE TFEU

534. The principal provision in the TFEU designed to eliminate barriers to the free movement of goods that are of a non-fiscal nature, is Article 34 of the TFEU. It is one of the key components in the establishment of the common market in which market forces can operate freely without the restraints of national barriers. In its jurisprudence, the Court of Justice has developed this provision into a broad anti-protectionist character. The vigour of Article 34 of the TFEU lies in its remarkable scope since it covers tangible physical non-fiscal barriers to trade (e.g. Member States which impose formalities at the frontier) and extends to more subtle, non-fiscal barriers to trade (e.g. the presence of a manufacturer in a given Member State in order to be able to market its product(s)).

535. Article 34 of the TFEU could even apply if Member States fail to adopt adequate measures to prevent private individuals creating obstacles to the free movement of goods. Such kind of inactivity of a Member State, where it refrains from adopting the measures required to deal with obstacles to the free movement of goods due to action of private individuals, can be found in the *Commission v*

French Republic case.[613] The Court of Justice in this case held France responsible for the actions of national farmers seeking to restrict the import of agricultural goods from neighbouring Member States by intercepting lorries transporting these goods and/or by destroying their loads. Since Member States are obliged to ensure the free movement of goods in their territory by taking adequate measures for the purposes of preventing any restriction due to the acts of private individuals, the non-intervention of the French authorities against these acts was considered to be an infringement of Article 34 of the TFEU.[614]

536. In some exceptional circumstances, if a Member State bears the heavy burden of proof, it may legitimately maintain some barriers to the free movement of goods. Nonetheless, the scope of unilateral Member State's action that may permissibly restrain trade has been steadily reduced through various harmonisation initiatives of the Union.[615]

537. Article 34 of the TFEU states that *"[q]uantitative restrictions on imports and all measures having equivalent effect shall be prohibited between Member States"*. Under Article 35 of the TFEU a similar worded prohibition applies to quantitative restrictions on exports. A distinct difference between Article 34 of the TFEU and Article 35 of the TFEU is that the latter basically only applies to measures which discriminate against goods.

538. This principle can be found in the *Groenveld* case[616] where the Court of Justice held that Article 35 of the TFEU *"concerns national measures which have as their specific object or effect the restriction of patterns of exports and thereby the establishment of a difference in treatment between domestic trade of a Member State and its export trade in such a way as to provide a particular advantage for national production or for the domestic market of the state in question at the expense of the production or of the trade of other Member States"*.[617] This narrow interpretation of Article 35 of the TFEU can be clarified by several reasons. One is that, if the scope of Article 35 of the TFEU is too broad, it would encompass restrictions which have no bearing on intra-Union trade.[618] In some cases[619] the

---

613 Case C-265/95, *Commission of the European Communities v French Republic* [1997] ECR I-06959.

614 *Ibid.*, paragraph 31 and Commission Staff Working Document, Free Movement of Goods – Guide to the application of Treaty provisions governing Free Movement of Goods (Articles 28–30 EC) of 12 May 2009, SEC(2009) 673, p. 11 and 12.

615 S. WEATHERILL and P. BEAUMONT, *EU law –The essential guide to legal workings of the European Union* (London, Penguin Books, 1993), p. 500–501.

616 Case 15/79, *P.B. Groenveld BV v Produktschap voor Vee en Vlees* [1979] ECR 3409.

617 *Ibid.*, paragraph 7.

618 European Commission, *Guide to the concept and practical application of Articles 28–30 EC*, January 2001, p. 35 and 36.

619 Case 155/80, *Sergius Oebel* [1981] ECR 1993.

Court of Justice omitted the last requirement of the *Groenveld* principle *"at the expense of the production or of the trade of other Member States"* and in other cases[620] it did not refer to the requirement *"for providing particular advantage for national production"*.

539. Nonetheless, the general approach followed by the Court of Justice seems to be that Article 35 of the TFEU catches trade barriers which create a difference in treatment between trade within a Member State and exports, and have an actual and specific effect on exports.[621]

540. In the *Meroni* case[622] the Court of Justice held that the prohibition laid down in the Article 34 of the TFEU is mandatory and explicit and its implementation does not require any subsequent intervention of the Member States or Union institutions. The prohibition of Article 34 of the TFEU has therefore direct effect and creates individual rights which national courts must protect.[623] The Court also recognised that the provisions of Article 35 of the TFEU are directly applicable and that they confer on individuals rights which the courts of Member States must protect.[624] This means that individuals can invoke these provisions on the free movement of goods before a national court. If a national court considers that a measure is contrary to Articles 34 and 35 of the TFEU, it may refuse to apply these measures.[625]

541. The typical example of a quantitative restriction or quota system involves the situation where state A imposes a numerical limit on the imports of a certain product of state B in order to protect the domestic production of the product in question. Such kind of trade restrictions is, compared to fiscal barriers as discussed above, more prejudicial to free trade because they impose, rather than simply rendering the import of a certain product more expensive, an absolute ban on trade beyond the stipulated ceiling of imports. The Court of Justice emphasises in its jurisprudence that both a partial restraint and an absolute ban on import constitute quantitative restrictions caught by Article 34 of the TFEU.[626] Even a hidden or covert quota system may be considered a quantitative restriction.[627]

620 Case C-388/95, *Kingdom of Belgium v Kingdom of Spain* [2000] ECR 3123.
621 European Commission, *Guide to the concept and practical application of Articles 28–30 EC*, January 2001, p. 36.
622 Case 74/76, *Iannelli & Volpi SpA v Ditta Paolo Meroni* [1977] ECR 557.
623 *Ibid.*, paragraph 13.
624 Case 83/78, *Pigs Marketing Board v Raymond Redmond* [1978] ECR 2374, paragraph 66.
625 European Commission, *Guide to the concept and practical application of Articles 28–30 EC*, January 2001, p. 55.
626 Case 2/73, *Riseria Luigi Geddo v Ente Nazionale Risi* [1973] ECR 865, paragraph 7.
627 European Commission, *Guide to the concept and practical application of Articles 28–30 EC*, January 2001, p. 9.

(a) Dassonville formula

542. The evolution of Article 34 of the TFEU is primarily attributable to the case law of the Court of Justice which shaped this provision into a very useful and flexible instrument against national measures that create barriers to the free movement of goods.[628] Since the adoption of the *Procureur du Roi v Dassonville* case[629] the scope and meaning of Article 34 of the TFEU have been defined by stating that *"all trading rules enacted by Member States which are capable of hindering, directly or indirectly, actually or potentially, intra-Community trade are to be considered as measures having an effect equivalent to quantitative restrictions"*.[630] This definition has been cited continuously in almost unchanged terms. Only in some judgments of the Court of Justice the words "all trading rules" have been replaced by the words "all rules" or "all measures".[631]

543. The striking feature in this landmark case is that the Court of Justice in its interpretation of Article 34 of the TFEU does not concentrate on the form of the national practice, but on the effect of that practice. The precise legal form of a practice is not of direct relevance. Thus, two aspects are important *"the emphasis on the effect of a measure having, or importantly, merely potentially having, effect on trade and the fact that there is no need for discriminatory intent"*.[632] The formulation of the Court of Justice, the so-called *Dassonville* formula, makes clear that one must look to the effects of a measure rather than to its aim in order to decide whether the measure falls under Article 34 of the TFEU.[633] As a side note, the Court of Justice, by making the scope of Article 34 of the TFEU very broad, also grants itself equally broad powers to supervise national measures.[634]

---

628 S. WEATHERILL and P. BEAUMONT, *EU law -The essential guide to legal workings of the European Union* (London, Penguin Books, 1993), p. 500–501.

629 Case 8–74, *Procureur du Roi v Benoît and Gustave Dassonville* [1974] ECR 837.

630 *Ibid.*, paragraph 5.

631 D. CHALMERS, G. DAVIES and G. MONTI, *European Union Law* (Cambridge, Cambridge University press 2010), p. 748 which refers to the Case C-88/07, *Commission of the European Communities v Kingdom of Spain* [2009] ECR I-1353; Case C-319/05, *Commission of the European Communities v* Federal Republic of Germany [2007] ECR I-9811 and Joined Cases C-158/04 and C-159/04, *Alfa Vita Vassilopoulos AE, formerly Trofo Super-Markets AE and Carrefour Marinopoulos AE v Elliniko Dimosio, Nomarchiaki Aftodioikisi Ioanninon* [2006] ECR I-8135. See also L. W. GORMLEY, "The Definition of Measures having Equivalent Effect" in A. ARNULL, P. EECKHOUT and T. TRIDIMAS, *Continuity and Change in EU Law Essays in Honour of Sir Francis Jacobs* (Oxford, Oxford University Press, 2008), p. 189–204 and P. OLIVIER and M. JARVIS, *Free Movement of Goods in the European Community* (London, Sweet & Maxwell, 2003), p. 91–214.

632 European Commission, *Guide to the concept and practical application of Articles 28–30 EC*, January 2001, p. 9.

633 S. WEATHERILL and P. BEAUMONT, *EU law -The essential guide to legal workings of the European Union* (London, Penguin Books, 1993), p. 500–501.

634 D. CHALMERS, G. DAVIES and G. MONTI, *European Union Law* (Cambridge, Cambridge University press 2010), p. 748.

544. In the *Cassis de Dijon* case[635] the Court of Justice confirms that Article 34 of the TFEU could also catch national measures which apply equally to domestic and imported goods. Consequently, there is no need for any discriminatory element in order to be caught under Article 34 of the TFEU. On the basis of this ruling of the Court it can be concluded that Article 34 of the TFEU does not only apply to national measures which discriminate against imported goods, but also to those which in law seem to apply equally to both domestic and imported goods, however in practice they are more burdensome for imported ones. Such kind of particular burden stems from the fact that the imported goods are in fact required to comply with two sets of rules (one laid down by the importing Member State and the other by the exporting one).[636]

545. Since the adoption of this ruling of the Court of Justice, it has become a general rule that products and services are primarily subject to the laws of their origin and not to the laws or further requirements of their destination unless mandatory requirements apply.[637] This approach by the Court of Justice mediates between different laws and has been described as the principle of tolerance. It enables states to conduct trade while allowing Member States to maintain their national laws and avoiding harmonisation measures.[638] However, this approach of mutual recognition has been criticised as it tends to sacrifice non-economic concerns (e.g. regulatory race to the bottom is feared), but also from a free trade perspective since it appears to be ineffective in practice. Thus, mutual recognition has not turned out to be the legal panacea. On the contrary, it seems to leave many obstacles to movement in place.[639]

---

635 Case 120/78, *Rewe-Zentral AG v Bundesmonopolverwaltung für Branntwein* [1979] ECR 649.

636 European Commission, *Guide to the concept and practical application of Articles 28–30 EC*, January 2001, p. 13 and 14 and Case 120/78, *Rewe-Zentral AG v Bundesmonopolverwaltung für Branntwein* [1979] ECR 649, paragraphs 14 and 15.

637 Case C-288/89, *Stichting Collectieve Antennevoorziening Gouda and others v Commissariaat voor de Media* [1991] ECR I-4007.

638 C. JOERGES and J. NEYER, *Deliberative Supranationalism Revisted*, EUI Working Paper No. 2006/20 (Florence, European University Institute, 2006) and K. NICOLAIDIS and G. SHAFFER, "Managed Mutual Recognition Regimes: Governance Without Global Government", *Law and Contemporary Problems* (2005), p. 317. For a further discussion see also S. K. SCHMIDT (ed.), "Mutual Recognition as a New Mode of Governance", *Journal of European Public Policy* (2007), p. 667–825.

639 D. CHALMERS, G. DAVIES and G. MONTI, *European Union Law* (Cambridge, Cambridge University press 2010), p. 764–766; K. ALTER and S. MEUNIER-AITSAHALIA, "Judicial Politics in the European Community: European Integration and the Pathbreaking Cassis de Dijon Decision, *Comparative Political Studies* (1994), p. 535; Commission Report to the Council, Parliament and ESC of 23 July 2002, *Second Biennial Report on the Application of the Principle of Mutual Recognition in the Single Market*, COM(1992)419 and J. PELKMANS, "Mutual Recognition in Goods: On Promises and Disillusions", *Journal of European Public Policy* (2007), p. 699–716.

546. This way of interpreting Article 34 of the TFEU and in particular the recourse to the flexible notion of measures having equivalent effect, has proven to be a powerful means in order to challenge protectionist practices of Member States.[640] Consequently, a national requirement for a prior import authorisation even as a pure formality will be caught by Article 34 of the TFEU.[641]

547. With respect to measures which have no specific cross-border impact, but have an impact on trade or economic activity, the Court of Justice generally seems to be reluctant to extend Article 34 of the TFEU. Nonetheless, some recent rulings of the Court of Justice appear to put this approach in question and apply Article 34 of the TFEU if access to a market is hindered.[642] If this approach would lead to the situation in which Article 34 of the TFEU will be applied to any measures that reduce sales, even without unequal effect or discrimination, its application would become very broad. It would require new limits to the notion of a restriction to movement for example by requiring that only direct obstacles to movement or only significant or substantial obstacles to movement are prohibited. Such concepts have not yet been developed by the Court of Justice. It is argued in some legal doctrine that such an extension of Article 34 of the TFEU could potentially raise constitutional problems.[643]

548. Thus, it is uncertain whether such a true "market access" approach with respect to Article 34 of the TFEU will ever, or should ever, be applied. It remains to be seen what the Court of Justice will decide in this context. Nonetheless, it should be emphasised that the Court's ruling in the *Commission v Italy* case[644] was a Grand Chamber Judgment which perhaps contains the underlying

---

640 S. WEATHERILL and P. BEAUMONT, *EU law –The essential guide to legal workings of the European Union* (London, Penguin Books, 1993), p. 503–505.

641 *Ibid.*

642 Joined Cases C-158/04 and C-159/04, *Alfa Vita Vassilopoulos AE, formerly Trofo Super-Markets AE and Carrefour Marinopoulos AE v Elliniko Dimosio, Nomarchiaki Aftodioikisi Ioanninon* [2006] ECR I-8135; Case C-142/05, *Åklagaren v Percy Mickelsson and Roos* [2008] ECR I-7841, Case C-265/06, *Commission of the European Communities v Portuguese Republic* [2008] ECR I-2245 and Case C-110/05, *Commission of the European Communities v Italian Republic* [2009] ECR I-519.

643 D. CHALMERS, G. DAVIES and G. MONTI, *European Union Law* (Cambridge, Cambridge University press 2010), p. 752 which refers to G. DAVIES, *Nationality Discrimination in the European Internal Market* (The Hague, Kluwer Law International, 2003), p. 93–115 and J. SNELL, *Goods and Services in EC Law A Study of the Relationship Between the Freedoms* (Oxford, Oxford University Press, 2002), 251 p. See also G. STRAETMANS, "The Outer Limits of Free Movement of Goods" in M. BULTERMAN, L. HANCHER, A. MCDONNELL and H. SEVENSTER (eds.), *Views of European Law from the Mountain – Liber Amicorum Piet Jan Slot* (The Netherlands, Kluwer Law International, 2009), p. 91–105.

644 Case C-110/05, *Commission of the European Communities v Italian Republic* [2009] ECR I-519.

message that Article 34 of the TFEU will undergo another expansionary phase the coming years.[645]

(b) Keck formula

549. Beside the *Dassonville* formula, reference to the *Keck* formula which originates from the *Keck* case[646] must also be made.[647] In this case, the Court of Justice made a significant distinction as to which kind of measures must be deemed to be a hindrance to trade and made some limitations to the scope of the term *"measures having equivalent effect"*. The Court of Justice stated that: *"in view of the increasing tendency of traders to invoke Article 30 of the Treaty as a means of challenging any rules whose effect is to limit their commercial freedom even where such rules are not aimed at products from other Member States, the Court considers it necessary to re-examine and clarify its case-law on this matter"*.[648] This means that the origin and intention of the Court of Justice of re-examining the case law seem *inter alia* necessary in order to limit the flow of cases which aim at challenging key pillars of national welfare and social provisions which were never intended to interfere with the free movement provisions of the Union.[649]

550. The Court of Justice held that: *"contrary to what has previously been decided, the application to products from other Member States of national provisions restricting or prohibiting certain selling arrangements is not such as to hinder directly or indirectly, actually or potentially, trade between Member States within the meaning of the Dassonville judgment (Case 8/74 [1974] ECR 837), so long as those provisions apply to all relevant traders operating within the national territory and so long as they affect in the same manner, in law and in fact, the marketing of domestic products and of those from other Member States. [...]*

---

645 D. CHALMERS, G. DAVIES and G. MONTI, *European Union Law* (Cambridge, Cambridge University press 2010), p. 752.

646 Joined Cases C-267/91 and C-268/91, *Criminal proceedings against Bernard Keck and Daniel Mithouard* [1993] ECR I 06097.

647 The reasoning of the *Keck* formula seems to be present in preceding cases such as Case 75/81, *Joseph Henri Thomas Blesgen v Belgian State* [1982] ECR 1211; Case 148/85, *Direction générale des impôts et procureur de la République v Marie-Louise Forest, née Sangoy, and Minoterie Forest SA* [1986] ECR 3349 and Case C-23/89, *Quietlynn Limited and Brian James Richards v Southend Borough Council* [1990] ECR I-3059. An outline on case law of Article 34 of the TFEU (ex article 28 of the EC Treaty) before the *Keck* judgment can be found in the opinion of Advocate General Jacobs in the Case C-412/93, *Société d'Importation Edouard Leclerc-Siplec v TF1 Publicité SA and M6 Publicité SA*, paragraphs 23–33.

648 Joined Cases C-267/91 and C-268/91, *Criminal proceedings against Bernard Keck and Daniel Mithouard* [1993] ECR I 06097, paragraph 14.

649 European Commission, *Guide to the concept and practical application of Articles 28–30 EC*, January 2001, p. 14 and C. BARNARD, "Filling the Remaining Pieces into the Goods and Persons Jigsam", *European Law Review* (2001), p. 50.

Provided that those conditions are fulfilled, the application of such rules to the sale of products from another Member State meeting the requirements laid down by that State is not by nature such as to prevent their access to the market or to impede access any more than it impedes the access of domestic products. Such rules therefore fall outside the scope of Article 30 of the Treaty".[650]

551. This judgment means that national measures which lay down requirements to be met by goods continue to be treated under the *Cassis de Dijon* case[651] and therefore fall *per se* within the scope of Article 34 of the TFEU. However, if it concerns selling arrangements, they will only fall within the scope of Article 34 of the TFEU on the condition that the party which invokes a violation can prove that these selling arrangements introduce a discrimination on the basis of the origin of products, either in law or in fact. When measures are manifestly discriminatory, discrimination occurs in law.

552. It is more complex to establish discrimination in fact. Besides, what types of measures are concerned with the characteristics of the products is relatively easier to comprehend than what types of measures constitute selling arrangements. The Court of Justice includes in the list of selling arrangements measures relating to the conditions and methods of marketing[652], measures which relate to the place of the sale of goods or restrictions regarding by whom goods may be sold[653], measures which relate to the time of sale of goods[654] and measures which relate to price controls.[655]

2. Meaning of Article 36 of the TFEU

553. *Prima facie* breaches of Articles 34 and 35 of the TFEU can be justified under Article 36 of the TFEU which states that *"[t]he provisions of Articles 34*

---

650 Joined Cases C-267/91 and C-268/91, *Criminal proceedings against Bernard Keck and Daniel Mithouard* [1993] ECR I 06097, paragraphs 16 and 17.

651 Case 120/78, *Rewe-Zentral AG v Bundesmonopolverwaltung für Branntwein* [1979] ECR 649.

652 Case C-412/93, *Société d'Importation Edouard Leclerc-Siplec v TF1 Publicité SA and M6 Publicité SA* [1995] ECR I-179, paragraph 22.

653 Case C-391/92, *Commission of the European Communities v Hellenic Republic* [1995] ECR I-1621 paragraph 15.

654 Joined Cases C-401/92 and 402/92, *Criminal proceedings against Tankstation 't Heukske vof and J. B. E. Boermans* [1994] ECR I-2199 paragraph 14 and Joined cases C-418/93, C-419/93, C-420/93, C-421/93, C-460/93, C-461/93, C-462/93, C-464/93, C-9/94, C-10/94, C-11/94, C-14/94, C-15/94, C-23/94, C-24/94 and C-332/94, *Semeraro Casa Uno Srl v Sindaco del Comune di Erbusco (C-418/93), Semeraro Mobili SpA v Sindaco del Comune di Erbusco (C-419/93), RB Arredamento Srl v Sindaco del Comune di Stezzano (C-420/93), Città Convenienza Milano Srl v Sindaco del Comune di Trezzano sul Naviglio (C-421/93), Città Convenienza Bergamo Srl v Sindaco del Comune di Stezzano (C-460/93), Centro Italiano Mobili Srl v Sindaco del Comune di Pineto (C-461/93) [...]* [1996] ECR I-2975.

655 European Commission, *Guide to the concept and practical application of Articles 28–30 EC*, January 2001, p. 14–16.

and 35 shall not preclude prohibitions or restrictions on imports, exports or goods in transit justified on grounds of public morality, public policy or public security; the protection of health and life of humans, animals or plants; the protection of national treasures possessing artistic, historic or archaeological value; or the protection of industrial and commercial property. Such prohibitions or restrictions shall not, however, constitute a means of arbitrary discrimination or a disguised restriction on trade between Member States".

554. This provision offers a list of potential justifications, but also issues a firm warning that any purported justification will be scrutinised carefully in order to make sure that the justification is genuine. The Court of Justice made clear in its case law that it was unwilling to extend the scope of available justifications of Article 36 of the TFEU.[656] In the *Campus Oil*[657] case the Court declared that *"the purpose of Article 36 of the Treaty [*read Article 36 of the TFEU*] is not to reserve certain matters to the exclusive jurisdiction of the Member States; it merely allows national legislation to derogate from the principle of the free movement of goods to the extent to which this is and remains justified in order to achieve the objectives set out in the Article".*[658]

3. Mandatory requirements

555. Beside these derogations on the basis of Article 36 of the TFEU, the case law of the Court of Justice provides for the so-called mandatory requirements (e.g. environmental protection and consumer protection) on which Member States could rely in order to justify their national measures which impede cross-border trade. Whereas the derogations of Article 36 of the TFEU are exhaustive, the concept of mandatory requirements is an open one. However, it is generally

656 Case 113/80, *Commission of the European Communities v Ireland* [1981] ECR 1625, paragraph 8; Case C-265/95, *Commission of the European Communities v French Republic* [1997] ECR I 6959 and Case C-120/95, *Nicolas Decker v Caisse de maladie des employés privés* [1998] ECR I 1831.

657 Case 72/83, *Campus Oil Limited and others v Minister for Industry and Energy and others* [1984] ECR 2727, paragraph 32.

658 For a further analysis of this provision see C. BARNARD, *The substantive Law of the EU The Four Freedoms* (Oxford, Oxford University Press, 2007), p. 64–90; A. ARNULL, A. DASHWOOD, M. DOUGAN, M. ROSS, E. SPAVENTA and D. WYATT, *Wyatt and Dashwoods's European Union Law* (London, Sweet & Maxwell, 2006) p. 616–630; S. WEATHERILL and P. BEAUMONT, *EU law –The essential guide to legal workings of the European Union* (London, Penguin Books, 1993), p. 525–564 and C. BARNARD, "Derogations, Justifications and the Four Freedoms: Is State Interest Really Protected?" in C. BARNARD and O. ODUDU (eds.), *The Outer Limits of European Union Law* (Oxford, Hart Publishing, 2009), p. 273–305.

contended that the Court of Justice will not accept restrictions to the free movement provisions which relate to non-economic interests.[659]

556. The concept of mandatory requirements as a non-exhaustive list of protected interests has been laid down by the Court of Justice in its *Cassis de Dijon* case.[660] In this case, the Court of Justice held that the mandatory requirements relate *"in particular to the effectiveness of fiscal supervision, the protection of public health, the fairness of commercial transactions and the defence of the consumer"*.[661] Until recently, it was well-established case law that these mandatory requirements could not be relied on to justify national rules which were not applicable to domestic and imported products without distinction.[662] Consequently, from a theoretically point of view, discriminatory measures cannot be justified on grounds others than those covered by Article 36 of the TFEU.[663]

557. This approach of the Court of Justice on mandatory requirements has been criticised in some legal doctrine in which it is argued that such separation between Article 36 of the TFEU and the mandatory requirements is artificial. Nonetheless, the Court of Justice seems to have found ways to overcome this separation without renouncing its earlier practice.[664] Such an approach has been applied in the *Walloon Waste* case[665] which concerned a measure that prohibited the storage, tipping, or dumping in Wallonia of waste originating in another Member State or in a region of Belgium other than Wallonia.

558. With respect to the issue of whether the measure could be justified by imperative requirements of environmental protection, the Court of Justice

659 L. HANCHER and S. JANSSEN, "Shared Competences and Multi-Faceted Concepts – European Legal Framework for Security of Supply", in B. BARTON, C. REDGWELL, A. RØNNE and D. ZILLMAN (eds.), *Energy Security – Managing Risk in a Dynamic Legal and Regulatory Environment* (Oxford, Oxford University press, 2004), p. 108.

660 Case 120/78, *Rewe-Zentral AG v Bundesmonopolverwaltung für Branntwein* [1979] ECR 649.

661 *Ibid.*, paragraph 8.

662 Case 113/80, *Commission of the European Communities v Ireland* [1981] ECR 1625, paragraph 11; Case 207/83, *Commission of the European Communities v United Kingdom of Great Britain and Northern Ireland* [1985] ECR 1201, paragraph 22 and Joined Cases C-1/90 and C-176/90, *Aragonesa de Publicidad Exterior SA and Publivía SAE v Departamento de Sanidad y Seguridad Social de la Generalitat de Cataluña* [1991] ECR I-4151, paragraph 13.

663 Commission staff working document, *Free Movement of Goods – Guide to the application of Treaty provisions governing Free Movement of Goods* (Articles 28–30 EC) of 12 May 2009, SEC(2009) 673, p. 42.

664 Another example in the jurisprudence of the Court of Justice beside the *Walloon Waste* case is the case C-320/03, *Commission of the European Communities v Republic of Austria* [2005] ECR I-9871.

665 Case C-2/90, *Commission of the European Communities v Kingdom of Belgium* [1992] ECR 4431.

argued that it was true that imperative requirements could be taken into account only in the case of measures which applied without distinction to both domestic and imported products.[666] However, in assessing whether or not the barrier in question is discriminatory, account had to be taken of the particular nature of waste, of the principle under Article 191(2) of the TFEU that environmental damage should be remedied at source, and of the principles of self-sufficiency and proximity set out in the Basel Convention of 22 March 1989 on the control of transboundary movements of hazardous wastes and their disposal. Having regard to the differences between waste produced in different places and to the connection of the waste with its place of production, the Court of Justice concluded in this case that the contested measures cannot be regarded as discriminatory.[667]

559. Moreover, in some legal doctrine it is argued that there are indications that in its case law, the Court of Justice is reconsidering its earlier case law and treating mandatory requirements the same way as the Article 36 of the TFEU justifications.[668] This approach can be found in cases such as the *Dusseldorp* case[669] and the *Ahner-Waggon* case[670] where the Court relied on imperative requirements in which it was at least doubtful whether the measure could be considered as applicable without distinction.[671]

4. Proportionality test

560. As emphasised above, national measures restricting the free movement of goods provisions of the TFEU have been justified narrowly by the Court of Justice on the basis of Article 36 of the TFEU or the so-called mandatory

---

666 Opinion Advocate General Jacobs of 26 October 2000, Case C-379/98 *PreussenElektra AG v Schhleswag AG, in the presence of Windpark Reußenköge III GmbH and Land Schleswig-Holstein*, p. 28.

667 Case C-2/90, *Commission of the European Communities v Kingdom of Belgium* [1992] ECR 4431, paragraphs 34–36.

668 Commission staff working document, *Free Movement of Goods – Guide to the application of Treaty provisions governing Free Movement of Goods* (Articles 28–30 EC) of 12 May 2009, SEC(2009) 673, p. 42 which refers to P. OLIVIER and M. JARVIS, *Free Movement of Goods in the European Community* (London, Sweet & Maxwell, 2003), p. 234–275.

669 Case C-203/96, *Chemische Afvalstoffen Dusseldorp BV and Others v Minister van Volkshuisvesting, Ruimtelijke Ordening en Milieubeheer* [1998] ECR 4075, paragraphs 44 and 49.

670 Case C-389/96, *Aher-Waggon GmbH v Bundesrepublik Deutschland* [1998] ECR 4473, paragraph 19.

671 Opinion Advocate General Jacobs of 26 October 2000, Case C-379/98 *PreussenElektra AG v Schhleswag AG, in the presence of Windpark Reußenköge III GmbH and Land Schleswig-Holstein*, p. 28 and 29 and see *Supra*.

requirements.[672] Moreover, these measures must also respect the principle of proportionality.

561. In order to comply with the proportionality test three criteria must be fulfilled: (a) the measure must be suitable to protect the interests requiring protection (the so-called suitability test); (b) the measure must be necessary for attaining the objective(s) pursued (the so-called necessity test) and (c) the intended or actual result of the restriction caused by the measure on intra-Union trade must not be disproportionate (the so-called proportionality test *sensu stricto*).[673] In order to give the impression that only under exceptional circumstances, restrictions to the free movement provisions of the TFEU can be allowed, the Court of Justice has been very cautious in applying the proportionality test. It primarily focuses in its case law on the suitability and necessity tests of which the latter is the most important. The suitability test in practice seems to cause rarely any problems since it only requires a causal link between the measure in question and its intended objective or result.[674]

562. In several cases, such as the *Commission v Germany* case,[675] the Court of Justice described the proportionality test in the following manner: "*it must be ascertained not only whether the means which they employ are suitable for the purpose of attaining the desired objectives but also whether those means do not go beyond what is necessary for that purpose (see Case C-284/95 Safety Hi-Tech [1998] ECR I-4301, paragraph 57)*".[676] The application of this proportionality test

---

672 P. OLIVIER and M. JARVIS, *Free Movement of Goods in the European Community* (London, Sweet & Maxwell, 2003), p. 215–384.

673 J. SCHWARZE J, *European Administrative Law* (London, Sweet and Maxwell, 2006), p. 708; C. BARNARD, *The substantive Law of the EU The Four Freedoms* (Oxford, Oxford University Press, 2007), p. 64–90; D. CHALMERS, G. DAVIES and G. MONTI, *European Union Law* (Cambridge, Cambridge University press 2010), p. 766–773 and *Supra*.

674 H. BJØRNEBYE, *Investing in EU energy security – Exploring the regulatory approach to tomorrow's electricity production*, Dissertation submitted to the faculty of law, University of Oslo, 2009, p. 86. See also H. BJØRNEBYE, *Investing in EU Energy Security. Exploring the regulatory approach to tomorrow's electricity production* (The Netherlands, Kluwer Law International, 2010), 456 p. This author furthermore stated that: "*[t]he insignificance of the suitability test can also be explained by the fact that it will, in most cases, be subsumed by the more far-reaching necessity test. If a Member State is unable to demonstrate any causal relationship between a national measure and the objective that the measure is contended to pursue, it is hardly likely to be possible to argue that the measure in question is necessary for the promotion of the stated objective.*"

675 Case C-463/01, *Commission of the European Communities v Federal Republic of Germany* [2004] ECR I-11705.

676 *Ibid.*, paragraph 78; similar descriptions can be found in Case C-309/02, *Radlberger Getränkegesellschaft mbH & Co. and S. Spitz KG v Land Baden-Württemberg* [2004] ECR I-11763, paragraph 79 and Joined cases C-158/04 and C-159/04, *Alfa Vita Vassilopoulos AE, formerly Trofo Super-Markets AE and Carrefour Marinopoulos AE v Elliniko Dimosio, Nomarchiaki Aftodioikisi Ioanninon* [2006] ECR I-8135, paragraph 22. H. BJØRNEBYE, *Investing in EU energy security – Exploring the regulatory approach to tomorrow's electricity*

by the Court of Justice often results in the conclusion that the effect on Union trade of national measures is more restrictive than strictly necessary for the pursued objectives. Nonetheless, the proportionality test will always need to be carried out in the light of the particular merits of the case at hand.[677]

563. Before discussing more in detail the energy case law of the Court of Justice in the light of Article 36 of the TFEU, it should be noted that even if a measure is justifiable under one of the derogations of Article 36 of the TFEU it must not, according to the second part of Article 36 of the TFEU, *"[...] constitute a means of arbitrary discrimination or a disguised restriction on trade between Member States"*. This second part is designed to avoid abuse on the part of the Member States.

564. The Court of Justice has emphasised in its case law that the function of the second sentence is to: *"prevent restrictions on trade based on the grounds mentioned in the first sentence of [Article 36 of the TFEU] from being diverted from their proper purpose and used in such a way as either to create discrimination in respect of goods originating in other Member States or indirectly to protect certain national products"*.[678]

5. Non-fiscal barriers and energy

(a) Energy is a good in the sense of the Treaty provisions

565. The analysis above has discussed in detail the meaning of the free movement of goods provisions of the TFEU that are of a non-fiscal nature. Whether these provisions of Articles 34–36 of the TFEU can be applied to the forms of energy such as electricity, gas, oil, etc. depends on the question whether these forms can be qualified as a good in the sense of these Treaty provisions. In

---

production, Dissertation submitted to the faculty of law, University of Oslo, 2009, p. 85 and 86. See also H. BJØRNEBYE, *Investing in EU Energy Security. Exploring the regulatory approach to tomorrow's electricity production* (The Netherlands, Kluwer Law International, 2010), 456 p.

677 H. BJØRNEBYE, *Investing in EU energy security – Exploring the regulatory approach to tomorrow's electricity production*, Dissertation submitted to the faculty of law, University of Oslo, 2009, p. 86 and 87. See also H. BJØRNEBYE, *Investing in EU Energy Security. Exploring the regulatory approach to tomorrow's electricity production* (The Netherlands, Kluwer Law International, 2010), 456 p.

678 Commission staff working document, *Free Movement of Goods – Guide to the application of Treaty provisions governing Free Movement of Goods* (Articles 28–30 EC) of 12 May 2009, SEC(2009) 673, p. 37 and Case 34/79, *Regina v Maurice Donald Henn and John Frederick Ernest Darby* [1979] ECR 3795, paragraph 21 and Joined Cases C-1/90 and C-176/90, *Aragonesa de Publicidad Exterior SA and Publivía SAE v Departamento de Sanidad y Seguridad Social de la Generalitat de Cataluña* [1991] ECR I-4151, paragraph 20.

the *Costa v ENEL* case[679] the Court of Justice accepted that electricity may fall within the scope of Article 36 of the TFEU.[680] Thus, the Court implicitly decided that electricity is a good. In the *Almelo* case[681] the Court of Justice explicitly stated that in Union and national laws, it is accepted that electricity constitutes a good within the meaning of Article 34 of the TFEU. Under the Union's nomenclature (code CN 27.16) electricity is regarded as a good.[682]

566. Not only electricity has been regarded as a good within the meaning of the TFEU, but also natural gas and other petroleum products. In the *Campus Oil* case[683] and the *Commission v Greece* case[684] the Court of Justice applied the Articles 34–37 of the TFEU to measures concerning oil.

(b) The application of Article 36 of the TFEU in the energy field

567. The foregoing clarified that electricity, gas and oil constitute goods for the purposes of Title I and II in Part III of the TFEU and thus, also for Articles 34–37 of the TFEU which are part of those Titles. According to these provisions and the case law of the Court of Justice, any existing quantitative restriction as well as measures having equivalent effect on trade between Member States, must be prohibited.

568. In a number of cases relating to energy aspects, the derogations on the free movement of goods set forth in Article 36 of the TFEU have been discussed. This provision lists the defences that a Member State could use in order to justify its national measure which impedes cross-border trade. The Court of Justice emphasised consistently in its case law that this list of derogations in Article 36 of the TFEU must be interpreted narrowly.[685]

569. The first relevant energy case in the light of Article 36 of the TFEU is the *Campus Oil* case[686] in which the Court of Justice allowed restrictions to the free movement provisions of the Treaty. In this case, the Court of Justice for the first

---

679 Case 6/64, *Flaminio Costa v E.N.E.L.* [1964] ECR 1251.

680 *Ibid.*, paragraph 11.

681 Case C-393/92, *Municipality of Almelo and others v NV Energiebedrijf Ijsselmij* [1994] ECR I 1477.

682 *Ibid.*, paragraph 28 and see *Supra*.

683 Case 72/83, *Campus Oil Limited and others v Minister for Industry and Energy and others* [1984] ECR 2727, paragraphs 12–20.

684 Case C-347/88, *Commission of the European Communities v Hellenic Republic* [1984] ECR I 4747, paragraph 50.

685 Case 46/76, *W. J. G. Bauhuis v The Netherlands State* [1977] ECR 5; Case 72/83, *Campus Oil Limited and others v Minister for Industry and Energy and others* [1984] ECR 2727 and Case C-120/95, *Nicolas Decker v Caisse de maladie des employés privés* [1998] ECR I-1831.

686 Case 72/83, *Campus Oil Limited and others v Minister for Industry and Energy and others* [1984] ECR 2727.

time provided a concrete example in which it balanced the exceptions of Article 36 of the TFEU and issues of an economic nature. Moreover, this judgment was one of the first decisions that indicated to which extent a Member State could rely on the concept of public security to justify its restrictions on trade. Thus, the ruling of the Court of Justice in the *Campus Oil* case was not only relevant for lawyers in the energy law field.[687]

570. The *Campus Oil* case[688] came before the Court of Justice in 1983 following a request of the High Court of Ireland for a preliminary ruling. This case concerned a 1982 order of the Irish minister for industry and energy according to which importers of any of the various petroleum products had to purchase a proportion of their requirements (up to a maximum of 35% of the importer total requirements of petroleum products and 40% of its requirements of each type of petroleum product) at fixed prices through the Irish National Petroleum Corporation. *Campus Oil* was one of the trading companies in petroleum products that was established in Ireland and affected by the 1982 order. The purchase obligation is intended to guarantee the continued activity of the refiners and to ensure a minimum supply of petroleum products at all times. Six Irish undertakings trading in petroleum products disputed this purchase obligation before the Irish Court. The latter decided to request the Court of Justice whether this purchase obligation constitutes a restriction on imports and consequently a breach of Article 34 of the TFEU.

571. The Court of Justice stated that *"the obligation placed on all importers to purchase a certain proportion of their supplies of a given product from a national supplier limits to that extent the possibility of importing the same product, it thus has a protective effect by favouring national production and, by the same token, works to the detriment of producers in other Member States, regardless of whether or not the raw materials used in the national production in question must themselves be imported. [...]*

687 H. BJØRNEBYE, *Investing in EU energy security – Exploring the regulatory approach to tomorrow's electricity production*, Dissertation submitted to the faculty of law, University of Oslo, 2009, p. 66. See also H. BJØRNEBYE, *Investing in EU Energy Security. Exploring the regulatory approach to tomorrow's electricity production* (The Netherlands, Kluwer Law International, 2010), 456 p. See also K. MORTELMANS, "Case note to Campus Oil", *Common Market Law Review* (1984), p. 687–713; P. OLIVER, "A review of the Case law of the Court of Justice on Articles 30 to 36 EEC in 1984", *Common Market Law Reviw* (1985), p. 301–328; D. VANDERMEERSCH, "Restrictions on the Movement of Oil In and Out the European Community: The Campus Oil and Bulk Oil Cases", *Journal of Energy & Natural Resources Law* (1987), p. 31–54 and A. J. GUNST, "Energy Trade in the European Common Market – Free Movement, Exceptions and Regulatory Inaction", *Journal of Energy & Natural Resources Law* (2003), p. 447–469.

688 Case 72/83, *Campus Oil Limited and others v Minister for Industry and Energy and others* [1984] ECR 2727.

The answer to the High Court's first question is therefore that Article 30 of the EEC Treaty must be interpreted as meaning that national rules which require all importers to purchase a certain proportion of their requirements of petroleum products from a refinery situated in the national territory constitute a measure having equivalent effect to a quantitative restriction on imports".[689]

572. Contrary to the views expressed by the Irish government, the Court of Justice, as the Advocate General Sir Gordon Slynn's in its opinion[690], made clear that the Irish purchase obligation undoubtedly constitutes a measure which has an equivalent effect to a quantitative restriction on imports and therefore in breach with Article 34 of the TFEU.[691]

573. The next question which had to be answered by the Court of Justice was to which extent an exception from the scope of Article 34 of the TFEU could be justified under Article 36 of the TFEU. The Irish government and the Irish National Petroleum Corporation argued that *"it is for the Member States to determine, for the purposes of Article 36 [*now Article 36 of the TFEU*], and in particular with regard to the concept of public security, their interests that are to be protected and the measures to be taken to that end. They contend that Ireland's heavy dependence for its oil supplies on imports from other countries and the importance of oil for the life of the Country make it indispensable to maintain refining capacity on the national territory, thereby enabling the national authorities to enter into long-term delivery contracts with the countries producing crude oil. [...]"*[692]

574. The Court of Justice firstly acknowledged in paragraph 27 of its judgment that recourse to Article 36 of the TFEU is no longer justified if harmonisation measures have been adopted in order to ensure the protection of one of the interests included in Article 36 of the TFEU. It furthermore stated that *"national*

689 *Ibid.*, paragraphs 16 and 20.

690 Opinion of Advocate General Sir Gordon Slynn in Case 72/83, *Campus Oil Limited and others v Minister for Industry and Energy and others* at p. 2760–2761.

691 Case 72/83, *Campus Oil Limited and others v Minister for Industry and Energy and others* [1984] ECR 2727, paragraph 16 and 20; H. BJØRNEBYE, *Investing in EU energy security – Exploring the regulatory approach to tomorrow's electricity production*, Dissertation submitted to the faculty of law, University of Oslo, 2009, p. 68. See also H. BJØRNEBYE, *Investing in EU Energy Security. Exploring the regulatory approach to tomorrow's electricity production* (The Netherlands, Kluwer Law International, 2010), 456 p. P. OLIVER, "A review of the Case law of the Court of Justice on Articles 30 to 36 EEC in 1984", *CMLR* 1985, p. 308. The author D. VANDERMEERSCH, as correctly emphasised by H. BJØRNEBYE (see the reference in this footnote) in its article, "Restrictions on the Movement of Oil In and Out the European Community: The Campus Oil and Bulk Oil Cases", *Journal of Energy and Natural Resources Law* (1987), Vol 5 No 1, p. 38, perceives the purchasing obligation of the Irish minister for industry and energy as *"an outright quantitative restriction on imports."*

692 Case 72/83, *Campus Oil Limited and others v Minister for Industry and Energy and others* [1984] ECR 2727, paragraph 22.

measures such as those provided for in the 1982 order cannot therefore be justified unless supplies of petroleum products to the Member State concerned are not sufficiently guaranteed by the measures taken for that purpose by the Community Institutions".[693] The Court of Justice continued by ruling that: "*the existing Community rules give a Member State whose supplies of petroleum products depend totally or almost totally on deliveries from other countries certain guarantees that deliveries from other Member States will be maintained in the event of a serious shortfall in proportions which match those of supplies to the market of the supplying state. However, this does not mean that the Member State concerned has "an unconditional assurance" that supplies will "in any event" be maintained at least at a level sufficient to meet its minimum needs. In those circumstances, the possibility for a Member State to rely on Article 36 to justify appropriate complementary measures at national level cannot be excluded, even where there exist Community rules on the matter*".[694]

575. Compared to previous judgments, the Court of Justice seems to apply a more rigid approach towards Union measures in the *Campus Oil* case and a less rigorous test to determine the adequacy of Union legislation in subsequent judgments which referred to the *Campus Oil* case.[695] It took the view that a real danger could still occur in the event of a crisis despite of the existence of some precautionary measures on a Union level and measures adopted in response to the International Energy Agency because there was no "unconditional assurance" that the Irish government could maintain its supplies at a level to meet its minimum needs.[696]

576. It remains questionable whether Union measures will ever provide "*an unconditional assurance*" that the interest of Article 36 of the TFEU in "*any event*" will not be endangered. Therefore, it needs to be analysed on a case-by-case basis whether given Union measures such as the Third Electricity and Gas Directives provide for a full harmonisation of national laws in order to determine

693 *Ibid.*, paragraph 27 and H. BJØRNEBYE, *Investing in EU energy security – Exploring the regulatory approach to tomorrow's electricity production*, Dissertation submitted to the faculty of law, University of Oslo, 2009, p. 69. See also H. BJØRNEBYE, *Investing in EU Energy Security. Exploring the regulatory approach to tomorrow's electricity production* (The Netherlands, Kluwer Law International, 2010), 456 p.; Case 5/77, *Carlo Tedeschi v Denkavit Commerciale s.r.l.* [1977] ECR 1555, paragraph 35 and Case 35/76, *Simmenthal SpA v Ministère des finances italien* [1976] ECR 1871.

694 Case 72/83, *Campus Oil Limited and others v Minister for Industry and Energy and others* [1984] ECR 2727, paragraph 31, emphasis added.

695 H. BJØRNEBYE, *Investing in EU energy security – Exploring the regulatory approach to tomorrow's electricity production*, Dissertation submitted to the faculty of law, University of Oslo, 2009, p. 70 which referred to the Case C-124/95, *The Queen, ex parte Centro-Com Srl v HM Treasury and Bank of England* [1997] ECR I 81, paragraph 46.

696 P. D. CAMERON, *Competition in Energy Markets – Law and Regulation in the European Union* (second edition) (Oxford, Oxford University Press, 2007), p. 548.

whether recourse to Article 36 of the TFEU remains possible (e.g. *Commission v Italy* case relating to appliances burning gaseous fuels).[697, 698]

(i) Relationship between energy security and the concept of public security of Article 36 of the TFEU

577. After the Court of Justice held in the *Campus Oil* case that recourse to Article 36 of the TFEU was possible, it determined to which extent the grounds for exception in Article 36 of the TFEU could justify the Irish purchase obligation.

The Irish government maintained that the public security exemption of Article 36 of the TFEU was applicable to its purchase obligation. Before the Court of Justice expressed its view on the public security interests as a ground for exemption it stated, contrary to the views of the Advocate General Sir Gordon Slynn[699], that the public policy interests of Article 36 of the TFEU were not pertinent in the *Campus Oil* case. The Court of Justice then ruled with respect to the public security interests that: "*it should be stated in this connection that petroleum products, because of their exceptional importance as an energy source in the modern economy, are of fundamental importance for a country's existence since not only its economy but above all its institutions, its essential*

---

697 See Case C-112/97, *Commission of the European Communities v Italian Republic* [1999] ECR 1821, paragraph 53 in which the Court of Justice held in paragraphs 52–55 that "*[i]n any event, it contends, reliance on Article 36 of the Treaty cannot be precluded where the particular interest at issue is not sufficiently safeguarded by Community measures because it concerns situations which are not provided for in the harmonising directives (Case 72/83 Campus Oil and Others v Minister for Industry and Energy [1984] ECR 2727). It states, finally, that the possibility for a Member State to resort to Article 36 of the Treaty is expressly provided for in Article 100a thereof, on the basis of which the Directive was adopted. It must be pointed out in that regard that, according to well-established case-law, where Community directives provide for the harmonisation of the measures necessary to ensure the protection of animal and human health and establish Community procedures to check that they are observed, recourse to Article 36 is no longer justified and the appropriate checks must be carried out and the measures of protection adopted within the framework outlined by the harmonising directive (Case 5/77 Tedeschi v Denkavit [1977] ECR 1555, paragraph 35). In this case it has already been stated in paragraph 32 of this judgment that the Directive has completely harmonised the essential requirements which gas appliances must satisfy. As is clear from the fifth recital in the preamble to the Directive, requirements as to health and safety rank among them.*"

698 H. BJØRNEBYE, *Investing in EU energy security – Exploring the regulatory approach to tomorrow's electricity production*, Dissertation submitted to the faculty of law, University of Oslo, 2009, p. 70–71. See also H. BJØRNEBYE, *Investing in EU Energy Security. Exploring the regulatory approach to tomorrow's electricity production* (The Netherlands, Kluwer Law International, 2010), 456 p.

699 Opinion of Advocate General Sir Gordon Slynn in Case 72/83, *Campus Oil Limited and others v Minister for Industry and Energy and others* [1984] ECR 2727 which argued that the maintenance of essential oil supplies could be categorised as a public security interest since "*it is vital to the stability of and cohesion of the life of the modern State*". He furthermore stated that if the Court of Justice would come to a different conclusion with respect to the "*public security interest*", the more general category of "*public policy*" could also be applied by the Court of Justice.

public services and even the survival of its inhabitants depend upon them. An interruption of supplies of petroleum products, with the resultant dangers for the country's existence, could therefore seriously affect the public security that Article 36 allows states to protect".[700]

578. The Court of Justice in the foregoing defined the concept of public security and its link with the concept of security of supply. This reasoning of the Court of Justice made clear that the concept of public security refers in Article 36 of the TFEU to *"all its institutions, its essential public services and even the survival of its inhabitants"* and is not limited to the military security of a Member State.[701] Both the Court of Justice and the Advocate General Sir Gordon Slynn have shown in this case that the concept of public security in Article 36 of the TFEU is not only applicable to situations of external and internal security.[702]

579. It furthermore ruled, contrary to the view of the Commission, that *"it is true that, as the Court has held on a number of occasions, most recently in its judgments of 9 June 1982 (Case 95/81 Commission v Italy (1982) ECR 2187, Article 36 refers to matters of non-economic nature. A Member State cannot be allowed to avoid the effects of measures provided for in the Treaty by pleading the economic difficulties caused by the elimination of barriers to intra-community trade, however, in the light of the seriousness of the consequences that an interruption in supplies of petroleum products may have for a country's existence, the aim of ensuring a minimum supply of petroleum products at all times is to be regarded as transcending purely economic considerations and thus as capable of constituting an objective covered by the concept of public security".*[703] With this ruling the Court of Justice accepts that the aim of ensuring a minimum supply of petroleum products constitutes an objective which is covered by the concept of public security. Moreover, a measure which pursuits legitimate interests, but also has an economic effect, does not preclude the recourse to Article 36 of the TFEU.

580. The latter finding in the light of the former judgments of the Court of Justice, in which it emphasised that the derogations of Article 36 of the TFEU all

---

700 Case 72/83, *Campus Oil Limited and others v Minister for Industry and Energy and others* [1984] ECR 2727, paragraph 34.

701 H. BJØRNEBYE, *Investing in EU energy security – Exploring the regulatory approach to tomorrow's electricity production*, Dissertation submitted to the faculty of law, University of Oslo, 2009, p. 72–75. See also H. BJØRNEBYE, *Investing in EU Energy Security. Exploring the regulatory approach to tomorrow's electricity production* (The Netherlands, Kluwer Law International, 2010), 456 p.

702 *Ibid.*, p. 75 and D. VANDERMEERSCH, "Restrictions on the Movement of Oil In and Out the European Community: The Campus Oil and Bulk Oil Cases", *Journal of Energy & Natural resources Law* (1987), p. 43.

703 Case 72/83, *Campus Oil Limited and others v Minister for Industry and Energy and others* [1984] ECR 2727, paragraph 35.

relate to non-economic interests[704], is remarkable since the effects of the Irish purchase obligation are, arguably, primarily economic in nature. It could be argued that the measures which aim to ensure the security of supply of energy are not exclusively non-economic because energy is indispensable for the well functioning and general economy of a country.

The approach of the Court of Justice in the *Campus Oil* case seems to overcome its former judgments by ruling that the economic effects of a measure from a Member State are irrelevant if the objectives of that measure transcend purely economic considerations and are based upon a legitimate ground for exemption under the TFEU.[705] Consequently, if a measure is primarily motivated by economic considerations, a Member State could argue on the basis of the *Campus Oil* case that the principle objective is public security. The given Member State will not only have to demonstrate that the national measure will likely address a real threat to the public security, but also that the measure is proportionate to the legitimate objectives it pursues.[706]

581. Furthermore, with respect to the scope of the *Campus Oil* case one commentator argued that it was highly unlikely that the Court of Justice would extent it to other products than oil such as gas and electricity.[707] On the contrary, it could be argued that ensuring a sufficient minimum supply of electricity is vital for the functioning of the institutions of a Member State, its essential public services and even the survival of its inhabitants. Subsequent case law, which will be discussed below, highlighted that the commentators which argued that the Court of Justice would not extend its scope of the *Campus Oil* case to other products than oil, got it wrong.[708]

582. In the later 1990 case of *Commission v Hellenic Republic* case[709] the Court of Justice was asked again to rule on the relationship between energy security and the concept of public security of Article 36 of the TFEU. It concerned a somewhat similar case as the *Campus Oil* case in which the government of

[704] See *Supra*.

[705] H. BJØRNEBYE, *Investing in EU energy security – Exploring the regulatory approach to tomorrow's electricity production*, Dissertation submitted to the faculty of law, University of Oslo, 2009, p. 73. See also H. BJØRNEBYE, *Investing in EU Energy Security. Exploring the regulatory approach to tomorrow's electricity production* (The Netherlands, Kluwer Law International, 2010), 456 p.

[706] See *Supra*.

[707] Editorial Comments to Campus Oil, *ECLR* (1984), p. 255.

[708] See *Infra* and H. BJØRNEBYE, *Investing in EU energy security – Exploring the regulatory approach to tomorrow's electricity production*, Dissertation submitted to the faculty of law, University of Oslo, 2009, p. 75. See also H. BJØRNEBYE, *Investing in EU Energy Security. Exploring the regulatory approach to tomorrow's electricity production* (The Netherlands, Kluwer Law International, 2010), 456 p.

[709] Case C-347/88, *Commission of the European Communities v Hellenic Republic* [1990] ECR I-4747.

Greece defended the exclusive right to import and market a quantity of petroleum products corresponding to 65% of the requirements of the domestic market. Under the law, distribution companies were required to obtain their suppliers from Greek public-sector refineries to the extent of a percentage of the requirements of the domestic market corresponding to the unadjusted part of the marketing monopoly. The government of Greece acknowledged that the rationale behind this provision is to secure an outlet for the products of its public sector refineries.[710]

583. The Commission started, against these exclusive import and marketing rights for petroleum products in Greece, an enforcement action before the Court of Justice because Greece had failed to fulfil its obligations under Articles 34, 35 and 37, (1) of the TFEU. The government of Greece argued that the exclusive rights of the public sector refineries can be justified on the grounds of public security. It stated that *Greece's special geopolitical situation makes it essential to adopt measures designed to ensure that the country has a regular supply of crude oil and petroleum products.*[711] That aim could, according to the government of Greece, only be achieved by maintaining public-sector refineries in operation and to require from the distribution companies that they obtain a proportion of their supplies from those refineries until such time as it is possible for them to market their products at competitive prices.

584. The Court of Justice recalled its ruling in the *Campus Oil* case and stated that it held in this case that *"a Member State which is totally or almost totally dependant on imports for its supplies of petroleum products may rely on grounds of public security within the meaning of Article 36 of the Treaty* [Article 36 of the TFEU] *for the purpose of requiring importers to cover a certain proportion of their needs by purchases from a refinery situated in its territory at prices fixed by the competent ministry on the basis of the costs incurred in the operation of that refinery, if the production of the refinery cannot be freely disposed of at competitive prices on the market concerned".*[712] However, the Court of Justice stated that Greece in this case had not shown that if the State's rights with regard to the import and marketing of petroleum products were not maintained in force, the public sector refineries would be unable to dispose of their products on the market at competitive prices and thereby ensure their continued operation. This

---

710 K. TALUS, *Vertical Natural Gas Transportation capacity, upstream commodity contracts and EU Competition law*, Dissertation submitted to the faculty of law, University of Helsinki, 2010, p. 61 and 62. See also K. TALUS, *Vertical Natural Gas Transportation Capacity, Upstream Commodity Contracts and EU Competition Law* (The Netherlands, Kluwer Law International, 2010), 321 p.

711 Case C-347/88, *Commission of the European Communities v Hellenic Republic* [1990] ECR I-4747, paragraph 47.

712 *Ibid.*, paragraph 48.

means that the national measures of Greece did not pass the proportionality requirements of Article 34 of the TFEU.

585. Greece did not only lose this case before the Court of Justice, but also a similar case 10 years later.[713] This case concerned a Greek system on the compulsory maintenance of emergency stocks of petroleum products to be kept in Greece and to be stored by the petroleum marketing companies in tanks owned or leased by them outside the refineries. Since 1 January 1996, these marketing companies were entitled to transfer their obligation, in whole or in part, to refineries in Greece where they bought products during the previous calendar year. The transfer may be made up to a total quantity equal to the volume of products of each category delivered to the marketing companies during a 90-day period in the previous year by each of the refineries with which they do business.

586. The Commission argued that the obligation imposed upon the marketing companies to buy petroleum products from refineries established in Greece in order to be able to transfer their storage obligation, constitutes an obstacle to the free movement of goods. According to the Commission that obligation in practice, clearly discriminates in favour of national products to the detriment of foreign ones. The import of petroleum products into Greece, while not being prohibited, is strongly discouraged since, if the marketing companies were to obtain their supplies in other Member States, they would lose the possibility of transferring their storage obligation to the refineries.[714]

587. The government of Greece argued that even if the system in question constitutes an obstacle to the free movement of goods, that obstacle is justified by a public interest objective in particular, the security of supply of petroleum products which benefits from the derogation provided by Article 36 of the TFEU. It furthermore stated that this objective could not be achieved by less restrictive measures.[715]

588. The Court of Justice ruled that this Greek system on the compulsory maintenance of emergency stocks of petroleum products constitutes a measure having equivalent effect to a quantitative restriction. It noted that the Greek

---

[713] Case C-398/98, *Commission of the European Communities v Hellenic Republic* [2001] ECR I-7915.

[714] *Ibid.*, paragraphs 19–21 and K. TALUS, *Vertical Natural Gas Transportation capacity, upstream commodity contracts and EU Competition law*, Dissertation submitted to the faculty of law, University of Helsinki, 2010, p. 62 and 63. See also K. TALUS, *Vertical Natural Gas Transportation Capacity, Upstream Commodity Contracts and EU Competition Law* (The Netherlands, Kluwer Law International, 2010), 321 p.

[715] Case C-398/98, *Commission of the European Communities v Hellenic Republic* [2001] ECR I-7915, paragraphs 19–21.

government was right when it argued that the maintenance on national territory of a stock of petroleum products allowing continuity of supplies to be guaranteed constitutes a public security objective within the meaning of Article 36 of the TFEU. Nonetheless, since the arguments of the government of Greece are purely of an economic nature, they could not serve as a justification for a quantitative restriction within the meaning of Article 36 of the TFEU. The Court of Justice furthermore stated, with reference to the opinion of the Advocate General D. Ruiz-Jarabo Colomer, that the objective of public security could be achieved by less restrictive measures without it being necessary to make the transfer of the storage obligation to refineries established in Greece conditional upon the obligation to obtain supplies of petroleum products from those refineries.[716]

589. The cases above seem to indicate that it is uncertain whether the Court of Justice in subsequent cases would continue to apply the approach taken by the Court in the *Campus Oil* case and allow Member States on the basis of the concept of security of supply to derogate from the basic free movement provisions of the Treaty.

590. Moreover, the Advocate General Jacobs in his opinion in the *PreussenElektra* case[717] emphasised that in his view recourse to Article 36 of the TFEU in respect of security of supply as the Court of Justice admittedly held in the *Campus Oil* case is no longer possible because the *"electricity directive provides for types of measures necessary to ensure security of supply. Moreover wind as an energy source is not yet as important for the modern economy as petroleum products"*. He also noted that the special economic role of petroleum products was a decisive factor in the Court's rather exceptional judgment in the *Campus Oil* case.[718] This opinion of the Advocate General Jacobs clearly raises doubts whether the same approach on security of supply even with petroleum products would apply today due to the considerable amount of legislation on a European level covering the energy security of supply.[719] Nonetheless,

716 *Ibid.*, paragraphs 22–31 and K. TALUS, *Vertical Natural Gas Transportation capacity, upstream commodity contracts and EU Competition law*, Dissertation submitted to the faculty of law, University of Helsinki, 2010, p. 63. See also K. TALUS, *Vertical Natural Gas Transportation Capacity, Upstream Commodity Contracts and EU Competition Law* (The Netherlands, Kluwer Law International, 2010), 321 p.

717 Case C-379/98, *PreussenElektra AG v Schhleswag AG, in the presence of Windpark Reußenköge III GmbH and Land Schleswig-Holstein* [2001] ECR I-2099.

718 Opinion of Advocate General Jacobs of 26 October 2000 in the Case C-379/98, *PreussenElektra AG v Schhleswag AG, in the presence of Windpark Reußenköge III GmbH and Land Schleswig-Holstein*, paragraph 209.

719 Council Directive 2009/119/EC of 14 September 2009 imposing an obligation on Member States to maintain minimum stocks of crude oil and/or petroleum products (from 2012), *OJ* L 265/9; Council Directive 2006/67/EC of 24 July 2006 imposing an obligation on Member States to maintain minimum stocks of crude oil and/or petroleum products, *OJ* L 217/209;

petroleum products remain very important for the modern economy of Member States.

(ii) Concept of security of supply in the Golden Share cases

591. The concept of security of supply has not only been applied in the free movement cases, but also in the *Golden Share* cases[720] of the Court of Justice.[721] These cases concerned the free movement of capital and in some cases the freedom of establishment which were restricted by the measures of the Member States. The Court of Justice ruled in these cases whether Member States could retain the power to intervene in decisions taken by privatised companies which are active in strategically important sectors and, if so, to which extent and on what conditions. These cases, and in particular the cases that related to undertakings in the energy sector, are relevant for the present topic since they seem to extend the concept of public security and the scope of exemptions on the grounds of public security to new energy products.[722]

K. TALUS, *Vertical Natural Gas Transportation capacity, upstream commodity contracts and EU Competition law*, Dissertation submitted to the faculty of law, University of Helsinki, 2010, p. 61. See also K. TALUS, *Vertical Natural Gas Transportation Capacity, Upstream Commodity Contracts and EU Competition Law* (The Netherlands, Kluwer Law International, 2010), 321 p. P. D. CAMERON, *Competition in Energy Markets – Law and Regulation in the European Union* (second edition) (Oxford, Oxford University Press, 2007), p. 549.

720 Case C-367/98, *Commission of the European Communities v Portuguese Republic* [2002] ECR I-4731; Case C-483/99, *Commission of the European Communities v French Republic* [2002] ECR I-4781; Case C-503/99, *Commission of the European Communities v Kingdom of Belgium* [2002] ECR I-4809; Case C-463/00, *Commission of the European Communities v Kingdom of Spain* [2003] ECR I-4581; Case C-98/01, *United Kingdom of Great Britain and Northern Ireland* [2003] ECR I-4641; Case C-174/04, *Commission of the European Communities v Italian Republic* [2005] ECR I-4933; Joined Cases C-282/04 and C-283/04, *Commission of the European Communities v Kingdom of the Netherlands* [2006] ECR I-9141 and Case C-112/05, *Commission of the European Communities v Federal Republic of Germany* [2007] ECR I-8995; Case C-274/06, *Commission of the European Communities v Kingdom of Spain* [2008] ECR I-26; Case C-326/07, *Commission of the European Communities v Italian Republic* [2009]ECR I-2291 and Joined Cases C-463/04 and C-464/04, *Federconsumatori, Adiconsum, ADOC, Ercole Pietro Zucca (C-463/04) and Associazione Azionariato Diffuso dell'AEM SpA, Filippo Cuccia, Giacomo Fragapane Pietro Angelo Puggioni, Annamaria Sanchirico, Sandro Sartorio (C-464/04) v Comune di Milano* [2007] ECR I-10419.

721 See also with respect to the Pax Electrica I in Belgium T. VANDEN BORRE, "Via nieuwe en fragmentarische maatregelen in het Federale energiebeleid naar een globale visie?" in K. DEKETELAERE (ed.), *Jaarboek Energierecht 2005–2006* (Antwerp, Intersentia, 2007), p. 102–108.

722 H. BJØRNEBYE, *Investing in EU energy security – Exploring the regulatory approach to tomorrow's electricity production*, Dissertation submitted to the faculty of law, University of Oslo, 2009, p. 76–78. See also H. BJØRNEBYE, *Investing in EU Energy Security. Exploring the regulatory approach to tomorrow's electricity production* (The Netherlands, Kluwer Law International, 2010), 456 p.

592. The most relevant case in this respect was the *Commission v Belgium* case[723] in which the Court of Justice accepted that the security of supply concept could justify restrictions to the free movement of establishment. In the other *Golden Share* cases the Court of Justice refused to accept the various attempts to justify restrictions on the free movement of capital on public security considerations. It recognised that the public security considerations could justify an obstacle to the free movement of goods, capital and establishment. However, the requirements of such public security considerations must be interpreted strictly, so that their scope cannot be determined unilaterally by each Member State without any control by the Union institutions. Consequently, as the Court of Justice ruled in the *Église de Scientologie* case[724], public security can be relied on only if there is a genuine and sufficiently serious threat to a fundamental interest of society. Thus, the scope of public security and security of supply to justify restrictions to the free movement provisions of the TFEU is narrow and must fulfil the proportionality and necessity test.[725]

593. In the *Commission v Belgium* case[726] the Commission brought an action before the Court of Justice against the national provisions vesting the Kingdom of Belgium Golden Shares in *Société nationale de transport par canalisations* (hereinafter SNTC) and *Société de distribution du gaz SA* (hereinafter Distrigaz). The SNTC was the owner of a system of lines and conduits that constitute major infrastructures for the domestic conveyance of energy products, whilst Distrigaz comprised, *inter alia*, infrastructures for the domestic conveyance and storage of gas, including unloading and cross-border facilities. The Golden Shares entitled the Belgian State to oppose any transfer, use as security or change in the intended destination of lines and conduits or of certain other strategic assets and certain management decisions regarded as contrary to the guidelines for the country's energy policy. According to the Commission these powers of the Belgian State are contrary to the Treaty provisions on the free movement of capital.[727]

---

723 Case C-503/99, *Commission of the European Communities v Kingdom of Belgium* [2002] ECR I-4809.

724 Case C- 54/99, *Association Eglise de scientologie de Paris and Scientology International Reserves Trust v The Prime Minister* [2000] ECR I-1335, paragraph 17.

725 K. TALUS, *Vertical Natural Gas Transportation capacity, upstream commodity contracts and EU Competition law*, Dissertation submitted to the faculty of law, University of Helsinki, 2010, p. 64. See also K. TALUS, *Vertical Natural Gas Transportation Capacity, Upstream Commodity Contracts and EU Competition Law* (The Netherlands, Kluwer Law International, 2010), 321 p.

726 Case C-503/99, *Commission of the European Communities v Kingdom of Belgium* [2002] ECR I-4809.

727 *Ibid.*, paragraphs 19 and 28 and H. BJØRNEBYE, *Investing in EU energy security – Exploring the regulatory approach to tomorrow's electricity production*, Dissertation submitted to the faculty of law, University of Oslo, 2009, p. 77 and 78. See also H. BJØRNEBYE, *Investing in*

594. The Belgian government argued that any restrictions to the free movement of capital or free movement of establishment due to the national provisions in question are justified by the exception of public security and by overriding requirements of the general interest. Moreover, these provisions in relation to the objective pursued by them fulfil the criteria of necessity and proportionality. It furthermore emphasised that both companies occupy a strategic position as regards the country's energy supplies, especially in view of Belgium's dependence on foreign energy resources.[728]

595. The Court of Justice ruled that: "*the objective pursued by the legislation at issue, namely the safeguarding of energy supplies in the event of a crisis, falls undeniably within the ambit of a legitimate public interest. Indeed, the Court has previously recognised that the public-security considerations which may justify an obstacle to the free movement of goods include the objective of ensuring a minimum supply of petroleum products at all times (Campus Oil, [...], paragraphs 34 and 35). The same reasoning applies to obstacles to the free movement of capital, inasmuch as public security is also one of the grounds of justification referred to in Article 73d(1)(b) of the Treaty* [Article 65(1) (b) of the TFEU*]*".[729] The Court of Justice recognised that security of supply interests could provide legitimate grounds for justification.

596. The Court continued by stating that: "*however, the Court has also held that the requirements of public security, as a derogation from the fundamental principle of free movement of capital, must be interpreted strictly, so that their scope cannot be determined unilaterally by each Member State without any control by the Community institutions. Thus, public security may be relied on only if there is a genuine and sufficiently serious threat to a fundamental interest of society (see, in particular, Église de scientologie, cited above, paragraph 17)*".[730] The Court of Justice continued to ascertain whether the provisions in question enable the Belgian State to ensure a minimum level of energy supplies in the event of a genuine and serious threat and whether they are proportional. It concluded that the provisions in question meet these requirements and are justified by the objective of guaranteeing energy supplies in the event of a crisis.[731]

EU Energy Security. Exploring the regulatory approach to tomorrow's electricity production (The Netherlands, Kluwer Law International, 2010), 456 p.

728 Case C-503/99, *Commission of the European Communities v Kingdom of Belgium* [2002] ECR I-4809, paragraphs 26–30.

729 *Ibid.*, paragraph 46. In the Case C-483/99, *Commission of the European Communities v French Republic* [2002] ECR I-4781 and Case C-463/00, *Commission of the European Communities v Kingdom of Spain* [2003] ECR I-4581, the Court of Justice took a similar view.

730 Case C-503/99, *Commission of the European Communities v Kingdom of Belgium* [2002] ECR I-4809, paragraph 47.

731 *Ibid.*, paragraphs 48–55.

597. This *Commission v Belgium* case[732] has clarified that the public security exemption introduced by the *Campus Oil* case not only applies to petroleum products, but also to gas supplies. Moreover, the Court of Justice held in the *Commission v Spain* case[733] that: "*as regards the three other undertakings concerned, which are active in the petroleum, telecommunications and electricity sectors, it is undeniable that the objective of safeguarding supplies of such products or the provision of such services within the Member State concerned in the event of a crisis may constitute a public-security reason (see, for similar situations, Commission v France, paragraph 47, and Commission v Belgium, paragraph 46) and therefore may justify an obstacle to the free movement of capital*".[734]

598. In the *Commission v Portugal* case[735] the Court of Justice, recently illustrated the scope and limits of the ground of public security in order to justify an obstacle to the free movement of capital. It highlighted that: "*it is undisputed that requirements of public security must, in particular as a derogation from the fundamental principle of the free movement of capital, <u>be interpreted strictly</u>, so that their scope cannot be determined unilaterally by each Member State without any control by the institutions of the European Union. Thus, public security may be relied on only if there is a genuine and sufficiently serious threat to a fundamental interest of society (see, in particular, Case C-54/99 Église de scientologie [2000] ECR I-1335, paragraph 17, and Case C-171/08 Commission v Portugal, paragraph 73)*".[736]

599. The Portuguese Republic argued that: "*since the risk of serious threats to the security of energy supply cannot be excluded and since such threats are by definition sudden and, in the majority of cases, unforeseeable, it is the duty of the Member State concerned to ensure that adequate mechanisms are put in place to enable it to react rapidly and effectively to guarantee that the security of that supply is not interrupted*". According to the Court this argument is not entirely without merit. However, it ruled that: "*since the Portuguese Republic has done no more than raise that ground relating to the security of the energy supply, without stating clearly the exact reasons why it considers that the special rights at issue, considered either individually or as a whole, would make it possible to prevent such an interference with a fundamental interest of society, a justification based on public security cannot be upheld in the present case*". This suggests that if exact raisons and acceptable ones are provided the Court of Justice might accept the use of the public security ground in order to justify an obstacle to the free

[732] *Ibid.*

[733] Case C-463/00, *Commission of the European Communities v Kingdom of Spain* [2003] ECR I-4581.

[734] *Ibid.*, paragraph 71.

[735] Case C-543/08, *European Commission v Portuguese Republic*, not yet published.

[736] *Ibid.*, paragraph 85, emphasis added.

movement provisions of the TFEU. Since these reasons were lacking the Court ruled that a justification based on public security could not be upheld in this case.

600. Moreover, as discussed above, the provisions of national law must pass the proportionality test. In that regard, the Court of Justice in the *Commission v Portugal* case[737] stated that: *"the exercise of the special rights which the holding of golden shares in the share capital of EDP confers on the Portuguese State is not subject to any specific and objective condition or circumstance". [...] "Thus, such uncertainty constitutes serious interference with the free movement of capital in that it confers on the national authorities, as regards the use of such rights, a latitude so discretionary in nature that it cannot be regarded as proportionate to the objectives pursued (see, to that effect, Case C-326/07 Commission v Italy, paragraph 52)".*[738]

601. Thus, the scope of the public security exemption can be extended to other energy products such as gas, electricity, but also to other sectors such as telecommunications. Furthermore, the Court of Justice seems to accept that Member States can adopt measures, contrary to the free movement provisions of the TFEU, which aimed at ensuring the security of supply, but can be justified on the basis of public security grounds.[739] Nonetheless, these measures still need to pass the necessity, proportionality test and should be based on exact and acceptable reasons.

602. It can be concluded, that the *Golden Share* cases has proven that the Court of Justice interpreted the public security and security of supply grounds strictly since it refused in most cases that these grounds could justify restrictions to the free movement provisions of the TFEU. The case law seems to suggest that the objectives and means of a Golden Share scheme have been accepted by the Court of Justice and that the decisive factor, to convince the Court of the necessity of these schemes, was the level of precision and detail.[740] In order to ensure that

---

737 Case C-543/08, *European Commission v Portuguese Republic*, not yet published.

738 *Ibid.*, paragraphs 90–92.

739 H. BJØRNEBYE, *Investing in EU energy security – Exploring the regulatory approach to tomorrow's electricity production*, Dissertation submitted to the faculty of law, University of Oslo, 2009, p. 80 and 81. See also H. BJØRNEBYE, *Investing in EU Energy Security. Exploring the regulatory approach to tomorrow's electricity production* (The Netherlands, Kluwer Law International, 2010), 456 p.

740 This can be derived from the Case C-326/07, *Commission of the European Communities v Italian Republic* [2009] ECR I-2291, paragraphs 71–73; Case C-543/08, *European Commission v Portuguese Republic*, not yet published and Case C-274/06, *Commission of the European Communities v Kingdom of Spain* [2008] ECR I-26, paragraphs 45 and 46 and 52 and 53. In this case in paragraph 50 a distinction was also made between an *ex poste* and *ex ante* objection, the Court held that: *"[e]n second lieu, le régime institué par la DA 27 n'est pas assimilable au régime examiné dans l'arrêt Commission/Belgique, précité. Ce dernier constitue,*

restrictive national measures did not go further than what is necessary to satisfy the security of energy supply objective, the Court of Justice applied the proportionality and necessity tests.

603. It is clear from the foregoing that the intensity of the review undertaken, or as in the *PreussenElektra* case[741] the decision on whether to undertake a proportionality and necessity assessment, are both essential for the outcome of a case before the Court of Justice. It will determine a Member State's margin of discretion when it addresses its national security of supply concerns.[742] In addition to the security of supply arguments, the Court of Justice accepted in its case law, in particular the *PreussenElektra* case[743], that environmental grounds could also justify the adoption of discriminatory measures. These environmental grounds will be discussed in the next part of this section.

(c) Environmental grounds as a mandatory requirement in the energy field

604. The protection of the environment is not listed in Article 36 of the TFEU. However, the Court of Justice recognised in its case law that certain obstacles to the free movement provisions of the TFEU must be accepted in so far as they could be regarded as necessary in order to satisfy imperative requirements.[744] Moreover, according to settled case law, and in particular the *Commission v Denmark* case[745], the protection of the environment is one of the mandatory

en effet, un régime d'opposition a posteriori aux décisions adoptées par une société. En revanche, la DA 27, en ce qu'elle suspend l'exercice des droits de vote et en permet la non-reconnaissance ou la limitation, produit ses effets avant même que la société ait adopté une décision, c'est à dire sans que soit établi un risque, même potentiel, d'atteinte à la sécurité de l'approvisionnement en énergie". Thus, the Belgian Golden Share scheme contains an *ex poste* objection on the other hand the Spanish Golden Share scheme contains an *ex ante* objection. According to the Court of Justice the latter scheme will not allow to verify whether the public security was endangered or not and K. TALUS, *Vertical Natural Gas Transportation capacity, upstream commodity contracts and EU Competition law,* Dissertation submitted to the faculty of law, University of Helsinki, 2010, p. 65–68. See also K. TALUS, *Vertical Natural Gas Transportation Capacity, Upstream Commodity Contracts and EU Competition Law* (The Netherlands, Kluwer Law International, 2010), 321 p.

741 Case C-379/98, *PreussenElektra AG v Schhleswag AG, in the presence of Windpark Reußenköge III GmbH and Land Schleswig-Holstein* [2001] ECR I-2099.

742 H. BJØRNEBYE, *Investing in EU energy security – Exploring the regulatory approach to tomorrow's electricity production*, Dissertation submitted to the faculty of law, University of Oslo, 2009, p. 85–92. See also H. BJØRNEBYE, *Investing in EU Energy Security. Exploring the regulatory approach to tomorrow's electricity production* (The Netherlands, Kluwer Law International, 2010), 456 p.

743 Case C-379/98, *PreussenElektra AG v Schhleswag AG, in the presence of Windpark Reußenköge III GmbH and Land Schleswig-Holstein* [2001] ECR I-2099.

744 Case C-120/78, *Rewe-Zentral AG v Bundesmonopolverwaltung für Branntwein* [1979] ECR 649.

745 Case C-302/86, *Commission of the European Communities v Kingdom of Denmark* [1988] ECR 4607.

requirements which may limit the application of Article 34 of the TFEU.[746] The scope of this mandatory requirement has been clarified by the Court of Justice in its *PreussenElektra* case.[747]

(i) PreussenElektra case

605. This case concerned the German Feeding-in Act 1990 (*Stromeinspeisungsgesetz*) which was designed to promote the use of electricity from renewable energy sources and required that the supply undertakings (both in public and in private ownership) purchased all electricity generated within their area of supply from renewable energy sources such as water, wind, sun and biomass at a fixed price. This minimum price was calculated on the basis of the average nationwide sales price for electricity which in effect was set in order to provide a subsidy to generators of renewable electricity. Over time, the level of subsidy has been raised. The Commission kept a close eye on these developments and proposed a number of amendments which would make the *Stromeinspeisungsgesetz* compatible with the State aid rules as such. It proposed a reduction of the minimum purchase price for electricity from wind to 75% of the average price, the adoption of a rule according to which the purchase price was calculated on the basis of avoided costs and/or a limitation of the support mechanism in time and/or according to electricity generated.[748]

606. In order to implement the First Electricity Directive, Germany adopted on 24 April 1998 the *"Gesetz zur Neuregelung des Energiewirtschaftsrechts"* (Law reforming the energy supply industry) which led to the adoption in paragraph 4 of the *Stromeinspeisungsgesetz*, headed as before *"hardship clause"*, of a new compensation mechanism which is worded as follows: "*1. In so far as the kilowatt hours to be compensated for exceed 5% of the total kilowatt hours supplied by the electricity supply undertaking through its network during a calendar year, the upstream network operator shall be obliged to reimburse the electricity supply undertaking in respect of the supplementary costs resulting from the kilowatt hours exceeding that share. In the case of upstream network operators, the burden constituted by the right to reimbursement within the meaning of the first sentence also forms part of those supplementary costs. If there is no such operator, the obligation laid down in the first sentence of Paragraph 2 ceases, as regards*

[746] Opinion Advocate General Jacobs of 26 October 2000 in the Case C-379/98, *PreussenElektra AG v Schhleswag AG, in the presence of Windpark Reußenköge III GmbH and Land Schleswig-Holstein*, p. 27 and Case C-302/86, *Commission of the European Communities v Kingdom of Denmark* [1988] ECR 4607, paragraph 9.

[747] Case C-379/98, *PreussenElektra AG v Schhleswag AG, in the presence of Windpark Reußenköge III GmbH and Land Schleswig-Holstein* [2001] ECR I-2099.

[748] A. JOHNSTON, K. NEUHOFF, D. FOUQUET, M. RAGWITZ and G. RESCH, "The proposed new EU renewables directive: interpretation, problems and prospects", *EEELR* (2008), p. 131 and 132.

electricity supply undertakings in the circumstances referred to in the first and second sentences, at the beginning of the first calendar year after those circumstances arose, in the case of installations not yet essentially completed at that time. In the case of wind turbines, the relevant time is the installation of the mast and the rotor. [...]".[749]

607. Whether this paragraph 4 of the *Stromeinspeisungsgesetz* constitutes State aid within the meaning of Article 107(1) of the TFEU has been discussed in great detail by the Advocate General Jacobs in his opinion and the Court of Justice.[750] It is sufficient for my analysis to note that both concluded that the German support mechanism for renewable energy producers does not constitute State aid because there was no direct or indirect transfer of State resources. The support for the renewable energy producers came from the utilities.[751]

608. Another issue in the *PreussenElektra* case, which is more relevant in this context, is whether the German support mechanism for renewable energy sources constitutes a quantitative restriction on imports or a measure having equivalent effect within the meaning of Article 34 of the TFEU. The purchase obligation of electricity generated from renewable energy sources in the *Stromeinspeisungsgesetz* on German network operators within their area might reduce the demand for electricity generated in other Member States. Therefore, the Court of Justice was asked whether the purchase obligation could be classified as a measure incompatible with Article 34 of the TFEU.

609. The Advocate General Jacobs in his opinion very easily concluded that the German support mechanism for renewable energy sources constituted a measure having an effect equivalent to a quantitative restriction.[752] The *Stromeinspeisungsgesetz* favoured the marketing of electricity of German origin to the detriment of imported electricity. In this respect, Schleswag, one of the plaintiffs, claimed that it had been offered electricity generated from renewable energy sources in Sweden at a relatively low price. However, it had been forced to

749 Case C-379/98, *PreussenElektra AG v Schhleswag AG, in the presence of Windpark Reußenköge III GmbH and Land Schleswig-Holstein* [2001] ECR I-2099, paragraph 10.

750 Opinion Advocate General Jacobs of 26 October 2000 in the Case C-379/98, *PreussenElektra AG v Schhleswag AG, in the presence of Windpark Reußenköge III GmbH and Land Schleswig-Holstein*, p. 14–24.

751 A. JOHNSTON, K. NEUHOFF, D. FOUQUET, M. RAGWITZ and G. RESCH, "The proposed new EU renewables directive: interpretation, problems and prospects", *EEELR* (2008), p. 132; B. DELVAUX, "The EC State Aid Regime regarding Renewables: Opportunities and Pitfalls", *EELR* (2003), p. 103–112 and M. BRONCKERS and R. VAN DER VLIES, "The European court's *PreussenElektra* judgement: tensions between EU principles and national renewable energy initiatives", *E.C.L.R.* (2001), p. 458–468.

752 Opinion Advocate General Jacobs of 26 October 2000 in the Case C-379/98, *PreussenElektra AG v Schhleswag AG, in the presence of Windpark Reußenköge III GmbH and Land Schleswig-Holstein*, p. 25.

decline this offer due to its obligation to purchase all the wind-generation electricity generated within its area of supply. Since the purchase obligation was in principle prohibited by Article 34 of the TFEU, the Advocate General Jacobs analysed whether this infringement could be justified. In his view, recourse to Article 36 of the TFEU on either in respect of security of supply as in the *Campus Oil* case[753] was not possible.[754]

610. He stressed that wind as an energy source is not yet as important for the modern economy as petroleum products. Furthermore, according to the Advocate General Jacobs the special economic role of petroleum products was a decisive factor in the Court's rather exceptional judgment in the *Campus Oil* case. Thus, Advocate General Jacobs concluded that recourse to grounds such as public security and security of supply in order to justify restrictions to the free movement provisions was not an option.[755]

611. The Advocate General Jacobs subsequently analysed whether the second ground of justification, namely the protection of the environment relied on by the interveners, the Commission and government of Germany, could justify the restrictions. Notwithstanding the discrimination caused by the *Stromeinspeisungsgesetz* on imported electricity, he proceeded to analyse whether this could be justified under the Treaty in the interest of the environmental protection. In this context, he first criticised the reasoning of the Court of Justice in the *Walloon waste* case[756] and emphasised *"that it is desirable that even directly discriminating measures can sometimes be justified on grounds of environmental protection"*.[757] Moreover, Advocate General Jacobs stressed that the Court of Justice should clarify its position on whether directly discriminatory measures can be justified by imperative requirements in order to provide the necessary legal certainty.[758]

---

753 Case 72/83, *Campus Oil Limited and others v Minister for Industry and Energy and others* [1984] ECR 2727.

754 Opinion Advocate General Jacobs of 26 October 2000 in the Case C-379/98 *PreussenElektra AG v Schhleswag AG, in the presence of Windpark Reußenköge III GmbH and Land Schleswig-Holstein*, p. 26 and 27 and A. JOHNSTON, K. NEUHOFF, D. FOUQUET, M. RAGWITZ and G. RESCH, "The proposed new EU renewables directive: interpretation, problems and prospects", *EEELR* (2008), p. 133 and 134.

755 Opinion Advocate General Jacobs of 26 October 2000 in the Case C-379/98 *PreussenElektra AG v Schhleswag AG, in the presence of Windpark Reußenköge III GmbH and Land Schleswig-Holstein*, p. 26 and 27.

756 Case C-2/90, *Commission of the European Communities v Kingdom of Belgium* [1992] ECR 4431.

757 Opinion Advocate General Jacobs of 26 October 2000 in the Case C-379/98, *PreussenElektra AG v Schhleswag AG, in the presence of Windpark Reußenköge III GmbH and Land Schleswig-Holstein*, p. 27–29.

758 A. JOHNSTON, K. NEUHOFF, D. FOUQUET, M. RAGWITZ and G. RESCH, "The proposed new EU renewables directive: interpretation, problems and prospects", *EEELR* (2008), p. 133 and 134.

612. In his view, the Court of Justice should apply a more flexible approach in respect of the imperative requirement of environmental protection. He supports this view on the basis of the fact that the amendments to the Treaties agreed in Amsterdam show an increased concern for the environment, although Article 36 of the TFEU itself had not been amended. Furthermore, according to Advocate General Jacobs in this context Article 11 of the TFEU, which provides that *"environmental protection requirements must be integrated into the definition and implementation of the Union policies and activities, in particular with a view to promoting sustainable development"*, is also of particular importance. This provision is not merely pragmatic, but also imposes legal obligations. He also stated that environmental measures of Member States are liable to be found inherently discriminatory because they are based on principles, reflected also in Article 191, (2) of the TFEU, which provide that *"environmental damage should as a priority be rectified at source"*. Advocate General Jacobs found that if environmental measures can only be justified on the basis of the environmental protection requirement where they are applicable without distinction, this would risk the very purpose of the national measure. Thus, if such environmental measures necessarily have a discriminatory impact, the possibility that they may be justified should not be excluded according to Advocate General Jacobs.[759]

613. On the above mentioned assumption that the environmental protection requirement could justify restrictions to the free movement provisions, the Advocate General Jacobs analysed whether *Stromeinspeisungsgesetz* could pass the proportionality test. He concluded that since the *Stromeinspeisungsgesetz* tried to rectify the pollution caused by greenhouse gas emissions resulting from conventional generation of electricity it failed to satisfy the proportionality test because electricity generated from renewables sources outside Germany would reduce the pollution caused by greenhouse gas emissions to the same extent. Finally, with respect to the Commission's argument stating that the measure was also justified due to the possible loss of energy through transmission over long distances, the Advocate General Jacobs referred the assessment of this argument to the national court.[760]

---

759 Opinion Advocate General Jacobs of 26 October 2000 in the Case C-379/98, *PreussenElektra AG v Schhleswag AG, in the presence of Windpark Reußenköge III GmbH and Land Schleswig-Holstein*, p. 27–29 and A. JOHNSTON, K. NEUHOFF, D. FOUQUET, M. RAGWITZ and G. RESCH, "The proposed new EU renewables directive: interpretation, problems and prospects", *EEELR* (2008), p. 133 and 134.

760 A. JOHNSTON, K. NEUHOFF, D. FOUQUET, M. RAGWITZ and G. RESCH, "The proposed new EU renewables directive: interpretation, problems and prospects", *EEELR* (2008), p. 133 and 134 and Opinion Advocate General Jacobs of 26 October 2000 in the Case C-379/98, *PreussenElektra AG v Schhleswag AG, in the presence of Windpark Reußenköge III GmbH and Land Schleswig-Holstein*, p. 29.

614. The Court of Justice with respect to the question whether the German scheme in the *Stromeinspeisungsgesetz* was compatible with Article 34 of the TFEU, surprisingly held that these rules are compatible with the free movement provision of the Treaty. In this context, it once again reiterated the *Dassonville* formula[761] and recalled its reasoning in the *Campus Oil* case[762] and *Du Pont de Nemours Italiana* case.[763] It continued by stating that the *Stromeinspeisungsgesetz* *"expressly state[s] that the purchase obligation imposed on electricity supply undertakings applies only to electricity produced from renewable energy sources within the scope of that statute and within the respective supply area of each undertaking concerned, [...]"*.[764]

615. The Courts of Justice thus held that the *Stromeinspeisungsgesetz* was capable, at least potentially, of hindering intra-Union trade. In the author's view and other doctrine this is a remarkable conclusion.[765] The purchase obligation at issue in the *Stromeinspeisungsgesetz* undoubtedly favours the marketing of electricity generated in Germany to the detriment of imported electricity and therefore the conclusion in the author's view should have been that it was in principle prohibited by Article 34 of the TFEU.

616. The Court of Justice continued its reasoning and held that in order to determine whether the purchase obligation at issue is nevertheless compatible with the Article 34 of the TFEU, account must be taken of the aim of the provision in question and the particular features of the electricity market. It stressed that the *Stromeinspeisungsgesetz* is intended to promote the use of renewable energy sources for generating electricity which is useful for the protection of the environment in so far as it contributes to the reduction of greenhouse gas emissions which is one of the priority objectives of the Union and its Member States. Furthermore, according to the Court of Justice, the policy is also designed to protect the health and life of humans, animals and plants. Besides, Article 191, (2) of the TFEU and Article 11 of the TFEU require that environmental protection is integrated into the definition and implementation of other Union policies. It noted that the first Electricity

761 Case 8–74, *Procureur du Roi v Benoît and Gustave Dassonville* [1974] ECR 837.

762 Case 72/83, *Campus Oil Limited and others v Minister for Industry and Energy and others* [1984] ECR 2727.

763 Case C-21/88, *Du Pont de Nemours Italiana SpA v Unità sanitaria locale N° 2 di Carrara* [1990] ECR 889.

764 Case C-379/98, *PreussenElektra AG v Schhleswag AG, in the presence of Windpark Reußenköge III GmbH and Land Schleswig-Holstein* [2001] ECR I-2099, paragraph 71.

765 A. JOHNSTON, K. NEUHOFF, D. FOUQUET, M. RAGWITZ and G. RESCH, "The proposed new EU renewables directive: interpretation, problems and prospects", *EEELR* (2008), p. 134 and F. JACOBS, "The role of the Court of Justice of the European Union in the Protection of the Environment", *Journal of Environmental Law* (2006), p. 185–205.

Directive expressly referred in its preamble to the environmental objectives of the Union.[766]

617. The Court of Justice also held that due to the nature of electricity, it is difficult to determine its origin and in particular the source of energy from which it is generated once it has been put into the transmission or distribution system. In order to strengthen this view, the Court of Justice also referred to the Commission's proposal which stated that a system of certificates of origin for electricity generated from renewable sources should be established in order to make trade in that type of electricity reliable and possible in practice.[767]

618. This reasoning was striking since the Court of Justice only three years earlier in the *Outokumpu* case[768] dismissed a similar contention of the Finnish government. In this case the Court of Justice was asked to assess the compatibility of the Finnish taxation scheme for electricity with Union law. It was a Finnish electricity generator who brought an action before a national judge against a Finnish law (No 1473/94 of 29 December 1994 on excise duty on certain sources of energy) which established the levying of excise duties determined on the basis of the method of generation for electricity generated in Finland.[769] The Finnish government argued that due to the characteristics of electricity, the origin and method of generation cannot be determined once it has entered the network. Consequently, according to the Finnish government the imported electricity cannot have the advantage of differential rates. The Court of Justice ruled with reference to its previous case law[770] that practical difficulties cannot justify the application of internal taxation which discriminates against products from other Member States.[771]

---

766 Case C-379/98, *PreussenElektra AG v Schhleswag AG, in the presence of Windpark Reußenköge III GmbH and Land Schleswig-Holstein* [2001] ECR I-2099, paragraphs 72–78 and H. BJØRNEBYE, *Investing in EU energy security – Exploring the regulatory approach to tomorrow's electricity production*, Dissertation submitted to the faculty of law, University of Oslo, 2009, p. 92–96. See also H. BJØRNEBYE, *Investing in EU Energy Security. Exploring the regulatory approach to tomorrow's electricity production* (The Netherlands, Kluwer Law International, 2010), 456 p.

767 A. JOHNSTON, K. NEUHOFF, D. FOUQUET, M. RAGWITZ and G. RESCH, "The proposed new EU renewables directive: interpretation, problems and prospects", *EEELR* (2008), p. 134–137.

768 Case C-213/96, *Outokumpu Oy* [1998] ECR I-1777.

769 See *Supra*.

770 Case C-375/95, *Commission of the European Communities v Hellenic Republic* [1998] ECR I-1777, paragraph 47.

771 Case C-213/96, *Outokumpu Oy* [1998] ECR I-1777, paragraphs 37 and 38.

619. In some legal doctrine it is argued that the apparent contradictions between the *Outokumpu* case[772] and *PreussenElektra* case[773] can be clarified and that the ruling of the Court of Justice in the *PreussenElektra* judgment did not necessarily overrule the former *Outokumpu* judgment. According to this doctrine, the German scheme in the *Stromeinspeisungsgesetz* did not only pursue a national interest which should be protected, but also a Union interest in a balanced development. It subsidised the regional generation of green electricity of which the additional costs were redistributed amongst all the German consumers. To achieve this goal in Germany, the provisions in the *Stromeinspeisungsgesetz* were necessary. The Finish scheme was adopted in order to ensure that green energy could compete with imported as well as with domestic non-green electricity. This severe rule in Finland was on the contrary not necessary to achieve its goal.[774]

620. The Court of Justice in the *PreussenElektra* case[775] finally concluded on the basis of its reasoning summarised above that "*in the current state of Community law concerning the electricity market, legislation such as the amended Stromeinspeisungsgesetz is not incompatible with Article 30 of the Treaty [Article 34 of the TFEU]*".[776]

621. In this *PreussenElektra* judgment, the Court of Justice seems to succeed on the basis of a rather tortuous reasoning to apply the environmental protection grounds either by finding that the provisions in the *Stromeinspeisungsgesetz* were not discriminatory or simply avoided that issue. The judgment did not contain a thorough analysis of whether the provisions in the *Stromeinspeisungsgesetz* are of a discriminatory nature or whether these provisions are disproportionate. Should it not be a logical and necessary first step for the Court of Justice to assess whether a measure is directly discriminatory or not?[777] In the author's view, this is the case. Moreover, such an assessment cannot take into account whether the provision at stake is appropriate or desirable for policy reasons.

---

[772] *Ibid.*

[773] Case C-379/98, *PreussenElektra AG v Schhleswag AG, in the presence of Windpark Reußenköge III GmbH and Land Schleswig-Holstein* [2001] ECR I-2099.

[774] A. JOHNSTON, K. NEUHOFF, D. FOUQUET, M. RAGWITZ and G. RESCH, "The proposed new EU renewables directive: interpretation, problems and prospects", *EEELR* (2008), p. 134–137 and H. BJØRNEBYE, *Investing in EU energy security – Exploring the regulatory approach to tomorrow's electricity production*, Dissertation submitted to the faculty of law, University of Oslo, 2009, p. 96 and 97. See also H. BJØRNEBYE, *Investing in EU Energy Security. Exploring the regulatory approach to tomorrow's electricity production* (The Netherlands, Kluwer Law International, 2010), 456 p.

[775] Case C-379/98, *PreussenElektra AG v Schhleswag AG, in the presence of Windpark Reußenköge III GmbH and Land Schleswig-Holstein* [2001] ECR I-2099.

[776] *Ibid.*, paragraph 81.

[777] F. JACOBS, "The Role of the Court of Justice of the European Union in the protection of the environment", *Journal of Environmental Law* (2006), p. 185–205.

(ii) *Integration principle and environmental protection as a mandatory requirement*

622. With respect to uncertainties in the Court's case law when applying environmental protection as a mandatory requirement, Advocate General Jacobs suggested in his opinion that the Court of Justice should apply a more favourable treatment and to set this mandatory requirement apart from all others. In favour of this approach he referred to the amendments of the Treaties agreed in Amsterdam and in particular the adoption of Article 11 of the TFEU of the so-called integration principle which imposes legal obligations. According to the Advocate General Jacobs this provision provides that special account should be taken of environmental concerns in interpreting the provisions of the TFEU, including the free movement provisions.[778]

623. Some authors argue that recourse to this provision might cause some difficulties since it is unclear whether Article 11 of the TFEU derogates from other Treaty provisions and whether this provision should be considered *lex specialis*? They conclude that Article 11 of the TFEU imposes only an obligation on the Union legislator and therefore, it would be difficult to interpret this provision as a limitation on the free movement rights in the absence of Union harmonisation legislation.[779] In my view, the Advocate General Jacobs did not suggest that the integration principle should prevail over the free movement provisions of the TFEU. However, these environmental concerns should be taken into account and consequently, lead to a choice for the least environmentally harmful option.[780]

624. The integration principle refers to the external integration of environmental protection requirements or the integration of environmental objectives in other policy sectors. However, it also has two other dimensions. One of them requires that environmental measures do not cause shifts in pollution from one medium to another and can be found in the Industrial Emission Directive.[781, 782] The other dimension of this principle is inserted into Article 191, (3) of the TFEU and requires from the Union when it prepares its policy on the environment that it takes into account *"available scientific and*

778 *Ibid.*, p. 190–191.

779 A. JOHNSTON, K. NEUHOFF, D. FOUQUET, M. RAGWITZ and G. RESCH, "The proposed new EU renewables directive: interpretation, problems and prospects", *EEELR* (2008), p. 136 and 137.

780 See *Infra*.

781 Directive 2008/1/EC of the European Parliament and of the Council of 15 January 2008 concerning integrated pollution prevention and control (codified version), *OJ* L 24/8 (hereinafter Industrial Emission Directive).

782 J. DE CENDRA DE LARRAGÁN, *Distributional choices in EU climate change law and policy – Towards a principled approach*, Dissertation submitted to the faculty of law, Maastricht University, 2010, p. 100.

technical data, environmental conditions in the various regions of the Union, the potential benefits and costs of action or lack of action and the economic and social development of the Union as a whole and the balanced development of its regions".

625. In practice, the Commission complies with this provision through the adoption of an integrated impact assessment in which it analyses the environmental, economic and social impacts of its proposals. In relation to the integration principle there are a number of issues such as the relation between the objective of environmental protection and other objectives in the TFEU, the legal enforceability of the principle[783], the implementation of the principle in secondary legislation[784] and the contents of the principle of external integration.[785]

626. In this context, it is relevant to have a closer look at the first issue and in particular the relation between environmental requirements and other requirements in the TFEU. This principle is designed to ensure that if decisions are taken on a European level even when commercial policy is involved or other fields such as energy, development aid, trade and external relations, etc., the protection of the environment is at least taken into consideration. Potential

---

783 *Ibid.*, p. 100 and see J. H. JANS and H. H. B. VEDDER, *European Environmental Law* (Groningen, Europa Law Publishing, 2008), p. 21 and in this context the case law of the Court of Justice Case in C-300/89, *Commission of the European Communities v Council of the European Communities* [1991] ECR I-2867; Case C-513/99, *Concordia Bus Finland Oy Ab, formerly Stagecoach Finland Oy Ab v Helsingin kaupunki and HKL-Bussiliikenne* [2002] ECR I-7213; Case C-341/95, *Gianni Bettati v Safety Hi-Tech Srl* [1998] ECR I-4355; Case C-293/97, *The Queen v Secretary of State for the Environment and Ministry of Agriculture, Fisheries and Food, ex parte H.A. Standley and Others and D.G.D. Metson and Others* [1999] ECR I-2603; Case C-209/98, *Entreprenørforeningens Affalds/Miljøsektion (FFAD) v Københavns Kommune* [2000] ECR I-3743 and Case C-6/03, *Deponiezweckverband Eiterköpfe v Land Rheinland-Pfalz* [2005] ECR I-2753.

784 Examples of secondary legislation on a Community level into which the integration principle has been inserted are the Council Directive 85/337/EEC of 27 June 1985 on the assessment of the effects of certain public and private projects on the environment, *OJ* L 216/40 (the so-called EIA-Directive) which has been amended in 1997, 2003 and 2009 and Directive 2001/42/EC of the European Parliament and of the Council of 27 June 2001 on the assessment of the effects of certain plans and programmes on the environment *OJ* L 197/30 (the so-called EIA-Directive).

785 See for further analysis of this issue of the integration principle A. EPINEY, "Environmental Principles" in R. MACRORY, *Reflections on 30 years of EU Environmental law – a high level of protection* (Groningen, Europa Law Publishing, 2006), p. 19–39; N. DHONDT, *Integration of Environmental Protection into other EC Policies: legal theory and practice* (Groningen, Europa Law Publishing, 2003), 532 p.; L. KRÄMER, *EC Environmental law* (London, Sweet & Maxwell, 2007), 513 p. and J. H. JANS and H. H. B. VEDDER, *European Environmental Law* (Groningen, Europa Law Publishing, 2008), 496 p. and T. SCHUMACHER, "The environmental integration clause in Article 6 of the EC Treaty: Prioritising Environmental Protection", *Environmental Law Review* (2001), p. 29–43 and See M. G. DOHERTY, "Judicial Use of Principles of EC Environmental Policy", *Environmental Law Review* (2000), p. 254–256.

conflicts between the protection of the environment and, for example, free movement of goods which need to be solved cannot be inferred from the integration principle as such.[786] Such kind of conflicts should be resolved on the basis of the proportionality principle.[787]

627. The Court of Justice in its *ADBHU* case[788] seems to apply this approach since it allowed restrictions to trade as long as these measures are not discriminatory and do not entail restrictions that go beyond what is strictly necessary for the protection of the environment.[789] In line with this approach this would mean that if an objective could be achieved in different ways, the integration principle would entail a preference for the least environmentally harmful option.[790]

628. With respect to the legal enforceability of the integration principle, the Court of Justice in its case law seems to attach a legal value to this principle. In the *Chernobyl I* case[791] and *TiO_2* case[792] it used the principle to justify environmental legislation on the basis of Treaty provisions than other Article 192 of the TFEU and to broaden the objectives of the other Treaty provisions with environmental objectives (e.g. in the context of commercial policy and internal market policy). The Court of Justice stated in the *Chernobyl I* case that: *"articles 130r and 130s are intended to confer powers on the Community to undertake specific action on environmental matters. However, those articles leave intact the powers held by the Community under other provisions of the Treaty, even if the measures to be taken under the latter provisions pursue at the same time any of the objectives of environmental protection.*

Moreover, that interpretation is confirmed by the second sentence of Article 130r(2), pursuant to which "environmental protection requirements shall be a component of the Community's other policies". That provision, which reflects the principle whereby all Community measures must satisfy the requirements of environmental protection, implies that a Community measure cannot be part of

786 J. DE CENDRA DE LARRAGÁN, *Distributional choices in EU climate change law and policy – Towards a principled approach*, Dissertation submitted to the faculty of law, Maastricht University, 2010, p. 101 and T. SCHUMACHER, "The environmental integration clause in Article 6 of the EC Treaty: Prioritising Environmental Protection", *Environmental Law Review* (2001), p. 29–43.

787 J. H. JANS and H. H. B. VEDDER, *European Environmental Law* (Groningen, Europa Law Publishing, 2008), p. 17.

788 Case 240/83, *Procureur de la République v Association de défense des brûleurs d'huiles usagées* [1985] ECR 531.

789 *Ibid.*, paragraph 15.

790 J. H. JANS and H. H. B. VEDDER, *European Environmental Law* (Groningen, Europa Law Publishing, 2008), p. 17 and 18.

791 Case C-62/88, *Hellenic Republic v Council of the European Communities* [1990] ECR I-1527.

792 Case C-300/89, *Commission of the European Communities v Council of the European Communities* [1991] ECR I-2867.

Community action on environmental matters merely because it takes account of those requirements".[793]

629. In its TiO_2 case, the Court of Justice confirms this view by stating that: *it must be observed in the first place that, pursuant to the second sentence of Article 130r(2) of the Treaty, "environmental protection requirements shall be a component of the Community's other policies". That principle implies that a Community measure cannot be covered by Article 130s merely because it also pursues objectives of environmental protection.*[794] The integration principle seems to allow the Council to take into account environmental objectives without the need to invoke an additional legal basis allowing the adoption of environmental measures.[795]

630. The Advocate General Jacobs also suggested that the Court of Justice should give a broader interpretation to the protection of health of animals, humans and plants and include in environmental protection in Article 36 of the TFEU.[796] It is indeed true that legal rules and especially Treaty provisions should be interpreted as a living and evolving text which needs to be adapted to a changing context because due to their nature, they are more difficult to amend. He also argued that environmental protection should be distinguished from all other mandatory requirements because environmental measures are inherently liable to be found discriminatory and should therefore be capable of being justified on grounds of environmental protection.[797]

---

[793] Case C-62/88, *Hellenic Republic v Council of the European Communities* [1990] ECR I-1527, paragraph 19 and 20.

[794] Case C-300/89, *Commission of the European Communities v Council of the European Communities* [1991] ECR I-2867, paragraph 22.

[795] J. H. JANS and H. H. B. VEDDER, *European Environmental Law* (Groningen, Europa Law Publishing, 2008), p. 17–20. For a further discussion on the legal status of the integration principle see J. H. JANS and H. H. B. VEDDER, *European Environmental Law* (Groningen, Europa Law Publishing, 2008), p. 16–23; S. MAHMOUDI, "Integration of Environmental Considerations into Transport" in R. MACRORY (ed.), *Reflections on 30 Years of EU Environmental Law: A high level of protection* (Groningen, Europa Law Publishing, 2005), p. 185–196; R. WILLIAMS, "Community Development Cooperation Law, Sustainable Development and the Convention on Europe: From Dislocation to Consistency?", *Yearbook of European Environmental Law* (2005), p. 303–375; N. DHONDT, "Integration of Environmental Protection into the EC Energy Policy, *Yearbook of European Environmental Law* (2005), p. 247–302; G. MARIN DURAN and E. MORGERA, "Towards Environmental Integration in EC External Relations? A comparative analysis of selected association agreements", *Yearbook of European Environmental Law* (2006), p. 179–210 and H. H. B. VEDDER, *Competition Law and Environmental Protection in Europe – Towards Sustainability?* (Groningen, Europa Law Publishing, 2003), 478 p.

[796] F. JACOBS, "The Role of the Court of Justice of the European Union in the protection of the environment", *Journal of Environmental Law* (2006), p. 185–205 and see also *Supra*.

[797] *Ibid.*, p. 185–205 and see also *Supra*.

631. It is clear that the Court of Justice in its *PreussenElektra* case[798] unfortunately missed the opportunity to clarify its position whether directly discriminatory measures can be justified by mandatory requirements and to establish legal certainty. This was not only the case in the *PreussenElektra* judgment, but also in the *Commission v Austria* case[799] which concerned an Austrian ban of heavy lorries on a particular section of a motorway on grounds of environmental protection. Also in this case, the Court of Justice decided to sidestep the issue once again. The author agrees with the view of the Advocate the General Jacobs that the outcome of the *PreussenElektra* judgment should not be criticised since it enhances environmental protection. However, the means by which the Court of Justice came to its conclusions are regrettable. The Court of Justice seems to assess measures aiming to protect the environment favourably. Nonetheless, this should not be to the detriment of legal certainty. In conclusion of the foregoing, it seems that the Court of Justice has environmentally-friendly credentials, however, it has until recently failed to provide in its case law an adequate conceptual basis for its approach.[800]

6. Conclusion

632. The previous analysis discussed both the *Dassonville* and *Keck formula* of the Court of Justice which shaped Article 34 of the TFEU into an efficient instrument against national measures that created non-fiscal barriers to the free movement of goods. It has demonstrated that under Article 36 of the TFEU and the so-called mandatory requirements national measures restricting free movement of goods can be justified by the Court of Justice.

633. With respect to the energy field, this analysis concluded on the basis of case law of the Court of Justice that electricity, gas and oil constitute goods for the purposes of Title I and II in Part III of the TFEU and thus, also for Articles 34–37 of the TFEU which are part of those Titles. With respect to the application of Article 36 of the TFEU in the energy field the author argued on the basis of case law of the Court of Justice that it is uncertain whether the Court in subsequent cases would continue to apply the approach taken by the Court in the *Campus Oil* case and allow Member States on the basis of the concept of security of supply to derogate from the basic free movement provisions of the Treaty. The analysis of the *Golden Share* cases in this context proved that the Court of Justice interpreted the public security and security of supply grounds strictly.

---

798 Case C-379/98, *PreussenElektra AG v Schhleswag AG, in the presence of Windpark Reußenköge III GmbH and Land Schleswig-Holstein* [2001] ECR I-2099.

799 Case C-320/03, *Commission of the European Communities v Republic of Austria* [2005] ECR I-9871.

800 F. JACOBS, "The Role of the Court of Justice of the European Union in the protection of the environment", *Journal of Environmental Law* (2006), p. 185–205.

634. The protection of the environment is not listed in Article 36 of the TFEU. However, as discussed above, the Court of Justice recognised in its case law that certain obstacles to the free movement provisions of the TFEU can be justified on the basis of the protection of the environment.

635. The analysis of the most prominent case law of the Court of Justice has proven that both security of supply and environmental protection arguments, on certain conditions, could justify, in energy, national measures which restrict the free movement of goods.

3. INTERNAL MARKET PROVISIONS

(a) *Meaning of Articles 114 and 115 of the TFEU*

636. The Union has also intervened in the energy markets by secondary legislation. Due to the lack of any specific chapter or title on energy, until recently, this was primarily on the basis of Article 114 of the TFEU which empowers the Union and sets out the legal procedures for the adoption of measures to achieve the internal market. Even this is no longer explicitly mentioned in Article 114 of the TFEU. This provision still applies by way of derogation from Article 115 of the TFEU. Article 115 of the TFEU gives a general competence to the Council, acting unanimously, to issue directives for the approximation of laws, regulations or administrative provisions of the Member States that directly affect the establishment or functioning of the common market.[801]

637. This means that Article 115 of the TFEU is applicable to those provisions which are explicitly excluded by paragraph 2 of Article 114 of the TFEU, such as fiscal provisions, those relating to the free movement of persons and to the rights and interests of employed people. It also applies to those issues which are not linked to the establishment or functioning of the internal market, but covered by the broader notion of the common market.[802]

---

[801] For further reading see D. CHALMERS, G. DAVIES and G. MONTI, *European Union Law* (Cambridge, Cambridge University press 2010), p. 674–712; A. DASHWOOD, "The Limits of European Community Powers", *European Law Review* (1996), p. 113; N. N. SHUIBHNE (ed.), *Regulating the Internal Market* (Cheltenham, Edward Elgar, 2006), p. 370; S. WEATHERILL, "Why Harmonise?" in T. TRIDIMAS and P. NEBBIA (eds.), *European Union Law for the Twenty-First Century* (Oxford and Portland, Hart Publishing, 2004), 465 p. and C. JOERGES and R. DEHOUSSE (eds.), *Good Governance in Europe's Integrated Market* (Oxford, Oxford University Press, 2002), 348 p.

[802] P. CRAIG and G. DE BÚRCA, *EU law text, cases and materials* (Oxford, Oxford University Press, 2003), p. 1184.

638. Article 114 of the TFEU is a residual provision, since it only applies where otherwise provided in the TFEU. This means that in areas such as the common agricultural policy (Article 43 of the TFEU), the right of establishment (Article 50 of the TFEU), the common transport policy (Article 91 of the TFEU), with a specific legal basis provided by the TFEU, this legal basis should be used in preference to Article 114 of the TFEU for the adoption of measures designed to attain the internal market.

639. Whereas Article 115 of the TFEU only allows the adoption of directives, Article 114 of the TFEU authorises the Council to adopt measures which includes directives, but also for example regulations. However, in practice, the Council has shown a preference for using directives for the adoption of measures on the basis of Article 114 of the TFEU. It is also worth nothing that Article 114 of the TFEU, contrary to Article 115 of the TFEU, requires the ordinary legislative procedure of Article 294 of the TFEU for the enactment of approximation measures.[803]

(b) Case law of the Court of Justice interpreting Article 114 of the TFEU

640. Article 114 of the TFEU is broadly framed. However, the Court of Justice has confirmed in its case law that this provision does not give the Union legislator a general competence to regulate the internal market. The exercise of this competence is specific[804] and thus bound to the principle of conferral of powers. The leading case of the Court of Justice interpreting Article 114 of the TFEU is the first *Tobacco Advertising* case.[805] According to the Court of Justice the measures referred to in Article 114 of the TFEU *"are intended to improve the conditions for the establishment and functioning of the internal market"*.[806] It furthermore stated that *"if a mere finding of disparities between national rules and of the abstract risk of obstacles to the exercise of fundamental freedoms or of distortions of competition liable to result there from were sufficient to justify the choice of Article 100a as a legal basis, judicial review of compliance with the*

[803] A. ARNULL, A. DASHWOOD, M. DOUGAN, M. ROSS, E. SPAVENTA and D. WYATT, *Wyatt and Dashwoods's European Union Law* (London, Sweet & Maxwell, 2006) p. 61–66.

[804] Case C-376/98, *Germany v. European Parliament and Council* [2000] ECR I-8419, paragraphs 83 and 107.

[805] *Ibid.* and D. CHALMERS, G. DAVIES and G. MONTI, *European Union Law* (Cambridge, Cambridge University press 2010), p. 687. See also J. SNELL, "Who Has Got the Power: Free Movement and Allocation of Competences", *YEL* (2003), p. 323–351; T. K. HERVEY, "Up in Smoke?" Community (Anti) Tobacco Law and Policy", *European Law Review* (2001), p. 101–125; S. WEATHERILL, "Competence Creep and Competence Control", *YBEL* (2004), p. 1–55 and M. KUMM, "Constitutionalising Subsidiarity in Integrated Markets: The Case of Tobacco Regulation in the European Union", *European Law Journal* (2006), p. 503–533.

[806] Case C-376/98, *Germany v. European Parliament and Council* [2000] ECR I-8419, paragraphs 83 and 107, paragraph 83.

proper legal basis might be rendered nugatory. The Court would then be prevented from discharging the function entrusted to it by Article 164 of the EC Treaty (now Article 220 EC) of ensuring that the law is observed in the interpretation and application of the Treaty".[807]

Prior to the Court's ruling in the first *Tobacco Advertising* case the Union institutions applied a generous concept to regulate the internal market as if they had the general power to regulate that market. The mere abstract risk to the exercise of fundamental freedoms or distortions of competition, or a mere finding of disparities between national rules, was sufficient to justify measures of harmonisation on the basis of one or more internal market provisions of the Treaty.[808]

641. The Court of Justice in the first *Tobacco Advertising* case ruled that Article 114 of the TFEU can be used to adopt measures on a European level if those measures contribute to the elimination of obstacles to the exercise of the fundamental freedoms. Article 114 of the TFEU could not only be used to adopt measures which eliminate actual obstacles to trade, but also to address future obstacles that might result *from the multifarious development of national laws, the emergence of such obstacles must be "likely" and the measure in question must be "designed to prevent them".*[809, 810] Moreover, the Court of Justice ruled that Article 114 of the TFEU can also serve as a legal basis to remove distortions of competition arising from the diverse national laws. This second ground for which Article 114 of the TFEU can be used, overlaps to a certain extent with the first one since any obstacle to trade will likely distort competition. The distortions must also be appreciable.[811] It furthermore ruled in its *Titanium Dioxide* case[812] that Union legislator could also adopt measures

---

807 *Ibid.*, paragraph 83. See for further reading D. KHANNA, "The Defeat of the European Tobacco Advertising Directive: A blow for Health", *YEL* (2001), p. 113 and D. WYATT, "Constitutional significance of the Tobacco Advertising Judgement of the Court of Justice of the European Union" and G. DE BÚRCA, "The Political Will versus Constitutional Limits", *The ECJ's Tobacco Advertising Judgment*, ECLS Occasional Paper, No 5 (Cambridge, Cambridge University Press, 2001), p. 9–35 (www.cels.law.cam.ac.uk/publications/occasional%20papers/Paper_5.pdf).

808 D. WYATT QC, "Community Competence to Regulate the Internal Market" in M. DOUGAN and S. CURRIE (eds.), *50 years of the European Treaties looking back and thinking forward* (Oxford, Hart Publishing, 2009), p. 94–101.

809 Case C-491/01, The *Queen v Secretary of State for Health, ex parte British American Tobacco (Investments) Ltd and Imperial Tobacco Ltd* [2002] ECR I-11453, paragraph 61 which refers to Case C-350/92, *Kingdom of Spain v Council of the European Union* [1995] ECR I-1985, paragraph 35 and Case C-376/98, *Germany v. European Parliament and Council* [2000] ECR I-8419, paragraph 86.

810 C. BARNARD, *The substantive Law of the EU The Four Freedoms* (Oxford, Oxford University Press, 2007), p. 574.

811 Case C-300/89, *Commission of the European Communities v Council of the European Communities* [1991] ECR I-2867, paragraph 15.

812 *Ibid.*

for future distortions, if the distortions were likely to arise from diverse national rules.[813]

642. The Court's ruling in the first *Tobacco Advertising* case appears to contain both elements which are supportive with respect to the powers of the Union law-maker to regulate the internal market, and elements which are restrictive. The Court's conclusion that harmonisation measures can be based upon Article 114 of the TFEU if the measures in question make "some" contribution to the internal market, falls in the former category.

643. One of the elements which restrict the powers of the Union legislator is the requirement that the Union measures regulating the internal market make a "positive contribution" to the internal market. Another requirement is that the distortions of competition or obstacles to trade are "appreciable". This implies that the mere existence of disparities between national laws cannot be regarded in itself as leading to the conclusion that distortion of competition exists which justifies the adoption of harmonisation measures on a European level.[814] However, due to the lack of any quantitative assessment by the Court of Justice, the meaning of the term "appreciable" remains uncertain. In some legal doctrine it is argued that the ambiguity of this term prevents Article 114 of the TFEU to become a general power to harmonise national rules.[815] However, the appreciability requirement will not always be an adequate limit since there are matters which cause very appreciable distortions of competition (for example the existence of different languages), but which we do not expect to be harmonised yet.[816]

644. Subsequent cases of the Court of Justice, such as the *Swedish Match* case[817] and the second *Tobacco Advertising* case[818] appear to apply a different and more relaxed approach with respect to the use of Article 114 of the TFEU. It

813 C. BARNARD, *The substantive Law of the EU The Four Freedoms* (Oxford, Oxford University Press, 2007), p. 575.

814 D. WYATT QC, "Community Competence to Regulate the Internal Market" in M. DOUGAN and S. CURRIE (eds.), *50 years of the European Treaties looking back and thinking forward* (Oxford, Hart Publishing, 2009), p. 114. This author furthermore discussed whether on the basis of subsequent case law and due to the lack of any quantitative assessment by the Court of Justice, the Court of Justice is adopting a low threshold for the concept of appreciability of distortions of competition see pages 127–130.

815 D. CHALMERS, G. DAVIES and G. MONTI, *European Union Law* (Cambridge, Cambridge University press 2010), p. 694 and A. DASHWOOD, "The limits of European Community Powers", *European Law Review* (1996), p. 113.

816 *Ibid.*

817 Case C-210/03, *Swedish Match AB and Swedish Match UK Ltd v Secretary of State for Health* [2004] ECR I-11893.

818 Case C-380/03, *Federal Republic of Germany v European Parliament and Council of the European Union* [2006] ECR I-11573.

upheld a number of uses of Article 114 of the TFEU, but also referred to some key parts of its ruling in the first *Tobacco Advertising* case.

645. In the *Swedish Match* case the Court of Justice concluded that Article 114 of the TFEU provides a Treaty basis to adopt a Union-wide ban of products. In this case the Court of Justice seems to accept that if obstacles exist in the trade of a product due to the fact that some Member States prohibit that product and other Member States do not, Article 114 of the TFEU provides an adequate legal basis to prohibit the product in all Member States. Obstacles to trade are removed by prohibiting the product. This approach contradicts with the rationale on the elimination of obstacles to trade in the first *Tobacco Advertising* case.

646. In some legal doctrine it is argued that the approach of the Court of Justice in the *Swedish Match* case extends the powers of the Union in the internal market to subject matters which make no contribution to the internal market and fail to comply with the subsidiarity requirement.[819] In the second *Tobacco Advertising* case the Court of Justice incorporated the approach of the *Swedish Match* case. It concluded, that if there are obstacles to trade, or if it is likely that such obstacles will emerge in the future, Article 114 of the TFEU could serve as a legal basis to adopt measures which *"may consist in requiring all the Member States to authorise the marketing of the product or products concerned, subjecting such an obligation of authorisation to certain conditions, or even provisionally or definitively prohibiting the marketing of a product or products".*[820, 821]

647. In the *Österreichischer Rundfunk* case[822] the Court of Justice has ruled in this context that Article 114 of the TFEU *"as legal basis does not presuppose the existence of an actual link with free movement between Member States in every situation referred to by the measure founded on that basis. As the Court has previously held (see Case C-376/98 Germany v Parliament and Council [2000] ECR I-8419, paragraph 85, and Case C-491/01 British American Tobacco (Investments) and Imperial Tobacco [2002] ECR I-11453, paragraph 60), to justify recourse to Article 100a of the Treaty as the legal basis, what matters is that the*

819 D. WYATT QC, "Community Competence to Regulate the Internal Market" in M. DOUGAN and S. CURRIE (eds.), *50 years of the European Treaties looking back and thinking forward* (Oxford, Hart Publishing, 2009), p. 115–121.

820 Case C-380/03, *Federal Republic of Germany v European Parliament and Council of the European Union* [2006] ECR I-11573, paragraphs 41–43.

821 D. WYATT QC, "Community Competence to Regulate the Internal Market" in M. DOUGAN and S. CURRIE (eds.), *50 years of the European Treaties looking back and thinking forward* (Oxford, Hart Publishing, 2009), p. 121. Emphasis added.

822 Joined Cases C-465/00, C-138/01 and C-139/01, *Rechnungshof v Österreichischer Rundfunk and Others and Christa Neukomm and Joseph Lauermann v Österreichischer Rundfunk* [2003] ECR I-4989.

measure adopted on that basis must actually be intended to improve the conditions for the establishment and functioning of the internal market".[823]

648. The Court of Justice furthermore accepted that Article 114 of the TFEU could serve as an appropriate legal basis to create a Union body. It ruled that *"nothing in the wording of Article 95 EC implies that the addressees of the measures adopted by the Community legislature on the basis of that provision can only be the individual Member States. The legislature may deem it necessary to provide for the establishment of a Community body responsible for contributing to the implementation of a process of harmonisation in situations where, in order to facilitate the uniform implementation and application of acts based on that provision, the adoption of non-binding supporting and framework measures seems appropriate.*

It must be emphasised, however, that the tasks conferred on such a body must be closely linked to the subject-matter of the acts approximating the laws, regulations and administrative provisions of the Member States".[824, 825]

649. This *ENISA* case and the *Smoke Flavouring* case[826] have recently defined the meaning and scope of approximation. According to these cases Article 114 of the TFEU cannot only be used for measures which actually harmonise matters, but also for measures that contribute to the harmonising process without engaging in it as such.[827]

---

823 *Ibid.*, paragraph 41.

824 Case C-217/04, *United Kingdom of Great Britain and Northern Ireland v European Parliament and Council of the European Union (hereinafter ENISA case)* [2006] ECR I-3771, paragraphs 44 and 45.

825 C. BARNARD, *The substantive Law of the EU The Four Freedoms* (Oxford, Oxford University Press, 2007), p. 577–579.

826 Case C-66/04, *United Kingdom of Great Britain and Northern Ireland v European Parliament and Council of the European Union* [2005] ECR I-10553 in which the Court of Justice ruled in paragraphs 45–50, that *"the expression 'measures for the approximation' in Article 95 EC the authors of the Treaty intended to confer on the Community legislature a discretion, depending on the general context and the specific circumstances of the matter to be harmonised, as regards the harmonisation technique most appropriate for achieving the desired result, in particular in fields which are characterised by complex technical features. [...] Such an interpretation of Article 95 EC is also borne out by the fact that, according to their very wording, paragraphs 4 and 5 of that article recognise that the Commission has power to adopt harmonisation measures. The reference to that power of the Commission in those paragraphs, read in conjunction with paragraph 1 of that article, indicates that an act adopted by the Community legislature on the basis of Article 95 EC, in accordance with the codecision procedure referred to in Article 251 EC, may be limited to defining the provisions which are essential for the achievement of objectives in connection with the establishment and functioning of the internal market in the field concerned, while conferring power on the Commission to adopt the harmonisation measures needed for the implementation of the legislative act in question".* Emphasis added.

827 D. CHALMERS, G. DAVIES and G. MONTI, *European Union Law* (Cambridge, Cambridge University press 2010), p. 692.

650. To conclude this part, one may wonder whether there is a need to put some outer limits on the competence of the Union to regulate the internal market since currently the Union law-maker appears to enjoy in part *de jure* and *de facto* the competence to regulate the internal market which amounts to a competence to regulate society, provided that the measures in question contribute to a certain extent to the internal market.[828]

(c) *Article 114 and energy*

651. Given the broad scope of Article 114 of the TFEU, the European legislator very often used this provision for the adoption of secondary legislation in different areas on a European level. In particular in the energy field, the First Electricity and Gas Directives, both amended by the Second Electricity and Gas Directives and subsequently amended, by the Third Electricity and Gas Directives were based on Article 114 of the TFEU and gradually introduced an opening process of the electricity and gas markets.

These were not the only measures of secondary legislation in the energy field which used Article 114 of the TFEU as a legal basis for their adoption; examples are Directive 2005/89/EC, Directive 2009/125/EC, etc.[829]

(d) *Conclusion*

652. This part has discussed the meaning of Article 114 of the TFEU on the basis of the legal doctrine and case law of the Court of Justice. It has emphasised that the Court of Justice has ruled that although Article 114 of the TFEU is

---

828 D. WYATT QC, "Community Competence to Regulate the Internal Market" in M. DOUGAN and S. CURRIE (eds.), *50 years of the European Treaties looking back and thinking forward* (Oxford, Hart Publishing, 2009), p. 135 and 136. In this context reference could be made to the debate whether it would be appropriate to adopt a European Contract Code on the basis of Article 114 of the TFEU (ex Article 95 of the EC Treaty). Such a common code would likely facilitate interstate contracts and business to some extent. In some legal doctrine it is argued that this would be disproportionate since the benefits to trade do not justify the cultural and social costs. See D. CHALMERS, G. DAVIES and G. MONTI, *European Union Law* (Cambridge, Cambridge University press 2010), p. 694 which refers to G. DAVIES, "Subsidiarity: The Wrong Idea, in the Wrong Place, at the Wrong Time", *Common Market Law Review* (2006), p. 63; S. WEATHERILL, "Why Object to the Harmonisation of Private Law by the EC?", *European Review of Private Law* (2004), p. 633; S. WEATHERILL and S. VOGENAUER (eds.), *The Harmonisation of European Contract Law: Implications for European Private Laws, Business and Legal Practise* (Oxford, Hart Publishing, 2006), 259 p.; P. VAN DEN BERGH, "Forced Harmonisation of Contract Law in Europe: Not to be Continued" in S. GRUNDMANN and J. STUYCK (eds.), *An Academic Green Paper on European Contract Law* (The Hague, Kluwer, 2002), p. 245–264 and A. HARTKAMP and E. HONDIUS *et al.*, *Towards a European Civil Code* (The Hague, Kluwer Law International, 2004), 868 p.

829 See *Supra*.

broadly framed, it does not provide the Union legislator with a general competence to regulate the internal market.

With respect to the energy field, Article 114 of the TFEU has been the legal basis for various energy measures of secondary legislation due to the lack of any specific powers in the energy field. The recourse to this provision was for many years a sufficient legal basis in order to establish a large amount of secondary legislation in the energy field. Chapter 5 will demonstrate whether Article 114 of the TFEU is still relevant as a legal basis for the adoption of secondary legislation in the energy field even after the adoption of the new energy title in the Lisbon Treaty.

4. ARTICLE 113 OF TFEU

(a) *Meaning of Article 113 of the TFEU*

653. The Treaty provisions with respect to the abolition of fiscal obstacles, such as custom duties and discriminatory taxation, and charges with equivalent effects are covered by Articles 28, 30 and 110 of the TFEU and have been discussed above.[830]

654. However, the mere application of these provisions in the TFEU is insufficient for the abolition of fiscal obstacles. TFEU therefore gives the Council, under Article 113 of TFEU the power to adopt provisions for the harmonisation of legislation concerning turnover taxes, excise duties and other forms of indirect taxation to the extent that such harmonisation is necessary to ensure the establishment and the functioning of the internal market and to avoid distortion of competition. Such harmonisation measures must be adopted by the Council acting unanimously in accordance with a special legislative procedure and after consulting the European Parliament and the Economic and Social Committee.[831]

(b) *Article 113 of the TFEU and energy*

655. With respect to energy, the Union framework dealing with the harmonisation of the taxation of energy products, adopted on the basis of Article 113 of TFEU, is Directive 2003/96/EC. This directive has regulated the taxation of energy and electricity in the Union since 2004. It therefore establishes a general framework for the taxation of energy products and electricity, and

---

830 *Ibid.*

831 K. J. M. MORTELMANS, "Chapter VIII The Functioning of the Internal Market: The Freedoms" in P. J. G. KAPTEYN, A. M. MCDONNELL, K. J. M. MORTELMANS and C. W. A. TIMMERMANS (eds.), *The Law of the European Union and the European Communities*, fourth edition (The Netherlands, Kluwer Law International, 2008), p. 616.

leaves Member States with a wide range of flexibility.[832] The Commission recently adopted a proposal amending Directive 2003/96/EC.[833]

656. One of the main difficulties with respect to Article 113 of TFEU is the unanimity requirement which results in very long discussions on a European level before fiscal harmonisation measures are adopted, and if a compromise is eventually reached, it is often a poor one which allows Member States a wide range of flexibility.

(c) Conclusion

657. This part has provided a brief discussion with respect to Article 113 of the TFEU which allows the Council to adopt provisions for the harmonisation of legislation concerning turnover taxes, excise duties and other forms of indirect taxation. This provision served as the legal basis for the adoption of Directive 2003/96/EC which harmonised the taxation of energy products. Whether this will continue to be the case will be discussed in chapter 5.

5. THE ENVIRONMENTAL LAW PROVISIONS IN THE TFEU

(a) Introduction

658. For many years, the Union adopted its regulatory framework for the energy sector on the basis of the general non-energy provisions of the TFEU. Beside the internal market provisions of the TFEU, as discussed above, the European legislator also used the environmental provisions as a legal basis to adopt measures in the energy field. Title XX of the TFEU, comprising Articles 191 to 193, establishes the Union's competence in the environmental field.

659. The objective of this part is to examine the environmental provisions of the TFEU used to adopt secondary legislation. This analysis provides a clear understanding of the environmental provisions in the TFEU which is important in order to determine to which extent the European legislator can continue to adopt energy measures on the basis of this provision.

---

832 E. D. CROSS, B. DELVAUX, L. HANCHER, P. J. SLOT, G. VAN CALSTER and W. VANDENBERGHE, "EU Energy Law (chapter 5)" in M. M. ROGGENKAMP, A. RØNNE, C. REDGWELL and I. DEL GUAYO (ed.), *Energy Law in Europe. National, EU and International Regulation*, 2nd ed. (Oxford, Oxford University Press, 2007), p. 321 and 327.

833 See *Supra*.

660. One of the shared competences between the European Union and its Member States is environmental policy and law. Consequently, in the absence of Union legislation, Member States are free to adopt legislation in the environmental field provided that these measures are compatible with the general rules of the TFEU. However, if the Union has acted in the environmental field, Union legislation should be considered as supreme and binding upon future and past actions of the Member States.[834]

661. Nonetheless, a Member States' competence to deal with a given environmental issue will not end once the Union has adopted an environmental measure to tackle this issue. The possibility for a Member State to act will depend on the content of the Union measure and on the legal basis on which this Union measure is grounded.[835]

(b) Present environmental competences of the Union

662. The Environmental Title XX of the TFEU, comprising Articles 191 to 193, establishes the Union's competence in the environmental field. The pursued objectives of the Union policy on the environment are defined in paragraph 1 of Article 191 of the TFEU:

(a) preserving, protecting and improving the quality of the environment;
(b) protecting human health;
(c) prudent and rational utilisation of natural resources;
(d) promoting measures on an international level to deal with regional or worldwide environmental problems, and in particular combating climate change.

i) Objectives of the environmental policy

663. The main significance of paragraph 1 of Article 191 TFEU is that it determines the range of Union competences in the environmental field. Paragraph 1 obtained this significance via Article 192 of the TFEU as this provision enables the Union to act in order to achieve the objectives mentioned in paragraph 1 of Article 191 TFEU.

664. One could also deduce from paragraph 1 of Article 191 TFEU two obligations. On the one hand, this provision seems to oblige the Union and its

---

834 K. DEKETELAERE (ed.), "European 'Environmental Law" in *International Encyclopedia of Laws* (The Hague, London and Boston, Kluwer Law International, 2000), p. 27 and L. KRÄMER, *EC Environmental law* (London, Sweet & Maxwell, 2007), p. 102–104.

835 L. KRÄMER, *EC Environmental law* (London, Sweet & Maxwell, 2007) p. 126 and see *Infra*.

institutions to adopt an environmental policy and on the other hand, it seems to contain an obligation on the Union and its institutions to take the necessary measures to fulfil one of the objectives enumerated in paragraph 1 of Article 191 TFEU. The latter obligation would mean that if one of the goals listed in paragraph 1 of Article 191 TFEU can only be achieved in a satisfactory way on a Union level, the Union's institutions are not longer free to adopt measures or not. However, this must be analysed on a case-by-case basis and in connection with a particular goal mentioned in paragraph 1 of Article 191 TFEU (ex 174 of the EC Treaty).[836] The next part will have a closer look at the various objectives cited paragraph 1 of Article 191 TFEU.

1. Preserving, protecting and improving the quality of the environment

665. This first objective mentioned in paragraph 1 of Article 191 TFEU (ex 174 of the EC Treaty) is fairly generally formulated as the term "environment" is given no further definition in the TFEU itself. Due to this lack of any definition of the term "environment", it is impossible to determine with absolute certainty what should be understood by an environmental policy within the Union. However, this has the advantage that the concept of environment is sufficiently flexible to be adapted to new developments and new needs. This objective also seems to include the protection of animal health, fauna and flora.[837]

666. The formulation *"preserving, protecting and improving"* of this objective is worded broadly and in a flexible way.[838] Consequently, this indicates that there is no restriction to a certain type of environmental measure which can be adopted. The nature of these environmental measures can be preservative, repressive,

[836] A. EPINEY, "Environmental Principles" in R. MACRORY (ed.), *Reflections on 30 Years of EU Environmental Law: A high level of protection* (Groningen, Europa Law Publishing, 2005), p. 24 and 25.

[837] J. H. JANS and H. H. B. VEDDER, *European Environmental Law* (Groningen, Europa Law Publishing, 2008), p. 26 and 27. Examples of secondary legislation in this respect are Council Directive 92/43/EEC of 21 May 1992 on the conservation of natural habitats and of wild fauna and flora, *OJ* L 206/7 and Council Directive 1999/22/EC of 29 March 1999 relating to the keeping of wild animals in zoos, *OJ* L 94/24. Please note that the Court of Justice confirmed in its case law, Case C-189/01, *H. Jippes, Afdeling Groningen van de Nederlandse Vereniging tot Bescherming van Dieren and Afdeling Assen en omstreken van de Nederlandse Vereniging tot Bescherming van Dieren v Minister van Landbouw, Natuurbeheer en Visserij* [2001] ECR I-5689, paragraphs 71–79, that animal welfare cannot be regarded as a general principle of Community law. Nonetheless, the legal basis for animal welfare can be found in Article 38 of the TFEU (ex Article 32 of the EC Treaty) et seq.

[838] Reference to this objective and the human health objective can be found in the preamble of Directive 2006/7/EC of the European Parliament and of the Council of 15 February 2006 concerning the management of bathing water quality and repealing Directive 76/160/EEC, *OJ* L 64/37.

curative, etc.[839] It includes schemes or measures to prevent pollution, improve the state of the environment and to combat or clean up pollution.[840]

2. Protecting human health

667. The main question which rises is whether this objective is a wider concept than protecting *"public"* health. The latter seems to be more limited since it indicates to provide for measures which protect in a given society the collective health interests of people. On the contrary, the objective protecting *"human"* health is broader, as it encompasses not only the collective health interests, but also the interests of certain individuals or a group in a given society. In such cases, the principle of subsidiarity must be taken into account.

668. The Court of Justice, however in the *Criminal proceedings against Frans-Nederlandse Maatschappij voor Biologische Producten BV* case[841] made the distinction between these concepts somewhat blurred. This case concerned the application of Article 36 of the TFEU to measures restricting the importation of plant protecting products. Article 36 of the TFEU, as we discussed above, refers to *"the protection of health and life of humans, animals or plants"* and not *"public"* health. Nonetheless, the court equates the two concepts by stating in this judgment that: *"[I]n that respect, it is not disputed that the national rules in question are intended to protect public health and that they therefore come within the exception provided by Article 36* [Article 36 of the TFEU]*"*.[842, 843] The fact that the Treaty mentions the protection of human health as a separate objective of the Union's objectives in the environmental field allows the Union to adopt measures only with a subsidiary link to the environment.[844]

3. Prudent and rational utilisation of natural resources

669. This objective of paragraph 1 of Article 191 TFEU refers to *"prudent and rational utilisation of natural resources"* proven to be useful for the adoption of

839 J. H. JANS and H. H. B. VEDDER, *European Environmental Law* (Groningen, Europa Law Publishing, 2008), p. 26–29.

840 M. M. T. A. BRUS, R. R. RAIMOND and T. G. DRUPSTEEN, "Balancing National and European Competence in Environmental Law", *Connecticut Journal of International Law* (1993–1994), p. 647.

841 Case 272/80, *Criminal proceedings against Frans-Nederlandse Maatschappij voor Biologische Producten BV*, [1981] ECR 3277.

842 *Ibid.*, paragraph 13, emphasis added.

843 J. H. JANS and H. H. B. VEDDER, *European Environmental Law* (Groningen, Europa Law Publishing, 2008), p. 29 and 30.

844 M. M. T. A. BRUS, R. R. RAIMOND and T. G. DRUPSTEEN, "Balancing National and European Competence in Environmental Law", *Connecticut Journal of International Law* (1993–1994), p. 648 and D. VANDERMEERSCH, "The Single European Act and the Environmental Policy of the European Community, *European Law Review* (1987), p. 414.

energy efficiency legislation and other energy-related legislation and programmes.[845] It provides that one of the objectives of the common policy on the environment is to ensure that natural resources are used in a prudent and rational way. This third objective with respect to the management of natural resources indicates that it is an objective in itself which does not depend on whether or not, it is necessary to the quality of the environment.[846]

670. With respect to the concept of natural resources, a distinction can be made between renewable (e.g. wind, water, solar, etc.) and non-renewable natural resources (e.g. oil, natural gas, uranium, etc.). In some doctrine it is argued that nature conservation, soil protection, waste disposal, policy on urban areas, water management, energy saving, etc. fall under the management of natural resources.[847] Under principle 2 of the Stockholm Declaration of the United Nations Conference on the Human Environment[848], natural resources of the earth include air, water, land, flora and fauna. On the basis of the foregoing it can be concluded that the meaning of natural resources is very broad and consequently, this objective has a wide scope.[849]

671. The phrase "*prudent use*" means that the Union in its common environmental policy must aim to prevent pollution of natural resources and take measures to preserve their quality. If the purpose of a measure is to protect renewable natural resources, such a measure must aim at encouraging moderate use, thereby allowing their regular renewal. The phrase "*rational use*" means that the Union in its common environmental policy must aim at avoiding all kinds of waste and to use only the necessary quantity of natural resources to meet the present needs, in order to ensure that the needs of future generations are not compromised. The emphasis in this phrase is the adoption of measures to preserve the quality of natural resources. If the purpose of a measure is to protect

---

845 See Regulation No 2422/2001 of the European Parliament and of the Council of 6 November 2001 on a Community energy efficiency labelling programme for office equipment, *OJ* L 332/1; Directive 2003/30/EC of the European Parliament and of the Council of 8 May 2003 on the promotion of the use of biofuels or other renewable fuels for transport, *OJ* L 123/42; Directive 1999/94/EC of the European Parliament and of the Council of 13 December 1999 relating to the availability of consumer information on fuel economy and CO_2 emissions in respect of the marketing of new passenger cars, *OJ* L 12/16 and Decision No 1230/2003/EC of the European Parliament and of the Council of 26 June 2003 adopting a multiannual programme for action in the field of energy: "Intelligent Energy – Europe", *OJ* L 176/29.

846 M. M. T. A. BRUS, R. R. RAIMOND and T. G. DRUPSTEEN, "Balancing National and European Competence in Environmental Law", *Connecticut Journal of International Law* (1993–1994), p. 648 and H. G. SEVENSTER, *Milieubeleid en Gemeenschapsrecht: het interne juridische kader en de praktijk* (Deventer, Kluwer, 1992), p. 99.

847 H. G. SEVENSTER, *Milieubeleid en Gemeenschapsrecht: het interne juridische kader en de praktijk* (Deventer, Kluwer, 1992), p. 100.

848 www.unep.org/Documents.Multilingual/Default.asp?documentid=97&articleid=1503.

849 J. H. JANS and H. H. B. VEDDER, *European Environmental Law* (Groningen, Europa Law Publishing, 2008), p. 30 and 31.

non-renewable natural resources, such a measure must aim at preventing their rapid exhaustion.[850]

672. The inclusion of this objective in the EC Treaty was accompanied with the Declaration No 9 annexed to the Single European Act which stated that the Union's activities in the sphere of the environment may not interfere with national policies on the exploitation of energy sources. This provision in the declaration means that if the Union is exercising its competence in the environmental field, it is not allowed to take any decision which could have an impact on a Member State policy regarding the exploitation of energy sources. In the author's view, it is clear that this restriction on the Union's competences is limited to the *"exploitation"* of energy resources and not the *"use"* of energy resources. Consequently, the value of this restriction seems to be limited. Moreover in some doctrine it is argued that the restriction of the Union's competence is limited to the environmental field. In this doctrine it is therefore stated that it does not necessarily mean that the Union on the basis of another Treaty provision could perhaps take measures which interfere with a Member State's policy regarding the exploitation of energy sources.[851]

673. The Treaty on European Union did not confirm this declaration. On the contrary, it amended Article 192 of the TFEU by inserting paragraph 2 which provides that the Council may in the interest of its environmental policy, acting unanimously, adopt measures which may interfere in the field of the energy policy defined by the Member States.[852] This provision will be discussed in more detail below.

4. Promoting measures on an international level to deal with regional or worldwide environmental problems

(a) Meaning of this provision in Article 191 of the TFEU

674. It is obvious that environmental problems (e.g. water pollution or air pollution) do not stop at the borders of the Members States. However, whether or not the Union's competences in the environmental field were limited in a territorial sense was unclear in the beginning. The present wording of Article 191 of the TFEU and the explicit reference *"to promote measures at international level to deal with regional or worldwide environmental problems, and in particular*

850 Opinion Advocate General LÉGER of 16 May 2000 in the Case C-36/98, *Kingdom of Spain v Council of the European Union*, paragraphs 73–81.

851 N. SCHRIJVER, *Sovereignty over natural resources. Balancing rights and duties* (Cambridge, Cambridge University Press, 1997), 484 p.

852 Opinion Advocate General LÉGER of 16 May 2000 in the Case C-36/98, *Kingdom of Spain v Council of the European Union*, paragraphs 96–101.

combating climate change" allows the Union to achieve extraterritorial objectives in the environmental field. Nonetheless, several aspects of this objective are still unclear.

675. Does this provision intend to exclude unilateral measures since most of the current Union's extraterritorial measures aim at protecting the environment of all the Member States as well as third countries? Does the reference to *"regional or worldwide environmental problems"* mean that measures intended to protect the environment of one or only a few third countries are excluded? For example, if the import of non-sustainable biomass has specific consequences for the environment of one or a few Member States, it is debatable whether this amounts to a regional or worldwide environmental problem. The author is in favour of interpreting this environmental objective of paragraph 1 of the Article 191 of the TFEU rather broadly and not to exclude measures directed at protecting the environment in only one state or a few states.

676. In this context reference could be made to the rulings of the Court of Justice in the *Kramer* and *Drift-Net* cases[853] in which the Court ruled that: *"it follows from these provisions taken as a whole that the Community has at its disposal, on the internal level, the power to take any measures for the conservation of the biological resources of the sea, measures which include the fixing of catch quotas and their allocation between the different Member States. It should be made clear that, although Article 5 of Regulation No 2141/70 is applicable only to a geographically limited fishing area, it none the less follows Article 102 if the act of accession, from Article 1 of the said regulation and moreover from the very nature of things that the rule-making authority of the Community ratione materiae also extends – in so far as the Member States have similar authority under public international law – to fishing on the high seas"*.[854] This ruling clarifies that the Union can exercise powers outside the direct territory of the Member States. In this case the issue was whether the Union had the authority to regulate the fishing on high seas. However, such authority is conditional *on the subject matter of the functional jurisdiction falling within the material sphere of operation of the relevant Treaty provisions, and on the provisions themselves not containing any restriction limiting the territorial sphere of operation to the territory of the Member States.*

677. This condition means that the Union has the same rule-making authority in matters within its jurisdiction as that conferred on the Member State under

[853] Joined Cases 3, 4 and 6/76, *Cornelis Kramer and others* [1976] ECR 1279 and Case C-405/92, *Etablissements Armand Mondiet SA v Armement Islais SARL* [1993] ECR I-6133.

[854] Joined cases 3, 4 and 6/76, *Cornelis Kramer and others* [1976] ECR 1279, paragraphs 30–33, emphasis added.

international law. In the *Drift-Net* case, in which a measure which prohibited the use of driftnets was disputed on the grounds that the Union had no power to adopt such kind of measures, the Court of Justice confirmed its *Kramer* judgment.[855]

678. If we apply this approach of the Court of Justice to the environmental field, this would mean that the Union has extraterritorial authority in so far as the Member States are competent under international law to protect the environment outside their own territories and to the extent that the subject matter falls within scope of Article 191 of the TFEU. Nonetheless, in this context international law, including the provisions of the WTO needs to be taken into account. Consequently, any measure on the basis of Article 191 of the TFEU which brings the Union in conflict with its obligations under international law must be excluded.[856] On the basis of this analysis the next part will discuss the scope of the recently adopted Directive 2008/101/EC.

(b) Directive 2008/101/EC and Article 191 of the TFEU

(i) Background

679. In the 21st century air transport has become an integral part of society which enables passengers and freight to travel over long distances and contributes to European and global integration. However, aviation also has a negative side since it contributes to climate change.[857] Moreover, over the years air travel has become cheaper and thus the greenhouse gas emissions from aviation increase even faster.[858] For example, someone flying from London to New York and back seems to generate roughly the same level of greenhouse gas emissions as the average person in the EU does by heating their home for a whole year. Direct emissions from the aviation sector account for about 3% of the EU's

855 Case C-405/92, *Etablissements Armand Mondiet SA v Armement Islais SARL* [1993] ECR I-6133, paragraph 12 in which also reference was made to the Case C-61/77, *Commission of the European Communities v Ireland* [1978] ECR 417; Case C-258/99, *Commission of the European Communities v Kingdom of Spain* [1991] ECR 3977 and Case C-286/90, *Anklagemyndigheden v Peter Michael Poulsen and Diva Navigation Corp.* [1992] ECR I-6019.

856 J. H. JANS and H. H. B. VEDDER, *European Environmental Law* (Groningen, Europa Law Publishing, 2008), p. 34 and 35.

857 Proposal for a Directive of the European Parliament and of the Council amending Directive 2003/87/EC so as to include aviation in the scheme for greenhouse gas emission allowance trading within the Community, COM(2006)818, p. 2 and Commission staff working document of 20 December 2006, Accompanying document to the Proposal for a Directive of the European Parliament and of the Council amending Directive 2003/87/EC so as to include aviation in the scheme for greenhouse gas emission allowance trading within the Community Impact Assessment of the inclusion of aviation in the scheme for greenhouse allowance trading within the Community, SEC(2006)1684, p. 7.

858 See http://ec.europa.eu/clima/policies/transport/aviation/index_en.htm.

total greenhouse gas emissions of which the large majority comes from international flights (flights between two Member States or between a Member State and a third country).

680. Emissions from the aviation sector are higher than from other sectors, such as refineries and steel production, covered by the ETS of the EU. Unfortunately, the rapid growth in the aviation sector's greenhouse gas emissions contrasts with the success of many other sectors of the economy in reducing emissions. Moreover, without action, the growth in emissions from flights from EU airports will by 2012 cancel out more than a quarter of the 8% emission reduction the EU-15 must achieve to reach its Kyoto Protocol target. By 2020, greenhouse gas emissions of the aviation sector are forecasted to be more than twice the present levels.

681. Policies to address climate change have currently not required any substantial contribution from the aviation sector. The greenhouse gas emissions from domestic flights are covered by the Kyoto Protocol's reduction or limitation targets for developed countries. However, the vast majority of flights being international are not covered by the Kyoto Protocol. Instead, the Kyoto Protocol with respect to the international aviation emissions placed an obligation on its parties to pursue the limitation or reduction of greenhouse gases through the International Civil Aviation Organization.[859]

682. In order to address the growing climate change impact of aviation, the Directive 2008/101/EC, which amended Directive 2003/87/EC, was adopted in order to impose a cap on CO_2 emissions from all international flights by including emissions from aviation activities in the Union scheme.[860]

(ii) Main features of the Directive 2008/101/EC

683. Under Directive 2008/101/EC, and beginning on the 1st January 2012, all flights flying in, or out of a European Union airport will be covered under the ETS, unless these activities are included in the limited number of categories exempted under Annex I of the directive.[861] Directive 2008/101/EC covers aviation operators without discrimination on the basis of nationality. Consequently, all carriers, EU and non-EU, which fly to and from the EU fall under the obligations of the Directive 2008/101/EC. For the year 2012 the total quantity of allowances to

---

[859] *Ibid.*

[860] Proposal for a Directive of the European Parliament and of the Council amending Directive 2003/87/EC so as to include aviation in the scheme for greenhouse gas emission allowance trading within the Community, COM(2006)818, p. 2.

[861] See Article 3(a) and 3(c) of the Directive 2003/87/EC.

be allocated to aircraft operators[862] shall be equivalent to 97% of the average greenhouse gases emitted in the years 2004 to 2006. In 2013, the total quantity of allowances to be allocated shall be equivalent to 95%.[863] These allowances will be distributed for free or through auctioning to the aircraft operators. Initially, 15% of these allowances will be auctioned in a process which will be open to anyone and 85% of these allowances will be allocated free of charge.[864]

684. Aircraft operators who performed aviation activities covered by Directive 2008/101/EC are obliged to surrender after the conclusion of each calendar year a number of allowances equal to their total emissions in that year.[865] The total number of allowances which an aircraft operator must surrender is determined by the emissions of flights landing and departing in the European Union, calculated by means of fuel consumption. It is important to note that for the Directive 2008/101/EC it is irrelevant whether the fuel is consumed over the territory of one of the Member States of the Union or elsewhere. The Directive 2008/101/EC with these provisions, consequently does not set emission limits for aircraft operators nor does it regulate the construction, manoeuvring, design or any other aspect of the operation or navigation of an air carrier.

685. Another feature of the Directive 2008/101/EC which should be mentioned is the way in which the scheme can be adapted due to developments in third countries or at a global scale. This is addressed in Article 25a of the Directive 2003/87/EC of which the first paragraph states that *"[w]here a third country adopts measures for reducing the climate change impact of flights departing from that country which land in the Community, the Commission, after consulting with that third country, and with Member States within the Committee referred to in Article 23(1), shall consider options available in order to provide for optimal interaction between the Community scheme and that country's measures" [...].*[866]

(iii) Validity of Directive 2008/101/EC

686. The High Court of Justice of England and Wales, Queen's Bench Division requested the Court of Justice by order of 8 July 2010 to give a preliminary ruling

---

862 Under Article 3(o) of the Directive 2003/87/EC means aircraft operator *the person who operates an aircraft at the time it performs an aviation activity listed in Annex I or, where that person is not known or is not identified by the owner of the aircraft, the owner of the aircraft.*

863 See Article 3(c) and 3(d) of the Directive 2003/87/EC.

864 Legal service of the Council of the European Union, draft written observations to the president and the members of the Court of Justice of the European Union in the Case C-366/10, p. 7.

865 Article 12 of the Directive 2003/87/EC.

866 Legal service of the Council of the European Union, draft written observations to the president and the members of the Court of Justice of the European Union in the Case C-366/10, p. 7 and 8.

on questions with respect to the validity of Directive 2003/87/EC, as amended by Directive 2008/101/EC in order to include aviation activities in the scheme for greenhouse gas emission allowance trading within the Union. This reference for a preliminary ruling has been done in the course of proceedings before the High Court of Justice of England and Wales, Queen's Bench Division between The Air Transport Association of America, American Airlines Inc., Continental Airlines Inc., United Airlines, Inc. which are the applicants and The secretary of State for Energy and Climate Change of the United Kingdom of Great Britain and Northern Ireland which is the defendant.

687. More specifically, the Court of Justice has been sought to rule on a number of questions such as whether certain rules of international law (for example principle of customary law, Chicago convention, Kyoto Protocol and Open Skies Agreement) can be relied upon in order to challenge the validity of Directive 2003/87/EC, as amended by Directive 2008/101/EC. Is the amended directive invalid, if and insofar as it applies the ETS to those parts of flights (either generally or by aircraft registered in third countries) which take place outside the airspace of EU Member States? In the view of the applicants such an application of the Directive 2003/87/EC as amended by Directive 2008/101/EC would contravene one or more of the invoked principles of customary international law; is the Amended Directive invalid, if and insofar as it applies the Emissions Trading Scheme to those parts of flights (either generally or by aircraft registered in third countries) which take place outside the airspace of EU Member States: (a) as contravening Articles 1, 11 and/or 12 of the Chicago Convention and (b) as contravening Article 7 of the Open Skies Agreement?; etc.[867]

688. The next part will only make a comment with respect to the validity of Directive 2003/87/EC as amended by Directive 2008/101/EC, namely the territorial scope of this Directive, despite all the questions referred to the Court of Justice for a preliminary ruling since they fall outside the scope of this book,.

689. This territorial scope is defined in Article 355 of the TFEU and Article 52 of the TEU which also refers to Article 355 of the TFEU. In essence, the territorial field of application of the Treaties is constituted by the Member States of the European Union. Consequently, under international law, the Treaties apply to all areas which fall within the jurisdiction and under the sovereignty of the Member States. Moreover, the application of the Treaties extends to the airspace and maritime waters which fall within the jurisdiction and under the sovereignty of

---

[867] Case C-366/10, Reference for a preliminary ruling from High Court of Justice Queen's Bench Division (Administrative Court) (United Kingdom) made on 22 July 2010 – *The Air Transport Association of America, American Airlines, Inc., Continental Airlines, Inc., United Airlines, Inc. v The Secretary of State for Energy and Climate Change*, OJ C260/6.

the Member States. It could happen that the application of Union law also has extraterritorial effects.[868] Also case law of the Court of Justice clarified in the *Kramer* and *Drift-Net*-case, as discussed above, that the Union can exercise powers outside the direct territory of the Member States. It has the same rule-making authority in matters within its jurisdiction as that conferred on the Member State under international law.[869]

690. On the basis of the analysis above, it must be questioned whether Member States are competent under international law, to protect the environment outside there territory in such a way that they can oblige aircraft operators to surrender on an annual basis a certain number of allowances on the basis of the fuel which is consumed over their territory or elsewhere.

691. The Court of Justice in its judgment of the Case C-366/10 argued that the: "*European Union legislation may be applied to an aircraft operator when its aircraft is in the territory of one of the Member States and, more specifically, on an aerodrome situated in such territory, since, in such a case, that aircraft is subject to the unlimited jurisdiction of that Member State and the European Union (see, by analogy, Poulsen and Diva Navigation, paragraph 28)*". Moreover, according to the Court: "*It is only if the operator of such an aircraft has chosen to operate a commercial air route arriving at or departing from an aerodrome situated in the territory of a Member State that the operator, because its aircraft is in the territory of that Member State, will be subject to the allowance trading scheme*".[870]

692. With respect to the fact that the calculation of emission allowances to be surrendered is based on the whole international flight the Court of Justice argued that the: "*European Union policy on the environment seeks to ensure a high level of protection in accordance with Article 191(2) TFEU, the European Union legislature may in principle choose to permit a commercial activity, in this instance air transport, to be carried out in the territory of the European Union only on condition that operators comply with the criteria that have been established by the European Union and are designed to fulfil the environmental protection objectives which it has set for itself, in particular where those objectives follow on from an international agreement to which the European Union is a signatory, such as the Framework Convention and the Kyoto Protocol*".[871]

---

868 K. LENAERTS, and P. VAN NUFFEL, in BRAY, R. (ed.), *Constitutional Law of the European Union* (London, Thomson Sweet & Maxwell, 2005), p. 351–357.

869 See *Supra*.

870 Case C-366/10, Reference for a preliminary ruling from High Court of Justice Queen's Bench Division (Administrative Court) (United Kingdom) made on 22 July 2010 – *The Air Transport Association of America, American Airlines, Inc., Continental Airlines, Inc., United Airlines, Inc. v The Secretary of State for Energy and Climate Change* [2011] ECR I-0000, paragraphs 124 and 127.

871 *Ibid.*, paragraph 128.

693. On the basis of the foregoing the Court of Justice concluded that the EU had competence to adopt Directive 2008/101/EC even with the inclusion within the EU emission trading scheme of parts of flights that take place outside the territory of the EU.[872]

ii) Principles of the environmental policy

694. Paragraph 2 of Article 191 of the TFEU sets out the principles on which the Union's environmental policy is based:

(a) high level of protection principle;
(b) precautionary principle;
(c) prevention principle;
(d) source principle;
(e) polluter pays principle;
(f) safeguard clause.

1. HIGH LEVEL OF PROTECTION PRINCIPLE

695. Paragraph 2 of Article 191 of the TFEU states that the *"Union policy on the environment shall aim at a high level of protection taking into account the diversity of situations in the various regions"*. This principle is one of the most important substantive principles of the Union's environmental policy. As the Court of Justice held in its Safety Hi-Tech case[873] *"whilst it is undisputed that Article 130r(2) of the Treaty [*now paragraph 2 of Article 191 of the TFEU*] requires Community policy in environmental matters to aim for a high level of protection, such a level of protection, to be compatible with that provision, does not necessarily have to be the highest that is technically possible. […]*.[874]

696. It is apparent from the text of paragraph 3 of Article 114 of the TFEU that this principle is not only directed at the Commission, but also at the European Parliament and Council. Due to the wording *"seek to achieve"* in paragraph 3 of Article 114 of the TFEU, it remains uncertain whether this principle can be invoked in a court of law to challenge the legality of the environmental measures adopted by the Council.[875] The limited enforceability of this principle can also be

[872] See also Opinion Advocate General KOKOTT of 6 October 2011 in the Case C-366/10, Reference for a preliminary ruling from High Court of Justice Queen's Bench Division (Administrative Court) (United Kingdom) made on 22 July 2010 – *The Air Transport Association of America, American Airlines, Inc., Continental Airlines, Inc., United Airlines, Inc. v The Secretary of State for Energy and Climate Change*, paragraphs 145–160.

[873] Case C-284/95, *Safety Hi-Tech Srl v S. & T. Srl.* [1998] ECR I-4301. Emphasis added.

[874] *Ibid.*, paragraph 49, emphasis added.

[875] G. VAN CALSTER and K. DEKETELAERE, "Amsterdam, the Intergouvernmental Conference and Greening the EU Treaty", *European Environmental Law Review* (1998), p. 15

derived from the wording of the text in paragraph 2 of Article 191 of the TFEU which states that the *"Union policy on the environment shall aim at a high level of protection"*.[876]

2. PRECAUTIONARY PRINCIPLE

697. The primary purpose of the precautionary principle mentioned in paragraph 2 of Article 191 of the TFEU is to prevent serious or irreversible harm to the environment. It entails that preventive action should be taken in response to threats of environmental harm at an early stage, including in situations of scientific uncertainty. Under this principle the benefit of any doubt has been given to nature or *in dubio pro natura*.[877]

698. The most commonly cited formulation of the precautionary principle is the one of Principle 15 in the Rio Declaration of 1992 which states that: *"[i]n order to protect the environment, the precautionary approach shall be widely applied by States according to their capabilities. Where there are threats of serious or irreversible damage, lack of full scientific certainty shall not be used as a reason for postponing cost-effective measures to prevent environmental degradation"*.[878] This interpretation of the precautionary principle has also been endorsed by the Commission in its Communication on this principle which provides further guidance on how this principle should be applied.[879]

---

and L. KRÄMER, *EC Environmental Law* (London, Sweet and Maxwell, 2007), p. 11–12. The latter author argues that if a Commission "proposal" is not based on a high level of environmental protection, the European Parliament has under Article 263 of the TFEU (ex Article 230 of the EC Treaty) a right of action against the Commission for its failure. It is, as mentioned by the authors J. H. JANS and H. H. B. VEDDER, indeed unclear how a *"proposal of the Commission"* can be regarded an *"act"* under Article 263 of the TFEU (ex Article 230 of the EC Treaty) see J. H. JANS and H. H. B. VEDDER, *European Environmental Law* (Groningen, Europa Law Publishing, 2008), p. 37.

876 J. H. JANS and H. H. B. VEDDER, *European Environmental Law* (Groningen, Europa Law Publishing, 2008), p. 35–37. Emphasis added.

877 A. TROUWBORST, "International nature conservation law and the adaptation of biodiversity to climate change: a mismatch?", *Journal of Environmental Law* (2009), p. 419.

878 See also the preamble to the Convention on biological diversity which states that: *"where there is a threat of significant reduction or loss of biological diversity, lack of full scientific certainty should not be used as a reason for postponing measures to avoid or minimize such a threat"* and Article 3(3) of the 1992 Framework Convention on Climate Change (UNFCCC) which states that *"[t]he Parties should take precautionary measures to anticipate, prevent or minimize the causes of climate change and mitigate its adverse effects. Where there are threats of serious or irreversible damage, lack of full scientific certainty should not be used as a reason for postponing such measures, taking into account that policies and measures to deal with climate change should be cost-effective so as to ensure global benefits at the lowest possible cost."*

879 Communication of 2 February 2000 from the Commission on the precautionary principle, COM(2000)1, 28p.

699. According to this communication, the Commission considers that the Union has the right to establish the level of protection of the environment, of human, animal or plant health, it deems appropriate. It is furthermore specified that in the situations where action is deemed necessary, the measures taken on the basis of the precautionary principle should be proportionate to the chosen level of protection, non-discriminatory in their application, consistent with similar measures already taken, based on an examination of the potential benefits and costs or lack of action, subject to review in the light of new scientific data and capable of assigning responsibility for producing the scientific evidence necessary for a more comprehensive risk assessment.[880] The precautionary principle on the basis of this Communication has been described as a risk management tool which is part of a risk assessment framework. According to this approach precautionary action should only be taken after an objective quantitative risk assessment by experts.[881]

700. Also the Court of Justice applied the precautionary principle in various cases by ruling: *"where there is uncertainty as to the existence or extent of risks to human health, the institutions may take protective measures without having to wait until the reality and seriousness of those risks become fully apparent"*.[882] In the past the Union could easily use the first sentence of paragraph 3 of Article 191 of the TFEU, which states that in preparing its policy on the environment, the Union shall into take account *"available scientific and technical data"*, as a ground for not acting until there was sufficient data proving certain undesirable environmental effects. Currently, with the precautionary principle and the case law of the Court of Justice such an interpretation would not be accepted.[883]

701. The Court of Justice furthermore ruled that the inadequate nature of available scientific data does not preclude the Union or a Member State from

---

880 *Ibid.*, p. 2 and 3.

881 N. DE SAEDELEER, *Environmental Principles From Political Slogans to Legal Rules* (Oxford, Oxford University Press, 2002), p. 110 and 111.

882 Case C-180/96, *United Kingdom of Great Britain and Northern Ireland v Commission of the European Communities* [1998] ECR I-2265, paragraph 99; Case C-157/96, *The Queen v Ministry of Agriculture, Fisheries and Food, Commissioners of Customs & Excise, ex parte National Farmers' Union, David Burnett and Sons Ltd, R. S. and E. Wright Ltd, Anglo Beef Processors Ltd, United Kingdom Genetics, Wyjac Calves Ltd, International Traders Ferry Ltd, MFP International Ltd, Interstate Truck Rental Ltd and Vian Exports Ltd* [1998] ECR I-2211, paragraph 63; Case T-76/96 R, *The National Farmers' Union, International Traders Ferry Ltd, UK Genetics, RS & EM Wright Ltd and Prosper De Mulder Ltd v Commission of the European Communities* [1996] ECR II-815, paragraph 63; Case T-199/96, *Laboratoires pharmaceutiques Bergaderm SA and Jean-Jacques Goupil v Commission of the European Communities* [1998] ECR II-2805 and Case T-70/99 R, *Alpharma Inc. v Council of the European Union* [1999] ECR II-2027.

883 J. H. JANS and H. H. B. VEDDER, *European Environmental Law* (Groningen, Europa Law Publishing, 2008), p. 38.

adopting protective measures.[884] Nonetheless, the Court did not seem to give a *carte blanche* either in the sense that the burden of proof is reversed unreservedly[885], nor will mere hypothetical risks suffice for adopting measures.[886]

702. Consequently, protective measures seem to be acceptable only if a risk assessments has been carried out which is as complete as possible given the particular circumstances of the given case, from which it is apparent that the implementation of such protective measures in the light of the precautionary principle is necessary in order to ensure that there is no danger for the environment and human health.[887]

3. Prevention principle

703. The Union's policy on environment shall be based on the prevention principle which allows action to be taken in order to protect the environment at an early stage. It requires, *inter alia*, that measures are taken to prevent damage occurring at all and not take measures after damage has occurred. Instead of

---

884 Case C-236/01, *Monsanto Agricoltura Italia SpA and Others v Presidenza del Consiglio dei Ministri and Others* [2003] ECR I-8105, paragraph 112 which referred to the Case T-13/99, *Pfizer Animal Health SA v Council of the European Union* [2002] ECR II-3305, paragraph 160 and 162 and Case T-70/99, *Alpharma Inc. v Council of the European Union* [2002] ECR II-3495, paragraph 160 and 162.

885 Case C-314/99, *Kingdom of the Netherlands v Commission of the European Communities* [2002] ECR I-5521.

886 Case T-392/02, *Solvay Pharmaceuticals* v *Council* [2003] ECR II-4555, paragraph 129 and Case T-229/04, *Kingdom of Sweden v Commission of the European Communities* [2007] ECR II-2437.

887 J. H. JANS and H. H. B. VEDDER, *European Environmental Law* (Groningen, Europa Law Publishing, 2008), p. 38 and 39. For further analysis on the precautionary principle see: T. O'RIORDAN and J. CAMERON (eds.), *Interpreting the Precautionary principle* (London, Earthscan, 1994), 316 p.; J. MORRIS (ed.), *Rethinking Risk and the Precautionary Principle* (Oxford, Buttherworth-Heinmann, 2000) 294 p.; G. CORCELLO, "La perspective communautaire du principe de précaution", *Revue du Marché Commun et de l'Union Européenne* (2001), p. 447–454; J. SCOTT and E. VOS, "The Juridification of Uncertainty: Observations of the Ambivalence of the Precautionary Principle within the EU and the WTO" in C. JOERGES and R. DEHOUSSE (eds.), *Good Governance in Europe's Integrated Market* (Oxford, Oxford University Press, 2002), p. 25; W. T. DOUMA, *The Precautionary Principle: its Application in International, European and Dutch Law* (PhD., Groningen, 2003), 472 p.; E. FISHER, J. JONES and R. VON SCHOMBERG (eds.), *Implementing the Precautionary Principle: Perspectives and Prospects* (Cheltenham, Edward Elgar, 2006), 352 p.; T. O'RIORDAN, J. CAMERON and A. JORDAN (eds.), *Reinterpreting the Precautionary principle* (London, Cameron May, 2001), 284 p.; A. DOYLE and T. CARNEY, "Precaution and Prevention: Giving effect to Article 130r Without Direct Effect", *European Environmental Law Review* (1999), p. 44–47; N. DE SAEDELEER, *Environmental Principles From Political Slogans to Legal Rules* (Oxford, Oxford University Press, 2002), p. 91–226 and N. DE SAEDELEER, "The Precautionary Principle in EC Health and Environmental Law", *European Law Review* (2006), p. 139–172.

combating the effects of environmental nuisances, the Union should avoid them. It furthermore requires that in all planning and decision-making projects the environmental impact of these projects is considered as soon as possible.[888]

704. The fact that this principle was inserted into the Treaty has allowed the Court of Justice in the Joined cases C-175/98 and C-177/98[889] to interpret the provisions of the Waste framework Directive[890] in favour of the environment. The Court of Justice has ruled in these joined cases that *"Article 4 of Directive 75/442, adopted on the basis of Article 130s of the EC Treaty (now, after amendment, Article 175 EC), is intended to implement the principles of precaution and preventive action contained in the second sentence of the second paragraph of Article 130r of the Treaty. By virtue of those principles, it is for the Community and the Member States to prevent, reduce and, in so far as is possible, eliminate from the outset, the sources of pollution or nuisance by adopting measures of a nature such as to eliminate recognised risks. [...]*

In so far as waste, even waste which is stored temporarily, can cause serious harm to the environment, the provisions of Article 4 of Directive 75/442, which are intended to implement the principle of precaution, also apply to temporary storage".[891]

705. In many sectorial European Directives the prevention principle has been implemented such as the EIA Directive[892], ambient air quality directives[893] and waste directives.[894, 895] Also the Industrial Emission Directive is an important instrument to implement the principle. In particular Article 3 of this Directive

888 M. M. T. A. BRUS, R. R. RAIMOND and T. G. DRUPSTEEN, "Balancing National and European Competence in Environmental Law", *Connecticut Journal of International Law* (1993–1994), p. 650 and J. H. JANS and H. H. B. VEDDER, *European Environmental Law* (Groningen, Europa Law Publishing, 2008), p. 40–41.

889 Joined Cases C-175/98 and C-177/98, *Criminal proceedings against Paolo Lirussi (C-175/98) and Francesca Bizzaro (C-177/98)* [1991] ECR I-6881.

890 Council Directive 75/442/EEC of 15 July 1975 on waste, *OJ* L 194/39 which shall be repealed with effect from 12 December 2008 by Directive 2008/98/EC of the European Parliament and of the Council of 19 November 2008 on waste and repealing certain Directives, *OJ* L 312/3.

891 Joined Cases C-175/98 and C-177/98, *Criminal proceedings against Paolo Lirussi (C-175/98) and Francesca Bizzaro (C-177/98)* [1991] ECR I-6881, paragraphs 51 and 53.

892 Council Directive 85/337/EEC of 27 June 1985 on the assessment of the effects of certain public and private projects on the environment, *OJ* L 216/40 (the so-called EIA Directive) which has been amended in 1997, 2003 and 2009.

893 For example: Directive 2002/3/EC of the European Parliament and of the Council of 12 February 2002 relating to ozone in ambient air, *OJ* L 67/14 and Directive 2004/107/EC of the European Parliament and of the Council of 15 December 2004 relating to arsenic, cadmium, mercury, nickel and polycyclic aromatic hydrocarbons in ambient air, *OJ* L 23/3, etc.

894 For example: Directive 94/62/EC of the European Parliament and Council of 20 December 1994 on packaging and packaging waste, *OJ* L 365/10, etc.

895 J. H. JANS and H. H. B. VEDDER, *European Environmental Law* (Groningen, Europa Law Publishing, 2008), p. 40–42.

stipulates that Member States must adopt the necessary measures which provide that the competent authorities ensure that installations are operated in such a way that *"(a) all the appropriate preventive measures are taken against pollution, in particular through application of the best available techniques; (b) no significant pollution is caused; [...]; (d) energy is used efficiently; (e) the necessary measures are taken to prevent accidents and limit their consequences; [...]"*.[896] This principle furthermore calls for pollution control at the source rather than attempting to cure after the facts the harmful effects of emissions.[897, 898]

4. Source principle

706. The Union's policy on environment shall be based on the source principle. On the basis of this principle environmental damage should, as a priority, be rectified at its source. This principle complements the prevention principle.[899] It implies that damage to the environment should preferably not be prevented by applying an end-of-pipe technology. Besides, it prefers environmental quality standards instead of emission standards for water and air pollution.

707. In this context reference should be made to the Case C-2/90[900] in which the Court of Justice clarified the meaning of the source principle by stating that: *"imperative requirements can indeed be taken into account only in the case of measures which apply without distinction to both domestic and imported products (see inter alia the judgment in Joined Cases C-1/90 and C-176/90 Aragonesa de Publicidad Exterior and Publivía [1991] ECR I-4151). However, in assessing*

---

896 Another example of secondary legislation which implements this principle is the Council Directive 96/82/EC of 9 December 1996 on the control of major-accident hazards involving dangerous substances, *OJ* L 10/13 which stipulates in Article 5(1) that *"Member States shall ensure that the operator is obliged to take all measures necessary to prevent major accidents and to limit their consequences for man and the environment."*

897 N. DE SAEDELEER, "Environmental Principles – From Political Slogans to Legal Rules", *European Journal of International Law* 2005, p. 164. For a further analyses of the principle of prevention see N. DE SAEDELEER, *Environmental Principles From Political Slogans to Legal Rules* (Oxford, Oxford University Press, 2002), p. 61–90 and A. DOYLE and T. CARNEY, "Precaution and Prevention: Giving effect to Article 130r Without Direct Effect", *European Environmental Law Review* (1999), p. 44–47.

898 In particular the Third Environmental Action Programme focused on the prevention principle, *OJ* C46/1. The principle can have its full effect if according to the programme the following conditions are met: (a) the requisite knowledge and information must be improved and made readily available to decision-makers and all interested parties, including the public; (b) it is necessary to formulate and introduce procedures for judgment which will ensure that the appropriate facts are considered early in the decision-making process relating to any activity likely to affect the environment significantly and (c) the implementation of adopted measures must be monitored to ensure their correct application and their adaptation if circumstances or new knowledge should so require.

899 C. ZACKER, "Environmental Law of the European Economic Community: New Powers Under the Single European Act", *International and Comparative Law Review* (1991), p. 266.

900 Case C-2/90, *Commission of the European Communities v Kingdom of Belgium* [1992] ECR 4431.

whether or not the barrier in question is discriminatory, account must be taken of the particular nature of waste. The principle that environmental damage should as a matter of priority be remedied at source, laid down by Article 130r (2) of the Treaty as a basis for action by the Community relating to the environment, entails that it is for each region, municipality or other local authority to take appropriate steps to ensure that its own waste is collected, treated and disposed of; it must accordingly be disposed of as close as possible to the place where it is produced, in order to limit as far as possible the transport of waste".[901]

708. Not only the Case C-2/90 is relevant in this respect, but also the *Sydhavnens Sten & Grus* case[902] in which the Court of Justice ruled that *"[a]s regards the justification based on the protection of the environment, in particular the principle referred to in Article 130r(2) of the Treaty that environmental damage should as a priority be rectified at source, it must be pointed out that the protection of the environment cannot serve to justify any restriction on exports, particularly in the case of waste destined for recovery (see, to that effect, Case C-203/96 Dusseldorp and Others [1998] ECR I-4075, paragraph 49).*[903] *That is so a fortiori where, as in the case before the national court, environmentally non-hazardous building waste is involved"*. This ruling seems to indicate that Member States are allowed to impose export restrictions on waste if this is necessary for the protection of the environment.[904]

5. Polluter pays principle

709. According to this principle the polluter should bear the cost of measures to reduce it. This principle is one of the cornerstones of the European environmental policy even before it was incorporated into the Treaty. Since the polluter pays principle is an economic rule for cost allocation its source lies in the theory of externalities. It charges polluters the external costs arising from their pollution. When the polluter takes responsibility for all costs arising from pollution, internalisation is complete. However, the internalisation is incomplete when part or all the costs are shifted to the Union.

710. The meaning of the polluter pays principle experienced a gradual shift. In the beginning it was an instrument to prevent distortion of competition, later it became the basis to internalise and prevent chronic pollution. Subsequently, it

---

901 *Ibid.*, paragraph 34, emphasis added.

902 Case C-209/98, *Entreprenørforeningens Affalds/Miljøsektion (FFAD) v Københavns Kommune* [2000] ECR I-3743 and see also Case C-2/90, *Commission of the European Communities v Kingdom of Belgium* [1992] ECR I-4431 in which the Court of Justice in paragraph 34 clarifies the meaning of the source principle.

903 *Ibid.*, paragraph 48.

904 J. H. Jans and H. H. B. Vedder, *European Environmental Law* (Groningen, Europa Law Publishing, 2008), p. 42 and 43.

formed the basis to guarantee the integrated reparation of damage.[905] Some doctrine argued that the polluter pays principle is essential for the adoption of a preventive environmental protection policy as this principle makes it possible to provide funds to carry out such policy and, if necessary, to modify behaviour. However, the principle also appears to contain the idea that the right to pollute could be obtained if a monetary equivalent of the environmental cost is granted.[906]

6. Safeguard clause

711. Paragraph 2 of Article 191 of the TFEU furthermore stipulates that "*harmonisation measures answering environmental protection requirements shall include, where appropriate, a safeguard clause allowing Member States to take provisional measures, for non-economic environmental reasons, subject to a procedure of inspection by the Union*". This provision allows Member States to adopt measures protecting the environment in cases or urgency. This kind of safeguard clause can be found in different directives.[907]

712. According to paragraph 3 of Article 191 of the TFEU the Union in preparing its policy on the environment, shall take account of: (a) available scientific and technical data; (b) environmental conditions in the various regions of the Union; (c) the potential benefits and costs of action or lack of action and (d) the economic and social developments of the Union as a whole and the balanced development of its regions.

(a) Available scientific and technical data

713. This criterion under the Single European Act was to ensure that the European Union would only act if sufficient scientific data were available to prove that a given product or activity would have a harmful effect on the environment. However, as discussed with the precautionary principle, a different approach is more appropriate.[908] Any kind of provisional, indicative and

---

905 N. DE SAEDELEER, *Environmental Principles From Political Slogans to Legal Rules* (Oxford, Oxford University Press, 2002), p. 21.

906 *Ibid.*, p. 33 and 34 and 59 and 60.

907 See for example 11 of the Council Directive 91/414/EEC of 15 July 1991 concerning the placing of plant protection products on the market, *OJ* L 230/1 which states in its paragraph 1 that: "*Where a Member State has valid reasons to consider that a product which it has authorized or is bound to authorize under Article 10 constitutes a risk to human or animal health or the environment, it may provisionally restrict or prohibit the use and/or sale of that product on its territory. It shall immediately inform the Commission and the other Member States of such action and give reasons for its decision*". Another example is Article 32 of the Directive 98/8/EC of the European Parliament and of the Council of 16 February 1998 concerning the placing of biocidal products on the market, *OJ* L 123/1. Emphasis added and J. H. JANS and H. H. B. VEDDER, *European Environmental Law* (Groningen, Europa Law Publishing, 2008), p. 45 and 46.

908 See *Supra*.

tentative scientific data could now be considered as sufficient to take measures on a European level.

(b) Environmental conditions in the various regions of the Union

714. This criterion of paragraph 3 of Article 191 of the TFEU entails on the basis of the environment in a given region a differentiated environmental policy approach. In practice, this criterion can, for example, be found in Article 9 of the Water Framework Directive[909] with respect to the recovery of costs of water services which states that: *"[...] Member States may in so doing have regard to the social, environmental and economic effects of the recovery as well as the geographic and climatic conditions of the region or regions affected"*.

715. It seems, furthermore, to give preference to environmental quality objectives rather than emission limits. However, as discussed above, from the point of view of the source principle, there is a preference for emission standards rather than environmental quality standards. When environmental measures need to be adopted it is up to the Union's legislator to evaluate the merits of both approaches. Finally, it should be noted that this criterion could also be applied differently and lead to additional protective measures in order to conserve those areas in which the environmental quality is high.[910]

(c) Potential benefits and costs of action or lack of action

716. Since environmental action on a Union level does not only produce (should produce) benefits for the environment, but also entails costs for Member States (administrative organisation, enforcement, etc.) and private actors (manufactures and importers of goods and products which harm the environment), this criterion requires that the potential benefits and costs of action or lack of action, must be assessed. This criterion appears to be closely linked to the principle of proportionality and therefore adding little. The Industrial Emission Directive contains an example of this criterion in its definition of "best available techniques" by defining this concept as *"the most effective and advanced stage in the development of activities and their methods of operation which indicate the practical suitability of particular techniques for providing in principle the basis for emission limit values designed to prevent and, where that is not practicable, generally to reduce emissions and the impact on the environment as a whole: [...] (b) "available techniques" means those developed on a scale which allows implementation in the relevant industrial sector, under*

---

909 Directive 2000/60/EC of the European Parliament and of the Council of 23 October 2000 establishing a framework for Community action in the field of water policy, *OJ* L 327/1.

910 J. H. JANS and H. H. B. VEDDER, *European Environmental Law* (Groningen, Europa Law Publishing, 2008), p. 47 and 48.

economically and technically viable conditions, taking into consideration the costs and advantages, whether or not the techniques are used or produced inside the Member State in question, as long as they are reasonably accessible to the operator; […]".[911, 912]

(d) Economic and social developments of the Community as a whole and the balanced development of its regions

717. This provision is an elaboration of the principle mentioned in Article 27 of the TFEU and states that: *"[w]hen drawing up its proposals with a view to achieving the objectives set out in Article 26, the Commission shall take into account the extent of the effort that certain economies showing differences in development will have to sustain for the establishment of the internal market and it may propose appropriate provisions. […]"*. It allows the adoption of differentiated environmental policies or so-called multi-speed environmental policy depending on the economic and social development of certain regions.[913] The element of economic and social development which allows Member States to derogate from European standards could also be translated in terms of financial support by the Union for those Member States.[914, 915]

---

911 See Article 2(12) of the Industrial Emission Directive, emphasis added.

912 J. H. JANS and H. H. B. VEDDER, *European Environmental Law* (Groningen, Europa Law Publishing, 2008), p. 48 and 49.

913 See for example Article 6, (5) of the European Parliament and Council Directive 94/62/EC of 20 December 1994 on packaging and packaging waste, *OJ* L 365/10 which stipulates that "*Greece, Ireland and Portugal may, because of their specific situation, i.e. respectively the large number of small islands, the presence of rural and mountain areas and the current low level of packaging consumption, decide to […]*"(emphasis added) and see also the preamble, consideration 13 of the Directive 2000/60/EC of the European Parliament and of the Council of 23 October 2000 establishing a framework for Community action in the field of water policy, *OJ* L 327/1.

914 See for example the preamble of the Council Directive 92/43/EEC of 21 May 1992 on the conservation of natural habitats and of wild fauna and flora, *OJ* L 206/7 (hereinafter the Habitats Directive) which provides that: *"[w]hereas it is recognized that the adoption of measures intended to promote the conservation of priority natural habitats and priority species of Community interest is a common responsibility of all Member States; whereas this may, however, impose an excessive financial burden on certain Member States given, on the one hand, the uneven distribution of such habitats and species throughout the Community and, on the other hand, the fact that the "polluter pays" principle can have only limited application in the special case of nature conservation"* emphasis added; and Article 8 of the Habitats Directive which stipulates the possibility of a Community co-financing. Another example is paragraph 5 of Article 192 of the TFEU (ex Article 175 of the EC Treaty) which states that *"[w]ithout prejudice to the principle that the polluter should pay, if a measure based on the provisions of paragraph 1 involves costs deemed disproportionate for the public authorities of a Member State, such measure shall lay down appropriate provisions in the form of temporary derogations, and/or financial support from the Cohesion Fund set up pursuant to Article 177 [of the TFEU]."*

915 J. H. JANS and H. H. B. VEDDER, *European Environmental Law* (Groningen, Europa Law Publishing, 2008), p. 49 and 50.

(e) External competences of the Union in the environmental field

718. Finally, paragraph 4 of 191 of the TFEU contains the external competences of the Union in the environmental field. Since the entry into force of the Single European Act on 1 July 1987 the Union has, on the basis of paragraph 4 of Article 191 of the TFEU, the explicit competence to conclude international environmental agreements, binding to the institutions of the Union and on Member States.[916] Before the adoption of paragraph 4 in Article 191 of the TFEU, the Union cooperated in the environmental field with third countries on the basis of the implied powers doctrine.

719. This implied powers doctrine, developed by the Court of Justice, acknowledges that external competences could derive from internal competences in two situations. Firstly, in the situation where the Union has already adopted internal measures on the basis of its internal competences.[917] Secondly, in the situation where the Union has not yet adopted any internal measures, but where the exercise of external powers is necessary to obtain the objectives of the Union. This situation has been acknowledged by the Court of Justice in *Opinion 1/76*[918] in which it held that "*[...] the Court has already had occasion to state, most recently in its judgment of 14 July 1976 in joined cases 3, 4 and 6/76, Cornelis Kramer and Others, (1976) ECR 1279, that authority to enter into international commitments may not only arise from an express attribution by the Treaty, but equally may flow implicitly from its provisions. The Court has concluded inter alia that whenever Community law has created for the institutions of the Community powers within its internal system for the purpose of attaining a specific objective, the Community has authority to enter into the international commitments necessary for the attainment of that objective even in the absence of an express provision in that connexion*".[919]

720. This situation has in particular been important for the declaration made by the Union in the United Nations Framework Convention on Climate change.[920] In Annex C of this Convention, the Union and its Member States declared that "*the commitment to limit anthropogenic CO_2 emissions set out in Article 4(2) of the Convention will be fulfilled by the Community as a whole*

[916] See Article 216, (2) of the TFEU (ex Article 300, (7) of the EC Treaty).

[917] Case C-22/70, *Commission of the European Communities v Council of the European Communities – European Agreement on Road Transport* [1971] ECR 263.

[918] Opinion 1/76 of the Court of Justice of 26 April 1977 on the *Draft Agreement establishing a European laying-up fund for inland waterway vessels* [1977] ECR 741.

[919] *Ibid.*, paragraph 3, emphasis added. See also Joined Cases 3, 4 and 6/76, *Cornelis Kramer and others* [1976] ECR 1279, paragraph 19/20; Opinion 2/00 of the Court of Justice of 6 December 2001 on the *Cartagena Protocol* [2001] ECR 9713 and Case C-459/03, *Commission of the European Communities v Ireland* [2006] ECR I-4635.

[920] United Nations Framework Convention on Climate Change – Declarations, *OJ* L 33/13.

through action by the Community and its Member States, within the respective competence of each". When this Convention was concluded, there was on a Union level no internal legislation in force on which the competence to implement the commitment set out in Article 4, (2) of the Convention could be based. Consequently, the conclusions of the Court of Justice in the *Cornelis Kramer and others* case[921] and its opinion 1/76 have been implicitly applied by the Council which based its competence on paragraph 1 of Article 192 of the TFEU.[922]

721. The first subparagraph of paragraph 4 of Article 191 of the TFEU states that: *"[w]ithin their respective spheres of competence, the Union and the Member States shall cooperate with third countries and with the competent international organisations. The arrangements for Union cooperation may be the subject of the agreements between the Union and the third parties concerned"*. It is clear from this provision that the material scope of the external competences of the Union in the environmental field is determined by the objectives contained in paragraph 1 of Article 191 of the TFEU and the principles referred to in paragraph 2 of Article 191 of the TFEU.

722. The competence of the Union in the environmental field must be regarded as a shared competence which implies that the Union and the Member States are allowed to legislate and adopt legally binding acts in the environmental field. Nonetheless, Member States shall only exercise their competence to the extent that the Union has not exercised, or has decided to cease exercising, its competence. In this respect, reference should be made to the second subparagraph of paragraph 4 of Article 191 of the TFEU which states that: *"[t]he previous subparagraph shall be without prejudice to Member States' competence to negotiate in international bodies and to conclude international agreements"*.

723. The adoption of this second subparagraph, inserted by the Single European Act, caused confusion and could be contrary to the existing case law, and in particular the *Commission v Council* case or the so-called *ERTA* (European Agreement on Road Transport) case[923], be interpreted in such a way that a Member State could still negotiate international agreements in the environmental field even if the Union had already adopted internal measures in that field. In order to avoid such an interpretation, the Final Act of the Single European Act added a declaration which stated that the second subparagraph of paragraph 4 of the Article 191 of the TFEU does not affect the principles

---

921 Joined Cases 3, 4 and 6/76, *Cornelis Kramer and others* [1976] ECR 1279.

922 J. H. JANS and H. H. B. VEDDER, *European Environmental Law* (Groningen, Europa Law Publishing, 2008), p. 58–60.

923 Case C-22/70, *Commission of the European Communities v Council of the European Communities – European Agreement on Road Transport* [1971] ECR 263.

resulting from the judgment handed down by the Court of Justice in the *ERTA* case. This declaration indicates clearly that the TFEU gives an express competence to the Union to conclude international agreements in the environmental field.[924]

724. In the *ERTA* case, the Court of Justice ruled that if the competence in internal matters has been transferred from the Member States to the Union, the implied competence of the Union is exclusive. Consequently, the exclusivity of the Union's external competence will depend on whether the Union legislator has taken measures internally or externally. This exclusivity implies that if the Union adopted internal measures, a Member State no longer has the right to adopt obligations which would affect those measures. With a total harmonisation in a given field, there will be a transfer of internal competence to the Union legislator which results in an exclusive external competence. Consequently, a Member State in that field will no longer have any competence of its own.

725. However, in the environmental field on a European level, a large part of the adopted measures contain minimum standards instead of a total harmonisation which expressly allows Member States to adopt more stringent measures. In this case, there can be no question of exclusive external competence of the Union. The Member States and the Union have to exercise this competence jointly.[925] Besides, Article 193 of the TFEU allows the Member States to adopt more stringent environmental standards than the European ones. This means that a wholly exclusive external Union competence in the environmental field is not possible.[926]

726. Before the *Opinion 2/00*[927] of the Court of Justice, the Commission and the Council had a different view on the most appropriate legal basis to adopt

---

924 J. H. JANS and H. H. B. VEDDER, *European Environmental Law*, 3rd ed. (Groningen, Europa Law Publishing, 2008), p. 62.

925 See for example the declaration concerning the competence if the European Community with regard to matters governed by the United Nations Convention on the law of the Sea of 10 December 1982 and the agreement of 28 July 1994 relating to the implementation of part XI of the Convention which states "2. Matters for which the Community shares competence with its Member States: "*[...] with regard to the provisions on maritime transport, safety of shipping and the prevention of marine pollution contained, inter alia, in Parts II, III, V, VII and XII of the Convention, the Community has exclusive competence only to the extent that such provisions of the Convention or legal instruments adopted in implementation thereof affect common rules established by the Community. When Community rules exist but are not affected, in particular in cases of Community provisions establishing only minimum standards, the Member States have competence, without prejudice to the competence of the Community to act in this field. Otherwise competence rests with the Member States*", emphasis added.

926 J. H. JANS and H. H. B. VEDDER, *European Environmental Law* (Groningen, Europa Law Publishing, 2008), p. 61–64.

927 Opinion 2/00 of the Court of Justice of 6 December 2001 on the *Cartagena Protocol* [2001] ECR 9713.

decisions to adhere to an international environmental agreement, convention or protocol. The Council and the Member States were of the opinion that Article 192 of the TFEU constituted the appropriate legal basis, whereas the Commission was in favour of paragraph 4 of Article 192 of the TFEU and 207 of the TFEU.

727. This controversy has been clarified in *Opinion 2/00* of the Court of Justice in which it was asked on which legal basis the Council decision to adhere to the Protocol of Cartagena should be based, according to the court: "*[*it*] already held (see Peralta, cited above, paragraph 57, and Safety Hi-Tech, cited above, paragraph 43), [*that*] Article 174 EC defines the objectives to be pursued in the context of environmental policy, while Article 175 EC constitutes the legal basis on which Community measures are adopted. It is true that Article 174(4) EC specifically provides that the arrangements for Community cooperation with non-member countries and international organisations may be the subject of agreements [...] negotiated and concluded in accordance with Article 300. However, in the present case, the Protocol does not merely establish arrangements for cooperation regarding environmental protection, but lays down, in particular, precise rules on control procedures relating to transboundary movements, risk assessment and management, handling, transport, packaging and identification of LMOs.*

Consequently, Article 175(1) EC is the appropriate legal basis for conclusion of the Protocol on behalf of the Community".[928]

728. Consequently, since practically all international environmental agreements, conventions and protocols do not merely establish arrangements for cooperation regarding environmental protection, the most appropriate legal basis is paragraph 1 of Article 192 of the TFEU and not paragraph 4 of Article 191 of the TFEU. Paragraph 4 of Article 191 of the TFEU will therefore only be used in rather exceptional circumstances.[929]

(c) Legal basis in the TFEU for environmental measures

729. While Article 191 of the TFEU defines the objectives to be pursued by the Union in the context of the environmental policy, Article 192 of the TFEU constitutes the legal basis on which environmental measures can be adopted. The first paragraph of Article 192 of TFEU stipulates that the European Parliament and the Council, acting in accordance with the ordinary legislative procedure of Article 294 of the TFEU, shall decide what action is to be taken by the Union in order to achieve the objectives of Article 191 of the TFEU in the environmental policy. The fact that the bulk of environmental measures are subject to the ordinary legislative procedure does not automatically lead to more

928 *Ibid.*
929 L. KRÄMER, *EC Environmental Law* (London, Sweet & Maxwell, 2007), p. 92–93.

environmentally-friendly legislation on a European level. Nonetheless, the currently ordinary legislative procedure is very welcomed.

730. Besides, from a procedural point of view there is no longer any difference between the choice of Article 114 of the TFEU and the first paragraph of Article 192 of the TFEU. Article 114 of the TFEU is broadly framed. However, the Court of Justice has confirmed that this provision does not give the Union legislator a general competence to regulate the internal market.[930, 931]

731. According to the second paragraph of Article 192 of the TFEU the Council, by way of derogation from the ordinary legislative procedure (first paragraph Article 192 of the TFEU) and without prejudice to the provisions of Article 114 of the TFEU, shall act unanimously in accordance with a special legislative procedure on a proposal from the Commission and after consulting the European Parliament, the Economic and Social Committee and the Committee of the Regions for the adoption of:

(a) provisions primarily of a fiscal nature;
(b) measures affecting:
  (i) town and country planning;
  (ii) quantitative management of water resources or affecting, directly or indirectly, the availability of those resources;
  (iii) land use, with the exception of waste management;
(c) measures significantly affecting a Member State's choice between different energy sources and the general structure of its energy supply.

i) Provisions primarily of a fiscal nature

732. The first uncertainty with respect to this provision is the meaning of the term *primarily*.[932] It remains unclear which measures can be qualified as primarily or not primarily of a fiscal nature. With regard to the interpretation of the term *fiscal nature* there is nothing in the Treaty to indicate how that term

930 See *Infra*.

931 J. H. JANS and H. H. B. VEDDER, *European Environmental Law* (Groningen, Europa Law Publishing, 2008), p. 53. See also K. DEKETELAERE, "European Environmental Tax Policy: Proposal for a Policy Vision and a Legal Framework", *EELR* (1996), p. 9–15 which argues that when measures have ecological purposes *"at stake, which are, moreover, of a primarily fiscal nature, Article 130S, second part, first line, EC Treaty seems to be the most advisable Treaty basis for these European environmental taxes"*. Moreover, this author suggests that *"depending on their contents, [...] "clean-up" measures can be based on the following articles of the EC Treaty: Article 130S(2) (ecological measures which are primarily of a fiscal nature); Article 130S(1) (ecological measures which are not primarily of a fiscal nature); [...]."*

932 *Ibid.*, p. 54.

should be construed. Not only Article 192 of the TFEU refers to this term, but also for example Article 30 of the TFEU. A similar term can be found in paragraph 2 of Article 114 of the EC Treaty which refers to *fiscal provisions*. With respect to the meaning of the latter term, the Court of Justice ruled in the *Commission v Council* case[933] that: "*by reason of their general character, those words cover not only all areas of taxation, without drawing any distinction between the types of duties or taxes concerned, but also all aspects of taxation, whether material rules or procedural rules*".[934]

733. Consequently, any legislative initiative which introduces a tax on CO_2 or other emissions clearly falls within the scope of paragraph 2 of Article 192 of the TFEU. Measures which introduce a CO_2 tax are of a fiscal nature and designed to achieve environmental goals. Since such measures fall within the scope of paragraph 2 of Article 192 of the TFEU, they can only be taken unanimously in accordance with a special legislative procedure.[935]

ii) Measures affecting town and country planning, quantitative management of water resources and land use

734. The meaning of the term *affecting* referred to in the second sentence of paragraph 2 of Article 192 of the TFEU is also unclear. However, the Court of Justice has provided some guidance in its ruling in the *Spain v Council* case[936] by stating that "*[...] measures relating to town and country planning and to land use with the exception of waste management and measures of a general nature. These are measures which, just like those based on Article 130s(1) of the Treaty, are intended to attain the objectives referred to in Article 130r of the Treaty, but which regulate the use of the territory of the Member States, such as measures relating to regional, urban or rural management plans or the planning of various projects concerning the infrastructure of a Member State*".[937]

---

933 Case C-338/01, *Commission of the European Communities v Council of the European Union* [2004] ECR I-4829.

934 *Ibid.*, paragraph 63.

935 J. H. JANS and H. H. B. VEDDER, *European Environmental Law* (Groningen, Europa Law Publishing, 2008), p. 54 and 55. These authors also discussed the unclear relationship between paragraph 2 of Article 192 of the TFEU (ex Article 175 of the EC Treaty) and Article 113 of the TFEU (ex Article 93 of the EC Treaty). They conclude that since the decision-making procedure of both provisions is similar, the practical importance between these two provisions is relative. However, as they emphasised correctly, the legal consequences are an important difference between both provisions. In particular, the legal consequences on the basis of Article 193 of the TFEU (ex Article 176 of the EC Treaty) which are attached to measures adopted on the basis of Article 192 of the TFEU (ex Article 175 of the EC Treaty).

936 Case C-36/98, *Kingdom of Spain v Council of the European Union* [2001] ECR I-779.

937 *Ibid.*

735. Furthermore, it should be noted that an exception of measures affecting land use is made for measures on waste management. Consequently, the unanimity requirement in accordance with the special legislative procedure shall not apply to measures dealing with waste management. In these cases, on the basis of the first paragraph of Article 192 of TFEU, the ordinary legislative procedure of Article 294 of the TFEU shall apply.[938]

iii) Measures significantly affecting a Member State's choice between different energy sources and the general structure of its energy supply

736. This third indent of the first subparagraph of the second paragraph of Article 192 of the TFEU means that only measures to protect the environment which *significantly* affect the choice made by a Member State between different sources of energy and the general structure of its energy supply must be adopted in accordance with the procedure laid down in paragraph 2 of Article 192 of the TFEU.[939]

737. In the first place, it is obvious that the threshold of "*measures significantly affecting*" is much higher than "*measures affecting*".[940] A Member State's choice between different energy sources means the power of a Member State to decide which kind of energy sources can be used to cover its demand for energy. With respect to the concept of energy sources three main categories fossil fuels, renewable sources and nuclear sources can be distinguished. The nuclear-powered sources are fusion and fission. The renewable energy sources are wind, solar, aerothermal, geothermal, hydrothermal and ocean energy, hydropower, biomass, landfill gas, sewage treatment plant gas and biogases. The fossil fuels are petroleum, coal and natural gas.

738. The general structure of its energy supply means the power of a Member State to organise the manner of its energy supply. Examples in this context are the Member States' choice to opt for a nuclear phase-out, to set a target for renewable energy sources, to replace all the coal fired power plants by gas fired power plants, not to allow the construction of coal-fired power plants anymore, etc.

739. This provision of Article 192 of the TFEU should be compared with the Declaration No 9 annexed to the Single European Act which states that "*[t]he*

938 J. H. JANS and H. H. B. VEDDER, *European Environmental Law* (Groningen, Europa Law Publishing, 2008), p. 55 and 56.

939 Opinion Advocate General LÉGER of 16 May 2000 in the Case C-36/98, *Kingdom of Spain v Council of the European Union*, paragraphs 96 and 97.

940 J. H. JANS and H. H. B. VEDDER, *European Environmental Law* (Groningen, Europa Law Publishing, 2008), p. 56 and 57. Emphasis added.

Conference confirms that the Community's activities in the sphere of the environment may not interfere with national policies regarding the exploitation of energy sources". The declaration means that if the Union is exercising its competence in the environmental field, it is not allowed to take any decision which could have an impact on a Member State's policy regarding the exploitation of energy sources. The Conference adopted this declaration because the Member States were concerned that, on the basis of the wording of that time of, Article 130r EEC Treaty (now Article 191 of the TFEU) in conjunction with Article 130s EEC Treaty (now Article 192 of the TFEU) might enable the Union to encroach upon their right to exploit some of their national resources which are important for their energy policy. On the basis of these provisions certain national resources such as oil, natural gas and uranium, which are important for the production of energy in most Member States, could have been the object of Union measures in order to achieve the environmental objective of *"prudent or rational utilisation of natural resources".* Such an interpretation of the Union competences in the environmental field was not an option for the Member States. Member States wish to this very day, to preserve wide powers in the energy field because measures in this field may have effects on their national economies.

740. The Treaty on European Union did not confirm this declaration; on the contrary, it amended Article 192 of the TFEU by inserting paragraph 2 which provides that the Union may, in the interest of its environmental policy, adopt measures on energy policy defined by the Member States. Measures in the interest of the environmental policy which interfere with the energy policy defined by Member States and which may have consequences in an area subject to the near-sovereign powers of the Member States must be adopted by the Council acting unanimously in accordance with the special legislative procedure.

741. Measures which could qualify as measures *significantly affecting a Member State's choice between different energy sources and the general structure of its energy supply* are for example the measures on a European level which oblige the Member States to construct wind turbines, hydroelectric dams, biomass plants, to phase out nuclear power plants in order to conserve non-renewable natural resources used to produce energy, such as natural gas, oil, coal or uranium. The rationale of these measures would be that due to the exploitation of these non-renewable natural resources they would be exhausted within a short period of time. Conversely, in order to safeguard water supplies the construction of hydro-electric dams, to safeguard the population of birds limit the construction of windmills or to limit emissions of greenhouse gases the construction of coal-fired power plants and gas-fired power plants, would be prohibited in certain areas of the Union.[941]

742. Under the second subparagraph of the second paragraph of Article 192 of the TFEU *"the Council, acting unanimously on a proposal from the Commission and after consulting the European Parliament and Social Committee of the Regions, may make the ordinary legislative procedure applicable to the matters referred to in the first subparagraph"*. This provision specifies that the Council, by means of self-delegation, could stipulate that measures which should be established under paragraph 2 of Article 192 of the TFEU can be adopted in accordance with the ordinary legislative procedure instead of the special legislative procedure.

743. The third paragraph of Article 192 of the TFEU states that *"general action programmes setting out priority objectives to be attained shall be adopted by the European Parliament and the Council, acting in accordance with the ordinary legislative procedure and after consulting the Economic and Social Committee and the Committee of the Regions. The measures necessary for the implementation of these programmes shall be adopted under the terms of paragraph 1 or 2, as the case may be"*. The most recent example of such a programme is the Sixth Environment Action Programme of the European Union 2002–2012.[942] This programme gives a broad look at the environmental challenges and provides for a strategic framework for the Commission's environmental policy up to 2012.[943] Such programmes on the environment can cover the whole environmental spectrum, but the measures in order to obtain the goals of these programmes must, depending on the subject matter, be adopted in accordance with the procedures mentioned in the first or second paragraph of Article 192 of the TFEU.

744. Paragraph 4 of Article 192 of the TFEU provides for the obligation for Member States to finance and implement the environmental policy of the Union.[944] Finally, paragraph 5 of Article 192 of the TFEU allows a temporary derogation or financial support from the Cohesion Fund[945] if a measure based on paragraph 1 of Article 175 involves disproportionate costs for a given Member State.[946]

---

941 Opinion Advocate General LÉGER of 16 May 2000 in the Case C-36/98, *Kingdom of Spain v Council of the European Union*, paragraphs 98–101.

942 Decision 1600/2002/EC of the European of the European Parliament and of the Council of 22 July 2002 laying down the Sixth Community Environment Action Programme, *OJ* L 242/1.

943 For further details see: http://ec.europa.eu/environment/newprg/index.htm.

944 See *Supra*.

945 The Cohesion Fund is set up pursuant to Article 177 of the TFEU (ex Article 161 of the EC Treaty).

946 See *Supra*.

(d) *Derogation clauses of the TFEU in the environmental field*

745. In the TFEU two provisions allow a stricter approach in the environmental field. The first legal basis in this respect is Article 193 of the TFEU which grants Member States the discretion to maintain or introduce more stringent national measures adopted pursuant to Article 192 of the TFEU, provided that these measures are not incompatible with the Treaty. The second provision is the so-called "Environmental Guarantee Clause" in paragraphs 4–6 of Article 114 of the TFEU which stipulates that the Commission on certain conditions could allow Member States to maintain under paragraph 4 of Article 114 of the TFEU or to enact under paragraph 5 of Article 114 of the TFEU, more stringent measures. This Environmental Guarantee Clause grants, contrary to Article 193 of the TFEU, the Commission – and not the Member States – the discretionary power to assess whether more stringent measures can be allowed, unless in the absence of a decision by the Commission.[947, 948]

746. Since these paragraphs allow a Member State to derogate from the principles of uniform application of Union's law and the unity of the market, they must be interpreted strictly. The General Court and Court of Justice confirmed that if a Member State invokes paragraphs 4–6 of Article 114 of the TFEU, it must prove that the conditions for application of that provision have been met.[949] Paragraphs 4–6 of Article 114 of the TFEU only allow Member States to adopt measures which aim to achieve a greater degree of protection than the harmonisation measure. This means that less stringent national standards derogating from the harmonisation measure will not be allowed.[950]

---

947 Article 114(6) of the TFEU.

948 The author Peter Pagh in his contribution "The Battle on Environmental Policy Competences – Challenging the Stricter Approach: Stricter Might Lead to Weaker Protection" states that 22 requests for using the clause have been decided by the Commission of which 15 cases requested to maintain stricter measures and 7 cases requested to enact more stringent measures. The Commission only approved in eights cases that the Member States could maintain stricter measures all in respect of the Council Directive 76/769/EEC of 27 July 1976 on the approximation of the laws, regulations and administrative provisions of the Member States relating to restrictions on the marketing and use of certain dangerous substances and preparations, *OJ* L 262/201. A conditional authorisation was granted in five cases often for a limited period of time. In one case the Commission found the request not admissible and in the last eight cases, the Commission rejected the applications of the given Member States. For further details see P. PAGH, "The Battle on Environmental Policy Competences – Challenging the Stricter Approach: Stricter Might Lead to Weaker Protection" in R. MACRORY (ed.), *Reflections on 30 years of EU Environmental law – a high level of protection* (Groningen, Europa Law Publishing, 2006), p. 5–7.

949 Case C-3/00, *Kingdom of Denmark v Commission of the European Communities* (or the so-called *Danish Foodstuff* case) [2003] ECR I-2643, paragraph 84 and Joined Cases T-366/03 and T-235/04, *Land Oberösterreich and Republic of Austria v Commission of the European Communities* [2005] ECR II-4005, paragraph 63.

950 J. H. JANS and H. H. B. VEDDER, *European Environmental Law* (Groningen, Europa Law Publishing, 2008), p. 111–112.

747. Member States which invoke paragraph 5 of Article 114 of the TFEU must show that there are circumstances specific to that Member State which justifies the more stringent measures. Consequently, as the General Court emphasised in the *Land Oberösterreich and Republic of Austria v Commission of the European Communities* case[951] that a simple statement of Member States stating that "*we want stricter environmental legislation*" will not pass the test.[952] The General Court in its *Dutch Diesel Engine* case[953] clarified this requirement further and stated that: "*[a]ny problem which arises in terms which are on the whole comparable throughout the Member States and which lends itself, therefore, to harmonised solutions at Community level is general in nature and is, consequently, not specific within the meaning of Article 95(5) EC.*

It is therefore necessary, in order correctly to interpret Article 95(5) EC, to envisage the requirement of national specificity of a problem essentially from the angle of the aptness or inaptness of the harmonisation of the applicable Community rules to confront adequately the difficulties encountered locally, since the established inaptness of those rules justifies the introduction of national measures".[954] Thus, the Member State must prove that due to the local nature of the problem, the adopted measures on a European level are insufficient. This requirement does not imply that the problem in the given Member State is the result of an environmental danger within that State alone; there is no need to prove that the problem is exclusive to that Member State.[955]

748. The requirement of a problem specific to that Member State cannot be found – at least not explicitly – contrary to paragraph 5 of Article 114 of the TFEU, in paragraph 4 of Article 114 of the TFEU. However, in the *Danish Foodstuff* case[956] the Court of Justice has ruled that "*it follows that neither the wording of Article 95(4) EC nor the broad logic of that article as a whole entails a requirement that the applicant Member State proves that maintaining the national provisions which it notifies to the Commission is justified by a problem specific to*

951 Joined Cases T-366/03 and T-235/04, *Land Oberösterreich and Republic of Austria v Commission of the European Communities* [2005] ECR II-4005. Please note that this judgment has been upheld by the Court of Justice in its judgment of the Joined Cases C-439/05P and C-454/05P, *Land Oberösterreich and Republic of Austria v Commission of the European Communities* [2007] ECR I-7141.

952 *Ibid.*, paragraph 65 in which the General Court stated that: "*[...] no scientific evidence was presented which showed that this area of Austria had unusual or unique ecosystems that required separate risk assessments from those conducted for Austria as a whole or for other similar areas of Europe.*"

953 Case T-182/06, *Kingdom of the Netherlands v Commission of the European Communities* [2007] ECR II-1983.

954 *Ibid.*, paragraphs 63 and 64.

955 J. H. JANS and H. H. B. VEDDER, *European Environmental Law* (Groningen, Europa Law Publishing, 2008), p. 116.

956 Case C-3/00, *Kingdom of Denmark v Commission of the European Communities* (or the so-called *Danish Foodstuff* case) [2003] ECR I-2643.

that Member State".[957] One might presume that on the basis of this ruling, the specific ground criterion is not necessary, if a Member State invokes paragraph 4 of Article 114 of the TFEU. However, the Court of Justice in the next paragraph of the same judgment states that: *"however, when a problem specific to the applicant Member State in fact exists, that circumstance can be highly relevant in guiding the Commission as to whether to approve or reject the notified national provisions. It is a factor which, in the present case, the Commission should have taken into account when it adopted its decision"*.[958] Thus, paragraph 4 of Article 114 of the TFEU does not require the existence of a specific problem to that Member State, but according to the Court of Justice it is *"highly relevant"* for the Commission.[959]

749. This "Environmental Guarantee Clause" has been rarely used and mainly in politically high profile cases by the Member States in order to demonstrate their environmental intentions. From an overall environmental perspective, the derogations on the basis of this clause did not seem to add any substantial element to the environmental protection.[960]

750. The other legal basis allowing a stricter approach in the environmental field is Article 193 of the TFEU.[961] Compared to the "Environmental Guarantee Clause", this derogation is not subject to prior review by the Commission. It follows from the case law of the Court of Justice that the consequences of taking more stringent measures, should be consistent with the pursued objective of the given measure.[962] Moreover, Article 193 of the TFEU does not allow Member States to adopt less stringent measures and it cannot be used to adopt measures in connection with European environmental legislation based on other

957 *Ibid.*, paragraph 59.

958 *Ibid.*, paragraph 60. Emphasis added.

959 J. H. JANS and H. H. B. VEDDER, *European Environmental Law* (Groningen, Europa Law Publishing, 2008), p. 116 and 117. These authors furthermore discussed in detail the role of the Commission under paragraph 6 of Article 114 of the TFEU (ex Article 95 of the EC Treaty), see J. H. JANS and H. H. B. VEDDER, *European Environmental Law* (Groningen, Europa Law Publishing, 2008), p. 117–121.

960 P. PAGH, "The Battle on Environmental Policy Competences – Challenging the Stricter Approach: Stricter Might Lead to Weaker Protection" in R. MACRORY (ed.), *Reflections on 30 years of EU Environmental law – a high level of protection* (Groningen, Europa Law Publishing, 2006), p. 7.

961 See H. SOMSEN, "Discretion in European Community Environmental Law: An analysis of ECJ Case Law", *C.M.L.R.* (2003), p. 1413–1453; Case C-379/92, *Peralta* [1994] ECR I-3453, paragraph 57; Case C-284/95, *Hi-Tech* [1998] ECR I-4301, paragraph 43 and Case C-86/03, *Greece v Commission* [2005] ECR I-10979, paragraph 88.

962 *Ibid.*, p. 8 and J. H. JANS and H. H. B. VEDDER, *European Environmental Law* (Groningen, Europa Law Publishing, 2008), p. 103 which refers to the Case C-232/97, *L. Nederhoff & Zn. v Dijkgraaf en hoogheemraden van het Hoogheemraadschap Rijnland* [1999] ECR I-6385, paragraph 58 and Case C-6/03, *Deponiezweckverband Eiterköpfe v Land Rheinland-Pfalz* [2005] ECR I-2753, paragraph 58.

provisions (such as Articles 43, 114 or 207 of the TFEU than Article 192 of the TFEU).

751. In the *Deponiezweckverband Eiterköpfe v Land Rheinland-Pfalz* case[963] the Court of Justice clarified the meaning of the concept of more stringent protective measures. It held that *"it follows that Article 176 EC and the Directive allow the Member States to introduce more stringent protection measures that go beyond the minimum requirements fixed by the Directive (see, to that effect, Fornasar and Others, paragraph 46, concerning Council Directive 91/689/EEC of 12 December 1991 on hazardous waste, OJ 1991 L 377, p. 20)".*[964] Consequently, Article 193 of the TFEU becomes applicable if the national measures are stricter than those required by the Union. Article 193 of the TFEU furthermore requires that these more stringent provisions are compatible with the TFEU. This provision implies that Member States when they adopt more stringent measures, still need to comply with for example the free movement provisions, the rules on competition or the taxation provisions. This can be illustrated by the ruling of the Court of Justice in the *Dusseldorp* case[965] in which the Court held that *"it must therefore be concluded that the object and effect of application of the principles of self-sufficiency and proximity to waste for recovery, such as oil filters, is to restrict exports of that waste and is not justified, in circumstances such as those in the present case, by an imperative requirement relating to protection of the environment or the desire to protect the health and life of humans in accordance with Article 36 of the Treaty. A Member State cannot therefore rely on Article 130t of the Treaty in order to apply the principles of self-sufficiency and proximity to such waste".*[966]

752. Another issue in this respect is whether these more stringent provisions adopted by the Member States on the basis of Article 193 of the TFEU should comply with the principle of proportionality. This was clarified by the Court of Justice in the *Deponiezweckverband Eiterköpfe v Land Rheinland-Pfalz case*[967] in which it stated that *"it is clear from the broad logic of Article 176 EC that, in adopting stricter measures, Member States still exercise powers governed by Community law, given that such measures must in any case be compatible with the Treaty. Nevertheless, it falls to the Member States to define the extent of the protection to be achieved.*

---

963 Case C-6/03, *Deponiezweckverband Eiterköpfe v Land Rheinland-Pfalz* [2005] ECR I-2753.

964 *Ibid.*, paragraph 32.

965 Case C-203/96, *Chemische Afvalstoffen Dusseldorp BV and Others v Minister van Volkshuisvesting, Ruimtelijke Ordening en Milieubeheer* [1998] ECR I-4075.

966 *Ibid.*, paragraph 49 and the Opinion of the Advocate General Ruiz-Jarabo Colomer of 30 November 2004 in the Case C-6/03, *Deponiezweckverband Eiterköpfe v Land Rheinland-Pfalz* [2005] ECR I-2753, paragraph 59.

967 Case C-6/03, *Deponiezweckverband Eiterköpfe v Land Rheinland-Pfalz* [2005] ECR I-2753.

In that context, in so far as it is a matter of ensuring that the minimum requirements laid down by the Directive are enforced, the Community principle of proportionality demands that measures of domestic law should be appropriate and necessary in relation to the objectives pursued.

In contrast, and inasmuch as other provisions of the Treaty are not involved, that principle is no longer applicable so far as it concerns more stringent protective measures of domestic law adopted by virtue of Article 176 EC and going beyond the minimum requirements laid down by the Directive.

As a result, the reply to the second question has to be that the Community-law principle of proportionality is not applicable so far as it concerns more stringent protective measures of domestic law adopted by virtue of Article 176 EC and going beyond the minimum requirements laid down by a Community directive in the sphere of the environment, inasmuch as other provisions of the Treaty are not involved".[968] This ruling of the Court of Justice made clear that more stringent national measures adopted on the basis of Article 193 of the TFEU, do not have to comply with the principle of proportionality.[969]

753. Finally, in this context it should be noted that there is a dispute in legal doctrine on whether a Member State could invoke Article 193 of the TFEU for the adoption of more stringent measures even when the issue in question has been totally harmonised by the Union's legislator.[970] In this respect the Court of Justice ruled in the *Fornasar* case[971] that: "*in that connection, it must be observed that the Community rules do not seek to effect complete harmonisation in the area of the environment. Even though Article 130r of the Treaty refers to certain Community objectives to be attained, both Article 130t of the EC Treaty (now Article 176 EC) and Directive 91/689 allow the Member States to introduce more stringent protective measures. Under Article 130r of the Treaty, Community policy on the environment is to aim at a high level of protection, taking into account the diversity of situations in the various regions of the Community".*[972] The Court of Justice in this judgment held that even with a total harmonisation by the Union's legislator, Member States are still allowed to adopt more stringent measures (*in casu* classifying as hazardous waste other than that featuring on the Hazardous Waste List).[973]

---

968 *Ibid.*, paragraphs 61–64, emphasis added.

969 J. H. JANS and H. H. B. VEDDER, *European Environmental Law* (Groningen, Europa Law Publishing, 2008), p. 106 and 107.

970 For a more detailed analyses of the different views in this respects see *Ibid.*, p. 107–110.

971 Case C-318/98, *Criminal proceedings against Giancarlo Fornasar, Andrea Strizzolo, Giancarlo Toso, Lucio Mucchino, Enzo Peressutti and Sante Chiarcosso* [2000], ECR 4785.

972 *Ibid.*, paragraph 46.

973 J. H. JANS and H. H. B. VEDDER, *European Environmental Law* (Groningen, Europa Law Publishing, 2008), p. 109 and 110.

754. However, this discussion seems to be re-opened since the judgment of the Court of Justice in the *Deponiezweckverband Eiterköpfe v Land Rheinland-Pfalz case*[974] in which it held that when Member States adopt more stringent measures, they exercise powers which *"are governed"* by Union law.[975] On the basis of this judgment some legal doctrine argues that if the Union legislator would not be granted the competence to restrict the Member States in maintaining or adopting more stringent environmental measures, it is unclear how these measures can then be *"governed"* by Union law.[976] On the basis of the analysis above, the next part discusses the legal basis of the adopted Renewable Directive 2009/28/EC.

(e) Renewable Directive 2009/28/EC and its legal bases Articles 192 and 114 of the TFEU

755. Given the benefits of the use of renewable energy sources, which include reducing greenhouse gas emissions, strengthening the security of energy supply, promoting technological development and innovation and providing opportunities for employment, the development of a market for energy from renewable sources is strongly stimulated on a European level.

756. Following the request of the European Council of March 2006, the Commission has proposed a European long-term vision for renewable energy sources in its *Renewable Energy Road Map*[977] on 10 January 2007. In this long-term vision, the Commission lays down a mandatory target of 20% for the share of renewable energy sources in the total energy consumption and of 10% for the share of renewable energy sources in the overall consumption of petrol and diesel in transport by 2020. Subsequently, in the framework of the further implementation of the renewed Lisbon Strategy for Growth and Jobs, the European Council of 8 and 9 March 2007 has called upon the EU Member States and Institutions to take measures that, in addition to strengthening the internal market and competitiveness, improving the framework conditions for innovation and for more research and education, also *develop a sustainable integrated European climate and energy policy.*

757. The Heads of State and Government reached an ambitious political consensus on the new European energy policy during the European Council of 8 and 9 March 2007. An integrated approach to climate and energy policy was put

---

[974] Case C-6/03, *Deponiezweckverband Eiterköpfe v Land Rheinland-Pfalz* [2005] ECR I-2753.

[975] *Ibid.*, paragraph 61.

[976] J. H. JANS and H. H. B. VEDDER, *European Environmental Law* (Groningen, Europa Law Publishing, 2008), p. 110.

[977] Communication of the Commission of 10 January 2007, *Renewable Energy Road Map – Renewable energies in the 21st century: building a more sustainable future*, COM(2006)848, 20 p.

forward, which must fully respect the Member States' choice of energy mix and their sovereignty over primary energy sources[978], and which will pursue the following three objectives: (i) increasing the security of supply; (ii) ensuring the competitiveness of European economies and the availability of affordable energy; (iii) promoting environmental sustainability and combating climate change.[979]

758. In addition to the third legislative package for the electricity and gas markets, the Commission proposed its climate and energy package on 23 January 2008. The European Parliament adopted this package on 17 December 2008, after which the Council gave its final approval on 6 April 2009. As a result, among other secondary legislation[980], the Renewable Energy Directive 2009/28/EC was adopted.

i) Introduction

759. The Renewable Energy Directive 2009/28/EC provides a common framework in order to achieve a total share of energy from renewable sources of 20% of the gross final consumption of energy in the EU by 2020. In order to reach this target, the directive prescribes mandatory national targets for the total share of energy from renewable sources. It must be stressed that these national overall targets are not limited to electricity produced from renewable energy sources as detailed in Directive 2001/77/EC[981], but relate to the total share of energy from renewable sources in the gross final consumption of energy and to the share of renewable energies in transport. From the Renewable Energy Directive 2009/28/EC, it can also be deduced that, as opposed to the production target as outlined in Directive 2001/77/EC[982], this directive provides a consumption target in renewable sources. Moreover, this target is mandatory.

760. In addition to this, the Renewable Energy Directive 2009/28/EC provides a legal framework for the statistical transfers between Member States, the realisation of joint projects between Member States and between Member States and third countries, the use of guarantees of origin, sustainability criteria for biofuels and bioliquids, etc. Before discussing some of the new provisions of the

978 See *Infra*.
979 See *Supra*.
980 See *Supra*.
981 Directive 2001/77/EC of the European Parliament and of the Council of 27 September 2001 on the promotion of electricity produced from renewable energy sources in the internal electricity market, *OJ* L 283/33 (hereinafter Directive 2001/77/EC).
982 K. VERHAEGEN, L. MEEUS, B. DELVAUX and R. BELMANS, "Electricity produced from renewable energy sources – what target are we aiming for?", *Energy Policy* (2007), p. 5576–5584.

Renewable Energy Directive 2009/28/EC the next part will analyse its Treaty basis.

ii) Legal basis of the Renewable Energy Directive 2009/28/EC

761. The Renewable Energy Directive 2009/28/EC has been based upon Article 192(1) of the TFEU in combination with Article 114 of the TFEU. According to the Commission's proposal a single legal basis was preferred. However, due to the fact that this directive contains provisions based on different parts of the TFEU, a dual legal basis seemed more appropriate.

762. According to case law of the Court of Justice, the choice of legal basis of the measure in question is not a subjective one, but must be based on objective factors such as the aim and the content of the measure in question which are amenable to judicial review.[983] Moreover, the main objective of a measure in question should be the decisive factor.[984] According to the Court of Justice, a general Treaty Article is a sufficient legal basis although in a subordinate manner the measure in question seeks to attain an aim pursued by a specific Treaty Article.[985] With respect to dual legal bases, the Court of Justice has ruled that only for measures which have several contemporaneous objectives which are indissolubly linked with each other without one being secondary and indirect with respect to the others, a dual legal basis must be applied.[986, 987]

763. The majority of the provisions of the Renewable Energy Directive 2009/28/EC are pursuing environmental objectives. Hence, it is not surprising that this directive was based on Article 192 of the TFEU since this provision gives, as discussed in the previous chapter, the European Union the power to adopt measures in the environmental field in order to, *inter alia*, preserve, protect and improve the quality of the environment, protect human health, make prudent and rational use of natural resources, etc. The use of Article 114 of the TFEU has been justified in the Commissions proposal by the fact that the Articles 17, 18 and 19 of the Renewable Energy Directive 2009/28/EC impose

983 Case 45/86, *Commission of the European Communities v Council of the European Communities* [1987] ECR 1493, paragraph 5 and Case C-300/89, *Commission of the European Communities v Council of the European Communities* [1991] ECR I-2867, paragraph 10.

984 Case C-377/89, *Kingdom of the Netherlands v European Parliament and Council of the European Union* [2001] ECR I-7079, paragraph 27.

985 *Ibid.*, paragraphs 27 and 28.

986 Case C-165/87, *Commission of the European Communities v Council of the European Communities* [1988] ECR 5545, paragraph 11.

987 Opinion of the Committee on legal affairs on the legal basis of the proposal for a directive of the European Parliament and of the Council on the promotion of the use of energy from renewable sources (COM(2008)0019 – C6-0046/2008 – 2008/0016(COD)) and See also *Supra* the analysis between Article 192 of the TFEU and Article 114 of the TFEU.

binding obligations with respect to the sustainability of biofuels and other bioliquids. The adopted sustainability criteria obviously by themselves aim for environmental protection, but the directive also prevents Member States from adopting measures that could obstruct the trade in raw materials or biofuels.

764. A complete harmonisation of biofuel sustainability criteria is aimed for with the adoption of the Renewable Directive 2009/28/EC. Such a harmonised approach must ensure Member States cannot individually adopt criteria that could constitute an obstacle to trade between Member States. Thus, according to the proposal of this directive, for these provisions of the Renewable Directive 2009/28/EC the creation of an internal market is the primary objective. Consequently, Article 114 of the TFEU is, according to the Commission, the proper legal basis for the adoption of measures which contain harmonised standards for biofuel sustainability.[988]

765. However, only one recital of the Renewable Energy Directive 2009/28/EC refers specifically to the internal market and states that "*since the measures provided for in Articles 17 to 19 also have an effect on the functioning of the internal market by harmonising the sustainability criteria for biofuels and bioliquids for the target accounting purposes under this Directive, and thus facilitate, in accordance with Article 17(8), trade between Member States in biofuels and bioliquids which comply with those conditions, they are based on Article 95 of the Treaty*". The other recitals and provisions of the Renewable Energy Directive 2009/28/EC, except for the provisions with respect to reporting, monitoring, amendments, etc., seem to fall under the scope of the objectives of Article 191 of the TFEU. Moreover, Articles 17 to 19 of the Renewable Directive 2009/28/EC, which according to the Commission should be adopted on the basis of Article 114 of the TFEU, are pursuing environmental objectives within the scope of Article 191 of the TFEU.

766. According to case law of the Court of Justice a general Treaty Article constitutes a sufficient legal basis.[989] Although in a subordinate manner, the measure in question also seeks to attain an aim pursued by a specific Treaty Article and that only measures which have several contemporaneous objectives which are indissolubly linked with each other without one being secondary and indirect with respect to the others, dual legal basis must be applied. Therefore, it could be argued that Article 192 of the TFEU should be the sole legal basis for the Renewable Energy Directive 2009/28/EC. Thus, there was no need to add

---

988 Proposal for a Directive of the European Parliament and of the Council on the promotion of the use of energy from renewable sources, COM(2008)19, p. 8.

989 See *Infra*.

Article 114 of the TFEU as a legal basis for the adoption of the Renewable Energy Directive 2009/28/EC.[990]

767. Nonetheless, the most striking aspect with respect to the legal basis of the Renewable Energy Directive 2009/28/EC is not the use of a dual legal basis, but the fact that the Commission has opted for Article 192(1) of the TFEU instead of Article 192(2) of the TFEU. To understand this remark we should discuss in more detail one of the basic features of the Renewable Energy Directive 2009/28/EC, namely the mandatory national overall target for the use of renewable energy sources.

1. Mandatory national overall target for the use of renewable energy sources

768. On the grounds of Article 3 of the Renewable Energy Directive 2009/28/EC, every Member State is required to ensure that its share of energy from renewable sources in the gross final consumption of energy is at least equal to its national overall target by 2020. This overall target is listed per Member State in Annex I, part A. This Annex I, part A demonstrates that the targets differ between Member States, as the Directive outlines that the potential for energy from renewable sources and the Member States' energy mix need to be taken into account.

769. The European overall target of 20% in renewable energy sources by 2020 has therefore been transferred into individual national targets through the Renewable Energy Directive 2009/28/EC. Whether these targets are adequate and fair is questionable; however this falls outside the scope of this book. Concretely, the European target has been divided by prescribing an equal increase of 5% in the share of energy from renewable sources for every Member State first. Subsequently, this share of energy was increased with an additional share of energy from renewable sources, weighted by the Member States' GDP. In order to enable the Member States to achieve the mandatory national targets by 2020, Article 3, paragraph 2 of the Renewable Energy Directive 2009/28/EC sets out that the Member States are required to take measures to ensure that their share of energy from renewable sources is equal to or larger than the share mentioned in the indicative trajectory. This indicative trajectory departs from

---

[990] Opinion of the Committee on legal affairs on the legal basis of the proposal for a directive of the European Parliament and of the Council on the promotion of the use of energy from renewable sources (COM(2008)0019 – C6-0046/2008 – 2008/0016(COD)) and See also *Supra* the analysis between Article 192 of the TFEU and Article 114 of the TFEU.

the base year 2005 and provides intermediate targets for various periods, as outlined in Annex I, part B.[991]

770. In view of reaching the mandatory national targets, it has to be stressed that Article 3, paragraph 1 of the Renewable Energy Directive 2009/28/EC explicitly mentions that the Member States are required to promote and encourage energy efficiency and energy saving. Improving energy efficiency is a main goal of the European Union and the target, as yet not mandatory, is to improve energy efficiency by 20% by 2020. According to the preamble, creating energy efficiency and energy saving policies in all sectors is one of the most effective methods by which Member States can increase their share of energy from renewable sources and as such more easily reach their national mandatory targets for energy from renewable sources at a lesser cost. The directive also stresses that the Member States must strive towards reducing the overall energy consumption in the transport sector and towards increased energy efficiency in this sector. Measures to this effect can include offering additional support to public transport, outlining energy efficient transport planning, increasing the share of electric cars and the production of more energy-efficient cars, smaller both in size and in engine capacity.

771. In order to reach the aforementioned national overall targets, Article 3, paragraph 3 of the Renewable Energy Directive 2009/28/EC determines on the one hand that Member States can apply support schemes. On the other hand, Articles 5 up to and including 11 of the Renewable Energy Directive 2009/28/EC allow joint projects between Member States and between Member States and third countries. Article 2, (k) of the Renewable Energy Directive 2009/28/EC defines a support scheme as *"any instrument, scheme or mechanism applied by a Member State or a group of Member States, that promotes the use of energy from renewable sources by reducing the cost of that energy, increasing the price at which it can be sold, or increasing, by means of a renewable energy obligation or otherwise, the volume of such energy purchased. This includes, but is not restricted to, investment aid, tax exemptions or reductions, tax refunds, renewable energy obligation support schemes including those using green certificates, and direct price support schemes including feed-in tariffs and premium payments"*. In addition to this, the third paragraph explicitly mentions that the Member States, without prejudice to Articles 87 and 88 of the EC Treaty have the right to, as per Articles 5 to 11 of the Renewable Energy Directive 2009/28/EC, support energy from renewable sources produced in a different Member State.

---

991 This trajectory establishes e.g. that the share of energy from renewable sources for the period 2011–2012 has to be at least equal to $S_{2005} + 0{,}20\ (S_{2020} - S_{2005})$, or for the period 2017–2018 at least $S_{2005} + 0{,}65\ (S_{2020} - S_{2005})$, whereby S_{2005} refers to the share of that Member State in 2005 as determined in Annex I, part A, and S_{2020} refers to the share of that Member State in 2020 as determined in Annex I, part A.

772. As yet, Member States continue to maintain different support schemes for energy from renewable sources on a national level, whereby mainly energy produced on their territory from renewable sources is eligible for support. In order to ensure a support scheme to function well, it is important that Member States have control over the scheme's costs and effects. Guaranteeing optimal functioning of the national support schemes is part of the outset of the Renewable Energy Directive 2009/28/EC according to its preamble, with the goal of maintaining investors' trust.

773. In addition, the Renewable Energy Directive 2009/28/EC aims at fostering cross-border support for renewable energy without affecting the national support schemes. It can be deduced that the Renewable Energy Directive 2009/28/EC does not opt to provide one single support scheme that will be applicable in all Member States. An important element in this directive is the new legal framework, which should stimulate cross-border cooperation in generating energy from renewable sources, but how this will be effectuated in practice is as yet not clear.

774. In addition to the aforementioned national overall targets for the share of energy from renewable sources, Article 3, paragraph 4 of the Renewable Energy Directive 2009/28/EC states that each Member State has to ensure that the share of energy from renewable sources in all forms of transport in 2020 will be at least 10% of the final consumption of energy in transport in that Member State.[992]

---

[992] In order to calculate this share, the guideline in Article 3, paragraph 4, second paragraph of the Renewable Energy Directive 2009/28/EC stipulates that: "*(a) for the calculation of the denominator, that is the total amount of energy consumed in transport for the purposes of the first subparagraph, only petrol, diesel, biofuels consumed in road and rail transport, and electricity shall be taken into account; (b) for the calculation of the numerator, that is the amount of energy from renewable sources consumed in transport for the purposes of the first subparagraph, all types of energy from renewable sources consumed in all forms of transport shall be taken into account; (c) for the calculation of the contribution from electricity produced from renewable sources and consumed in all types of electric vehicles for the purpose of points (a) and (b), Member States may choose to use either the average share of electricity from renewable energy sources in the Community or the share of electricity from renewable energy sources in their own country as measured two years before the year in question. Furthermore, for the calculation of the electricity from renewable energy sources consumed by electric road vehicles, that consumption shall be considered to be 2,5 times the energy content of the input of electricity from renewable energy sources*". Moreover, based on this Article the Commission can opt, by 31 December 2011, for a proposal that allows, subject to certain conditions, the whole amount of the electricity originating from renewable sources used to power all types of electric vehicles to be considered, and/or for a proposal for a methodology for calculating the contribution of hydrogen originating from renewable sources in the total fuel mix.

2. Mandatory target for the use of renewable energy sources and the legal basis of the Renewable Energy Directive 2009/28/EC

775. The foregoing clarified that the Renewable Energy Directive 2009/28/EC opted in favour of imposing mandatory targets for the use of renewable energy sources. These mandatory targets are used to offer certainty to investors and to promote the continuing development of technologies to generate energy from a myriad of renewable sources.

776. Given the fact that the Renewable Energy Directive 2009/28/EC with its mandatory targets obliges Member States to achieve a certain share of energy from renewable sources in their gross final consumption of energy and in all forms of transport by 2020, this directive undoubtedly affects a Member State's choice between different energy sources or the general structure of its energy supply. For example if a Member State decides to produce all its electricity from nuclear or gas-fired installations, the Renewable Energy Directive 2009/28/EC with its mandatory targets for the use of renewable energy sources undoubtedly affects such a Member State in its choice between different energy sources. Consequently, the Renewable Energy Directive 2009/28/EC should have been adopted on the basis of Article 192(2) of the TFEU instead of Article 192(1) of the TFEU.

777. Article 192(2) of the TFEU states that: "*[b]y way of derogation from the decision-making procedure provided for in paragraph 1 and without prejudice to Article 114, the Council acting unanimously in accordance with a special legislative procedure and after consulting the European Parliament, the Economic and Social Committee and the Committee of the Regions, shall adopt: [...]*

(c) measures significantly affecting a Member State's choice between different energy sources and the general structure of its energy supply.

The Council, acting unanimously on a proposal from the Commission and after consulting the European Parliament, the Economic and Social Committee and the Committee of the Regions, may make the ordinary legislative procedure applicable to the matters referred to in the first subparagraph".

778. The previous chapter of this book argued that a Member State's choice between different energy sources means the power of a Member State to decide which kind of energy sources can be used to cover its demand for energy. The general structure of its energy supply means the power of a Member State to organise the way of its energy supply. Examples in this context are a Member State's choice to opt for a nuclear phase-out, to set a target for renewable energy sources, to replace all coal-fired by gas-fired power plants, to prohibit the construction of new coal-fired power plants, etc.

779. The Commission has argued that renewable energy is a close substitute for conventional energy and is supplied through the same infrastructure and logistic systems. Moreover, all Member States already use renewable energy and they have already decided to increase its share. Therefore, according to the Commission, the Renewable Energy Directive 2009/28/EC will not significantly affect a Member States' choice between different energy sources or the general structure of their energy supply and thus, this directive should not be adopted on the basis of Article 192(2) of the TFEU.[993]

780. This reasoning of the Commission is in the author's view not correct. As discussed above, the Renewable Energy Directive 2009/28/EC sets mandatory targets for the overall share of energy from renewable sources in energy consumption and in transport which undoubtedly affects a Member State's choice between different energy sources and the general structure of its energy supply. One could argue that given the fact that Member States already use renewable energy and have decided to increase their share, the mandatory targets are "not" significantly affecting a Member State's choice between different energy sources and the general structure of its energy supply. However, this argument is not plausible because for some Member States, the Renewable Energy Directive 2009/28/EC has set mandatory targets of 20%, 25%, 30%, 38%, 40% and even 49% for the overall share of energy from renewable sources in their energy consumption.[994] These percentages are certainly significant.

781. Consequently, imposing legally binding targets for the overall share of energy from renewable sources in energy consumption and in transport on Member States is a violation of Article 192(2) of the TFEU. This directive should have been adopted with unanimity, under the so-called special legislative procedure[995], instead of qualified majority. The unanimity requirement enables Member States to use their veto right.

782. With the adoption of the Renewable Energy Directive 2009/28/EC, the European legislator did not only affect a Member State's choice between different energy sources and the general structure of its energy supply, but also imposed binding obligations with respect to the sustainability of biofuels and other bioliquids, and the binding target that the share of energy from renewable sources in all forms of transport in 2020 is at least 10% of the final consumption of energy in transport in that Member State.

---

[993] Proposal for a Directive of the European Parliament and of the Council on the promotion of the use of energy from renewable sources, COM(2008)19, p. 8.

[994] See Annex I, part A of the Renewable Directive 2009/28/EC.

[995] See Article 293–299 of the TFEU.

783. Given the impact of these binding obligations with respect to biofuels on (a) the use of the territory or the land use and (b) town and country planning of the Member States, such measures, according to Article 192(2) of the TFEU, should have been adopted with unanimity instead of a qualified majority.

iii) Conclusion

784. This part of the chapter has analysed the legal bases of the recently adopted Renewable Energy Directive 2009/28/EC. This directive provides a common framework in order to achieve a total share of energy from renewable sources of 20% of the gross final consumption of energy in the EU by 2020. In order to reach this target, the directive prescribes mandatory national targets for the total share of energy from renewable sources.

785. The Renewable Energy Directive 2009/28/EC is based upon Article 192(1) of the TFEU in combination with Article 114 of the TFEU. With respect to the question whether it was necessary to adopt the Renewable Directive 2009/28/EC on the basis of Article 114 of the TFEU it has been argued that there was no need to add Article 114 of the TFEU. This argument has been based upon case law of the Court of Justice which states that a general Treaty Article constitutes a sufficient legal basis although the measure in question also seeks to attain in a subordinate manner an aim sought by a specific Treaty Article.

786. With respect to Article 192(1) of the TFEU the author concluded that this is not the correct legal basis for the adoption of the Renewable Energy Directive 2009/28/EC. The fact that the Renewable Energy Directive 2009/28/EC imposes on Member States legally binding targets for the overall share of energy from renewable sources in energy consumption and in transport is a violation of Article 192(2) of the TFEU. Moreover, the Renewable Energy Directive 2009/28/EC has imposed various obligations with respect to biofuels and other bioliquids which affect Member States' land use and therefore needs to be based upon Article 192(2) of the TFEU. Consequently, this directive should have been adopted with unanimity under the special legislative procedure instead of a qualified majority.

787. Although, as discussed above, the legal basis of the Renewable Energy Directive 2009/28/EC is incorrect and should have been Article 192(2) of the TFEU, which requires unanimity, the Council has indeed adopted this directive unanimously. The fact that the Council adopted this directive in practice with unanimity makes it more difficult, or perhaps even impossible, to convince the Court of Justice of the European Union that the Renewable Energy Directive 2009/28/EC was adopted on the incorrect legal basis. The next part will discuss the so-called loophole Article 352(1) of the TFEU which has also served as a legal basis for the adoption of secondary legislation in the energy field.

6. LOOPHOLE ARTICLE 352(1) OF THE TFEU

(a) *Meaning of Article 352(1) of the TFEU*

788. The so-called loophole Article 352(1) of the TFEU or the so-called 'flexibility clause' has often been claimed to be a shortcut through which the Union has gained competence in areas which do not and should not belong to its competence.[996]

789. Historically, this provision has been used to extend the competences of the Union beyond the limits of the TFEU. In its Opinion 2/94 the Court of Justice gave some guidance as to the limits of the default mechanism in Article 352(1) of the TFEU by stating that this provision *"cannot serve as a basis for widening the scope of Community powers beyond the general framework created by the provisions of the Treaty as a whole and, in particular, by those that define the tasks and the activities of the Community".*[997] This means that a proposed measure is unlikely to qualify for adoption on the basis of Article 352(1) of the TFEU unless it falls, without being specifically authorised, within the scope of Article 3 of the TEU and Articles 3, 4, 5, 6, 8 and 119 of the TFEU.

790. The Court of Justice has also stressed in its case law that the use of Article 352(1) of the TFEU as the legal basis for a measure is only justified where no other provision of the Treaty gives the Union institutions the necessary power to adopt the relevant measure.[998] Consequently, a measure on the basis of Article 352(1) of the TFEU is only justified if no other Treaty provision gives the Union the necessary express or implied competence to act.

791. Finally, it should be noted that Article 352(1) of the TFEU does not contain any obligation for the Union to use this provision.[999] Moreover, it creates

---

[996] M. KURCZ, "La répartition des compétences au sein de l'Union européenne", *Revue du Droit de l'Union Européenne* (2005), p. 592.

[997] Opinion 2/94 of the Court of 28 March 1996, *Accession by the Community to the European Convention for the Protection of Human Rights and Fundamental Freedoms* [1996] ECR, p. I-1759.

[998] Case 45/86, *Commission of the European Communities v Council of the European Communities* [1987] ECR 1493, paragraph 13 and Case C-350/92, *Kingdom of Spain v Council of the European Union* [1995] ECR I-1985, paragraph 26 and Case C-271/94, *European Parliament v Council of the European Union* [1996] ECR I-1689.

[999] Case 22/70, *Commission v Council* [1971] ECR 263, paragraph 95.

a supplementary competence to achieve the Union objectives.[1000] However, the Court of Justice in its case law appears to apply Article 352(1) of the TFEU on occasions very creatively.[1001]

(b) *Article 352 of the TFEU and energy*

792. Before the adoption of the environmental provisions in the TFEU, Article 352(1) had been used to adopt not only secondary legislation in the environmental field, but also in the energy field.

7. THE TRANS-EUROPEAN NETWORKS PROVISION

(a) *Meaning of Articles 170–172 of the TFEU*

793. With the TEU, ex Article 129b, now Articles 170–172 of the TFEU on the promotion of trans-European Networks (TENs) in the areas of transport, telecommunications and energy infrastructure, was inserted.

794. In order to stimulate the completion of the internal market, and at a later stage economic and social cohesion, the Commission launched in the late 1980s and beginning of the 1990s the idea of developing trans-European networks. Some Member States such as Portugal, Austria, Finland, Sweden, Greece and the new Member States had a major interest in good transit connections.[1002] The idea behind these trans-European networks is explained in paragraph 1 of Article 170 of the TFEU which reads as follows: "*to help achieve the objectives referred to in Articles 26 and 174 and to enable citizens of the Union, economic operators and regional and local communities to derive full benefit from the setting-up of an area without internal frontiers, the Union shall contribute to the establishment and development of trans-European networks in the areas of transport, telecommunications and energy infrastructures*". Paragraph 2 of

---

1000 K. LENAERTS, and P. VAN NUFFEL, in BRAY, R. (ed.), *Constitutional Law of the European Union*, 2nd ed. (London, Thomson Sweet & Maxwell, 2005), p. 94 and 95.

1001 Joined Cases C-402/05 P and 415/05, *Yassin Abdullah Kadi and Al Barakaat International Foundation v Council of the European Union and Commission of the European Communities* [2008] ECR I-6351. See in this respect D. CHALMERS, G. DAVIES and G. MONTI, *European Union Law* (Cambridge, Cambridge University press 2010), p. 218 and 219.

1002 R. BARENTS and P. J. SLOT, "Chapter XII Sectoral Policies" in P. J. G. KAPTEYN, A. M. MCDONNELL, K. J. M. MORTELMANS and C. W. A. TIMMERMANS (eds.), *The Law of the European Union and the European Communities* (The Netherlands, Kluwer Law International, 2008), p. 1264. This author also referred to J. ERDMENGER in H. VON DER GROEBEN AND J. SCHWARZE (eds.), *Vertrag über die Europäische Union und Vertrag zur Gündung der Europäische Gemeinschaft, Kommentar EU-/EG-Vertrag* (Nomos, Baden-Baden, 2003), p. 1171–1195 and J. A. VINOIS, "Les réseaux transeuropéens: une nouvelle dimension donée au Marché unique", *Revue du Marché Unique Européen* (1993), p. 93–125.

Article 170 of the TFEU stresses that "*within the framework of a system of open and competitive markets, action by the Union shall aim at promoting the interconnection and interoperability of national networks as well as access to such networks. It shall take account in particular of the need to link island, landlocked and peripheral regions with the central regions of the Union*". This provision shows that the liberalisation directives in the electricity and gas markets which have, as an essential objective third party, access to transmission and distribution systems[1003] also fit into this general framework of trans-European networks.

795. In its White paper on Growth, Competitiveness and Employment, the Commission emphasises that the establishment of networks of the highest quality throughout the whole Union and beyond its frontiers is a priority task. The Commission furthermore clarifies that such a trans-European infrastructure is needed in order to travel safer, faster and at lower cost, make effective planning in Europe possible and to build bridges towards Eastern Europe. These goals should be achieved by removing regulatory and financial obstacles, get private investors involved in projects of European interest and to identify projects on the basis of the adopted (such as transport projects) or to be approved (such as energy) master plans.[1004]

796. According to paragraph 1 of Article 171 of the TFEU, the objectives of Article 170 of the TFEU could be achieved by establishing guidelines covering the objectives, priorities and broad lines of measures envisaged in the sphere of trans-European networks. Such guidelines shall identify projects of common interest. This provision furthermore provides that the Union shall implement any measure that may prove necessary to ensure the interoperability of the networks in particular in the field of technical standardisation. It also provides that the Union may support projects of common interest supported by the Member States through feasibility studies, loan guarantees or interest-rate subsidies.

797. In addition, the Union may also contribute to financing specific projects in Member States in the area of transport infrastructure through the Cohesion Fund set up in accordance with Article 177 of the TFEU. These guidelines and any other measure referred to in paragraph 1 of Article 171 of the TFEU shall,

---

1003 Case C-439/06, *Citiworks AG v Flughafen Leipzig/Halle GmbH, Bundesnetzagentur* [2008] ECR I-3919, paragraphs 43 and 44.

1004 White Paper of the European Commission of 5 December 1993 on Growth, Competitiveness, Employment – The Challenges and Ways Forward into the 21st Century, COM(93)700, p. 11 and P. J. SLOT, "Sectoral Policies" in P. J. G. KAPTEYN, A. M. MCDONNELL, K. J. M. MORTELMANS and C. W. A. TIMMERMANS (eds.), *The Law of the European Union and the European Communities*, fourth edition (The Netherlands, Kluwer Law International, 2008), p. 1264 and 1265.

according to paragraph 1 of Article 172 of the TFEU, be adopted by the European Parliament and the Council, acting in accordance with the ordinary legislative procedure and after consulting the Economic and Social Committee and the Committee of the Regions. For guidelines or projects of common interest which relate to the territory of a Member State, the approval of the concerned Member States will be required.

798. Pursuant to paragraph 2 of Article 171 of the TFEU Member States shall in liaison with the Commission coordinate among themselves the policies pursued on a national level which may have a significant impact on the achievement of the objectives referred to in Article 170 of the TFEU. The Commission may, in close cooperation with the Member States, take any useful initiative to promote such coordination. Under paragraph 3 of Article 171 of the TFEU, the Union is allowed to promote projects of mutual interest with third countries and to ensure the interoperability of networks.

(b) Secondary legislation on the basis of Articles 170–172 of the TFEU

799. On the basis of these Treaty provisions, the Council adopted a general TENs Regulation in 1995[1005], which has undergone several revisions.[1006] In 1996, as part of the Union moves towards the completion of the internal market, the trans-European Networks for Energy (TEN-E) were developed. At that time, the purpose of TEN-E was to provide a more political impulse to energy infrastructure investment focusing on the feasibility stage of gas and electricity network projects and in particular cross-border initiatives.

1005 Council Regulation (EC) No 2236/95 of 18 September 1995 laying down general rules for the granting of Community financial aid in the field of trans-European networks, *OJ* L 228/1, this Regulation has been amended several times, lastly by Regulation (EC) No 1159/2005 of the European Parliament and of the Council of 6 July 2005 amending Council Regulation (EC) No 2236/95 laying down general rules for the granting of Community financial aid in the field of trans-European networks, *OJ* L 191/16. See also Regulation (EC) No 680/2007 of the European Parliament and of the Council of 20 June 2007 laying down general rules for the granting of Community financial aid in the field of the trans-European transport and energy networks, *OJ* L 162/1; Decision 1364/2006/EC of the European Parliament and of the Council of 6 September 2006 laying down guidelines for trans-European energy networks and repealing Decision 96/391/EC and Decision No 1229/2003/EC of the European Parliament and of the Council of 26 June 2003 laying down a series of guidelines for trans-European energy networks and repealing Decision No 1254/96/EC, *OJ* L 176/11. See in this context also Council Directive 96/48/EC of 23 July 1996 on the interoperability of the trans-European high-speed rail system, *OJ* L 235/6 and Directive 2001/16/EC of the European Parliament and of the Council of 19 March 2001 on the interoperability of the trans-European conventional rail system, *OJ* L 110/1.

1006 For further details see E. D. CROSS, B. DELVAUX, L. HANCHER, P. J. SLOT, G. VAN CALSTER and W. VANDENBERGHE, "EU Energy Law (chapter 5)" in M. M. ROGGENKAMP, A. RØNNE, C. REDGWELL and I. DEL GUAYO (eds.), *Energy Law in Europe. National, EU and International Regulation* (Oxford, Oxford University Press, 2007), p. 300–303.

The EU in that period had no energy policy remit and no internal market. The network investment projects, mainly aiming at security of supply, where managed by national state-run companies. It was assumed that commercial interests would drive projects forward and therefore EU intervention in such projects in the implementation phases would not be necessary. Consequently, the TEN-E budget remained very low.[1007]

800. In the 1990s, the TEN-E policy has been developed and shaped through the TEN-E Guidelines[1008] and the corresponding TEN Financing Regulation.[1009] The 2006 TEN-E Guidelines listed approximately 550 TEN-E projects, eligible for Union co-financing, ranked in three categories: projects of European interest (42 in total); priority projects (240 in total) and projects of common interest (286 in total). These 2006 TEN-E Guidelines furthermore establish a framework for closer cooperation between Member States through, for example, a better exchange of information and specify that, when projects encounter significant delays or problems, a European coordinator may be appointed to the project. Such a coordinator should monitor progress and facilitate coordination of the implementation of the most critical identified priority projects.[1010, 1011]

801. Beside the 2006 TEN-E Guidelines, the TEN Financing Regulation sets the conditions for TEN-E funding. The budget of the TEN-E funding programme amounts to 155 million Euros for the period 2007–2013 or approximately 22

1007 It is currently annually some 22 million euros.

1008 Council Decision of 28 March 1996 laying down a series of measures aimed at creating a more favourable context for the development of trans-European networks in the energy sector, *OJ* L 161/154; Decision No 1229/2003/EC of the European Parliament and of the Council of 26 June 2003 laying down a series of guidelines for trans-European energy networks and repealing Decision No 1254/96/EC, *OJ* L 176/11 and Decision No 1364/2006/EC of the European Parliament and of the Council of 6 September 2006 laying down guidelines for trans-European energy networks and repealing Decision 96/391/EC and Decision No 1229/2003/EC, *OJ* L 262/1 (hereinafter the 2006 TEN-E guidelines).

1009 Regulation (EC) No 680/2007 of the European Parliament and of the Council of 20 June 2007 laying down general rules for the granting of Community financial aid in the field of the trans-European transport and energy networks, *OJ* L 162/1 (hereinafter the TEN Financing Regulation).

1010 In 2007 for example 4 coordinators were appointed for The French-Spanish connection (Mr. Mario Monti); Baltic and North Sea off-shore wind connections (Mr. Adamowitsch); The "Salzburgleitung": Austrian Power link Salzach neu – Tauern (Mr. Adamowitsch); The axis linking Caspian Sea countries and the Middle East to the European Union, including the Nabucco pipeline (Mr. van Aartsen) and The northern European power-link (Mr. Mielczarski).
See http://ec.europa.eu/energy/infrastructure/tent_e/coordinators_en.htm.

1011 Report from the Commission of 4 May 2010 to the European Parliament, the Council, the European Economic and Social Committee and the Committee of the regions on the implementation of the Trans-European Energy Networks in the period 2007–2009, Pursuant to Article 17 of Regulation (EC) 680/2007 and Articles 9(2) and 15 of Decision 1364/2006/EC, COM(2010)203, 11 p.

million Euros per year. The maximum co-financing rate is 50% for studies and 10% of eligible costs of works, in particular to projects of European interest. Nonetheless, in practice these co-financing rarely amounts to more than 0,01 – 1% of the total investment cost of a project.[1012]

802. At the moment, the TEN-E funding programme is in particular important for immature or risky projects and feasibility studies. However, this TEN-E funding programme does not contain a precise methodology to assess the additionality of its support. The current limited financial resources of the TEN-E programme appear to be adequate for studies on the remaining problematic interconnections. However, due to the shift to a low-carbon energy system which calls for a major change leading to massive investments in energy infrastructure the coming years, the TEN-E instrument needs to be reassessed in order to ensure its effectiveness. Besides, the current financial instruments allowed under the TEN-E Financing Regulation such as grants and interest rate rebates appear to be insufficient since the market rather needs innovative financial instruments such as guarantees or equity participation for risk mitigation.

803. Moreover, the TEN-E Financing Regulation currently does not allow the funding capital expenditures outside the EU, whereas large gas import infrastructure and related connection to upstream sources or electricity interconnections within third countries go well beyond the Union borders. The foregoing demonstrates that given the new policy environment in the energy field with ambitious goals the TEN-E Financing Regulation is not effective anymore.[1013]

804. The current objectives of the TEN-E policy are to (1) support the completion of the internal energy market in the EU, (2) reduce the isolation of less-favoured and island regions, (3) diversify and secure the energy supplies of the EU also through cooperation with third countries, (4) contribute to sustainable development and the protection of the environment (including the reduction of environmental risks due to the transportation of energy and renewable energy sources). The TEN-E policy framework currently includes

---

1012 *Ibid.*, p. 4 and 5 and Communication from the Commission of 17 November 2010 to the European Parliament, the Council, the European Economic and Social Committee and the Committee of the Regions, Energy infrastructure priorities for 2020 and beyond – A Blueprint for an integrated European energy network, COM(2010)677, p. 9 and 10.

1013 Communication from the Commission of 17 November 2010 to the European Parliament, the Council, the European Economic and Social Committee and the Committee of the Regions, Energy infrastructure priorities for 2020 and beyond – A Blueprint for an integrated European energy network, COM(2010)677, p. 9 and 10.

electricity, gas and olefin transmission networks. However, it does not include CO_2 transportation, neither oil pipeline infrastructure.[1014]

805. In April 2010, the Commission adopted a report on the implementation of the TEN-E framework in the period 2007–2009.[1015] This report concluded that the TEN-E has made a positive contribution to selected projects by giving them political visibility. The TEN-E label also appears to facilitate communication and collaboration with third parties. It helped them, for example, to get leverage funds from the financial market. However, in recent years the European energy policy framework has undergone a dramatic change and has set ambitious goals and binding targets on both greenhouse gas emissions and energy from renewable energy sources in order to improve competitiveness, fight against climate change and security of supply.[1016] In order to meet the objectives of this European energy policy framework such as the integration of renewable energy sources into the grid, the mitigation of the higher technological risks of low-carbon technologies, etc., an appropriate energy infrastructure development is needed.

806. This new policy environment in the energy field creates a challenge for the TEN-E policy framework which has neither the flexibility nor the resources in order to make a full contribution towards the delivery of the ambitious energy and climate objectives. Not only the dramatic change in the European energy policy framework, but also the evaluation of the current TEN-E policy framework, identified some shortcomings which require some changes. There seems to be a lack of focus and not enough clarity between the categorisation in the 2006 TEN-E Guidelines. The list of targeted projects is rigid whereby no technologies and projects can be included. The 2006 TEN-E Guidelines are based on a bottom-up approach and do not leave space for a top-down approach which could fill the identified infrastructure gaps. The existing TEN-E framework is not binding and does not contain any obligation on Member States or promoters to invest and construct infrastructure. It also lacks the mechanism to ensure that Member States grant projects of EU importance the same priority as their national projects. While there is a strong link between the EU's external energy

[1014] See the ENTSO-E, *Ten-Year Network Development Plan 2010–2020* (28 June 2010), 287 p. This plan can be consulted on https://www.entsoe.eu/fileadmin/user_upload/_library/SDC/TYNDP/TYNDP-final_document.pdf.

[1015] Report from the Commission of 4 May 2010 to the European Parliament, the Council, the European Economic and Social Committee and the Committee of the regions on the implementation of the Trans-European Energy Networks in the period 2007–2009, Pursuant to Article 17 of Regulation (EC) 680/2007 and Articles 9(2) and 15 of Decision 1364/2006/EC, COM(2010)203, 11 p.

[1016] See *Supra*.

policy and infrastructure development, there has been insufficient coordination between TEN-E and external aid programmes.[1017]

(c) *Conclusion*

807. It can be concluded that given the above-mentioned shortcomings in the TEN-E policy and its financing framework, it has demonstrated that due to the dramatic recent changes in the European energy policy framework, both the TEN-E policy and its financial means, are insufficient in meeting the goals of the new policy environment in the energy field.

8. MEANING OF ARTICLE 122 OF THE TFEU

808. According to Article 122, (1) of the TFEU the Council is allowed to decide on a proposal from the Commission, in a spirit of solidarity between Member States, upon measures appropriate to the economic situation in particular if severe difficulties arise in the supply of certain products, notably in the area of energy.

809. In the energy field this crisis mechanism of Article 122, (1) of the TFEU has already proven its practical relevance during the oil crises in the 1970s. During these oil crises, the Council applied the forerunner of this provision in the EEC Treaty in order to adopt measures securing the oil supply in the Union.[1018]

810. Under Article 122, (2) of the TFEU the Council, on a proposal from the Commission, is allowed to grant under certain conditions financial assistance to Member States that are in difficulties or are seriously threatened with severe difficulties due to natural disasters or exceptional occurrences beyond their control. When decisions are taken by the Council in this respect, the European Parliament shall be informed by the President of the Council.[1019]

1017 Communication from the Commission of 17 November 2010 to the European Parliament, the Council, the European Economic and Social Committee and the Committee of the Regions, *Energy infrastructure priorities for 2020 and beyond – A Blueprint for an integrated European energy network*, COM(2010)677, p. 9 and 10.

1018 F. AMTENBRINK, A. GEELHOED and S. KINGSTON, "Chapter × Economic, Monetary and Social Policy" in P. J. G. KAPTEYN, A. M. MCDONNELL, K. J. M. MORTELMANS and C. W. A. TIMMERMANS (eds.), *The Law of the European Union and the European Communities*, fourth edition (The Netherlands, Kluwer Law International, 2008), p. 909 and 910.

1019 *Ibid.*

9. IMPACT OF CASE LAW OF THE COURT OF JUSTICE ON THE ADOPTION OF SECONDARY LEGISLATION IN THE ENERGY FIELD

811. Over the past years, the institutions of the European Union, including the Commission, the European Parliament and the Council, have conducted various negotiations with regards to the third legislative package proposed by the Commission on 19 September 2007.[1020] This package proposes modifications to the existing electricity and gas directives and the existing electricity and gas regulations.[1021] In addition, the package includes a Proposal for a Regulation of the European Parliament and of the Council establishing an Agency for the Cooperation of Energy Regulators or ACER which was adopted on 13 July 2009.[1022]

812. The purpose of the next part is to demonstrate that not only Treaty provisions can influence the adoption of secondary legislation in the energy field, but also case law of the Court of Justice. This was the case with the adoption of ACER of which the powers remained rather limited. This is mainly due the so-called *Meroni doctrine* which entails that independent executive organs, unless a specific Treaty provision is available and the delegation remains limited to the aims and means provided by this provision, cannot be attributed discretionary powers, as this would render certain policy choices impossible for institutions that, based on the Treaty, carry the political responsibilities for such choices.

(a) Background

813. In order to address the lack of a regulatory framework for cross-border issues in the electricity and gas market and based on experiences with cooperation initiatives such as the European Regulators' Group for Electricity

1020 See http://ec.europa.eu/energy/gas_electricity/third_legislative_package_en.htm.

1021 Proposal for a Directive of the European Parliament and of the Council amending Directive 2003/54/EC concerning common rules for the internal market in electricity, COM(2007)528 (hereinafter Proposed Electricity Directive); Proposal for a Directive of the European Parliament and of the Council amending Directive 2003/55/EC concerning common rules for the internal market in natural gas, COM(2007)529 (hereinafter Proposed Gas Directive); Proposal for a Regulation of the European Parliament and of the Council Amending Regulation (EC) No 1228/2003 on conditions for access to the network for cross-border exchanges in electricity, COM(2007)531 (hereinafter Proposed Electricity Regulation); Proposal for a Regulation of the European Parliament and of the Council amending Regulation (EC) No 1775/2005 on conditions for access to the natural gas transmission networks, COM(2007)532 (hereinafter Proposed Gas Regulation).

1022 COM(2007)530 (hereinafter Proposal for a Regulation establishing the Agency); T. VANDEN BORRE, "Derde pakket voor de vrijmaking van de elektriciteitsmarkt: Quo Vadis?" in "Het Europese Energie – en Klimaatpakket", K. DEKETELAERE and B. DELVAUX (eds.), *Milieu – en Energierecht* (2008), p. 182–194 and L. HANCHER and A. DE HAUTECLOCQUE, "Manufacturing the EU Energy Markets: The Current Dynamics of Regulatory Practise", *Competition and Regulation in Network Industries* (2010), p. 311–321.

and Gas (ERGEG), the Electricity Regulatory Forum (Florence) and the Gas Regulatory Forum (Madrid), the Commission has decided that formal cooperation between national regulators should be set up. As a result, numerous structures were considered, along with their advantages and disadvantages, including the establishment of a network of national regulators, such as the network of Competition Authorities[1023], and the establishment of a cooperation form, such as in the case of the European banks. This analysis has led the Commission to conclude that the most suitable structure for an institution of this nature would be the establishment of an Agency, i.e. the establishment of the *Agency for the Cooperation of Energy Regulators* (hereinafter Agency).

814. According to the Commission proposal establishing the Agency, it, can as an independent entity apart from and outside of the Commission, take binding decisions to third parties with regard to specific technical issues and it can put forward proposals to the Commission. In first instance, on a European level, the Agency is required to coordinate and complete the regulatory tasks entrusted to national regulators. Hereby the Agency needs to create a framework in which the national regulatory bodies can cooperate (procedures, means of data exchange, etc.).[1024] Subsequently, the Agency will be required to monitor and evaluate the workings of the European networks for gas and electricity.[1025] The Agency will be involved in establishing the work programme, the investment plan and the technical and market codes, etc.[1026] Furthermore, the Proposal for a Regulation establishing the Agency provides for the solution that it can recommend to the Commission to make one of the technical or market codes legally binding.

815. In addition, the Agency will have its own decisional power with regard to a number of specific cross-border problems. Amongst others, the Agency will be authorised to make decisions concerning exemption procedures for new direct current interconnectors (Article 7 of the Proposed Electricity Regulation) and major new gas infrastructures (Article 22 of the Proposed Gas Directive)[1027], to determine which regulatory framework is applicable to infrastructures that cover more than one Member State's territory, and to take decisions on a number of specific technical questions in case this authority has been granted.[1028]

816. Finally, the Agency will advise the Commission on market regulation concerns, issue non-binding guidelines, etc. We can add to this that, in the

1023 Council Regulation (EC) No 1/2003 of 16 December 2002 on the implementation of the rules on competition laid down in Articles 81 and 82 of the Treaty, *OJ* L 1/1.
1024 Article 7 of the Proposal for a Regulation establishing the Agency.
1025 *Supra.*
1026 Article 6 of the Proposal for a Regulation establishing the Agency.
1027 *Supra.*
1028 Articles 7 and 8 of the Proposal for a Regulation establishing the Agency.

Commission's vision, the Agency will not be able to make normative decisions. The Commission will determine the further role of the Agency by means of binding guidelines. On the basis of the initial proposal of the Commission and the discussion above, it can be deduced that the tasks of the Agency will remain rather limited. Note that the costs of the Agency will be financed within the general budget of the European Communities, from remunerations for the treatment of an exemption decision in accordance with Article 8 of the Proposal for a Regulation establishing the Agency, and through voluntary contributions from Member States or their regulatory bodies.[1029]

817. Hereby it can already be noted that, based on the Commission's proposal, the decision power of the Agency will only be binding for specific technical situations that have explicitly been pointed out by directives, regulations or binding guidelines. The Agency's limited decision power is based on the so-called *Meroni doctrine*.[1030]

818. This doctrine entails that independent executive organs, unless a specific Treaty provision is available and the delegation remains limited to the aims and means provided for by this provision, cannot be attributed discretionary powers as this would render certain policy choices impossible for institutions that, based on the Treaty, carry the political responsibilities for these choices. The attribution of such discretionary power would impinge on the institutional balance that lies at the foundations of the Union.[1031] This balance implies that every institution of the European Union has the autonomy required to exercise its powers, that these powers cannot simply be transferred to other institutions or organs, and that the powers of the other institutions have to be respected.[1032]

819. The author deems the adoption of the Agency a first step in the right direction in order to tackle a number of cross-border issues with regards to the European electricity and gas markets. Nonetheless, it is remarkable that the powers of the proposed Agency remain rather limited. Based on the Commission's proposal, the Agency would mostly be granted an advisory mandate. It is unclear whether an Agency to which no more than such an advisory function is ascribed will truly be able to tackle a number of cross-border issues.[1033]

[1029] Articles 18 up until 22 of the Proposal for a Regulation establishing the Agency.

[1030] Case 9/56, *Meroni/High Authority* [1957–8] ECR 133.

[1031] K. LENAERTS and P. VAN NUFFEL, *Europees recht in hoofdlijnen* (Antwerpen, Maklu, 2008), p. 355–356; for a further discussion of this problematic, see *Infra*.

[1032] *Ibid.*, 245.

[1033] For a further analysis: C. F. ZIMMERMANN and K. TALUS, "Regulation of Electricity Markets at EU Level", *EEELR* (2008), p. 12–22 and S. NEVELING, "Europäiserung der Energieaufsicht? – Vorschläge von Kommission und ERGEG zur Neuordnung (Teil 1)", *Heft – IR Energie, Verkehr, Abfall*, Wasser (2007), p. 173–177.

(b) The Meroni doctrine and ACER

820. The Court, in its Judgment of 13 June 1958 *Meroni v. High Authority*, has judged that the Treaty establishing the European Coal and Steel Community (hereafter ECSC Treaty), on grounds of its Article 8, requires the High Authority: *"to ensure that the objectives set out in this Treaty are attained in accordance with the provisions thereof', and does not provide any power to delegate. However, the possibility of entrusting to bodies established under private law, having a distinct legal personality and possessing powers of their own, the task of putting into effect certain 'financial arrangements common to several undertakings' as mentioned in subparagraph (A) of Article 53 cannot be excluded"*.[1034]

821. It can therefore be deduced that the Court based the legal basis to delegate powers of the High Authority on Article 53 of the ECSC Treaty. Moreover, the Court deemed that: *"hence the power of the High Authority to authorize or itself to make the financial arrangements mentioned in Article 53 of the Treaty gives it the right to entrust certain powers to such bodies subject to conditions to be determined by it and subject to its supervision"*. In addition to this, the delegations on grounds of Article 53 *"are only legitimate if the High Authority recognizes them to be necessary for the performance of the tasks set out in Article 3 and compatible with this Treaty, and in particular with Article 65"*.

822. Finally, the Court judges that the High Authority, upon striving to reach the goals mentioned in Article 3 of the ECSC Treaty, will always face choices and will be required to prioritise specific powers, which requires and implies discretionary powers.[1035]

823. The delegation of powers can vary greatly. On the one hand, one can strive towards carefully defined executive powers, whereby their execution can be evaluated strictly on the basis of objective criteria as determined by the delegating body. On the other hand, the delegation can relate to discretionary powers, whereby a large amount of freedom of appreciation is presupposed and where the execution of these powers can amount to the execution of an economic policy. In the latter, the power to make choices, along with the responsibility for these choices, is transferred from the delegating to the other authority.[1036]

1034 Case 9/56, *Meroni/High Authority* [1957–8] ECR 133.

1035 K. LENAERTS, "Regulating the regulatory process: Delegation of powers in the European Union Community", *European Law Review* (1993), p. 32 and D. GERADIN, R. MUNOZ and N. PETIT, *Regulation through Agencies in the EU – A new paradigm of European Governance* (Cheltenham, Elgar, 2005), 280 p.

1036 Case 9/56, *Meroni/High Authority* [1957–8] ECR 133.

824. Based on the aforementioned elements, the Court concluded that the delegation of powers is only acceptable with executive powers that are clearly defined and subjected completely to the delegating authority's control. These limitations on the possibility to delegate powers are justified by the Court in view of respecting *"the balance of powers which is characteristic of the institutional structure of the Community"* (hereinafter institutional balance), which can be considered a fundamental guarantee provided by the Treaty, in particular to undertakings and associations of undertakings. As a result, delegating discretionary powers to other bodies than the ones the Treaty has established to execute and supervise these powers would impinge on this guarantee.[1037]

825. Firstly, it is important to stress that the *Meroni doctrine* dates from 1958 and was developed on the basis of the provisions of the ECSC Treaty.[1038] Taking into consideration the comprehensive modifications to the European Union's institutional framework, one might wonder whether the *Meroni doctrine*, and its reach, does not call for a thorough evaluation. Nevertheless, even though the doctrine was established in the framework of the ECSC Treaty, it is applicable to all European Treaties *mutatis mutandis.*[1039]

826. From what precedes, one may deduce that the Commission can delegate certain powers to other bodies, but that this delegation is subject to a number of limitations:

(a) a delegation of powers will only be permissible on the condition that the delegating authority has been attributed these powers itself;
(b) the delegation will only be acceptable with clearly defined executive powers, i.e. discretionary powers cannot be delegated;
(c) the delegating body will be obliged to supervise the execution of the delegated tasks and will be held responsible for this; and
(d) a delegation of powers cannot disrupt the institutional balance.[1040]

1037 K. LENAERTS, "Regulating the regulatory process: Delegation of powers in the European Union Community", *European law Review* (1993), p. 32.

1038 Note that the ECSC Treaty has expired on 23 July 2002, after its 50-year existence.

1039 K. LENAERTS, "Regulating the regulatory process: Delegation of powers in the European Union Community", *European Law Review* (1993), p. 32 and M. EVERSON, G. MAJONE, "Chapter VII: The problem of Delegation", in M. EVERSON, G. MAJONE, L. METCALFE and A. SCHOUT (eds.), "*The Role of specialized Agencies in Decentralising EU Governan*", Report presented to the Commission, 53.

1040 M. EVERSON, G. MAJONE, "Chapter VII: The problem of Delegation", in M. EVERSON, G. MAJONE, L. METCALFE and A. SCHOUT (eds.), "*The Role of specialized Agencies in Decentralising EU Governance*", Report presented to the Commission, p. 53.

827. The most important issue in this respect is the question of who will be held politically responsible for any given decision. Attributing decision-making powers to agencies cannot impinge on a European level and national level mechanisms of political control. One will not be able to replace these mechanisms of political control, included in amongst others the TFEU, by providing a representation of the Member States or the European institutions, such as the Commission, the Council and the European Parliament, in specific boards or organs within an agency. As a result, if it is opted not to alter the *Meroni doctrine*, a search for an institutional model for agencies, taking this doctrine into consideration, will have to be conducted.

828. It can be pointed out that establishing an agency based on Article 352 of the TFEU as opposed to a specific Treaty provision, such as Article 114 or 154 of the TFEU, will not solve the possible limitations as a result of the *Meroni doctrine*. The key difference between Article 352 of the TFEU and a specific Treaty provision for the establishment of a new instance with legal personality and with clearly defined powers, lies in the fact that the amount of powers will be more limited with a specific Treaty provision than when Article 352 of the TFEU is called upon. The decisional content of the executive powers delegated to another body will be limited to the same extent when this body is established on the basis of Article 352 of the TFEU as when it is based on a specific Treaty provision, as the institutional balance has to be respected as per the TFEU.[1041]

829. Supposing the *Meroni doctrine* may remain a barrier to the establishment of the Agency, with wide-ranging powers including discretionary powers, it would be theoretically possible, within the current institutional structure of the Union, to opt for a Treaty modification to Article 13 of the TEU and include agencies in this Article. Another option from a theoretical point of view is to provide the TEU with a specific legal basis on which an Agency could be established.

830. Another option that could be considered in order to conform to the *Meroni doctrine* and allow the Agency to be attributed wide-ranging powers is to stipulate that the Agency will give binding advice to the Commission with regard to the powers that can be qualified as discretionary. Subsequently, the Commission could base its decision on this advice or opt not to retain the Agency's advice, on the condition that this decision is motivated. The powers of the Agency not qualified as discretionary, i.e. the executive powers, could be carried out according to a predefined procedure.

[1041] K. LENAERTS, "Regulating the regulatory process: Delegation of powers in the European Union Community", *European Law Review* (1993), p. 32–34.

831. Hereby, it must be stressed that these executive powers can only be valid when they have been sufficiently clearly defined, i.e. when their boundaries have been unmistakably indicated.[1042] Executive powers of this nature include regulatory powers as well as the possibility to apply guidelines to special cases in accordance with individual decisions.[1043] In this manner, the author deems that the Commission will remain fully politically responsible for all decisions taken by the Agency and that the required institutional balance will remain respected.[1044]

(c) The adopted text on ACER

832. On 23 March 2009, the Czech Presidency and the responsible rapporteurs of the Committee on Industry, Research and Energy reached a compromise on the Third Legislative Package. This compromise was the object of the European Parliament vote on 21 April 2009 (second reading) which was backed by a large majority in the Parliament and contains the adoption of the new ACER Regulation, among other directives and regulations in the energy field.[1045]

833. The purpose of the Agency following Article 1 of the ACER Regulation is to assist the national regulatory authorities in exercising, on a Union level, the regulatory tasks performed in the Member States and if necessary to coordinate their action. This means that the Agency mainly has a supportive role to play on a European level. The name also makes clear that the Agency's purpose is to establish a European cooperation between the different national regulatory authorities on various cross-border issues. The Agency shall be a Union body with legal personality.[1046] Its tasks will be performed by an administrative board, a Board of Regulators, a Director with staff and a Board of Appeal.

834. These bodies of the Agency are allowed to issue opinions and recommendations addressed to the TSOs, regulatory authorities, European Parliament, Council and Commission. In specific cases (technical issues, some

---

1042 Case 291/86, *Central-Import Münster/Münster* [1988] 3679.

1043 K. LENAERTS and P. VAN NUFFEL, *Europees recht in hoofdlijnen* (Antwerpen, Maklu, 2008), p. 350; Case 41/69, *ACF Chemiefarma/Commissie* [1970] ECR 221 and Case 23/75, *Rey Soda/Cassa Conguaglio Zucchero* [1975] ECR 1279.

1044 H. VON ROSENBERG, "No ACE to win the trick – the proposed ACER and its influence on EC Competition law", *E.C.L.R.* (2008), p. 518 and 519 and K. LENAERTS, "Regulating the regulatory process: Delegation of powers in the European Union Community", *European Law Review* (1993), p. 32–36.

1045 Press release of 22 April 2009 of the European Parliament, "MEPs give green light to further liberalisation of EU electricity and gas markets" and MEMO/09/176 of 21 April 2009 of the European Commission, "Q&A on the European Parliament's vote on the Third Internal Energy Market Legislative Package."

1046 Until its premises are ready it will be hosted on Commission premises.

regulatory issues with respect to cross-border infrastructure and exemptions as provided for in Article 17 (5) of the Electricity Regulation 714/2009) referred to in Article 7, 8 and 9 of the ACER Regulation, the Agency is allowed to adopt individual decisions. The Agency can furthermore submit non-binding framework guidelines, for the development of network codes relating to the areas identified in the Commission's priority list, to the Commission in accordance with the Articles 6 of the Electricity Regulation 714/2009 and the Gas Regulation 715/2009.

i) The Agency's set of tasks

835. According to Article 5 of the ACER Regulation, the general task of the Agency is to provide an opinion or a recommendation to the European Parliament, the Commission or the Council upon their request, or on its own initiative, on any issue relating to the purpose for which it has been established. The concrete tasks of the Agency with respect to the cooperation of TSOs, the national regulatory authorities and the terms and conditions for access to and operational security of cross-border infrastructure have been further worked out in Article 6 up until and including Article 9 of the ACER Regulation.

836. Besides these tasks, the Agency will monitor on the basis of Article 11 of the ACER Regulation, in cooperation with the Commission, the Member States and the relevant national authorities, including national regulatory authorities, without prejudice to the competences of the competition authorities, the internal electricity and natural gas markets and in particular the retail prices of electricity, access to the network including access of electricity produced from renewable energy sources and compliance with consumer rights as laid down in the Electricity Directive 2009/72/EG and Gas Directive 2009/73/EG. In its monitoring work, the Agency may not duplicate or hamper monitoring by the Commission or national authorities, in particular national competition authorities. The results of the Agency's monitoring work will be published in an annual report which shall identify the possible barriers to the completion of the internal electricity and natural gas markets. When the Agency publishes its annual report, it is also allowed to submit an opinion to the European Parliament and the Commission on the measures that could be taken in order to remove the barriers referred to in its report.

837. Under Article 10 of the ACER Regulation, the Agency is obliged to consult market participants, extensively and at an early stage, i.e. TSOs, consumers, end-users and, where relevant, competition authorities in an open and transparent manner. It must ensure that the public and any interested parties have easy access to reliable information of its work. Furthermore, the Agency must publish on its website at least the agenda, the background documents and, where appropriate, the minutes of the meetings of the Administrative Board, the Board of Regulators and of the Board of Appeal.

838. With respect to the cooperation of TSOs, the Agency provides an opinion to the Commission on the draft statutes, list of members and the draft rules of procedures of the European Network of Transmission System Operators for Electricity (hereinafter ENTSO-E) and the ENTSO-G for gas.[1047] It monitors the execution of the tasks of the ENTSO-E and the ENTSO-G[1048], the regional cooperation between the TSOs in the electricity and gas sectors[1049] and the implementation of the Union-wide (ten-year) network-development plans. If it identifies, with the latter, inconsistencies between such a plan and its implementation, the Agency shall investigate the reasons for those inconsistencies and make recommendations to the TSOs, the national regulatory authorities or other competent bodies in order to implement the investments in accordance with the Union-wide ten-year network-development plans.

839. In the areas set out in the Articles 8 (6) of the Electricity Regulation 714/2009 and of the Gas Regulation 715/2009, such as rules for network security and reliability, network connection, third-party access, data exchange and settlement, interoperability, etc., the ENTSO-E and the ENTSO-G are allowed to adopt network codes. One of the Agency's tasks is to give an opinion on them and also on the draft annual work programme, the draft Union-wide (ten-year) network development plan and other relevant documents referred to in the Articles 8 (3) of the Electricity Regulation 714/2009 and of the Gas Regulation 715/2009 which were adopted by the ENTSO-E and ENTSO-G.[1050] If such a draft annual work programme or a draft Union-wide (ten-year) network-development plan of the Electricity Regulation 714/2009 and the Gas Regulation 715/2009 '*do not contribute to non-discrimination, effective competition and the efficient functioning of the market[s] or a sufficient level of cross-border interconnection open to third-party access, or do not comply with the relevant provisions*' of the new Electricity and Gas Directives and Regulations, the Agency shall provide a duly reasoned opinion as well as recommendations to ENTSO-E and ENTSO-G, the European Parliament, the Council and the Commission.[1051]

840. The Agency submits or reviews the non-binding framework guidelines to the Commission when it is requested to do so under the Articles 6(2) and 6(4) of the Electricity Regulation 714/2009 and the Gas Regulation 715/2009. From the reading of the Articles 6 of the Electricity Regulation 714/2009 and the Gas

1047 Article 6 (1) of the ACER Regulation.
1048 Article 6 (2) of the ACER Regulation.
1049 Article 6 (9) of the ACER Regulation.
1050 Article 6 (3) of the ACER Regulation.
1051 Article 6 (4) of the ACER Regulation.

Regulation 715/2009, it is clear that the Agency will not only give an opinion on the draft network codes from ENTSO-E and ENTSO-G, but that it will also be entirely involved in the adoption of these network codes.[1052] The network codes established by ENTSO-E and ENTSO-G need to be submitted to the Agency which will provide a reasoned opinion on these codes within a period of three months of the day of their receipt. If, in the view of the Agency, the network codes are in accordance with the relevant framework guideline it should submit them to the Commission. Furthermore, if the ENTSO-E and/or ENTSO-G have failed to develop a network code, the Agency, upon the request of the Commission, will prepare and submit a draft network code on the basis of the relevant framework guideline. When the ENTSO-E and/or the ENTSO-G have failed to implement a network code[1053] the Agency shall provide a duly reasoned opinion to the Commission.[1054]

841. The Agency will monitor and analyse the implementation of the network codes and Guidelines, adopted by the Commission on the basis of the Articles 6(11) of the Electricity Regulation 714/2009 and the Gas Regulation 715/2009, and their effect on harmonisation of applicable rules aimed at facilitating market integration as well as on non-discrimination, effective competition and the efficient functioning of the markets. The foregoing will be reported to the Commission. Finally, it must monitor the progress made by the projects to create new interconnection capacity. Compared to the amendment made by the European Parliament in its first reading which stated that: "*The Agency shall monitor the authorisation process of the construction of new cross-border capacities, and ensure the speeding up of this process within the confines of enhanced regional cooperation*", it is regrettable that the final text of the ACER Regulation does not contain a similar provision because currently one of the main shortcomings in the various Member States relates to the very long authorisation procedures (environmental permits, building permits, etc.) to construct new cross-border gas capacities and transmission lines.

842. With respect to the national regulatory authorities, the Agency has the power, on the basis of Article 7 (1) and (7) of the ACER Regulation, to adopt individual decisions on technical issues, the regulatory regime for electricity and natural gas infrastructure that connects or that might connect at least two Member States and, as a last resort, exemptions from the internal market rules for new electricity interconnectors and new gas infrastructure located in more

---

1052 Article 6 (4) of the ACER Regulation.

1053 It concerns a network code which has been elaborated under the Articles 8(2) of the Electricity Regulation 714/2009 and/or the Gas Regulation 715/2009 or a network code which has been established in accordance with Article 6 (1) to (10) of those regulations, but which was not adopted by the Commission under Article 6 (11) of those regulations.

1054 Article 6 (5) of the ACER Regulation.

than one Member State if those decisions are provided for by the new Electricity and Gas Directives and Regulations. It will furthermore provide a framework within which the national regulatory authorities can cooperate and promote cooperation between the national authorities on a regional and Union level.[1055]

843. At the request of the Commission or a regulatory authority, the Agency shall provide an opinion, based on matters of fact, on whether a decision taken by a regulatory authority complies with the Guidelines or the relevant provisions of the new Electricity and Gas Directives and Regulations. If a national regulatory authority does not comply with the opinion of the Agency within four months from the day of receipt, it shall inform the Commission and the Member State concerned accordingly. If in a specific case, a national regulatory authority encounters difficulties with the application of the Guidelines referred to in the new Electricity and Gas Directives and Regulations, it is allowed to ask the Agency for an opinion. This opinion needs to be delivered, following the consultation of the Commission, within three months upon the receipt of such a request.[1056]

844. With respect to cross-border infrastructure and in accordance with Article 8 of the ACER Regulation, the agency is allowed to decide upon the regulatory issues that fall within the competence of the national regulatory authorities, which may include the terms and conditions for access and operational security if one of the following conditions is met: *"(a) the competent national regulatory authorities have not been able to reach an agreement within the period of six months from when the case was referred to the last of those regulatory authorities or (b) upon a joint request from the competent national regulatory authorities"*. It should be noted that the competent national regulatory authorities may jointly request that the period mentioned in point (a) is extended by a period of up to six months and if the regulatory issues include exemptions on new interconnectors as referred to in Article 17 of the Electricity Regulation 714/2009 or the new infrastructure as referred to in Article 36 of the Gas Directive 2009/73/EG, the deadlines provided for in the ACER Regulation will not be cumulated with the deadlines provided for in the provisions of the Electricity Regulation 714/2009 or the Gas Directive 2009/73/EG.[1057]

845. The competence of the Agency on exemptions as provided for in Articles 17 (5) of the Electricity Regulation 714/2009 and 36 (4) of the Gas Directive, where the infrastructure concerned is located in the territory of more than one Member State, is also explicitly mentioned in Article 9 (1) of the ACER

1055 Article 7 (3) of the ACER Regulation.

1056 Article 7 (4), (5) and (6) of the ACER Regulation.

1057 Article 8 (5) of the ACER Regulation.

Regulation. It should be noted that, on the basis of Article 22 of the ACER Regulation, the Agency will request a fee for its exemption decision. This is the only fee which can be claimed by the Agency and its amount will be set by the Commission.

846. When the Agency prepares its decision in respect of the foregoing, it must consult the national regulatory authorities and concerned TSOs. Besides, the Agency must also be informed of the proposals and observations of all the concerned TSOs. If on the basis of the foregoing, a case has been referred to the Agency, it should provide its decision within a period of 6 months from the day of referral and if necessary, the Agency may provide an interim decision to ensure that the security of supply or operational security of the infrastructure in question is protected.[1058]

847. In the aforementioned situations where the Agency becomes competent to adopt terms and conditions, these terms and conditions should include in accordance with Article 8 (2) of the ACER Regulation: (a) a procedure for capacity allocation; (b) a time frame for allocation; (c) shared congestion revenues and (d) the levying of charges on the users of the new direct current interconnectors in the meaning of Article 17 (1) (d) of the Electricity Regulation 714/2009 or the major new gas infrastructure, i.e. interconnectors, LNG and storage facilities in the meaning of Article 36 (1) (d) of the Gas Directive 2009/73/EG. Article 8 (4) of the ACER Regulation furthermore stipulates that the Commission can adopt Guidelines that clarify in which situations the Agency becomes competent to decide upon the terms and conditions for access to an operational security of cross-border infrastructure.[1059]

848. Beside the tasks mentioned above, the Agency will, on the basis of Article 9 (2) of the ACER Regulation, also provide an opinion on the decisions of national regulatory authorities with respect to the certification of TSOs within the meaning of the second subparagraph of Articles 3 (1) of the Electricity Regulation 714/2009 and the Gas Regulation 715/2009. Finally, pursuant to Article 18 of the Electricity Regulation 714/2009 or Article 23 of the Gas Regulation 715/2009, the Agency may be, under clearly defined circumstances outlined by the Commission in Guidelines, commissioned with additional tasks which do not involve decision-making powers.

1058 Articles 8 (1) and 8 (3) of the ACER Regulation.

1059 Article 8 (4) of the ACER Regulation. It should be noted that these Guidelines need to be adopted in accordance with the regulatory procedure mentioned in Article 32 (2) of the ACER Regulation.

ii) Analysis and conclusion

849. Compared to the amendments made by the European Parliament in its first reading the most important changes in the finally adopted text of the ACER Regulation are that the Agency lost its power to approve various draft measures (such as network codes, annual work programme, the Union-wide (ten-year) network-development) which need to be taken by the ENTSO-E and the ENTSO-G, its powers to issue guidelines which establish basic, clear and objective principles for the harmonisation of network rules and its powers to issue binding guidelines to assist regulatory authorities and market players in sharing good practices. Furthermore, its role to monitor the European Union's gas storage requirements and to coordinate the national crisis management mechanisms on a Union level and the communications in this respect between the EU and third-country operators was not adopted in the ACER Regulation.

850. The Agency's powers to take action and remove any barriers to trade and even to force individual TSOs by means of financial penalties to comply with its decisions, has also not been preserved in the ACER Regulation. Given the fact that the Agency does not have decision-making powers, except in a limited number of cases, it seems reasonable for it not to impose any effective sanction in order to remove barriers to trade. Nonetheless, it remains questionable, even if this was not the purpose of the ACER Regulation, whether we need an independent European regulator for electricity and gas markets on a European level in order to ensure non-discrimination, the efficient functioning of the market and effective competition of European markets.[1060]

851. It is clear from the foregoing that the Agency on the basis of the new ACER Regulation has mainly been given an advisory role in order to handle cross-border issues in the Electricity and gas markets on a European level. The consequences of this lack of decision-making power are currently unclear.

852. The aim of this part with respect to ACER was to elaborate on recent developments and the workings and powers of the Agency that is established. It is clear that the adoption of the Agency was a first step in the right direction in order to tackle a number of cross-border issues with regards to the European electricity and gas market. Nonetheless and unfortunately, the powers of the proposed Agency remain rather limited. The Agency has mainly been given an advisory role in order to handle the cross-border issues in European electricity

[1060] See *Infra*.

and gas markets. ACER will at least in the first years mostly fulfil the role that ERGEG has played so far.[1061]

853. This is mainly due to the so-called *Meroni doctrine* which entails that independent executive organs, unless a specific Treaty provision is available and the delegation remains limited to the aims and means specified in this provision, cannot be attributed discretionary powers, as this would render certain policy choices impossible for institutions that, on the basis of the Treaty, carry the political responsibilities for such choices.

854. If this doctrine remains one of the obstacles to attribute discretionary powers, in the author's view there is a need for a thorough evaluation of the principles of the *Meroni doctrine*. Otherwise Europe will continue to be deprived of a real regulator in the energy field on a European level. The fact that the Agency could not be attributed discretionary powers due to the *Meroni doctrine* clearly demonstrates the limits of the competences of the European Union and calls for alternative options to overcome these barriers.

E. CONCLUSION

855. For many years, as discussed above, the Union has been adopting secondary legislation in the energy sphere, *inter alia*, on the basis of Articles 113, 114, 122, 192 and 352 of the TFEU.[1062] Consequently, Union competence

---

1061 L. HANCHER and A. DE HAUTECLOCQUE, "Manufacturing the EU Energy Markets: The Current Dynamics of Regulatory Practise", *Competition and Regulation in Network Industries* (2010), p. 321.

1062 For example measures taken under Article 113 of the TFEU (ex Article 93 EC Treaty) include Directive 2003/96/EC of 27 October 2003 restructuring the Community framework for the taxation of energy products and electricity, *OJ* L 283/51; measures taken under Article 114 of the TFEU (ex Article 95 EC Treaty) include Directive 96/92/EEC of 19 December 1996 concerning common rules for the internal market in electricity, *OJ* L 27/20; Directive 2003/54/EC of 26 June 2003 concerning the common rules for the internal market in electricity and repealing Directive 96/92/EC, *OJ* L 176/37; Regulation No 1228/2003 of 26 June 2003 on conditions for access to the network for cross-border exchanges in electricity, *OJ* L 20/24; Directive 98/30/EC of 22 June 1998 concerning the common rules for the internal market in natural gas, *OJ* L 204/1; Directive 2003/55/EC of 26 June 2003 concerning the common rules for the internal market in natural gas and repealing Directive 98/30/EC, *OJ* L 176/57; Regulation No 1775/2005 of 28 September 2005 on conditions for access to the natural gas transmission networks, *OJ* L 289/1; Directive 2005/89/EC of 18 January 2006 concerning measures to safeguard security of electricity supply and infrastructure investment, *OJ* L 33/22; measures taken under Article 122 of the TFEU (ex Article 100 of the EC Treaty) include Directive 68/414/EEC of 20 December 1968 imposing an obligation on Member States of the EEC to maintain minimum stocks of crude oil, *OJ* L 308/14; Directive 73/238/EEC of 24 July 1973 on measures to mitigate the effects of difficulties in the supply of crude oil and petroleum products, *OJ* L 228/1; Directive 98/93/EC of 14 December 1998 amending Directive 68/414/EEC imposing an obligation on Member States of the EEC to maintain minimum stocks of

undoubtedly existed in the energy policy area. However, this Union competence was not based on explicit energy policy provisions, but on other explicit or more general provisions of the Treaty.

856. This chapter has firstly given an historical overview of the Treaty context with respect to energy. It highlighted that over the past years the Treaties were amended several times. However, before the adoption of the Lisbon Treaty, these amendments with respect to the energy field were rather limited. This part was followed by an analysis of the principles of attribution, subsidiarity and proportionality which govern the existence and exercise of Union powers and applied them to the energy sector. It can be concluded, that, although the Treaties did not confer any specific powers to the European legislator to adopt measures in the energy field, recourse to more general legal bases in the TFEU appeared to be sufficient in order to comply with the principle of attribution. With respect to the subsidiarity principle, it seems that most of the energy measures will pass the subsidiarity test.[1063] However, in order to determine whether a given energy measure complies with the principle of subsidiarity, the European legislator must always analyse this on a case-by-case basis. The third and final principle discussed in this chapter was the principle of proportionality. It has been highlighted that the European legislator often applies a standard reference. Nonetheless, it should be assessed whether each measure complies with the three-tier proportionality test.

857. The next part of chapter 4 discussed the meaning and application of the Articles 106(1), (2) and 345 of the TFEU in order to determine to which extent they could hamper the application of some of the TFEU provisions, and secondary legislation, to the electricity and gas markets. It has concluded on the basis of case law of the Court of Justice that even with the adoption of the electricity and gas directives, Member States seem to retain a substantial discretion with respect to the adoption of public service obligations in the energy field and with respect to which services could be classified as services of general

---

crude oil and/or petroleum products, *OJ* L 358/100; Directive 96/57/EC of 3 September 1996 on energy efficiency requirements for household electric refrigerators, freezers and combinations thereof, *OJ* L 236/36; measures taken under Article 172 of the TFEU (ex Article 156 of the EC Treaty) include Decision No 1229/2003 of 26 June 2003 laying down a series of guidelines for trans-European energy networks and repealing Decision No 1254/96/EC, *OJ* L 176/11; measures taken under Article 192 of the TFEU (ex Article 175 of the EC Treaty) include, Directive 2001/77/EC of 27 September 2001 on the promotion of electricity produced from renewable energy sources in the internal electricity market, *OJ* L 283/33; Directive 2002/91/EC of 16 December 2002 on the energy performance of buildings, *OJ* L 1/65; Directive 2004/8/ EC of 11 February 2004 on the promotion of cogeneration based on a useful heat demand in the internal energy market and amending Directive 92/42/EEC, *OJ* L 52/50; etc.

[1063] J. H. JANS and H. H. B. VEDDER, *European Environmental Law* (Groningen, Europa Law Publishing, 2008), p. 12 en 13.

economic interest. With respect to Article 345 of the TFEU it argued that this provision does not prevent the Commission from adopting the ownership unbundling model in the energy sector.

858. This part was followed by an analysis of the free movement of goods provisions in the TFEU. These fundamental provisions have set boundaries which also apply to the electricity and gas markets. Moreover, both the European and national legislator must take into account these fundamental provisions when they adopt legislative measures in the energy field. This analysis contains a discussion of the fiscal barriers, non-fiscal barriers, Article 36 of the TFEU and the mandatory requirements. It has concluded that both the security of supply and environmental protection arguments, on certain conditions, could justify in energy measures restrictions to the free movement provisions of the TFEU.

859. After this discussion of the general provisions of the TFEU, this chapter analysed the legal bases used by the European legislator to intervene in the energy markets by secondary legislation. Due to the lack of any specific chapter, title or Article on energy, this secondary legislation was adopted on the basis of Article 114 of the TFEU, Title XX of the TFEU, comprising Articles 191–193, Article 113, Articles 170–172 of the TFEU, etc. The most prominent Articles with respect to the energy field are discussed in this part of the chapter. The analysis of these provisions allowed having a clear understanding of them. This understanding is important in order to determine to which extent the European legislator since the adoption of the Lisbon Treaty and the new energy title can continue to adopt measures in the energy field on the basis of these provisions.

860. The final part of chapter 4 has demonstrated that not only Treaty provisions can influence the adoption of secondary legislation, but also case law of the Court of Justice. It has concluded that mainly due to the so-called *Meroni doctrine* the Agency has mainly been given an advisory role in order to handle cross-border issues on a European level in both the electricity and gas markets. Moreover, this part has argued that if this doctrine remains one of the obstacles to attribute discretionary powers there is a strong need for a thorough evaluation of this doctrine.

861. Although, the TFEU does not contain an enabling provision putting forward the objectives, goals and procedures for Union action in the energy field, the European legislator has been adopting an enormous amount of secondary legislation in the energy field in the past years. Due to the recourse to more general legal provisions of the TFEU, the European legislator was to a certain extent unable to address the specific issues of the energy sector.[1064] Thus, this situation was not the most appropriate one.

862. Finally, with the adoption of the Treaty of Lisbon, after some 40 years or so, the Union legislature's activities in the energy field, have a legal basis for the adoption of a European energy policy. However, as discussed below, compared to the former situation, with respect to competences in the energy field, it seems that no major changes were adopted by the negotiators of the Lisbon Treaty. The purpose of the next chapter is to analyse in detail the most prominent changes to the TFEU provisions relevant for the energy sector since the adoption of the Lisbon Treaty.

1064 Policy Proposal by Jacques Delors, *Towards a European Energy Community: A Policy proposal*, study by S. ANDOURA, L. HANCHER and M. VAN DER WOUDE, p. 15. With respect to the specific issues of the energy sector see above Chapter 2.

CHAPTER 5

UNION COMPETENCES IN THE ENERGY FIELD AFTER THE ADOPTION OF THE LISBON TREATY

A. INTRODUCTION

863. Following the failed referenda on the draft Constitutional Treaty of Europe, a "period of reflection" on the future of Europe was launched to reconnect the citizens with the European project and to decide the fate of the Constitutional Treaty of Europe. At the European Council in June 2007, the EU leaders managed to overcome the institutional impasse and agreed on the outlines of a new Treaty which replaces the draft Constitutional Treaty of Europe.[1065]

864. This new Treaty, called the Treaty of Lisbon, was signed on 13 December 2007 and entered into force, after ratification by all Member States, on 1 December 2009. With the adoption of the Lisbon Treaty the proposed structure of the Constitutional Treaty of Europe is in practice highly recognisable in the consolidated version of the Treaties.

865. The TEU roughly corresponds with part I and IV of the Constitutional Treaty of Europe. The separate binding Charter of Fundamental Rights corresponds with part II of the Constitutional Treaty of Europe. This Charter of Fundamental Rights has furthermore been anchored through the reference to it in Article 6 of the TEU. Part III of the Constitutional Treaty of Europe corresponds *mutatis mutandis* with the new TFEU.[1066] In the final text of the TFEU, energy has been firmly enshrined as one of the eleven areas in which the Union shall share competence with the Member States. This means that, on the basis of Article 2 of the TFEU, the Union and the Member States can legislate and adopt legally binding acts in the energy field. However, the Member States

---

1065 Www.consilium.europa.eu/ueDocs/cms_Data/docs/pressData/en/ec/94932.pdf.

1066 T. CORTHAUT, "Plus ça change, plus c'est la même chose? A Comparison with the Constitutional Treaty", *Maastricht Journal of European and Comparative Law* (2008), p. 24 and 25.

shall exercise their competence to the extent that the Union has not exercised its competence or has decided to cease exercising its competence.[1067]

866. The most important innovation of the Lisbon Treaty in this context is the adoption of a specific chapter on energy which defines the key competences and overall objective of the Union's energy policy. This new energy title provides a legal basis for the adoption of measures which aim to ensure the general goals of the existing European energy policy such as the functioning of the internal market, the security of energy supply, energy efficiency.[1068] Before discussing all the aspects of this new energy title, the next part provides an overview of the most relevant changes for the energy sector in the TFEU since the adoption of the Lisbon Treaty.

B. RELEVANT CHANGES TO PROVISIONS OF THE TFEU

1. CATEGORIES OF UNION COMPETENCE

(a) Introduction

867. In addition to the question whether the Union has competence in a given area in accordance with the principles discussed above, it should be analysed what the nature of that competence is. This concerns more in particular, the different legal consequences for the national regulatory competences as a result of the existence or exercise of a Union competence. Depending on their relationship to the powers of the Member States, the Union competences can be divided into three principal categories among which (a) exclusive competence of the Union, (b) shared competence and (c) competence to support, coordinate or supplement actions of the Member States.[1069]

(b) Exclusive competences

868. To the first category of exclusive competences the regulatory authority to act solely belongs to the Union. In these areas the Member States have lost

---

[1067] See *Infra*.

[1068] T. CORTHAUT, "Plus ça change, plus c'est la même chose? A Comparison with the Constitutional Treaty", *Maastricht Journal of European and Comparative Law* (2008), p. 26 and 27.

[1069] A. ARNULL, A. DASHWOOD, M. DOUGAN, M. ROSS, E. SPAVENTA and D. WYATT, *European Union Law* (London, Sweet & Maxwell, 2006), p. 91 and A. DASHWOOD, "Relationship between Member States and EC/EU" in A. MCDONNELL (ed.), *A review of forty years of Community law – Legal developments in the European Communities and the European Union* (The Hague, Kluwer Law International, 2005), p. 51–55.

definitively and irreversibly their right to act autonomously. They can only adopt measures if, and so far as, the Union legislator authorises them to do so. In such a situation the Union legislator specifies according to which procedure and in which way the Member States are allowed to act.[1070] However, the Court of Justice held in the *Fisheries Conservation* case[1071] that the Member States could be implicitly authorised to act as a trustee of the common interest, provided that they adopt these measures in close consultation and cooperation with the Commission.[1072]

869. The exclusive Union competence is very much the exception which is evident because from a national perspective it involves a complete surrender of jurisdiction to the Union.[1073] The Court of Justice has recognised in its Opinion 1/75 that the Union legislator has, under Article 207 of the TFEU, the exclusive power to conduct a common commercial policy *vis-à-vis* third countries to trade goods.[1074]

870. Another example recognised as being one of the exclusive Union competences, is the competence to determine the conditions for fishing with a view to ensure protection of fishing grounds and the conservation of the

1070 K. LENAERTS, and P. VAN NUFFEL, in BRAY, R. (ed.), *Constitutional Law of the European Union, Constitutional Law of the European Union* (London, Thomson Sweet & Maxwell, 2005), p. 95–100 and A. ARNULL, A. DASHWOOD, M. DOUGAN, M. ROSS, E. SPAVENTA and D. WYATT, *European Union Law* (London, Sweet & Maxwell, 2006), p. 91 and 92.

1071 Case 804/79, *Commission of the European Communities v United Kingdom of Great Britain and Northern Ireland* [1981] ECR 1045.

1072 *Ibid.*, paragraph 30.

1073 D. CHALMERS, G. DAVIES and G. MONTI, *European Union Law* (Cambridge, Cambridge University Press, 2010), p. 207.

1074 Opinion 1/75 Draft OECD Understanding on Local Cost Standard [1975] ECR 1255 in which the Court of Justice stated that "*[…] In fact any unilateral action on the part of the Member States would lead to disparities in the conditions for the grant of export credits, calculated to distort competition between undertakings of the various Member States in external markets. Such distortion can be eliminated only by means of a strict uniformity of credit conditions granted to undertakings in the Community, whatever their nationality. It cannot therefore be accepted that, in a field such as that governed by the understanding in question, which is covered by export policy and more generally by the Common commercial policy, the Member States should exercise a power concurrent to that of the Community, in the Community sphere and in the international sphere, the provisions of Articles 113 and 114 [currently Article 207 of the TFEU] concerning the conditions under which, according to the Treaty, agreements on commercial policy must be concluded show clearly that the exercise of concurrent powers by the Member States and the Community in this matter is impossible. To accept that the contrary were true would amount to recognizing that, in relations with third countries, Member States may adopt positions which differ from those which the Community intends to adopt, and would thereby distort the institutional framework, call into question the mutual trust within the Community and prevent the latter from fulfilling its task in the defence of the Common interest*". See also Case 41/76, *Suzanne Criel, née Donckerwolcke and Henri Schou v Procureur de la République au tribunal de grande instance de Lille and Director General of Customs* [1976] ECR 1921, paragraph 32.

biological resources of the sea. Finally, the monetary policy for those Member States participating in the third stage of the Economic and Monetary Union and the introduction of the common customs tariff on the basis of Article 31 of the TFEU, preclude any parallel competence of the Member States and are exclusive competences of the Union legislator. If the Union has exclusive powers in a given area, any action by a Member State in that field is *a priori* in conflict with the TFEU. Furthermore, given the nature of these areas it would make no practical sense if the regulatory authority would be shared with the Member States.[1075]

871. Since the adoption of the Lisbon Treaty the TFEU and in particular the adoption of Article 3 of TFEU, it became crystal clear which the exclusive competences of the Union are. These exclusive competences comprise: (a) customs union, (b) establishment of the competition rules necessary for the functioning of the internal market; (c) monetary policy for the Member States whose currency is the Euro; (d) conservation of marine biological resources (fisheries policy) and (e) common commercial policy.

(c) Shared competences

872. The second main category of Union competences is the one of shared competences in which both the Union and its Member States are competent to regulate the relevant area. This is the most common and normal relationship between the Union and Member States powers in various fields. According to Article 4(2) of the TFEU these shared competences are (a) internal market; (b) social policy for the aspects defined in the TFEU; (c) economic, social and territorial cohesion; (d) agriculture and fisheries (except conservation of marine biological resources because this is an exclusive competence); (e) environment; (f) consumer protection; (g) transport; (h) trans-European networks; (i) energy: (j) area of freedom, security and justice and (k) common safety concerns in public health matters, for the aspects defined in the TFEU. This list of shared competences in Article 4(2) of the TFEU is not exhaustive.[1076]

873. In this model of cooperative federalism, a shared responsibility is granted to both the Union and its Member States in order to realise the common policy goals. The balance of responsibilities between the Union and its Member States for this category of competences is determined by the terms, limits and presence

---

[1075] K. LENAERTS, and P. VAN NUFFEL, in BRAY, R. (ed.), *Constitutional Law of the European Union, Constitutional Law of the European Union* (London, Thomson Sweet & Maxwell, 2005), p. 95–100 and A. ARNULL, A. DASHWOOD, M. DOUGAN, M. ROSS, E. SPAVENTA and D. WYATT, *European Union Law* (London, Sweet & Maxwell, 2006), p. 91 and 92.

[1076] See *Infra.*

of legislation on a European level.[1077] If the Union legislator has not yet exercised its shared regulatory competence, the Member States are allowed to exercise their competence as long as they comply with the primary Treaty provisions.[1078]

874. In a situation where the Union legislator has exercised its shared regulatory competences, the Member States are allowed to exercise their competence as long as they do not only comply with the primary provisions, but also with any obligation imposed by the relevant Union legislation.[1079] Such a relationship seems to be logical in situations where there is a need for uniformity; otherwise the harmonisation process would be robbed of much of its effect. This is in particular the case within the context of the internal market where distortions of competition and trade restrictions could arise due to maintenance of differing national regimes.[1080]

875. In the situation where the Union legislator has exercised its regulatory shared competences, the so-called pre-emptive effects could arise in which the Union legislator will occupy the relevant regulatory field and prevent Member States from exercising their own competence in that field. Unlike the principle of the precedence of Union law or the principle of supremacy, a doctrine called pre-emption has assumed relatively little interest in legal literature.

876. The doctrine of pre-emption governs matters of conflict between European and national measures and what the consequences are for these measures. Until recently, its main contents can be deduced from the Court of Justice's case law. According to the pre-emption doctrine, the national legislator is precluded from adopting new national legislative measures to the extent that they would be incompatible with Union legislation. This is the case when Union law covers the area in question exhaustively, or when national law would be in breach of Union law or when national law would constitute an obstacle to the smooth functioning of the internal market.[1081] When the pre-emption doctrine

1077 D. CHALMERS, G. DAVIES and G. MONTI, *European Union Law* (Cambridge, Cambridge University Press, 2010), p. 208 and R. SCHÜTZE, "Co-operative Federalism Constitutionalised: The Emergence of Complementary Competences in the EC Legal Order", *European Law Review* (2006), p. 168 and 169.

1078 See the *Cassis de Dijon* or the Case 120/78, *Rewe-Zentral AG v Bundesmonopolverwaltung für Branntwein* [1979] ECR 649.

1079 See the *Dim-Dip Headlights* case or the Case 60/86, *Commission of the European Communities v United Kingdom of Great Britain and Northern Ireland* [1988] ECR 3921.

1080 D. CHALMERS, G. DAVIES and G. MONTI, *European Union Law* (Cambridge, Cambridge University Press, 2010), p. 209.

1081 Case 106/77, *Amministrazione delle Finanze dello Stato v Simmenthal SpA* [1978] ECR 629, paragraphs 17, 18 and 21; Case C-158/91, *Criminal proceedings against Jean-Claude Levy* [1993] ECR I-4287, paragraph 9; Case 218/85, *Association comité économique agricole régional fruits et légumes de Bretagne v A. Le Campion* [1986] ECR 3513, paragraph 13; Case C-1/96, *The Queen v Minister of Agriculture, Fisheries and Food, ex parte Compassion in World*

is applied, the national legislator needs to decide whether it has competence to legislate at all, and if it does, whether and to what extent its competence is limited.

877. Since the adoption of the Lisbon Treaty, the doctrine of pre-emption has been codified for the first time in the Treaty and in particular in Article 2 of the TFEU.[1082] This provision states that: "*1. When the Treaties confer on the Union exclusive competence in a specific area, only the Union may legislate and adopt legally binding acts, the Member States being able to do so themselves only if so empowered by the Union or for the implementation of Union acts.*

2. When the Treaties confer on the Union a competence shared with the Member States in a specific area, the Union and the Member States may legislate and adopt legally binding acts in that area. The Member States shall exercise their competence to the extent that the Union has not exercised its competence. The Member States shall again exercise their competence to the extent that the Union has decided to cease exercising its competence.

3. The Member States shall coordinate their economic and employment policies within arrangements as determined by this Treaty, which the Union shall have competence to provide.

4. The Union shall have competence, in accordance with the provisions of the Treaty on European Union, to define and implement a common foreign and security policy, including the progressive framing of a common defence policy.

5. In certain areas and under the conditions laid down in the Treaties, the Union shall have competence to carry out actions to support, coordinate or supplement the actions of the Member States, without thereby superseding their competence in these areas.

Legally binding acts of the Union adopted on the basis of the provisions of the Treaties relating to these areas shall not entail harmonisation of Member States' laws or regulations.

6. The scope of and arrangements for exercising the Union's competences shall be determined by the provisions of the Treaties relating to each area".

878. This provision not only inserts the doctrine of pre-emption into the Treaty, it also sets out the different types of competences in the Treaties and for

---

Farming Ltd [1998] ECR I-1251, paragraph 41 and Case C-491/01, *The Queen v. Secretary of State for Health, ex. parte British American Tobacco ("BAT")* [2002] ECR I-11453, paragraph 77. See also E. D. CROSS, "Pre-emption of Member State Law in the European Economic Community: A Framework for Analysis", *Common Market Law Review* (1992), p. 450, 454 and 467–471; A. GOUCHA SOARES, "Pre-emption, Conflicts of Power and Subsidiarity", *European Law Review* (1998), p. 132–136 and 138 and R. SCHÜTZE, "Supremacy without Pre-emption? The Very Slowly Emergent Doctrine of Community Pre-emption", *Common Market Law Review* (2006), p. 1023–1024, 1032 and 1039.

1082 See *Infra*.

each of these competences the degree of intervention of the Union. Only in the first two paragraphs of Article 2 of the TFEU, covering the exclusive and shared competences, pre-emption applies as an instrument for precluding national legislation. The subsequent paragraphs of Article 2 of the TFEU seem to establish in the various fields the scope of the Union policy-making. They identify whether there should be a common policy, Union policy complementary to national policies or any Union policy at all. Thus, Article 2 of the TFEU not only sets out the relationship between the Union and its Member States, it also provides for the various models of integration.[1083]

879. It should be emphasised that contrary to the exclusive and supporting competences, the list of fields of shared competences of the Union is not exhaustive. This can be derived from the wording of Article 4(1) of the TFEU *"the Union shall share competence with the Member States where the Treaties confer on it a competence which does not relate to the areas referred to in Articles 3 and 6"*. Article 4(1) of the TFEU not only indicates that the list of fields of shared competences mentioned in the TFEU is not exhaustive, it also appears to indicate that the competences on the coordination of economic and employment policies of Article 5 of the TFEU and common foreign and security policy (hereinafter CFSP) referred to in Article 2(4) of the TFEU, belong to the category of fields which are shared competences.

880. With respect to competences on the coordination of economy and employment policies of Article 5 of the TFEU, it should be emphasised that the Union in this field only has a competence of coordination, while the Member States retain the competence to define the guidelines for these policies which means in essence the competence on substance. One could argue that this field should have been classified as "complementary" instead of "shared" competence. However, due to the coordination done by the Union in this field, the Convention decided that it was difficult to catalogue this competence as "complementary". Subsequent IGCs did not change this approach of the Convention. With regard to CFSP, it was considered that due to the specificity of the competences of the Union in this field, in which both the Union and its Member States are competent to adopt legally binding acts, it was politically difficult to make this field appear in one of the three categories of competences on a European level.

881. Nonetheless, the fact that a competence has been included in one or another category of competences cannot change its substance since this is defined in other provisions of the Treaties. The foregoing can be derived from paragraph 6 of Article 2 *"the scope of and arrangements for exercising the*

1083 D. CHALMERS, G. DAVIES and G. MONTI, *European Union Law* (Cambridge, Cambridge University Press, 2010), p. 206 and 207.

Union's competences shall be determined by the provisions of the Treaties relating to each area". Reference should also be made to Protocol 25, annexed to the Lisbon Treaty, on the exercise of shared competence which contains a sole Article which states: *"with reference to Article 2(2) of the Treaty on the Functioning of the European Union on shared competence, when the Union has taken action in a certain area, the scope of this exercise of competence only covers those elements governed by the Union act in question and therefore does not cover the whole area"*. The motives underlying this Protocol were to convince Member States which had difficulties accepting the "concept of shared competences". The main purpose of this Protocol is to confirm the existing legal situation and to avoid an unwanted "creeping" extension of the competences of the Union.[1084]

882. In this context it is also pertinent to note that the principle of precedence or primacy applies only if a conflict between a Union norm and a norm from the national legal order appears. In those cases the national norm needs to yield to the Union norm. Thus, pre-emption doctrine and the principle of primacy are interrelated. If placed in a time segment, the former precedes the latter.[1085] Before the adoption of the Treaty of Lisbon, the obligation to apply the principle of primacy of the Union law, the pre-emption doctrine and the Member States' requirement to implement the Union law, can all be traced back to the principle of cooperation or loyal cooperation[1086] as

---

1084 J.-C. PIRIS, *The Lisbon Treaty: A Legal and Political Analysis* (Cambridge, Cambridge University Press, 2010), p. 77 and 78 and P. CRAIG, "The Treaty of Lisbon, process, architecture and substance", *European Law Review* (2008), p. 144.

1085 For a further discussion on the principle of primacy see J.-C. PIRIS, *The Lisbon Treaty A Legal and Political Analysis* (Cambridge, Cambridge University Press, 2010), p. 79–82 and the declaration 17 concerning primacy which states that: *"[t]he Conference recalls that, in accordance with well settled case law of the Court of Justice of the European Union, the Treaties and the law adopted by the Union on the basis of the Treaties have primacy over the law of Member States, under the conditions laid down by the said case law. The Conference has also decided to attach as an Annex to this Final Act the Opinion of the Council Legal Service on the primacy of EC law as set out in 11197/07 (JUR 260): "Opinion of the Council Legal Service of 22 June 2007 – It results from the case-law of the Court of Justice that primacy of EC law is a cornerstone principle of Community law. According to the Court, this principle is inherent to the specific nature of the European Community. At the time of the first judgment of this established case law (Costa/ENEL,15 July 1964, Case 6/641[1]) there was no mention of primacy in the treaty. It is still the case today. The fact that the principle of primacy will not be included in the future treaty shall not in any way change the existence of the principle and the existing case-law of the Court of Justice"*. This declaration emphasises that even with the adoption of the Lisbon Treaty the legal situation has not changed.

1086 In the United States this principle is known as the fidelity principle which requires *"that each level and unit of government must act to ensure the proper functioning of the system of governance as a whole"* see D. HALBERSTAM, "The Political Morality of Federal Systems", *Virginia Law Review* (2004), p. 104.

provided for in Article 4(3) of the TEU[1087] as interpreted by the Court of Justice.[1088]

883. The principle of loyal cooperation requires Member States to facilitate the achievement of the Union's tasks and to abstain from any measure which could jeopardise the attainment of the objectives of the Treaty. The Member States are also required to carry out all measures to guarantee the scope and effectiveness of Union law.[1089] The Member States should not undertake any measures outside the Union legal order that would endanger the integrity and independence of the Union legal order.[1090] The principle of loyal cooperation is reciprocal[1091] applying to Member States and to Union institutions. Indisputably the principle of loyal cooperation is of great importance when interpreting Union law. Notwithstanding this, in order to enhance legal certainty in the division of competences between the Union and its Member States, it is advisable to give this principle a not too independent role.

884. Compared to its predecessor, Article 10 of the EC Treaty, Article 4(3) of the TEU introduces two new aspects. It firstly introduces the idea of "mutual respect" which means that each institution must not only support other

1087 This provision states that: *"[p]ursuant to the principle of sincere cooperation, the Union and the Member States shall, in full mutual respect, assist each other in carrying out tasks which flow from the Treaties. The Member States shall take any appropriate measure, general or particular, to ensure fulfilment of the obligations arising out of the Treaties or resulting from the acts of the institutions of the Union. The Member States shall facilitate the achievement of the Union's tasks and refrain from any measure which could jeopardise the attainment of the Union's objectives."*

1088 J. TEMPLE LANG, "Community Constitutional Law: Article 5 EEC Treaty", *Common Market Law Review* (1990), p. 647–648, 669–670 and 673–677; Case C-2/88, *Imm. J. J. Zwartveld and others* [1990] ECR I-3365, paragraph 17; Case C-275/00, *European Community, represented by the Commission of the European Communities v First NV and Franex NV* [2002] ECR I-10943, paragraph 49; Case C-339/00, *Ireland v Commission of the European Communities* [2003] ECR I-1175, paragraph 71; Case C-344/91, *Federal Republic of Germany v Commission of the European Communities* ECR 2004 I-2081, paragraph 79 and Case C-459/03, *Commission of the European Communities v Ireland* [2006] ECR I-4635, paragraph 174.

1089 Case C-265/95, *Commission of the European Communities v French Republic* [1997] ECR I-6959.

1090 Case C-2/88, *Imm. J. J. Zwartveld and others* [1990] ECR I-3365, paragraph 17; Case C-275/00, *European Community, represented by the Commission of the European Communities v First NV and Franex NV* [2002] ECR I-10943, paragraph 49; Case C-339/00, *Ireland v. Commission of the European Communities* [2003] ECR I-1175, paragraph 71; Case C-344/91, *Federal Republic of Germany v Commission of the European Communities* ECR 2004 I-2081, paragraph 79 and Case C-459/03, *Commission of the European Communities v Ireland* [2006] ECR I-4635, paragraph 174.

1091 Case C-339/00, *Ireland v. Commission of the European Communities* [2003] ECR I-1175, paragraph 72; Reasoned submission by Advocate General TIZZANO in Case C-339/00, *Ireland v Commission of the European Communities*, paragraph 73 and Case C-2/88, *Imm. J. J. Zwartveld and others* [1990] ECR I-3365.

institutions, but must also not transgress the prerogatives of other institutions. Secondly, according to this provision, the duty of cooperation upon the Union and its Member States applies to tasks which "flow from the Treaties". Compared to the previous duty of cooperation which applied to tasks arising from the fulfilment of Treaty obligations or actions taken by the institutions of the Union, this is a more open-ended duty of cooperation.[1092]

(d) Competence to support, coordinate or supplement

885. The third category of Union competence comprises the competence to support, coordinate or supplement actions of the Member States. The regulatory power of these competences remains with the Member States. The role of the Union under this competence could involve the adoption of guidelines, incentive measures or to facilitate the exchange of information on best practices. According to Article 6 of the TFEU, the areas of such supporting competence are (a) protection and improvement of human health; (b) industry; (c) culture; (d) tourism; (e) education, vocational training, youth and sport; (f) civil protection and (g) administrative cooperation. The actions of the Union in these fields are to complement the national policies. Furthermore, according to case law of the Court of Justice, Member States, even in areas of supporting competences must comply with the horizontal principles under the primary Union law (e.g. free movement of goods, services, persons and capital).[1093]

886. Beside these three categories of competences, the Treaties also contain some specific areas which do not fit into the threefold classification such as research, technical development, space, development of cooperation and humanitarian aid and can be found somewhere between shared and supporting competences. Compared to the shared competences, the need for uniformity, discussed above, in these areas appears to be a lesser prerequisite. This can be derived from Article 4(3) and (4) of the TFEU which states that the Union in some of these fields shall have competence to carry out activities and in particular to define and implement programmes or to conduct a common policy, however the exercise of that competence shall not result in Member States being prevented from exercising theirs.

887. It seems to imply that the relationship between the Union and its Member States in these fields should not be fixed *a priori*, but rather changeable. The Union has the competence to develop legislation in these fields. However, the

---

1092 D. CHALMERS, G. DAVIES and G. MONTI, *European Union Law* (Cambridge, Cambridge University Press, 2010), p. 223–227.

1093 See Case 293/83, *Gravier v City of Liège* [1985] ECR 593 and A. ARNULL, A. DASHWOOD, M. DOUGAN, M. ROSS, E. SPAVENTA and D. WYATT, *European Union Law* (London, Sweet & Maxwell, 2006), p. 95 and 96.

exercise of this competence does not prevent Member States from adopting their own legislation in these fields provided that the national provisions do not conflict with the European measures. Thus, the legal bases of these areas indicate that the Union has neither a shared nor a supporting competence. However, it appears that the Union has a complementary competence of its own in addition to the competence of the Member States in these fields.[1094] Other categories which do not fit into the threefold classification of competences are the coordination of economic and employment policies and CFSP.[1095]

(e) Since Lisbon, energy is a shared competence

888. With the adoption of the Lisbon Treaty, despite considerable political opposition[1096], "energy" has been explicitly recognised in Article 4(2), (i) of the TFEU as a new policy area among the general category of shared competences. On the basis of this provision both Union and Member States are competent to regulate in the energy area. Thus, if the Union legislator has not yet exercised its shared competence in the energy field, Member States remain free to adopt legislative measures in the energy field, provided that they comply with the primary Treaty provisions.

889. Where the Union has adopted legislative measures in the energy field, Member States remain free to exercise their competence provided that they comply with the primary provisions of the Treaty and any obligation imposed by the relevant secondary energy law provisions of the Union. It should be noted that Article 2(2) of the TFEU in effect embodies automatic pre-emption of Member State action in those situations where the Union has exercised its power in the energy area. Member States can only exercise their power in the energy area to the extent that the Union has not done so.[1097]

890. Given the extensive amount of secondary legislation in the energy field on a European level, the amount of shared powers held by the Member States in the energy area appears to be limited. Nonetheless, with respect to certain aspects in the energy area for example with respect to smart grids and meters Member States at the moment still hold a large amount of shared powers to adopt a legislative framework.

1094 J.-C. PIRIS, *The Lisbon Treaty: A Legal and Political Analysis* (Cambridge, Cambridge University Press, 2010), p. 75 and 76.

1095 See *Supra*.

1096 Policy Proposal by Jacques DELORS, *Towards a European Energy Community: A Policy Proposal* S. ANDOURA, L. HANCHER and M. VAN DER WOUDE (Notre Europe, March 2010), p. 11.

1097 P. CRAIG, "The Treaty of Lisbon, process, architecture and substance", *European Law Review* (2008), p. 144.

891. Whenever lines are drawn, as with the adoption of the Lisbon Treaty, there are inevitably borderline disputes. With respect to the energy field there might also be some borderline issues as to whether the actions/provisions in the energy field on a European level fall within one of the exclusive competences or fall within the shared competence.[1098] The European Commission and the European Parliament will generally qualify such borderline issues as exclusive competences. The Council and one or more Member States on the other hand will generally argue that they fall within the category of shared competences. The future case law of the Court of Justice might explain how in the energy field such borderlines issues need to be solved.

892. The way in which the competence in the energy field will be shared between the Union and its Member States, is determined by the relevant provisions that govern the energy field in the TFEU among which Article 194 of the TFEU.[1099] The way in which power is shared between the Union and its Member States in the various areas mentioned in the general category of shared competences (Article 4(2) of the TFEU) differs significantly. For example the sharing of competence in the energy area will differ from the complex competence sharing operating in the overall area of freedom, security and justice, where there are significant variations of power sharing.

893. The TFEU unfortunately does not contain any magic formula with a precise configuration on power sharing. This approach has been implicitly recognised by Article 2(6) of the TFEU which states that: *"[t]he scope of and arrangements for exercising the Union's competences shall be determined by the provisions of the Treaties relating to each area"*.[1100]

894. Thus, on the basis of the foregoing and in particular with respect to Article 194 of the TFEU[1101] and other provisions of the TFEU which are relevant for the energy area, the scope of the Union's shared power in the energy field appears to be broad. However, this is not the case for all the aspects in the energy area such as the Member State's right to determine the choice between different energy sources, the general structure of its energy supply, the conditions for exploiting its energy resources and measures of a primarily fiscal nature.

1098 P. CRAIG, "Competence: clarity, conferral, containment and consideration", *European Law Review* (2004), p. 323–344.

1099 See *Infra*.

1100 P. CRAIG, "The Treaty of Lisbon, process, architecture and substance", *European Law Review* (2008), p. 144.

1101 See *Infra*.

2. REVISED PROTOCOL ON THE APPLICATION OF THE PRINCIPLES OF SUBSIDIARITY AND PROPORTIONALITY

895. In addition to the Amsterdam Protocol, reference should be made to the revised Protocol on the application of the principles of subsidiarity and proportionality.[1102] This revised Protocol appears to reaffirm the procedural hurdles of the Amsterdam Protocol and to enhance the role of national parliaments as watchdogs of subsidiarity.[1103] It should be read together with the new Protocol on the role of national parliaments in the European Union.[1104]

(a) *Role of the national Parliaments*

896. The role of the national Parliaments on a European level was one of the four questions raised in the Declaration on the future of the Union adopted at the end of the Intergovernmental Conference of Nice in 2000. The process of European integration is based on the fact that Member States have accepted to exercise their competences in certain fields in common. This has led to a transfer of competences to institutions on a European level. Thus, it has deprived national parliaments in these fields the direct competences of on the adoption of legislative provisions.

897. Until 1979, members of the European Parliament were members of the national parliaments. This break led to the feeling that national parliaments transferred some of their legislative competences to their governments. The most important EU-decision-making institution on a European level is the Council, whereby the decision-making process on a European level is very much influenced by national governments instead of the national parliaments.[1105]

898. As a reaction, the Member States established in their parliaments, European affairs Committees which progressively strengthened the position of the Parliaments vis-à-vis their governments. They furthermore organised cooperation between themselves through the establishment in 1989 of the Conference of Community and European Affairs Committees of Parliaments of

---

1102 Protocol (No 2) to the Treaty on the Functioning of the European Union on the application of the principles of subsidiarity and proportionality (hereinafter revised Protocol).

1103 R. SCHUTZE, "Subsidiarity after Lisbon: reinforcing the safeguards of federalism?", *Cambridge Law Journal* (2009), p. 527 which refers to I. COOPER, "The Watchdogs of Subsidiarity: National Parliaments and the Logic of Arguing in the EU", *Journal of Common Market Studies* (2006), p. 281.

1104 Protocol (No 1) to the Treaty on the Functioning of the European Union on the role of national parliaments in the European Union.

1105 J.-C. PIRIS, *The Lisbon Treaty A Legal and Political Analysis* (Cambridge, Cambridge University Press, 2010), p. 122 and 123.

the EU, the so-called COSAC.[1106] The COSAC meetings in particular requested for years that the institutions on a European level, when they draft and adopt legislative provisions, should take their opinions into account.

899. The role of the national parliaments has been formally recognised by the EU Treaties.[1107] However, the most significant response to the demands of the national parliaments was recently in the adoption of the Lisbon Treaty. This Treaty conferred various rights upon the national parliaments such as the right to receive direct information from the European institutions and not through their national governments, but also the right to have a direct role in checking whether the EU institutions comply with the subsidiarity principle. Also in the area of freedom, security and justice, the national parliaments have been given a role in various evaluation and monitoring mechanisms.[1108] One of the reasons that resulted in a significant increase of the competences of national parliaments in the decision-making process on a European level is the fact that the European Parliament did not completely succeed in establishing itself as providing the solution for the democratic legitimacy required by the Union. However, the exercise of this role by the national parliaments will not be easy.[1109]

900. To fulfil this role effectively, national parliaments should be able to intervene early in the decision-making process, because many draft legislative acts on a European level (the so-called A items which will not be discussed anymore at the Council) have already been approved, on the level of the Council's working groups or Coreper, before their adoption by the Council. The provisions of the revised Protocol specify that the national parliaments have eight weeks to intervene in the decision-making process which is a very short deadline. Other means to ensure an effective application of these provisions of the Lisbon Treaty is to reinforce cooperation of the national parliaments through COSAC or otherwise. Some Member States still consider these mechanisms as insufficient and are looking for additional measures on a national level. In some Member States this is more a prerequisite than in others due to the fact that the control of national parliaments over the positions taken in the EU Council and the European Council by the representatives of their governments, for the time

[1106] COSAC is an acronym from the full name of this Conference in French which states: Conférence des Organes Spécialisés dans les Affaires Communautaire et européennes des parlements de l'Union européenne. Under the Lisbon Treaty the official name is now the Conference of Parliamentary Committees for Union Affairs.

[1107] See the Declaration of the role of national parliaments in the EU annexed to the Final Act of the Maastricht Treaty and the Protocols annexed to the Amsterdam and Nice Treaties.

[1108] See the Articles 70, 85 and 88 of the TFEU.

[1109] J.-C. PIRIS, *The Lisbon Treaty A Legal and Political Analysis* (Cambridge, Cambridge University Press, 2010), p. 123–125.

being, is different from one Member State to another according to their respective constitutional texts, practices and political cultures.[1110]

901. Nonetheless, it should be emphasised that since the adoption of the Lisbon Treaty national parliaments have undoubtedly obtained a deeper and tighter control over their government's positions on a European level. The new provisions ensure a direct and swift flow of information on issues on a European level which allows Member States to play a role, if they so wish. However, the actual balance of powers between national parliaments and their governments is organised differently in Member States. Thus, the effectiveness of any measure in this context to a large extent depends on the national political culture. Consequently, if there is a need to provide more democratic legitimacy in the Union, the answer should mainly be found on a national level.[1111]

902. The competences attributed to the national parliaments by the Lisbon Treaty can be found in Article 12 of TEU. Some of theses competences have been further developed in Protocol no. 1 on the role of the national parliaments in the EU and the revised protocol. The latter and its provisions with respect to the role of national parliaments will be discussed in more detail in the next section of this book.

903. According to Article 2 of the revised Protocol, the Commission must consult widely before proposing legislative acts. Furthermore, under Article 4 of this revised Protocol the Commission, European Parliament and Council[1112] must forward their draft legislative acts and amended drafts to national Parliaments. These draft legislative acts should contain a detailed statement concerning the proposed legislative acts which makes it possible to appraise compliance with the principle of subsidiarity and proportionality. Such a statement must contain some assessment of the proposal's financial impact and, in case of a directive, of its implications on the rules to be put in place by the Member States, including, where necessary, regional legislation. This statement must also contain qualitative and, wherever possible, quantitative indicators to

---

1110 The Danish model seems to be the most effective system currently in place. The German parliament has, due to the German Constitutional Court judgment of 30 June 2009, obtained new competences in this context. Furthermore, some national parliaments (e.g. in France and Portugal) have used the ratification process of the Lisbon Treaty in order to obtain an enhanced role in this context see J.-C. PIRIS, *The Lisbon Treaty: A Legal and Political Analysis* (Cambridge, Cambridge University Press, 2010), p. 125 and 126.

1111 J.-C. PIRIS, *The Lisbon Treaty: A Legal and Political Analysis* (Cambridge, Cambridge University Press, 2010), p. 125 and 126.

1112 It covers draft legislative acts originating from a group of Member States, the Court of Justice, the European Central Bank or the European Investment Bank.

substantiate the conclusion that a Union objective could be better achieved on a Union level.[1113]

904. Under Article 6 of the revised Protocol each national parliament or any chamber of a national parliament may, within eight weeks of the date of transmission of a draft legislative act, send a reasoned opinion stating why it considers that the proposed legislative act does not comply with the principle of subsidiarity.[1114] It should be noted that the possibility for national parliaments to submit opinions on new legislative proposals already exists since 2006. The Commission transmitted on its own initiative all new proposals to national parliaments and adopted a procedure for replying to their opinions.[1115]

905. In this context, in 2009 the Commission has received opinions from the two chambers of the Dutch parliament on the recast of the Directive 2002/91/EC of the European Parliament and of the Council of 16 December 2002 on the Energy Performance of Buildings.[1116] The First Chamber of the Dutch parliament was positive on the proposal. However, the Second Chamber argued that the recast breached the principle of subsidiarity, because in its opinion Member States should determine how and in which sectors national climate targets are to be achieved. Furthermore, during the legislative procedure, the European Parliament proposed various amendments which included the condition that from 2019 new buildings would have to be zero energy. Several Member States argued in the Council that the level of ambition was too high and considered that the definition of low and zero carbon buildings and quantitative targets raised subsidiarity issues. A compromise was reached in November 2009 according to which all new buildings should comply with high energy performance standards and obtain a significant share of their energy requirements from renewable energy sources. To achieve these objectives, it is up to the Member States to define the standards.[1117]

1113 Article 5 of the Protocol on the application of the principles of subsidiarity and proportionality and P. CRAIG, "The Treaty of Lisbon, process, architecture and substance", *European Law Review* (2008), p. 144.

1114 R. SCHUTZE, "Subsidiarity after Lisbon: reinforcing the safeguards of federalism?", *Cambridge Law Journal* (2009), p. 525–536 which refers to I. COOPER, "The Watchdogs of Subsidiarity: National Parliaments and the Logic of Arguing in the EU", *Journal of Common market Studies* (2009), p. 281 and see also G. BARRETT, "The king is dead, long live the king: the recasting by the Treaty of Lisbon of the provisions of the Constitutional Treaty concerning national parliaments", *European Law Review* (2008), p. 75.

1115 Report of 8 October 2010 from the Commission on subsidiarity and proportionality COM(2010)504, p. 5.

1116 *OJ* L 1/65.

1117 Report of 8 October 2010 from the Commission on subsidiarity and proportionality COM(2010)504, p. 7.

906. This example demonstrates in particular with respect to the energy field that national parliaments are able to submit opinions on new legislative proposals of the European legislator. Given the fact that energy is a very sensitive field for the Member States, national parliaments might submit in the future more opinions on proposals of secondary legislation in the energy field.

907. According to Article 7(1) of the revised Protocol, each national parliament shall have two votes. Where the national parliament is bicameral, each chamber shall have one vote. If the negative votes (called reasoned opinions) on the non-compliance of a draft legislative act with the principle of subsidiarity amount to one third or one quarter in the areas of freedom, security and justice of all the votes allocated to national parliaments, the proposed legislative act must be reviewed.[1118] This situation is the so-called "yellow card" mechanism, in soccer terms, because the Union legislator may decide to maintain, amend or withdraw the proposed legislative act.[1119] This yellow card mechanism was considered insufficient by the Netherlands. During the 2007 German Presidency[1120], the Netherlands negotiated and obtained an additional specific mechanism the so-called orange mechanism which only applies to the co-decision procedure.[1121]

908. In the case of an ordinary legislative procedure, this mechanism, under paragraph 3 of Article 7 of the revised Protocol, is strengthened since at least a simple majority of the votes allocated to the national parliaments is required to trigger the orange card mechanism. If the European Commission chooses to maintain the proposal, it will have to justify in a reasoned opinion why the proposed legislative act complies with the principal of subsidiarity. However, it will be politically difficult for the Commission to continue with the procedure for the adoption of a given proposal in the situation where a significant number of national parliaments found that the proposal violates the principle of subsidiarity.[1122] The reasoned opinion of the European Commission as well of the national parliaments must be submitted to the Union legislator (the European Parliament and the Council).

1118 P. KIIVER, "The Treaty of Lisbon, the National Parliaments and the principle of subsidiarity", *Maastricht Journal* (2008), p. 80.

1119 J.-C. PIRIS, *The Lisbon Treaty: A Legal and Political Analysis* (Cambridge, Cambridge University Press, 2010), p. 128 and R. SCHUTZE, "Subsidiarity after Lisbon: reinforcing the safeguards of federalism?", *Cambridge Law Journal* (2009), p. 527.

1120 This was at the end of the negotiation of the Intergovernmental Conference mandate at the European Council of June 2007.

1121 J.-C. PIRIS, *The Lisbon Treaty: A Legal and Political Analysis* (Cambridge, Cambridge University Press, 2010), p. 129.

1122 R. SCHUTZE, "Subsidiarity after Lisbon: reinforcing the safeguards of federalism?", *Cambridge Law Journal* (2009), p. 527 and J.-C. PIRIS, *The Lisbon Treaty: A Legal and Political Analysis* (Cambridge, Cambridge University Press, 2010), p. 129.

909. The Union legislator shall consider whether the proposed legislative act complies with the principle of subsidiarity. If a majority of the votes of the European Parliament or a majority of 55% of the members of the Council finds that the proposal violates the principle of subsidiarity, the proposed legislative act is rejected.[1123] If the European Commission decides to withdraw or, more significantly, amend its proposal, the procedure under Article 7(3) of the revised Protocol comes to an end. In this situation under paragraph 3 of Article 7 of the revised Protocol neither the European Parliament nor the Council will be involved.[1124]

910. This blocking provision is the key novelty of paragraph 3 of Article 7 of the revised Protocol. It includes for the first time in the European subsidiarity compliance-monitoring system, the possibility to put a proposal on a European Union level out of the action from the legislative agenda because it does not comply with the principle of subsidiarity. While this mechanism makes it from a procedural point of view easier for the European Parliament to block a proposal on a European Union level on subsidiarity grounds, it surprisingly makes it more difficult for the Council to reject a proposal rather on subsidiarity grounds than on a lack of substantive merits.[1125]

911. With the adoption of paragraph 3 in Article 7 of the revised Protocol, the drafters of the Treaty of Lisbon extended and reinforced the subsidiarity control mechanism. However, the drafters of the Lisbon Treaty and in particular of the Protocol, did not adopt, if we use the soccer term, a "red card" mechanism into the revised Protocol whereby the political decision on subsidiarity, ultimately remains with the European legislator. National parliaments do not get the final decision on whether a proposal violates the principle of subsidiarity. The European Parliament and the Council are the ultimate arbiters to decide on this issue. Some legal doctrine supported this idea because this would otherwise give national parliaments a veto right over proposals on a European Union level

[1123] R. SCHUTZE, "Subsidiarity after Lisbon: reinforcing the safeguards of federalism?", *Cambridge Law Journal* (2009), p. 527 and for a detailed analysis of paragraph 3 of Article 7 of the revised Protocol see G. BARRETT, "The king is dead, long live the king: the recasting by the Treaty of Lisbon of the provisions of the Constitutional Treaty concerning national parliaments", *European Law Review* (2008), p. 66–84.

[1124] G. BARRETT, "The king is dead, long live the king: the recasting by the Treaty of Lisbon of the provisions of the Constitutional Treaty concerning national parliaments", *European Law Review* (2008), p. 80 and P. KIIVER, "The Treaty of Lisbon, the National Parliaments and the principle of subsidiarity", *Maastricht Journal* (2008), p. 80 and 81.

[1125] R. SCHUTZE, "Subsidiarity after Lisbon: reinforcing the safeguards of federalism?", *Cambridge Law Journal* (2009), p. 527 and G. BARRETT, "The king is dead, long live the king: the recasting by the Treaty of Lisbon of the provisions of the Constitutional Treaty concerning national parliaments", *European Law Review* (2008), p. 80 and 81.

which is incompatible with the European Commission's constitutionally protected independence.[1126]

912. It can be concluded that, under Article 7(3) of the revised Protocol, national parliaments have received the procedural tools to make the adoption of proposals on a European Union level more difficult. However, these procedural tools do not necessarily entirely prevent the adoption of such proposals. With these procedural tools national parliaments are able to compel action from the European Commission and, if the situation remains intransigent, from the European Parliament and the Council, but they will not be able to determine the outcome of such an action.[1127]

913. Consequently, it seems that the Lisbon Treaty continues to proceduralise the principle of subsidiarity and gives the subsidiarity control mechanism some extra teeth with the adoption of paragraph 3 in Article 7 of the revised Protocol. However, the blockage of legislative proposals under Article 7(3) of the revised Protocol will likely be a highly exceptional and unusual situation. In the light of the foregoing it can be concluded that the Lisbon Treaty and its Protocol did not turn the principle of subsidiarity into a hard and efficient political safeguard of federalism.[1128]

(b) Compliance with subsidiarity under the revised Protocol

914. According to Article 1 of the revised Protocol each institution must comply with this principle. To ensure an effective application of the principle of subsidiarity each draft legislative act from the Commission, a group of Member States, the European Parliament, etc. must justify the relevance of their proposal with regard to the principle of subsidiarity.[1129]

1126 R. SCHUTZE, "Subsidiarity after Lisbon: reinforcing the safeguards of federalism?", *Cambridge Law Journal* (2009), p. 527; A. DASHWOOD, "The relationship between the Member States and the European Union/Community", *Common Market Law Review* (2004), p. 369; S. WEATHERILL states in this respect that *"a veto power vested in national Parliaments would distort the proper distribution of power and responsibility in the EU's complex but remarkably successful system of transnational governance by conceding too much to State control"* see S. WEATHERILL, "Using National Parliaments to improve Scrutiny of the limits of EU Action", *European Law Review* (2003), p. 909 and 910 and P. KIIVER, "The Treaty of Lisbon, the National Parliaments and the principle of subsidiarity", *Maastricht Journal* (2008), p. 81.

1127 G. BARRETT, "The king is dead, long live the king: the recasting by the Treaty of Lisbon of the provisions of the Constitutional Treaty concerning national parliaments", *European Law Review* (2008), p. 82 and 83.

1128 *Ibid.* p. 83 and 84 and R. SCHUTZE, "Subsidiarity after Lisbon: reinforcing the safeguards of federalism?", *Cambridge Law Journal* (2009), p. 527.

1129 Article 5 of the revised Protocol.

915. Article 5 of the revised Protocol stipulates that any legislative proposal should contain a detailed statement in order to appraise compliance with the subsidiarity principle. Such a statement should contain some assessment of the financial impact of the proposal and in the case of a directive, the implications for the rules which needs to be adopted by the Member States or, where necessary, the regional legislation. Furthermore, qualitative and, if possible, quantitative indicators should substantiate the reasons why a Union objective can be better achieved on a Union level.

916. The Amsterdam Protocol stated in this respect that the relevance of legislative proposals with regard to the principle of subsidiarity should be mentioned in the explanatory memorandum accompanying the legislative proposal.[1130] If the preamble of a measure lacks an adequate explanation of why it was considered necessary, this could be a ground to challenge that measure. In the Case C-233/94 the Advocate General Léger stated that *"All measures adopted by the Community should [...] indicate, either implicitly or explicitly, but in any event clearly, on what basis the authority concerned is acting – even if only to state, where this is the case, that the principle of subsidiarity does not come into play"*.[1131] However, the Court of Justice seemed to be rather easily satisfied because it held in the Case C-233/94 that it was not required to make an express reference to the principle of subsidiarity.[1132]

917. The revised Protocol and Amsterdam Protocol both stipulate that the European Commission must submit an annual report on the application of Article 5 of the TEU.[1133] It should be noted that the European Commission procedure for applying the subsidiarity principle is also set out in its impact assessment guidelines of which the revised version was adopted in January 2009. This version contains now two explicit sets of questions (e.g. on the presence of transnational aspects) which should help to improve the quality of the analyses in impact assessments.[1134]

1130 Amsterdam Protocol provision 9 and A. ARNULL, A. DASHWOOD, M. DOUGAN, M. ROSS, E. SPAVENTA and D. WYATT, *European Union Law* (London, Sweet & Maxwell, 2006), p. 100 and 101.

1131 Opinion of Advocate General Léger in Case C-233/94, Federal Republic of Germany v European Parliament and Council of the European Union, paragraph 90.

1132 Case C-233/94, *Federal Republic of Germany v European Parliament and Council of the European Union* [1997] ECR I-2405, paragraph 28.

1133 Amsterdam Protocol provision 9, fourth indent and Article 9 of the Protocol. The latest version is the report of 8 October 2010 from the Commission on subsidiarity and proportionality COM(2010) 504. For an analysis of the previous reports of the Commission see S. DOUGLAS-SCOTT, *Constitutional Law of the European Union* (Harlow, Pearson education Limited, 2002), p. 182 and 183.

1134 Impact Assessment Guidelines, SEC(2009)92 of 15 January 2009.

918. On the basis of the foregoing, it can be concluded that the Commission in practice at three key stages of the policy development process will check the compliance with the principle of subsidiarity: (a) a preliminary analysis in the roadmaps[1135] which are published for major initiatives when the Commission Work Programme has been agreed upon; (b) an analysis of subsidiarity as part of the impact assessment and (c) a justification in terms of subsidiarity in the recitals and explanatory memorandum of each draft legislative act.[1136]

919. The application of the principle of subsidiarity, as noted above, is not only the responsibility of the Commission, but also of the Council and the European Parliament. They must specifically address, during the ordinary course of the relevant procedures, whether the proposals of the Commission and any amendments to them, comply with the provisions of Article 5 of the TEU generally (i.e. the principles of subsidiarity, attribution and proportionality).

3. ARTICLE 122 OF THE TFEU AND THE LISBON TREATY

920. The Lisbon Treaty has included in this provision also the express appeal that the measures[1137] in this context should be taken "*in a spirit of solidarity between the Member States*". Another innovation of the Lisbon Treaty is the new reference to the supply of energy in Article 122 of the TFEU which states that: "[*w*]*ithout prejudice to any other procedures provided for in the Treaties, the Council, on a proposal from the Commission, may decide, in a spirit of solidarity between Member States, upon the measures appropriate to the economic situation, in particular if severe difficulties arise in the supply of certain products, notably in the area of energy* [...]". This reference has most likely a symbolic value.[1138]

921. Despite the reference to energy in Article 122 TFEU since the adoption of the Lisbon Treaty it must be emphasised that the relationship between Article 122 and 194 of the TFEU is unclear. In particular the decision-making procedure between both Articles varies considerably. While most measures will be adopted under Article 194 of TFEU on the basis of the ordinary legislative

1135 These roadmaps provide a preliminary description of a planned initiative of the Commission and aim to inform all the interested actors in a timely manner.

1136 Report of 8 October 2010 from the Commission on subsidiarity and proportionality COM(2010)543, p. 4 and see also Report of 10 June 2011 from the Commission on subsidiarity and proportionality COM(2011)344, 12 p.

1137 According to case law of the Court of Justice the Council is free to choose its form of measures see Case 5/73, *Balkan-Import-Export GmbH v Hauptzollamt Berlin-Packhof* [1973] ECR 1091, paragraph 18.

1138 T. CORTHAUT, "Plus ça change, plus c'est la même chose? A Comparison with the Constitutional Treaty", *Maastricht Journal of European and Comparative Law* (2008), p. 26 and 27.

procedure[1139], confers Article 122 of the TFEU, on the contrary, the decision-making power of the Council acting alone on a Commission proposal whereby the European Parliament will be excluded from the decision-making process.[1140]

4. CLIMATE CHANGE INSERTED IN ARTICLE 191 OF THE TFEU

922. With the adoption of the Treaty of Lisbon, the fight against climate change became explicit in Article 191 of the TFEU as one of the cornerstones of the Union's environmental policy. To a certain extent this seems to be symbolic because the EU has been for many years at the forefront of the fight against climate change. Moreover, the more sensitive provisions (e.g. primarily of a fiscal nature or which affect a Member State's choice between different energy sources) are also not dispensed from the unanimity requirement when it concerns provisions with respect to climate change.

923. Nonetheless, this insertion is welcome as it recognises and reinforces the commitment of the Union towards this global environmental problem. Climate change[1141] is a topic in which the Union wishes to be seen to provide something much more valuable than what Member States can provide. The fact that the words *"and in particular combating climate change"* has been added means that if climate change ceases to be politically less attractive, the Union's presence on an international level will not easily be sidestepped. Furthermore, it arguably prioritises climate change within the environmental objectives of the Union.[1142]

924. In this context, it is worth mentioning that the Union did not need an explicit reference in the TFEU in order to adopt measures to combat climate change. In the author's view this could be clarified by the fact that the concept of environment in paragraph 1 of Article 191 of the TFEU is a very flexible one which can be adapted to new developments and new needs for protection. However, since this reference there is no discussion anymore whether climate change on a European level falls under the environmental or energy field. It became clear that the Union measures to combat climate change fall under the environmental policy area of the EU.

1139 See *Infra*.

1140 Policy Proposal by Jacques Delors, *Towards a European Energy Community: A Policy proposal*, study by S. ANDOURA, L. HANCHER and M. VAN DER WOUDE, p. 99.

1141 Http://ec.europa.eu/climateaction/index_en.htm.

1142 M. LEE, "The environmental implications of the Lisbon Treaty", *Environmental Law Review* (2008), p. 131–138.

925. Moreover, the Treaty of Lisbon revised the infringement procedures in the Articles 260–262 of the TFEU. Consequently, Member States which might be daunted by the high costs to comply with the European legal framework combating climate change or the renewable targets on the basis of the Renewable Directive 2009/28/EC[1143], and which may be inclined to lag in meeting their climate targets, could make their defaulting on these targets an even more costly affair.[1144]

5. FOREIGN DIRECT INVESTMENT UNDER THE LISBON TREATY

926. With the coming into effect of the Lisbon Treaty, the scope of the common commercial policy is extended to the area of foreign direct investment. This can be derived from Article 207(1) of the TFEU which states that "*The common commercial policy shall be based on uniform principles, particularly with regard to changes in tariff rates, the conclusion of tariff and trade agreements relating to trade in goods and services, and the commercial aspects of intellectual property, foreign direct investment, the achievement of uniformity in measures of liberalisation, export policy and measures to protect trade such as those to be taken in the event of dumping or subsidies. The common commercial policy shall be conducted in the context of the principles and objectives of the Union's external action*".[1145] Moreover, Article 206 of the TFEU provides that the progressive abolition of restrictions on foreign direct investment is part of the establishment of the customs union.[1146]

1143 Directive 2009/28/EC of the European Parliament and of the Council of 23 April 2009 on the promotion of the use of energy from renewable sources and amending and subsequently repealing Directives 2001/77/EC and 2003/30/EC, *OJ* L 140/16.

1144 D. CHALMERS, G. DAVIES and G. MONTI, *European Union Law* (Cambridge, Cambridge University Press, 2010), p. 343–349 and T. CORTHAUT, "Plus ça change, plus c'est la même chose? A Comparison with the Constitutional Treaty", *Maastricht Journal of European and Comparative Law* (2008), p. 26. A further change to note is the adoption of a new definition of the qualified majority. Up to now, when the Council voted on the basis of a qualified majority, the number of votes attributed to each Member State was predetermined by the EC Treaty itself which provides for a scale ranging from 29 votes each for the four largest Member States to 3 votes for the smallest. This system of qualified majority will continue until November 2014, from then on the definition of the qualified majority by which a large number of acts will be adopted by the Council, unless the TFEU expressly requires unanimity, will be different. It will be a double majority which means that acts in order to be adopted must have the support of at least 55% of the members of the Council representing the participating Member States, comprising at least 65% of the population of these States (See Article 238, paragraph 3 of the TFEU).

1145 Emphasis added.

1146 T. EILMANSBERGER, "Bilateral Investment Treaties and EU Law" *Common Market Law Review* (2009), p. 394.

Since under Article 3(1) of the TFEU, both the customs union and common commercial policy fall within the exclusive competences of the Union, there will not be much debate on the exclusive nature of this new competence. However, the precise scope and content of this new competence for foreign direct investment, is unclear.[1147] This provision originates from the Constitutional Treaty of Europe. It reflected the developments in the WTO Doha negotiation round in that period of time. Under this provision the term 'foreign' seems to refer to investments from the EU into third countries and vice versa.

927. The term 'direct investment' is not new and can be found in Article 64 of the TFEU which is part of the free movement and capital provisions of the TFEU. In that context, the Court of Justice interpreted the term "direct investment" in its case law as "*investments of any kind undertaken by natural or legal persons and which serve to establish or maintain lasting and direct links between the persons providing the capital and the undertakings to which that capital is made available in order to carry out an economic activity*".[1148] This definition corresponds with the widely accepted definitions of 'direct investments' of the International Monetary Fund (hereinafter IMF) and Organisation for Economic Co-operation and Development (hereinafter OECD).

The Court of Justice also made a distinction between "'*direct' investments, namely investments in the form of participation in an undertaking through the holding of shares which confers the possibility of effectively participating in its management and control, and 'portfolio' investments, namely investments in the form of the acquisition of shares on the capital market solely with the intention of making a financial investment without any intention to influence the management and control of the undertaking*".[1149] This distinction can be transposed to the new competence for foreign direct investment in Article 207(1) of the TFEU. Given the distinction between 'direct' and 'portfolio' investment, it can be concluded that the notion of foreign direct investment does not encompass all forms of investments.[1150] The foregoing means that investment treaties covering those

---

1147 M. BUNGENBERG, "Going Global? The EU Common Commercial Policy After Lisbon" in C. HERMANN and J.P. TERHECHTE (eds.) *European Yearbook of International Economic Law* (2010), p. 143 and T. EILMANSBERGER, "Bilateral Investment Treaties and EU Law" *Common Market Law Review* (2009), p. 394.

1148 Case C-446/04, *Test Claimants in the FII Group Litigation v Commissioners of Inland Revenue* [2006] ECR I-11753, paragraph 180 and 181 and Case C-157/05, *Winfried L. Holböck v Finanzamt Salzburg-Land* [2007] ECR I-4051, paragraph 34.

1149 Case C-171/08, *European Commission v Portuguese Republic* not yet published, paragraph 49 and Case C-282/04 and C-283/04, *Commission of the European Communities v Kingdom of the Netherlands* [2006] ECR I-9141, paragraph 19.

1150 P. EECKHOUT, *EU External Relations Law* (Oxford, Oxford University Press, 2011), p. 62–64 and Communication from the Commission to the Council, the European Parliament, the European Economic and Social Committee and the Committee of the Regions, Towards a comprehensive European international investment policy, COM(2010)343, p. 2 and 3.

other forms of investment can only be fully substituted by EU investment agreements, if there is a further legal basis in the Treaties.[1151]

928. The scope of the new competence for foreign direct investment remains uncertain. Among international and European academics there are different views some narrower, some broader, but such an analysis would go beyond the scope of this book.[1152] Nonetheless, at least a number of matters which were previously regulated in investment treaties, concluded by Member States, fall under the exclusive competences of the Union since the entry into force of the Lisbon Treaty. This raises the question on the legal status of the existing investment treaties which Member States concluded prior to the entry into force of the Lisbon Treaty.[1153]

929. Given the absence of an explicit transitional regime in the TFEU clarifying the status of these investment treaties, the Commission adopted, at the time of writing, a proposal for a regulation which will authorise the continued existence of all investment agreements which are currently in force between Member States and third countries. The Commission states in its explanatory memorandum that the extent to which the existing investment agreements of Member States are incompatible with EU law can be subject to discussions. However, the Commission added that any legal uncertainty on the status and validity of these agreements should be avoided and therefore it proposed to maintain the continued existence of bilateral agreements relating to investment concluded between Member States and third countries.

930. It is indeed preferable to provide this kind of legal certainty, although it could be argued that it would be uncertain that under EU and international law the existing investment agreements became invalid since the entry into force of the Lisbon Treaty.[1154]

1151 P. EECKHOUT, *EU External Relations Law* (Oxford, Oxford University Press, 2011), p. 64.

1152 W. SHAN and S. ZHANG, "The Treaty of Lisbon: Half Way towards a Common Investment Policy" *The European Journal of International Law* (2011), p. 1060; M. BUNGENBERG, "Going Global? The EU Common Commercial Policy After Lisbon" in C. HERMANN and J.P. TERHECHTE (eds.) *European Yearbook of International Economic Law* (2010), p. 143–145 and T. EILMANSBERGER, "Bilateral Investment Treaties and EU Law" *Common Market Law Review* (2009), p. 394–397. For a discussion of the different interpretations on the scope of the new competence foreign direct investment see W. SHAN and S. ZHANG, "The Treaty of Lisbon: Half Way towards a Common Investment Policy" *The European Journal of International Law* (2011), p. 1060–1065.

1153 *Ibid.* and M. BUNGENBERG, "Going Global? The EU Common Commercial Policy After Lisbon" in C. HERMANN and J.P. TERHECHTE (eds.) *European Yearbook of International Economic Law* (2010), p. 145.

1154 Proposal for a Regulation of the European Parliament and of the Council establishing transitional arrangements for bilateral investment agreements between Member States and

931. On the basis of the foregoing it can be concluded that depending on the scope of this new exclusive competence of the Union on foreign direct investment, which is currently unclear as discussed above, it could become a very tactful tool for the Union to strengthen its external energy policy. However, this will only be possible if the oil and gas reserves are available for foreign direct investments from the Union since the main part of these reserves are preserved for development by national oil/gas companies.[1155]

6. CHANGES TO ARTICLE 352(1) OF THE TFEU

932. Article 352(1) of the TFEU states that *"if action by the Union should prove necessary, within the framework of the policies defined in the Treaties, to attain one of the objectives set out in the Treaties, and the Treaties have not provided the necessary powers, the Council, acting unanimously on a proposal from the Commission and after obtaining the consent of the European Parliament, shall adopt the appropriate measures. Where the measures in question are adopted by the Council in accordance with a special legislative procedure, it shall also act unanimously on a proposal from the Commission and after obtaining the consent of the European Parliament"*.

933. Since the adoption of the Lisbon Treaty some legal doctrine argued that on the basis of the wording of Article 352(1) of the TFEU the scope of Article 352 of the TFEU became compared to its predecessor, Article 308 of the EC Treaty, significantly wider.[1156] Under Article 308 of the EC Treaty the Union was granted legislative powers to realise "Union objectives in the course of the operation of the common market". The new Article 352(1) of the TFEU with the absorption of the Union in the Union applies to the Union objectives which are more extensive and ambitious than the Union ones. Moreover, this new provision no longer refers to the common market. However, the Court of Justice

third countries, COM(2010)344, p. 2 and 3 and P. EECKHOUT, *EU External Relations Law* (Oxford, Oxford University Press, 2011), p. 66 and 67.

1155 C. VAN DER LINDE, "External energy policy: Old fears and new dilemmas in a larger Union" in A. SAPIR (ed.), *Fragmented Power: Europe and the global economy* (Brussels, Bruegel, 2007), p. 286.

1156 See the judgment of the German Constitutional Court of 30 June 2009 which criticised the amended version of Article 352 of the TFEU as compared to its predecessor Article 308 of the EC Treaty. The Court stated in its paragraph 327 that this new version *"must lead to a new assessment of the provision. Article 352 of the TFEU is no longer restricted to the attainment of objectives in the context of the common market but makes reference to the policies defined in the Treaties"*. It furthermore stated in paragraph 328 that *"the newly worded provision makes it possible to substantially amend Treaty foundations of the European Union without mandatory participation of legislative bodies beyond the Member States' executive powers"*, see J.-C. PIRIS, *The Lisbon Treaty A Legal and Political Analysis* (Cambridge, Cambridge University Press, 2010), p. 121.

has for many years interpreted Article 308 of the EC Treaty (now Article 352(1) of the TFEU) as if the reference to the common market was not there.

934. Furthermore, the Treaty of Lisbon introduced a number of procedural and substantive constraints in order to limit the use of Article 352 of the TFEU. Beside the veto granted to national governments present in Article 352(1) of the TFEU and its predecessor an extra veto player has been introduced by the Lisbon Treaty in the form of the European Parliament when this provision is used as a legal basis.[1157] Its use is under Article 352(2) of the TFEU also subject to the information of all national parliaments. Besides the aforementioned procedural constraints the Lisbon Treaty added new substantive constraints with the adoption of Article 352(3) and (4) of the TFEU which provide that *"measures based on this Article shall not entail harmonisation of Member States' laws or regulations in cases where the Treaties exclude such harmonisation"* and *"this Article cannot serve as a basis for attaining objectives pertaining to the common foreign and security policy and any acts adopted pursuant to this Article shall respect the limits set out in Article 40, second paragraph, of the Treaty on European Union"*.[1158, 1159] Thus, it seems that since the adoption of the Lisbon Treaty the scope of Article 352 of the TFEU became compared to its predecessor, Article 308 of the EC Treaty, narrower.

935. In this context reference should be made to Declaration 41[1160] to the Lisbon Treaty of the IGC which declares that the objectives pursued by Article 352 TFEU are those mentioned in Article 3(2), (3) and (5) of the TEU.[1161] The IGC

1157 D. CHALMERS, G. DAVIES and G. MONTI, *European Union Law* (Cambridge, Cambridge University Press 2010), p. 214–216.

1158 The intent of the reference to the second paragraph of Article 40 of the TEU in Article 352(4) of the TFEU is that Article 352 of the TFEU cannot by used to adopt legislative acts in the area of the Common foreign and security policy. See also Declaration 41 to the Lisbon Treaty.

1159 D. CHALMERS, G. DAVIES and G. MONTI, *European Union Law* (Cambridge, Cambridge University press 2010), p. 214–216.

1160 Declaration 41 on Article 308 of the TFEU. Please note that this Declaration 41 is confusing since it refers to the numbering of the Treaties before they were renumbered by the Lisbon Treaty.

1161 Article 3(2), (3) and (5) of the TEU state that: "2. *The Union shall offer its citizens an area of freedom, security and justice without internal frontiers, in which the free movement of persons is ensured in conjunction with appropriate measures with respect to external border controls, asylum, immigration and the prevention and combating of crime. 3. The Union shall establish an internal market. It shall work for the sustainable development of Europe based on balanced economic growth and price stability, a highly competitive social market economy, aiming at full employment and social progress, and a high level of protection and improvement of the quality of the environment. It shall promote scientific and technological advance. It shall combat social exclusion and discrimination, and shall promote social justice and protection, equality between women and men, solidarity between generations and protection of the rights of the child. It shall promote economic, social and territorial cohesion, and solidarity among Member States. It shall respect its rich cultural and linguistic diversity, and shall ensure that Europe's cultural heritage is safeguarded and enhanced. [...] 5. In its relations with the wider world, the Union shall uphold and promote its values and interests and contribute to the protection of its citizens. It shall*

furthermore adopted Declaration 42[1162] with respect to Article 352 of the TFEU and provides that "*in accordance with the settled case law of the Court of Justice of the European Union, Article [352 of the TFEU], being an integral part of an institutional system based on the principle of conferred powers, cannot serve as a basis for widening the scope of Union powers beyond the general framework created by the provisions of the Treaties as a whole and, in particular, by those that define the tasks and the activities of the Union. In any event, this Article cannot be used as a basis for the adoption of provisions whose effect would, in substance, be to amend the Treaties without following the procedure which they provide for that purpose*".

936. This Declaration 42 of the IGC only seems to reiterate the adopted approach of the Court of Justice in its Opinion 2/94.[1163] However, the fact that this has been expressly included in a Declaration of the IGC emphasised by the governments of 27 Member States has a political significance not to be underestimated. It will be interesting to see how the Court of Justice will interpret this declaration and to which extent it will have an impact on the application of Article 352 of the TFEU.[1164] With the adoption of Article 194 of the TFEU, discussed below in the next part, the relevance of Article 352 in the future for the adoption of measures in the energy field will be rather limited.

C. THE NEW ARTICLE 194 OF THE TFEU

937. As far as known, the question of a specific energy policy provision has been up for consideration in several Inter-governmental conferences but to no avail, until the now buried Constitutional Treaty and its successor, the Lisbon Treaty. Thus, after some 40 years of the Union legislature's activities in the energy policy area, it seems that there could finally be a "*proper*" legal basis for the European energy policy. With the adoption of the Lisbon Treaty the new Energy title XXI has been inserted into the TFEU which reads as follows:

> "*TITLE XXI*
> *ENERGY*
> *Article 194*

contribute to peace, security, the sustainable development of the Earth, solidarity and mutual respect among peoples, free and fair trade, eradication of poverty and the protection of human rights, in particular the rights of the child, as well as to the strict observance and the development of international law, including respect for the principles of the United Nations Charter."

1162 Declaration 42 on Article 308 of the TFEU.

1163 Opinion 2/94 of the Court of 28 March 1996, *Accession by the Community to the European Convention for the Protection of Human Rights and Fundamental Freedoms* [1996] ECR I-1759.

1164 D. CHALMERS, G. DAVIES and G. MONTI, *European Union Law* (Cambridge, Cambridge University Press 2010), p. 216–218.

1. *In the context of the establishment and functioning of the internal market and with regard for the need to preserve and improve the environment, Union policy on energy shall aim, in a spirit of solidarity between Member States, to:*
   (a) ensure the functioning of the energy market;
   (b) ensure security of energy supply in the Union;
   (c) promote energy efficiency and energy saving and the development of new and renewable forms of energy; and
   (d) promote the interconnection of energy networks.
2. *Without prejudice to the application of other provisions of the Treaties, the European Parliament and the Council, acting in accordance with the ordinary legislative procedure, shall establish the measures necessary to achieve the objectives in paragraph 1. Such measures shall be adopted after consultation of the Economic and Social Committee and the Committee of the Regions.*
   Such measures shall not affect a Member State's right to determine the conditions for exploiting its energy resources, its choice between different energy sources and the general structure of its energy supply, without prejudice to Article 192(2)(c).
3. *By way of derogation from paragraph 2, the Council, acting in accordance with a special legislative procedure, shall unanimously and after consulting the European Parliament, establish the measures referred to therein when they are primarily of a fiscal nature".*

1. ARTICLE 194 OF THE TFEU AND THE OBJECTIVES OF EUROPEAN ENERGY POLICY

938. The energy objectives pursued by the Union are formulated in the first paragraph of Article 194 of the TFEU which read as follows: *"In the context of the establishment and functioning of the internal market and with regard for the need to preserve and improve the environment, Union policy on energy shall aim, in a spirit of solidarity between Member States, to:*
(a) ensure the functioning of the energy market;
(b) ensure security of energy supply in the Union;
(c) promote energy efficiency and energy saving and the development of new and renewable forms of energy; and
(d) promote the interconnection of energy networks".

939. With respect to the legally binding nature of these objectives of the European energy policy enumerated in the first paragraph of Article 194 of the TFEU, it can be concluded on the basis of similar provisions in the TFEU in the areas of industrial (Article 173 of the TFEU) and environmental (Article 191 of the TFEU) policy, that these objectives represent specific directives for action. Consequently, when there is a need for legislative provisions, the European Union is not only authorised to take action, but it also has an obligation to act under these circumstances. However, it must be noted that the ascertainment of "the need for legislative provisions" leaves a great deal of room for interpretation.

940. The enumeration of the objectives of the European energy policy is preceded by three guiding principles (a) the establishment and functioning of the internal market, (b) the need to preserve and improve the environment and (c) a spirit of solidarity between Member States. These guiding principles are to be taken heed of and the European legislator must unfold its energy policy within their framework. The sense and purpose of linking these guiding principles with the objectives of the European energy policy is questionable. Do these guiding principles subordinate or restrict the adoption of the energy policy on the basis of Article 194 of the TFEU?[1165]

941. Nonetheless, in my opinion the guiding principles are applicable to all the objectives of the European energy policy referred to in the first paragraph of Article 194 of the TFEU. Before the scope of these guiding principles is examined, the next part will first of all determine the meaning of the term Union policy on energy.[1166]

(a) Meaning of Union policy on energy

942. In order to determine the substance of the energy policy targets on a European level it is necessary to define the term Union policy on energy or the energy policy of the Union referred to in the first part of paragraph 1 of Article 194 of the TFEU. Since Article 194 of the TFEU does not contain any restrictions, it can be concluded that all activities directed at influencing the generation, transmission, distribution, supply and consumption of energy can be addressed by the European Union. However, the exercise of this competence is not unlimited as we discussed above in chapter 4.

943. Moreover, it is questionable whether the new energy title also applies to measures in the area of nuclear power. Before the adoption of the Lisbon Treaty, the EC Treaty contained Article 305(2) which stated that *"[t]he provisions of this Treaty shall not derogate from those of the Treaty establishing the European Atomic Energy Community"*. According to case law of the Court of Justice this provision means that, as far as the nuclear sector is concerned, the TFEU does

---

1165 See *Infra* and Policy Proposal by Jacques Delors, *Towards a European Energy Community: A Policy proposal*, study by S. ANDOURA, L. HANCHER and M. VAN DER WOUDE, p. 99.

1166 U. EHRICKE and D. HACKLÄNDER, "European Energy Policy on the Basis of the New Provisions in the Treaty of Lisbon" in A. BAUSCH and B. SCHWENKER (eds.), *Handbook Utility Management* (Berlin Heidelberg, Springer-Verlag, 2009), p. 744–746.

not apply whenever the EURATOM Treaty contains a derogating provision.[1167] The Court of Justice moreover ruled that where the EURATOM Treaty does not lay down any rules with respect to a matter which falls under the atomic energy sector such a matter falls under the TFEU.[1168] The provisions of the TFEU operate as, *lex generalis*, supplementing where necessary the EURATOM Treaty which is the *lex specialis*.[1169]

944. Since the adoption of the Lisbon Treaty both the TEU and TFEU no longer contain a similar provision as Article 305(2) of the EC Treaty. However, the Treaty of Lisbon contains the Protocol No 2 which deals with different aspects of the EURATOM Treaty.[1170] This Protocol provides for a number of crucial amendments to the EURATOM Treaty. Furthermore, it introduces the new Article 106A in the EURATOM Treaty of which the first paragraph lists the provisions of the TEU and TFEU that are to be applied to the EURATOM Treaty.[1171] This list contains no reference to the energy policy provisions in the TFEU, but deals mainly with procedural and institutional provisions. Nonetheless, paragraph 3 of the new Article 106A of the EURATOM Treaty states that *"the provisions of the Treaty on European Union and of the Treaty on the Functioning of the European Union shall not derogate from the provisions of this Treaty"*. In other words, this paragraph means that the EURATOM Treaty prevails except for those provisions referred to in the Protocol No 2.[1172]

1167 Joined Cases 27–58, 28–58 and 29–58, *Compagnie des hauts fourneaux et fonderies de Givors and others v High Authority of the European Coal and Steel Community* [1960] ECR 241; Joined Cases 188 to 190/80, *French Republic, Italian Republic and United Kingdom of Great Britain and Northern Ireland v Commission of the European Communities* [1982] ECR 2545, paragraphs 30 and 31 and Case C-18/94, *Barbara Hopkins and others v National Power plc and Powergen plc.* [1996] ECR 2281, paragraphs 11–24.

1168 Joined Cases 188 to 190/80, *French Republic, Italian Republic and United Kingdom of Great Britain and Northern Ireland v Commission of the European Communities* [1982] ECR 2545, paragraph 32 and Opinion 1/94, *Competence of the Community to conclude international agreements concerning services and the protection of intellectual property* [1994] ECR I-5267, paragraph 24.

1169 K. LENAERTS, and P. VAN NUFFEL, in BRAY, R. (ed.), *Constitutional Law of the European Union* (London, Thomson Sweet & Maxwell, 2005), p. 706 and 707 which refers to T. F. CUSACK, "A tale of two treaties: an assessment of the Euratom Treaty in relation to the EC Treaty", *Common Market Law Review* (2003), p. 117–142.

1170 Protocol No 2 Amending the Treaty establishing the European Atomic Energy Community.

1171 Article 7, Articles 9 to 9 F, Article 48(2) to (5), and Articles 49 and 49 A of the Treaty on European Union, Article 16 A, Articles 190 to 201b, Articles 204 to 211a, Article 213, Articles 215 to 236, Articles 238, 239 and 240, Articles 241 to 245, Articles 246 to 262, Articles 268 to 277, Articles 279 to 280 and Articles 283, 290 and 292 of the TFEU, and the Protocol on Transitional Provisions, shall apply to this Treaty.

1172 HARTLEY, *The Foundations of European Community Law. An Introduction to the Constitutional and Administrative Law of the European Community*, (Oxford, Oxford University Press, 2010), p. 96 and U. EHRICKE and D. HACKLÄNDER, "European Energy Policy on the Basis of the New Provisions in the Treaty of Lisbon" in A. BAUSCH and B. SCHWENKER (eds.), *Handbook Utility Management* (Berlin Heidelberg, Springer-Verlag, 2009), p. 745.

945. Thus, given the adoption of the Protocol No 2, the EURATOM Treaty will retain its special positions as *lex specialis* as opposed to the TEU and TFEU. However, this does not necessarily automatically rule out Article 194 of the TFEU as a legal basis for nuclear energy policy measures. Consequently, on a case-by-case basis it needs to be analysed whether Article 194 of the TFEU could serve as a legal basis for nuclear energy measures (for example a directive which contains for Member States legally binding targets of electricity from nuclear power plants due to security of supply reasons).

946. With respect to the concept of energy sources, it is my understanding that three main categories of fossil fuels, renewable sources and nuclear sources can be distinguished. The nuclear-powered sources are fusion and fission, however as we discussed in the previous paragraphs it should be asserted whether the measure in question falls within the EURATOM Treaty, the TEU or the TFEU. The renewable energy sources are wind, solar, aerothermal, geothermal, hydrothermal and ocean energy, hydropower, biomass, landfill gas, sewage treatment plant gas and biogases.[1173] The fossil fuels are petroleum, coal and natural gas.

(b) Meaning of "in the context of the establishment and functioning of the internal market and with regard for the need to preserve and improve the environment"

947. The first part of paragraph 1 of Article 194 of the TFEU indicates that the Union's policy in the energy field is part of the Union's most important task, the establishment of the internal market of which the essence is defined in Article 26 of the TFEU. Establishing the internal market is an autonomous and important goal of the European Union which has been legally sanctioned and enshrined in paragraph 3 of Article 3 of the TEU.

948. With respect to the words "in the context of" it can be noted that in English the meaning of these words is an open-ended formulation which does not necessarily lead to the conclusion that the energy policy goals are limited by the concept of the internal market. On the contrary, in the German, French, Spanish and Italian versions of these words, using the words *Rahmen*, *cadre*, *marco* and *quadro,* it establishes the outermost scope of a matter and thus, it embodies a spatial limitation whereby the concept of the internal market in Article 194 of the TFEU represents a barrier to the adoption of energy policy measures. In some doctrine it is argued that if the internal market concept would represent the outermost limit of the Union's energy policy measures, this would mean that the scope of the new Article 194 of the TFEU would not extend

[1173] Article 2(a) of the Renewable Directive 2009/28/EC.

beyond the measures that the internal market competence already permits. This interpretation would mean that the new energy competence of Article 194 of the TFEU would largely be of a declaratory nature.[1174]

949. The second guiding principle emphasises that the Union's energy policy must take into account the preservation and improvement of the environment. This provision refers to one of the objectives mentioned in paragraph 1 of Article 191 of the TFEU.[1175] The TFEU does not contain any definition of the term environment which makes this concept sufficiently flexible to be adapted to new developments and new needs for protection.[1176] Moreover, the environment appears to include the protection of animal health and fauna and flora.

In this context, the explicit reference to *"the need to preserve and improve the environment"* should be understood as an express condition for the Union's legislator when it adopts measures in the energy field. Thus, the Union's legislator should make sure that its legislative initiatives are not to the detriment of the environment. This explicit reference to the environment in Article 194 of the TFEU and the categorical placement of the energy title directly after the environmental title reveals the growing intertwinement of energy and environmental concerns.[1177]

(c) In a spirit of solidarity between Member States

950. The next part of paragraph 1 of Article 194 of the TFEU states that the objectives of the Union's energy policy should be achieved *"in the spirit of solidarity between Member States"*. These words appear to emphasise that solidarity between Member States in the energy field has become a basic feature of the EU membership. This solidarity should be taken into account by the Union's legislator when it adopts measures in the energy field. As a side note, this solidarity provision has been inserted by the Lisbon negotiators. It was not mentioned in paragraph 1 of Article III-256 of the Treaty establishing a Constitution for Europe.

951. Currently, each Member State remains responsible for its own energy mix and security. However, choices in the energy field made by one Member State (e.g. the choice of a Member State to rely largely or wholly on natural gas for

---

1174 U. EHRICKE and D. HACKLÄNDER, "European Energy Policy on the Basis of the New Provisions in the Treaty of Lisbon" in A. BAUSCH and B. SCHWENKER (eds.), *Handbook Utility Management* (Berlin Heidelberg, Springer-Verlag, 2009), p. 750 and 751.

1175 See *Supra*.

1176 See *Supra*.

1177 U. EHRICKE and D. HACKLÄNDER, "European Energy Policy on the Basis of the New Provisions in the Treaty of Lisbon" in A. BAUSCH and B. SCHWENKER (eds.), *Handbook Utility Management* (Berlin Heidelberg, Springer-Verlag, 2009), p. 752.

power generation) inevitably have an impact on the energy security of its neighbours and the Union as a whole.[1178] Thus, solidarity between Member States in the energy field has become one of the key areas of Union's energy policy (e.g. unilateral decision of Germany to shut down all its power plants by 2022) where action is necessary to address the challenges of the new energy era.

952. However, the meaning of this concept of solidarity remains uncertain since any definition or any guidance on how to apply it when the European legislator develops a new energy measure or policy is lacking.[1179] Besides, in Article 194 of the TFEU reference has been made to this new concept of solidarity in Article 2 of the TEU which states that *"[t]he Union is founded on the values of respect for human dignity, freedom, democracy, equality, the rule of law and respect for human rights, including the rights of persons belonging to minorities. These values are common to the Member States in a society in which pluralism, non-discrimination, tolerance, justice, "solidarity" and equality between women and men prevail"*[1180] and a number of other Articles in the TFEU such as Articles 67 and 80 of the TFEU (dealing with border control, immigration and asylum), Article 122 (dealing with economic policies) and Title VII, Article 222 of the TFEU (dealing with terrorist attacks, natural and manmade disasters).

953. Given the frequency of this term or concept of solidarity in a number of areas in the Treaties on a European level, it seems that this concept is in need of expansion.[1181] Since recently, the term of solidarity can also be found in secondary legislation and in particular the Gas Regulation No 994/2010 which makes reference to solidarity mechanisms and states in its recital 36 that Member States should devise measures to exercise solidarity.

954. The Court of Justice, in its case law, stated that the duty of solidarity is a principle which Member States have accepted by acceding to the Union.[1182] Moreover, it ruled that *"In permitting Member States to profit from the advantages*

1178 Green paper of 8 march 2006, *A European Strategy for Sustainable, Competitive and Secure Energy* of the Commission, COM(2006)105, p. 9.

1179 Policy Proposal by Jacques Delors, *Towards a European Energy Community: A Policy proposal*, study by S. ANDOURA, L. HANCHER and M. VAN DER WOUDE, p. 98.

1180 Emphasis added.

1181 U. EHRICKE and D. HACKLÄNDER, "European Energy Policy on the Basis of the New Provisions in the Treaty of Lisbon" in A. BAUSCH and B. SCHWENKER (eds.), *Handbook Utility Management* (Berlin Heidelberg, Springer-Verlag, 2009), p. 752.

1182 Cases C-413/04, *European Parliament v Council of the European Union* [2006] ECR I-11221, paragraph 68 and Case C-414/04, *European Parliament v Council of the European Union* [2006] ECR I-11279, paragraph 45 and Opinion of Advocate General Poiares Maduro of 21 June 2007 in the Case C-273/04, *Republic of Poland v Council of the European Union*, paragraph 51.

of the Community, the Treaty imposes on them also the obligation to respect its rules. For a State unilaterally to break, according to its own conception of national interest, the equilibrium between advantages and obligations flowing from its adherence to the Community brings into question the equality of Member States before Community Law and creates discrimination at the expense of their nationals, and above all of the nationals of the State itself which places itself outside the Community Rules. This failure in the duty of solidarity accepted by Member States by the fact of their adherence to the Community strikes at the fundamental basis of the Community legal order".[1183] This means that the duty of solidarity prohibits a Member State from upsetting the balance between the advantages and burdens of its membership of the Union based on its unilateral perception of its national interest.[1184] Thus, the Court of Justice paraphrases the duty of solidarity as the balance between advantages and burdens of a Member State's membership of the Union.[1185]

955. Solidarity between Member States in the energy area will most likely be relevant with respect to provisions of the security of supply and energy infrastructure. Until today, Member States have no legal obligation to provide mutual support to one another either in terms of short-term support in crisis situations or to cooperate in long-term infrastructure investments. On the contrary, Article 42 of the Third Electricity Directive and Article 46 of the Third Gas Directive allow Member States in the event of a sudden crisis in the energy market to relinquish their obligations under these directives.

956. Moreover, it remains uncertain how the principle of solidarity relates to the internal market provisions of the Treaties. The cornerstone of the internal market is competition and according to the classic liberal notions of competition it is the self-interest of an individual which drives its economic activities. The principle of solidarity, on the other hand, is characterised by shared identity and common interest. As a consequence, it could be argued that these principles are opposite.[1186] Hence, the adoption of binding energy solidarity measures could impede competition and therefore the European legislator should be careful when doing so. On the other hand, it must be noted that the European competition policy has been one of the most effective instruments of the

---

1183 Case 39/72, *Commission of the European Communities v Italian Republic* [1973] ECR 101, paragraphs 24 and 25 and Case 128/78, *Commission of the European Communities v United Kingdom of Great Britain and Northern Ireland* [1979], ECR 419, paragraphs 9 and 12.

1184 Opinion of Advocate General Poiares Maduro of 21 June 2007 in the Case C-273/04, *Republic of Poland v Council of the European Union*, paragraph 51.

1185 U. EHRICKE and D. HACKLÄNDER, "European Energy Policy on the Basis of the New Provisions in the Treaty of Lisbon" in A. BAUSCH and B. SCHWENKER (eds.), *Handbook Utility Management* (Berlin Heidelberg, Springer-Verlag, 2009), p. 752.

1186 *Ibid.*, p. 753.

integration. In this sense, competition policy has also promoted solidarity among Member States.[1187]

(d) Ensure the functioning of the energy market

957. The first objective formulated in paragraph 1 of Article 194 of the TFEU is fairly generally formulated. No further definition is given to the term energy markets in the TFEU which makes it impossible to determine with absolute certainty from the TFEU what might be understood by the concept "energy markets" and whether it refers to the "internal energy market".

958. The "internal energy market" concerns the internal cross-border issues within the European Union. While the term "energy markets" has a broader scope by covering both internal and external markets for the commodity "energy". Given the primary energy import dependency of most of the Member States, the functioning of the energy markets cannot be guaranteed without taking into account the external dimension of the energy markets.[1188] Moreover, the term market refers to a place where supply meets demand. However, it is questionable whether this term should be interpreted in a narrow sense or whether it refers to a very broad sense covering all the different aspects of supply and demand in all energy markets, and would fall under the regulatory authority of the European Union.

959. Given the foregoing, it is questionable whether the use of the words "energy markets" was on purpose or not. Regardless of the uncertainties with respect to the scope of energy markets. This term in paragraph 1 Article 194 of the TFEU should, in the author's view, be interpreted broadly and encompasses fuel (such as petroleum, natural gas, coal and uranium[1189]), renewable sources of energy, electricity and heat markets.

(e) Ensure security of energy supply in the Union

960. Under the second objective of paragraph 1 of Article 194 of the TFEU the Union's energy policy shall aim to *"ensure security of energy supply in the Union"*. Due to the global developments and the increasing dependence on imported

---

[1187] Policy Proposal by Jacques Delors, *Towards a European Energy Community: A Policy proposal*, study by S. ANDOURA, L. HANCHER and M. VAN DER WOUDE, p. 98 and U. EHRICKE and D. HACKLÄNDER, "European Energy Policy on the Basis of the New Provisions in the Treaty of Lisbon" in A. BAUSCH and B. SCHWENKER (eds.), *Handbook Utility Management* (Berlin Heidelberg, Springer-Verlag, 2009), p. 752 and 753.

[1188] See *Supra*. External measures on a European level with respect to energy security are for example the European Energy Charter Treaty, Energy Community Treaty, etc.

[1189] See *Supra* the interrelationship between the TEU, TFEU and EURATOM Treaty.

hydrocarbons[1190], the Union needs to take action to secure its energy future and protect its essential energy interests. It is therefore no surprise that the second objective of the Union's energy policy is to aim for the security of energy supply for the Union.

961. The TFEU does not provide a definition of security of energy supply which makes it also impossible to determine with absolute certainty from the TFEU what might be understood by this concept. Secondary legislation does not provide for an adequate definition of the concept of *"security of supply"* either although several directives refer to this concept. Some of these directives only define the concept of *"security"* as *"both security of supply of natural gas [electricity] and technical safety"*.[1191]

962. The Directive 2004/67/EC does not provide a definition, nor does the new Gas Regulation No 994/2010 which repealed this directive. Conversely, the Directive 2005/89/EC provides for a definition in Article 2, (b) which states that *"'security of electricity supply' means the ability of an electricity system to supply final customers with electricity, [...]"*.[1192] One of the issues with respect to this definition is whether supply is sufficient at any price or whether it must be affordable.[1193] The adoption of the Third

1190 With "business as usual" the EU's energy import dependence will jump from 50% of total EU energy consumption today to 65% in 2030. Reliance on imports of gas is expected to increase from 57% to 84% by 2030, of oil from 82% to 93%; see Communication from the Commission to the European Council and the European Parliament of 10 January 2007, *An energy policy for Europe*, COM(2007)1, p. 3.

1191 Article 2, (28) of the Directive 2003/54/EC of the European Parliament and the Council of 26 June 2003 concerning common rules for the internal market in electricity and repealing Directive 96/92/EC [2003] *OJ* L 176/37; Article 2, (32) of the Directive 2003/55/EC of the European Parliament and the Council of 26 June 2003 concerning common rules for the internal market in natural gas and repealing Directive 98/30/EC, *OJ* L 176/57; Article 2, (32) of the Directive 2009/73/EC of the European Parliament and of the Council of 13 July 2009 concerning common rules for the internal market in natural gas and repealing Directive 2003/55/EC, *OJ* L 211/94, etc.

1192 K. TALUS, *Vertical Natural Gas Transportation capacity, upstream commodity contracts and EU Competition law*, Dissertation submitted to the faculty of law, University of Helsinki, 2010, p. 54. See also K. TALUS, *Vertical Natural Gas Transportation Capacity, Upstream Commodity Contracts and EU Competition Law* (The Netherlands, Kluwer Law International, 2010), 321 p.

1193 H. BJØRNEBYE, "The Security of Electricity Supply and Infrastructure Directive: the Right Aims, but the Wrong Means" in B. DELVAUX, M. HUNT and K. TALUS (eds.) *EU Energy Law and Policy Issues* (Rixensart, Euronfidentiel, 2008), p. 173–188. This author compared the definition on *"security of electricity supply"* of the Directive of safeguarding security of electricity supply and infrastructure with the definition of *"energy security"* of the United Nations Development Programme. He emphasised that beside affordable energy prices also the concept of environmental protection is missing in the definition of the Directive of safeguarding security of electricity supply and infrastructure.

Package did not remedy this lack of a definition of the concept of security of supply.[1194]

963. Security of supply could be defined in several ways and any definition must take various factors into consideration. Its crucial elements could vary from country to country and depend on the time horizon. Thus, it is not a uniform concept. It could neither qualify as an absolute concept which can be fully guaranteed.[1195] Nonetheless, there are various attempts to define this concept. On a European level, the Commission stated that: "*the European Union's long-term strategy for energy supply security must be geared to ensuring, for the well-being of its citizens and the proper functioning of the economy, the uninterrupted physical availability of energy products on the market, at a price which is affordable for all consumers (private and industrial), while respecting environmental concerns and looking towards sustainable development, as enshrined in Articles 2 and 6 of the Treaty on European Union.*

Security of supply does not seek to maximise energy self-sufficiency or to minimise dependence, but aims to reduce the risks linked to such dependence. Among the objectives to be pursued are those balancing between and diversifying of the various sources of supply (by product and by geographical region)".[1196]

964. Not only the Commission, but also some academics made some suggestions to define the concept of security of supply which include:

"*A condition in which a nation and all, or most, of its citizens and businesses have access to sufficient resources at reasonable prices for the foreseeable future free from serious risk of major disruption of service. At different times, in different countries the crucial elements of energy security will vary*".[1197]

"*The ability of the energy industries, primarily in electricity and gas, to provide their services throughout the EU to a high standard and at a reasonable cost in a competitive, fully liberalised, pan European-market*".[1198, 1199]

1194 K. TALUS, *Vertical Natural Gas Transportation capacity, upstream commodity contracts and EU Competition law*, Dissertation submitted to the faculty of law, University of Helsinki, 2010, p. 54 and 55. See also K. TALUS, *Vertical Natural Gas Transportation Capacity, Upstream Commodity Contracts and EU Competition Law* (The Netherlands, Kluwer Law International, 2010), 321 p.

1195 *Ibid.*, p. 55. See also B. SAGA, "Introduction of Competition in Gas Markets: Effects on Contract Structures and Security of Supply" in L. HANCHER (ed.) *The European Energy Market: Reconciling Competition and Security of Supply* (Köln, Bundesanzeiger, 1995), p. 93.

1196 Green Paper of 29 November 2000, *Towards a European strategy for the security of energy supply*, COM(2000)769, 90 p.

1197 B. BARTON, C. REDGWELL, A. RØNNE and D. N. ZILLEMAN (eds.) "Introduction" in B. BARTON, C. REDGWELL, A. RØNNE and D. N. ZILLEMAN (eds.), *Energy Security – Managing Risk in a Dynamic Legal and Regulatory Environment* (Oxford, Oxford University Press, 2004), p. 5.

1198 P. D. CAMERON, *Competition in Energy Markets – Law and Regulation in the European Union* (second edition) (Oxford, Oxford University Press, 2007), p. 518.

1199 K. TALUS, *Vertical Natural Gas Transportation capacity, upstream commodity contracts and EU Competition law*, Dissertation submitted to the faculty of law, University of Helsinki,

965. These definitions all employ the risk perspective and are worded in a general way which makes them workable in various contexts.[1200] Instead of the risk perspective or threats approach, another approach towards the concept of security of supply is the reaction to the threat. Under this approach security of supply is defined as a non-linear reaction towards the threat.[1201] Other elements in these definitions are reasonable or affordable prices, uninterrupted physical availability of energy services, respecting environmental concerns. Short- and long-term trends and other factors such as the fight against terrorism (sabotage of gas pipelines, nuclear power plants, hydroelectric dams and other critical components of the energy infrastructure), global warming, natural disasters, wars and protection of the environment, also influence the concept of security of supply.

966. With respect to environmental protection, the concept of security of energy supply is in one sense in conflict with environmentally-based policies because the latter, for example, limit nuclear expansion and the burning of coal in power plants in order to reduce greenhouse gases.[1202] Due to the multi-faced nature of security of supply which involves legal, political and technical aspects and various trends, it remains very difficult or even impossible, to provide a comprehensive and uniform definition of security of supply.

967. Two distinctive approaches can be identified in the literature on the concept of security of supply. On the one hand, a security of supply definition

---

2010, p. 54 and 55. See also K. TALUS, *Vertical Natural Gas Transportation Capacity, Upstream Commodity Contracts and EU Competition Law* (The Netherlands, Kluwer Law International, 2010), 321 p.

1200 See also S. S. HAGHIGHI, *Energy Security – The External Legal Relations of the European Union with Major Oil- and Gas-Supplying Countries* (Oxford, Hart Publishing, 2007), p. 13–15 and A. BELYI, "New dimensions of energy security of the enlarging EU and their impact on relations with Russia", *European Integration* (2004), p. 351–369. The latter author distinguishes different dimensions of security of supply, namely a political dimension related to the reduction of energy import dependence, an economic dimension or the creation of a competitive market and the encouraging of energy price volatility and a normative dimension through the promotion of international institutions. S. S. HAGHIGHI, "Energy Security and the Division of Competences between the European Community and its Member States", *European Law Journal* (2008), p. 461–482. This author defines the concept of security of energy supply as "*the adequacy of energy supply at a reasonable price*". J. STERN, "Security of European Natural Gas Supplies: the Impact of Import Dependence and Liberalization", *RIIA Working Paper* (Royal Institute of International Affairs, London, 2002), p. 6.

1201 K. TALUS, *Vertical Natural Gas Transportation capacity, upstream commodity contracts and EU Competition law*, Dissertation submitted to the faculty of law, University of Helsinki, 2010, p. 55. See also K. TALUS, *Vertical Natural Gas Transportation Capacity, Upstream Commodity Contracts and EU Competition Law* (The Netherlands, Kluwer Law International, 2010), 321 p. A. BELYI, "New dimensions of energy security of the enlarging EU and their impact on relations with Russia", *European Integration* (2004), p. 354.

1202 B. BARTON, C. REDGWELL, A. RØNNE and D. N. ZILLEMAN (eds.) "Introduction" in B. BARTON, C. REDGWELL, A. RØNNE and D. N. ZILLEMAN (eds.), *Energy Security – Managing Risk in a Dynamic Legal and Regulatory Environment* (Oxford, Oxford University Press, 2004), p. 7–10.

which is very general and in which the threats to security of supply are seen as a *force majeure* type of event and on the other hand, a narrow approach which limits the relevant threats to strictly energy-related factors. The narrow approach compared to the force majeure approach appears to be, as discussed, more in line with the rulings of the Court of Justice.[1203]

968. Consequently, the second objective of paragraph 1 of Article 194 of the TFEU which states that the Union's energy policy shall aim to *"ensure security of energy supply in the Union"* should be interpreted in the light of a narrow approach. Nonetheless, it must be noted that in the author's view the concept of security of energy supply has become a very elastic concept and is often used to pass a number of policy choices.

(f) Promote energy efficiency and energy saving and the development of new and renewable forms of energy

969. This third objective of paragraph 1 of Article 194 of the TFEU which encompasses the promotion of *"energy efficiency and energy saving"* and the development of *"new and renewable forms of energy"* are main goals of the Union's energy policy. The TFEU does not contain any definition of the terms mentioned in the third objective. The term "Energy efficiency" refers to a higher yield in the unavoidable use of energy and the term "energy saving" refers to the elimination of energy waste. Both terms involve contributions of a quantitative nature.

970. The purpose of promoting *"energy efficiency and energy saving"* is to control and reduce energy consumption or the demand for energy. On the basis of this provision the Union's legislator will be competent to adopt measures which promote energy efficiency and energy saving in any sector (e.g. transport, building, products, appliances, energy generation and distribution, etc.). Although the terms *"energy efficiency" and "energy saving"* are often used

---

[1203] See *Supra* and also K. TALUS, *Vertical Natural Gas Transportation capacity, upstream commodity contracts and EU Competition law*, Dissertation submitted to the faculty of law, University of Helsinki, 2010, p. 59–69. See also K. TALUS, *Vertical Natural Gas Transportation Capacity, Upstream Commodity Contracts and EU Competition Law* (The Netherlands, Kluwer Law International, 2010), 321 p. H. BJØRNEBYE, *Investing in EU energy security – Exploring the regulatory approach to tomorrow's electricity production*, Dissertation submitted to the faculty of law, University of Oslo, 2009, p. 63–104. See also H. BJØRNEBYE, *Investing in EU Energy Security. Exploring the regulatory approach to tomorrow's electricity production* (The Netherlands, Kluwer Law International, 2010), 456 p. L. HANCHER and S. JANSSEN, "Shared Competences and Multi-Faced Concepts – European Legal Framework for Security of Supply" in B. BARTON, C. REDGWELL, A. RØNNE and D. N. ZILLEMAN (eds.), *Energy Security – Managing Risk in a Dynamic Legal and Regulatory Environment* (Oxford, Oxford University Press, 2004), p. 85–119.

interchangeably, it should be highlighted that there is a significant difference between them. *"Energy efficiency"* means that while an equivalent level of economic activity or service is maintained there will be less energy inputs used. *"Energy saving"* means an absolute decrease of energy consumption. Such a decrease can be done through behavioural changes, increased energy efficiency or decreased economic activities. Using the car less, heating a room less in the winter or enabling energy saving modes on a computer are examples of energy savings without energy efficiency improvements.[1204]

971. With respect to the Union's objective to aim for the *"development of new and renewable forms of energy"*, the term of renewable forms of energy encompasses, on the basis of the definitions in secondary legislation and in particular Directive 2009/28/EC, wind, solar, aerothermal, geothermal, hydrothermal and ocean energy, hydropower, biomass, landfill gas, sewage treatment plant, gas and biogases. However, this list cannot be seen as conclusive. It can be assumed that one of the decisive elements to be categorised as renewable energy sources is that new sources of renewable energy should overcome the disadvantages of fossil fuels of which the foremost are the production of greenhouse gases and threat to secure supply of energy due to the import dependency. This provision furthermore stipulates that the Union could adopt measures in order to develop new forms of energy which are not necessarily renewable.

(g) Promote the interconnection of energy networks

972. The fourth objective of the Union's energy policy shall aim to *"promote the interconnection of energy networks"*. On several occasions it has been emphasised that an increase of the interconnection between Member States will facilitate the inter-regional and cross-border transport of power and energy. To permit real competition and the creation of the single European energy market, additional interconnection capacity will be necessary. Interconnection is also a crucial mechanism for solidarity between Member States which has become a basic feature in the energy field since the adoption of the Lisbon Treaty.[1205] As a side note, this interconnection provision has been inserted by the Lisbon negotiators, but it was not mentioned in paragraph 1 of Article III-256 of the Treaty establishing a Constitution for Europe. The term networks in this context means in my view the network infrastructure in and between Member States

1204 Commission staff working document, Impact Assessment, Accompanying document to the Communication of the Commission to the European Parliament, the Council, the European Economic and Social Committee and the Committee of the Regions of 8 March 2011, *Energy Efficiency Plan 2011*, SEC(2011)277, p. 7 and 8.

1205 Communication from the Commission to the Council and the European Parliament of 10 January 2007, *Priority interconnection plan*, COM(2006)846, p. 3.

such as the network for cross-border exchanges, transmission network, interconnectors, pipelines, etc.

973. With respect to the promotion of the interconnection of networks the TFEU contains in addition to the provisions of paragraph 1 of Article 194 of the TFEU also the Articles 170–172 of the TFEU on the promotion of trans-European Networks (hereinafter TENs) in the areas of transport, telecommunications and energy infrastructure.[1206] In the 1990s, the trans-European Networks for Energy (hereinafter TEN-E) policy has been developed and shaped through the TEN-E Guidelines[1207] and the corresponding TEN Financing Regulation.[1208] The previous chapter highlighted that given the new policy environment in the energy field with ambitious goals, the TEN Financing Regulation and the TEN-E Guidelines are no longer effective.[1209]

974. Moreover, it must be admitted that the development of the TEN-E currently has been plodding along slowly. One of the main causes appears to be the fact that there are no standards in place with respect to the form and duration of the often prolonged ratification procedures in the Member States.[1210] Whether the new legal basis of Article 194 of the TFEU with respect to the promotion of the interconnection of networks will be able to overcome the current difficulties and shortcoming with the TEN-E framework remains to be seen.

(h) Conclusion

975. Before the adoption of the Lisbon Treaty, the European energy policy has been developed without any primary law objectives. The general considerations

---

1206 See *Supra*.

1207 Council Decision of 28 March 1996 laying down a series of measures aimed at creating a more favourable context for the development of trans-European networks in the energy sector, *OJ* L 161/154; Decision No 1229/2003/EC of the European Parliament and of the Council of 26 June 2003 laying down a series of guidelines for trans-European energy networks and repealing Decision No 1254/96/EC, *OJ* L 176/11 and Decision No 1364/2006/EC of the European Parliament and of the Council of 6 September 2006 laying down guidelines for trans-European energy networks and repealing Decision 96/391/EC and Decision No 1229/2003/EC, *OJ* L 262/1 (hereinafter the 2006 TEN-E guidelines).

1208 Regulation (EC) No 680/2007 of the European Parliament and of the Council of 20 June 2007 laying down general rules for the granting of Community financial aid in the field of the trans-European transport and energy networks, *OJ* L 162/1 (hereinafter the TEN Financing Regulation).

1209 See *Supra*.

1210 U. EHRICKE and D. HACKLÄNDER, "European Energy Policy on the Basis of the New Provisions in the Treaty of Lisbon" in A. BAUSCH and B. SCHWENKER (eds.), *Handbook Utility Management* (Berlin Heidelberg, Springer-Verlag, 2009), p. 749 and 750 which refers to B. MEIER-WEIGT, "Die Vorhaben von europäischem Interesse nach den Transeuropäischen Energienetze – TEN-E – Leitlinien und ihre Umsetzung in den europäischen Mitgliedsstaaten", *IR* (1/2007), S. 7 f.

to which it has oriented itself were a secure, economic (or at manageable prices) and sustainable (or as environmentally responsible as possible) supply of energy. To a certain extent these general considerations have been inserted, but not completely, into Article 194 of the TFEU. Nonetheless, it is also my understanding that these general considerations retain their validity as overriding goals and thus, the guiding principles and objectives mentioned in paragraph 1 of Article 194 of the TFEU are intermediate objectives of expediency in order to achieve the overriding goals.[1211]

976. The three guiding principles in the new energy title are (a) the establishment and functioning of the internal market, (b) the need to preserve and improve the environment and (c) a spirit of solidarity between Member States. These guiding principles should be taken into account and within their framework the European legislator must unfold its energy policy. Their sense, contents and purpose remain uncertain. Moreover, it is questionable whether these guiding principles subordinate or could even restrict the adoption of the energy policy on the basis of Article 194 of the TFEU.

977. With respect to the concept of energy sources, three main categories can be distinguished: fossil fuels, renewable sources and nuclear sources.

978. Moreover, as emphasised above, the meaning of some objectives of the new energy title is not always crystal clear. It is, for example, unclear what might be understood by the concept of energy markets and whether it refers to the internal energy market or not. Whether the new legal basis of Article 194 of the TFEU with respect to the promotion of the interconnection of networks will be able to overcome the current difficulties and shortcomings with the TEN-E framework remains questionable.

979. The main significance of paragraph 1 of Article 194 TFEU is that it determines the range of Union competences in the energy field. It obtained this significance via paragraph 2 of Article 194 TFEU since this provision enables the Union to adopt the measures necessary to achieve the objectives of paragraph 1. Finally, it can be noted that the negotiators of the Lisbon Treaty in some of the energy policy objectives, discussed above, have opted for the word "*ensure*" instead of "*promote*" which seems to imply that for those objectives, there is a stronger commitment for the Union's legislator.

[1211] U. EHRICKE and D. HACKLÄNDER, "European Energy Policy on the Basis of the New Provisions in the Treaty of Lisbon" in A. BAUSCH and B. SCHWENKER (eds.), *Handbook Utility Management* (Berlin Heidelberg, Springer-Verlag, 2009), p. 754.

2. THE LEGAL BASIS OF EUROPEAN ENERGY POLICY

(a) *Paragraph 2 of Article 194 TFEU*

980. The standard procedure to achieve the objectives of the European energy policy mentioned in paragraph 1 of Article 194 TFEU has been inserted into paragraph 2 of Article 194 TFEU which read as follows: "*Without prejudice to the application of other provisions of the Treaties, the European Parliament and the Council, acting in accordance with the ordinary legislative procedure, shall establish the measures necessary to achieve the objectives in paragraph 1. Such measures shall be adopted after consultation of the Economic and Social Committee and the Committee of the Regions.*

Such measures shall not affect a Member State's right to determine the conditions for exploiting its energy resources, its choice between different energy sources and the general structure of its energy supply, without prejudice to Article 192(2)(c)".

981. The first part of this paragraph stipulates that without prejudice to the application of the Treaties, the European Parliament and the Council, acting in accordance with the ordinary legislative procedure of Article 294 of the TFEU, shall establish the measures necessary to achieve the objectives of the European energy policy mentioned in paragraph 1. The use of the words "*without prejudice*" means that the provision laid down in paragraph 2 exists alongside the application of other provisions of the Treaties.

982. The second part of this paragraph stipulates that the measures adopted by the Union legislator "*shall not affect a Member State's right to determine the conditions for exploiting its energy resources, its choice between different energy sources and the general structure of its energy supply, without prejudice to Article 192(2)(c)*". These provisions are the restrictions on the energy competence of the European Union. They enumerate three rights of the Member States in the energy field which will not be affected by measures adopted by the Union. The scope of these three restrictions may become very relevant in the case of an overlap between, on the one hand, measures adopted on the basis of the energy policy objectives and on the other hand the rights of the Member States in the energy field. For example, a solidarity mechanism (e.g. releasing a certain amount of gas/oil reserves) adopted on the basis of the Union competence ensuring a secure energy supply might be on bad terms with a Member State's right to determine the conditions for exploiting its energy resources.[1212]

[1212] *Ibid.*, p. 755.

i) Member State's right to determine the conditions for exploiting its energy resources

983. This provision of paragraph 2 of Article 194 of the TFEU should be compared with the Declaration No 9 annexed to the Single European Act which states that *"[t]he Conference confirms that the Community's activities in the sphere of the environment may not interfere with national policies regarding the exploitation of energy sources"*.[1213] The meaning of this provision is in detail discussed above when Article 192 of the TFEU was analysed.[1214]

ii) Choice between different energy sources and the general structure of its energy supply, without prejudice to Article 192(2)(c)

984. The *"choice between different energy sources"* means the power of a Member State to decide which kind of energy sources can be used to cover its demand for energy.

985. The *"general structure of its energy supply"* refers to the power of a Member State to organise the manner of its energy supply. Thus, a Member State can decide which energy mix or sources it will use, but also which energy infrastructure it would apply. Examples in this context are a Member State's choice to opt for a nuclear phase-out, to set a target for renewable energy sources, to replace all the coal-fired power plants by gas-fired, no longer allow the construction of coal-fired power plants, etc.

986. Consequently, the Union legislator is allowed to adopt measures to achieve the objective mentioned in paragraph 1 of Article 194 of the TFEU, however such measures *shall not affect a Member State's right to determine the conditions for exploiting its energy resources, its choice between different energy sources and the general structure of its energy supply, without prejudice to Article 192(2)(c).* The reference to the third indent, of the first subparagraph, of the second paragraph, of Article 192 of the TFEU indicates that the Union, contrary to paragraph 2 of Article 194 of the TFEU, remains competent to adopt measures which *"significantly affect a Member State's choice between different energy sources and the general structure of its energy supply"*. However, such measures must be adopted by the Council acting unanimously in accordance with the special legislative procedure and cannot be enacted in ordinary legislative proceedings.

1213 Note that this part will also be discussed under the environmental law provisions of this book.

1214 See *Supra*.

987. The interpretation of the reference in Article 194 of the TFEU to Article 192(2)(c) of the TFEU is significant because various conclusions can be drawn from this reference. One conclusion could be that the reference is a matter of procedural restrictions which ensures that the environmental and energy law procedures are compatible. The right of veto for Member States provided for in accordance with Article 192(2)(c) of the TFEU could not be gotten around if the Union suggests measures which *"significantly affect a Member State's choice between different energy sources and the general structure of its energy supply"*.

988. This interpretation could be undermined by the adoption of the energy competence and the fact that paragraph 3 of Article 194 of the TFEU requires unanimity for measures in the energy field which are predominantly of a tax-related nature. The third paragraph of Article 194 of the TFEU deals with an exception to the adoption of measures on the basis of the ordinary legislative procedure.

989. It is striking that the negotiators of the Lisbon Treaty did not provide for a similar approach in paragraph 3 of Article 194 of the TFEU as with the environmental provisions of the TFEU by allowing the Council to act unanimously in accordance with the special legislative procedure to adopt measures affecting a Member State's right to determine the conditions for exploiting its energy resources, its choice between different energy sources and the general structure of its energy supply. However, the fact that the Lisbon Treaty negotiators did not adopt a similar approach might have been on purpose.

990. Moreover, aside from the methodological irregularity of the appearance of the unanimity requirement in the paragraphs 2 and 3 of Article 194 of the TFEU, the ambiguous wording in paragraph 3 *"by way of derogation from paragraph 2"* seems to argue that the reference to Article 192(2)(c) of the TFEU in the second paragraph is not a procedural restriction.

991. Under Article 192(2)(c) of the TFEU only measures which *"significantly affect a Member State's choice between different energy sources and the general structure of its energy supply"* need to be adopted by the Council acting unanimously in accordance with a special legislative procedure. On the contrary, the second paragraph of Article 194 of the TFEU contains no reference to the extent of the legal intervention. Such an open-ended formulation appears to imply that all measures, regardless of their gravity, fall under this provision.

992. The foregoing could lead to the conclusion that, contrary to the procedural restriction discussed above, the situations or concerns referred to in the second paragraph of Article 194 of the TFEU should be classified as situations or concerns for which on the part of the Union, a strict prohibition to intervene applies. This means that these situations or concerns are the exclusive competence of the Member States. However, this interpretation would mean that the Union could adopt extensive interventions in the energy sphere on the basis of the environmental competences of the TFEU which could not be covered by the genuine energy competence in the TFEU.[1215]

993. In the author's view, such a conclusion with respect to the interpretation of the reference in Article 194 of the TFEU to Article 192(2)(c) of the TFEU is incorrect. The correct interpretation of the reference is a matter of procedural restrictions. Consequently, the Member States' veto right provided for in accordance with Article 192(2)(c) of the TFEU cannot be avoided if the Union suggests measures which *"significantly affect a Member State's choice between different energy sources and the general structure of its energy supply"*. It must be highlighted that the foregoing clearly demonstrates that the new Article 194 of the TFEU is not well-written.

994. Regardless of the discussion mentioned above, it should be emphasised that the first subparagraph of the second paragraph of Article 192 of the TFEU does not allow the Union to adopt measures which affect a Member State's right to determine the conditions for exploiting its energy resources even if the Council acts unanimously. Thus, the adoption of the Lisbon Treaty established in Article 194 of the TFEU, contrary to the past, that the power to determine *"the conditions for exploiting energy resources"* remains an exclusive competence of the Member States.

(b) Third paragraph of Article 194 of the TFEU

995. According to the third paragraph of Article 194 of the TFEU by way of derogation from the ordinary legislative procedure (mentioned in the second paragraph Article 194 of the TFEU), the Council acting in accordance with a special legislative procedure shall unanimously and after consulting the European Parliament, establish the measures referred to therein when they are *"primarily of a fiscal nature"*. The environmental provisions contain, as discussed above, a similar provision. The first uncertainty with respect to this provision is the meaning of the term *primarily*.[1216]

996. With regards to the interpretation of the term *fiscal nature*, there is nothing in the Treaty to indicate how that term should be construed. Not only

---

1215 U. EHRICKE and D. HACKLÄNDER, "European Energy Policy on the Basis of the New Provisions in the Treaty of Lisbon" in A. BAUSCH and B. SCHWENKER (eds.), *Handbook Utility Management* (Berlin Heidelberg, Springer-Verlag, 2009), p. 756.

1216 See *Supra*.

Article 194 of the TFEU refers to this term, but also Article 30 of the TFEU and Article 192 of the TFEU.[1217] A similar term can be found in paragraph 2 of Article 114 of the TFEU which refers to *fiscal provisions*. With respect to the meaning of this latter term the Court of Justice ruled in the *Commission v Council* case[1218] that: "*by reason of their general character, those words cover not only all areas of taxation, without drawing any distinction between the types of duties or taxes concerned, but also all aspects of taxation, whether material rules or procedural rules*".[1219] Thus, measures in the energy field i.e. energy taxation measures (for example an injection tax for solar panels or other renewable energy source producing green electricity or an energy efficiency tax) which are primarily of a fiscal nature should be adopted according to paragraph 3 of Article 194 of the TFEU by the Council acting unanimously in accordance with the special legislative procedure.[1220]

997. In this context, it is not clear how paragraph 3 of Article 194 of the TFEU accords with the Council's power of harmonisation with respect to turnover taxes, excise duties and other forms of indirect taxation under Article 113 of the TFEU.[1221] Paragraph 3 of Article 194 of the TFEU does not seem to apply without prejudice to the provisions of Article 113 of the TFEU. Therefore, Article 113 of the TFEU had to be specifically excluded in a similar way as it has been done with respect to Article 114 of the TFEU.[1222]

3. DECLARATION 35 ON ARTICLE 194 OF THE TFEU

998. According to the Declaration 35 on Article 194 of the TFEU this Article "*does not affect the right of the Member States to take the necessary measures to ensure their energy supply under the conditions provided for in Article 347 TFEU*". According to Article 347 of the TFEU "*Member States shall consult each other with a view to taking together the steps needed to prevent the functioning of the internal market being affected by measures which a Member State may be called upon to take in the event of serious internal disturbances affecting the maintenance of law and order, in the event of war, serious international tension constituting a threat of war, or in order to carry out obligations it has accepted for the purpose of maintaining peace and international security*". This provision of the TFEU, together with the Articles 346 and 348 of the TFEU, allows Member States

1217 *Ibid.*

1218 Case C-338/01, *Commission of the European Communities v Council of the European Union* [2004] ECR I-4829.

1219 *Ibid.*, paragraph 63.

1220 See also *Infra*.

1221 See also *Infra*.

1222 See *Infra*.

themselves to deviate on grounds of internal or external security from their general Treaty obligations.

999. Thus, this declaration seems to indicate that Member States are allowed to adopt measures which deviate from the guiding principles and objectives of Article 194 of TFEU in order to ensure their energy supply. However, such energy supply measures should be adopted on the conditions provided for in Article 347 of the TFEU which requires Member States to consult each other in such a situation. Nonetheless, also in such a situation the Court of Justice in my opinion can review whether the adopted measure is manifestly disproportionate to the alleged risk to the energy supply of the Member State in question.[1223]

4. GAS REGULATION NO 994/2010 AND ARTICLE 194 OF THE TFEU

(a) *Background*

1000. In early 2006, 2008, 2009 and 2010, the EU witnessed gas supply interruptions. From 6–20 January 2009 the gas supplies were interrupted from Russia to the EU via Ukraine which affected, directly and indirectly, a majority of the Member States. Several times since the break-up of the Soviet Union, tensions between Russia and Ukraine and Belarus have raised due to the continuing difficulties between these states on the details of a new gas transit and supply regime.

1001. Vulnerability of the EU is not only to commercial disputes, but also to the risk of technical failures or accidents on major gas pipelines. The January gas crisis illustrated that the level of preparedness of the different Member States varied significantly or are not yet even developed. Although, Directive 2004/67/EC in its Article 8 provides that Member States should have adopted national emergency measures.

1002. If the January 2009 gas crisis continued for a longer period or spread to other parts of the EU its impact could have been much more devastating in economic, social and environmental terms. The scale of this gas crisis required an adequate response, however, a clear strategy and adequate instruments were lacking. The gas crisis highlighted different weaknesses in each stage of the

1223 Case C-414/97, *Commission of the European Communities v Kingdom of Spain* [1999] ECR I-5585, paragraph 22 and K. LENAERTS, and P. VAN NUFFEL, in BRAY, R. (ed.), *Constitutional Law of the European Union* (London, Thomson Sweet & Maxwell, 2005), p. 365–367.

energy security approach on an EU level and emphasised again the need for a common approach to an external energy policy.[1224]

1003. In 2009, the European Commission evaluated Directive 2004/67/EC and stated in its report that this directive is not sufficient to provide a timely response to a gas crisis which goes beyond the level that measures on a national level can mitigate. Moreover, an assessment of the real-time gas supply situation and potential responses within the EU are hindered due to a lack of transparency. Thus, the European Commission concluded in its report that Directive 2004/67/EC had to be revised.[1225] In response to the Russian-Ukrainian gas crisis in January 2009, the European Council and European Parliament also called for an accelerated revision of this directive. Following the report of the European legislator and the call of both the European Council and the Parliament, the Commission proposed a Gas Security Regulation which led to the adoption of the new Gas Security Regulation No 994/2010 of 20 October 2010.

(b) Main features of the Gas Regulation No 994/2010

1004. According to Article 1 of the Gas Security Regulation No 994/2010, this regulation establishes different provisions in order to safeguard the security of gas supply by allowing the implementation of exceptional measures when the market can no longer deliver the required gas supplies. This regulation provides, in a definition and attribution of the responsibilities among natural gas undertakings, Member States and the European Union with both preventive action and the reaction to concrete disruptions of the gas supply. Moreover, it adopts transparent mechanisms, in a spirit of solidarity, for the coordination of planning for, and response to, an emergency on a Member State, regional and European Union level.

1005. With these provisions the Gas Security Regulation No 994/2010 aims to strengthen the gas supply security by ensuring that all the Member States and their gas market players adopt well in advance effective measures in order to prevent and mitigate the consequences of potential gas supply disruptions. These measures should ensure that when the next crises in the gas supply hits, no

[1224] Commission staff working document of 16 July 2009 accompanying document to the proposal for a regulation of the European Parliament and of the Council concerning measures to safeguard security of supply and repealing Directive 2004/67/EC, *The January 2009 Gas supply disruption to the EU: an assessment*, COM(2009)977, p. 2–18.

[1225] Communication from the Commission to the European Parliament, the Council, the European Economic and Social Committee and the Committee of Regions of 13 November 2008 on the Directive 2004/76/EC of 26 April 2004 concerning measures to safeguard security of natural gas supply, COM(2008)769, p. 11.

Member State within the European Union is at risk of economic losses or worse.[1226]

1006. The Gas Security Regulation No 994/2010 is built upon two main approaches. On the one hand, a top-down approach which defines minimum gas supply security standards and on the other hand, a bottom-up approach which requires a full assessment of the risks affecting the security of gas supply.

1007. With respect to the top-down approach provides the Gas Security Regulation No 994/2010 with a common indicator, the so-called N-1 indicator, which defines a serious gas supply disruption. Under Article 6 of the Gas Security Regulation No 994/2010 (the so-called infrastructure standard) in the event of a disruption of the single largest infrastructure in a Member State, the Member States or the Competent Authority must ensure that the capacity of the remaining infrastructure, determined on the basis of the N-1 indicator, is able to satisfy the total gas demand during a day of exceptionally high gas demand occurring with a statistical probability of once in 20 years.[1227]

1008. Article 8 of the Gas Security Regulation No 994/2010 (the so-called supply standard) specifies that the Competent Authority requires from the natural gas undertakings to adopt measures which ensure gas supply to protected customers of the Member States in a number of situations. These situations are extreme temperatures during a 7-day peak period, any period of at least 30-days of exceptionally high gas demand and for a period of at least 30 days in case of the disruption of the single largest gas infrastructure.

1009. The bottom-up approach can be found in Article 9 of the Gas Security Regulation No 994/2010 which establishes a regular[1228] updated risk assessment by the Competent Authority on a national level and by the concerned Competent Authorities on a regional level.

1226 P. NOËL, "Ensuring success for the EU Regulation on gas supply security", *University of Cambridge (EPRG) & European Council on Foreign Relations (ECFR)*, 9 February 2010, p. 2 which refers to P. NOËL and S. FINDLATER, "A comment on the draft Regulation on security of gas supply", *EPRG, University of Cambridge*, 3 July 2009 and N. AHNER and S. RUESTER, "Challenges for European natural gas supply security regulation", *Network Industries Quarterly* (2010), p. 6 and 7.

1227 The proposal of the Gas Security Regulation initially stated that *"the remaining infrastructure (N-1) has the capacity to deliver the necessary volume of gas to satisfy total gas demand of the calculated area during a period of sixty days of exceptionally high gas demand during the coldest period."*

1228 For the first time and at the latest 18 months after the adoption of the Preventive Action and Emergency Plans and thereafter every 2 years before 30 September of the relevant year unless circumstances warrant more frequent updates.

(c) *Critical assessment of the Gas Security Regulation No 994/2010*

1010. In some doctrine it is argued, from an economic point of view, that an EU-wide unitary standard for the security of supply bears the risk for some Member States of being on a level of security which could be too high and consequently too expensive. Member States which already have a very secure supply of gas have no reason to adopt more stringent standards. Thus, the added value of the adopted N-1 indicator in Article 6 of the Gas Security Regulation No 994/2010 becomes questionable.

1011. It is uncertain whether this N-1 indicator, which is based upon the largest infrastructure, reflects the true risk situation in each Member State. The focus on the largest infrastructure of each Member State appears not to be the right approach since there are for some Member States other situations (e.g. a disruption of the gas supply from Russia via Ukraine and/or Belarus) which could be more sensible risks than a disruption of the single largest infrastructure.[1229]

1012. Therefore, it can be argued that preference should be given to a bottom-up approach, which is based on national and regional gas security assessments, instead of a top down approach. Standards for the security of gas supply should not be used as a tool in order to have a minimum level of security in all the Member States.[1230]

1013. Beside the objections from an economic point of view, there is also an objection from a legal point of view. Before analysing the legal objection, it should be emphasised that the legal instrument chosen by the European legislator, namely a regulation instead of a directive, is a very welcomed and adequate choice. The adoption of a regulation as a legal instrument on a European level ensures a rapid and uniform application of the intended gas security provisions.[1231]

1014. The Gas Security Regulation No 994/2010 is based upon the new energy title of the TFEU, Article 194(2). This paragraph stipulates that without prejudice to the application of the Treaties, the European Parliament and the Council, acting in accordance with the ordinary legislative procedure of Article 294 of the

---

1229 P. NOËL, "Ensuring success for the EU Regulation on gas supply security", *University of Cambridge (EPRG) & European Council on Foreign Relations (ECFR)*, 9 February 2010, p. 4 and 5 and N. AHNER and S. RUESTER, "Challenges for European natural gas supply security regulation", *Network Industries Quarterly* (2010), p. 6 and 7.

1230 *Ibid.*

1231 N. AHNER and S. RUESTER, "Challenges for European natural gas supply security regulation", *Network Industries Quarterly* (2010), p. 8 and see *Infra*.

TFEU, shall establish the measures necessary to achieve the objectives of the European energy policy mentioned in paragraph 1. According to its preamble, the Gas Security Regulation No 994/2010 has been adopted in accordance with the ordinary legislative procedure.

1015. The N-1 indicator, established in Article 6 of the Gas Security Regulation No 994/2010, is a security of gas supply standard which obliges Member States on the basis of its formula, as provided in point 2 of Annex I of the regulation, to adopt the necessary measures in order to ensure that the remaining infrastructure has the capacity to satisfy the total gas demand in accordance with the conditions stipulated in paragraph 1 of Article 6. This means that Member States are obliged on the basis of this N-1 indicator to maintain a level of gas supply security. The adoption of such a mandatory provision consequently makes national measures securing the gas supply redundant.[1232]

1016. However, one of the prerogatives of Member States on the basis of the TFEU is to decide on their fuel mix, the exploitation of their energy resources, etc.[1233] This can in particular be found in the second part of paragraph 2 of Article 194 of the TFEU which states that the measures adopted by the Union's legislator *"shall not affect a Member State's right to determine the conditions for exploiting its energy resources, its choice between different energy sources and the general structure of its energy supply, without prejudice to Article 192(2)(c)"*. The correct conclusion, as discussed above, with respect to this paragraph 2 is that measures on a European level which touch upon situations or concerns referred to in this paragraph should be classified as situations or concerns which need to be adopted with unanimity.

1017. One of the objectives of the Union's energy policy listed in the first paragraph of Article 194 of the TFEU is to *"ensure security of energy supply in the Union"*. Since the purpose of the Gas Security Regulation No 994/2010 is to establish a number of measures in order to safeguard the security of gas supply when the market can no longer deliver the required gas supplies, it can be concluded that this regulation falls within the scope of one of the Union's energy policy objectives. Moreover, this regulation can even be classified as measures which fall within the narrow approach of the concept of security of supply.[1234]

1018. However, this does not mean that the Union can exercise its competence with respect to the security of energy supply on the basis of Article 194(1) of the

1232 P. NOËL, "Ensuring success for the EU Regulation on gas supply security", *University of Cambridge (EPRG) & European Council on Foreign Relations (ECFR)*, 9 February 2010, p. 3.

1233 N. AHNER and S. RUESTER, "Challenges for European natural gas supply security regulation", *Network Industries Quarterly* (2010), p. 8.

1234 See *Supra*.

TFEU, unlimited. The EU when it exercises its competences in the energy field must, in particular, comply with the restrictions referred to in the second part of paragraph 2 of Article 194 of the TFEU.

1019. With respect to the adopted N-1 indicator, which obliges Member States to adopt measures ensuring that the remaining infrastructure has the capacity to satisfy the total gas demand, it could be argued that it falls under the situations or concerns of the second part of paragraph 2 of Article 194 of the TFEU. The N-1 indicator obliges Member States to maintain a minimum level of gas supply security. Given this mandatory security of gas supply standard, the right of a Member State to deal with a gas crisis itself and to decide which measures should be taken has been curtailed.

1020. A Member State's freedom to choose between measures has been limited due to the adoption of the Gas Security Regulation No 994/2010 (e.g. one of the Member States could decide, contrary to the mandatory security of gas supply standard, to deal with a gas crisis by adopting measures which don't apply to the gas sector).

1021. Thus, in the author's view this security of gas supply standard of the Gas Security Regulation No 994/2010 touches upon a Member State's choice between energy sources and, in particular, the general structure of its energy supply. Therefore, the Gas Security Regulation No 994/2010 should have been adopted under the special legislative procedure, requiring unanimity instead of the ordinary legislative procedure.

1022. However, it should be noted that although on the basis of this approach the legal basis of the Gas Security Regulation No 994/2010 is incorrect and should have been adopted with unanimity, the Council adopted this regulation with unanimity. The fact that the Council adopted this regulation in practice with unanimity makes it more difficult, or perhaps even impossible, to convince the Court of Justice of the European Union that the Gas Security Regulation No 994/2010 has been adopted on the incorrect legal basis.

5. CONCLUDING REMARKS WITH RESPECT TO THE NEW ENERGY TITLE

1023. Since the Lisbon Treaty entered into force, a new energy title has been included into the TFEU. Providing a legal basis for the adoption of measures which aim to ensure most of the goals of the existing European energy policy. However, with respect to the existing division of competences between the European Union and its Member States, nothing has been changed fundamentally. The Lisbon negotiators have adopted a carefully crafted compromise between the shared energy competence, between the Union and its Member States, and the national sovereignty over a number of energy situations or concerns referred to in the second paragraph of Article 194 of the TFEU. Hence, the new energy title in the TFEU can be seen as a mere codification of the existing practise in the energy field. Thus, the same pre-existing gaps, flaws and limitations remain in place.

1024. The author tends to conclude that, compared to the former situation, with respect to competences in the energy field, no major changes were adopted by the negotiators of the Lisbon Treaty, nor does this new energy title offer prospect of radical reform from the former situation.[1235] Moreover, this new energy title could even weaken the existing foundations of the European energy policy. It is for example not unthinkable that an undertaking would challenge a decision taken by the European Commission on the basis of Article 194 of TFEU for not taking into account one or more of the guiding principles.[1236]

D. ARTICLE 194 OF THE TFEU AND/OR OTHER PROVISIONS OF THE TFEU

1025. Although, the previous parts tried to provide a clear understanding of the meaning of Article 194 of the TFEU, there is still a question as to whether the adoption of this new energy title means that all the measures in the energy field need to be based upon this provision of the Treaty or may still be adopted on the basis of other Treaty provisions. The next part of this chapter will try to provide an answer to this question. In order to answer this question this part will start with an analysis of case law with respect to the choice between Article 192 of the TFEU and Article 114 of the TFEU. On the basis of this analysis some conclusions can be drawn with respect to Article 194 of the TFEU.

1. CHOICE BETWEEN ARTICLE 192 OF THE TFEU AND ARTICLE 114 OF THE TFEU

1026. Since the introduction of the co-decision procedure in the first paragraph of Article 192 of the TFEU, as discussed above, the choice between

1235 Policy Proposal by Jacques Delors, *Towards a European Energy Community: A Policy proposal*, study by S. ANDOURA, L. HANCHER and M. VAN DER WOUDE, p. 99.

1236 L. HANCHER and A. DE HAUTECLOCQUE, "Manufacturing the EU Energy Markets: The Current Dynamics of Regulatory Practise", *Competition and Regulation in Network Industries* (2010), p. 310.

Articles 114 and 192 of the TFEU from a procedural point of view will no longer make any difference. Nevertheless, a significant difference remains if the specific Union action in relation to the environment is based on Article 192 of the TFEU. Member States maintain the ability of adopting under Article 193 of the TFEU[1237] more stringent national measures than those provided for by the Union.

1027. It is clear that the purpose of Articles 191 to 193 of the TFEU is to serve as the legal basis for the adoption of a coherent framework for organising the environmental policy of the Union. However, it is equally clear that the provisions of Article 114 of the TFEU are intended to provide a legal basis for the adoption of measures for the approximation of the provisions laid down by national laws, regulations or administrative actions on the protection of the environment, which have as their object the establishment and functioning of the internal market. Consequently, it should be noted that there is a fine dividing line between the harmonisation measures on the basis of Article 114 of the TFEU and the specific Union action in the environmental field on the basis of Article 192 of the TFEU.[1238]

1028. The choice between Articles 114 and 192 of the TFEU, as the legal basis for Union action, has been the subject of some cases in the Court of Justice. On the basis of this case law we can conclude that the choice of a legal basis for a measure must be based on objective factors which include in particular the aim and content of the measure, as confirmed by the Court of Justice in its *Commission v Council of the European Communities* case[1239] in which it stated that: *"[i]t must be observed that in the context of the organization of the powers of the Community the choice of the legal basis for a measure may not be simply on an institution's conviction as to the objective pursued but must be based on objective factors which are amenable to judicial review (see the judgement in Case 45/86 Commission v Council [1987] ECR 1493, paragraph 11]. Those factors include in particular the aim and content of the measure"*.[1240] Consequently, the European legislator does not seem to be free to choose its legal basis.

1237 See *Infra*.

1238 A. ARNULL, A. DASHWOOD, M. DOUGAN, M. ROSS, E. SPAVENTA and D. WYATT, *Wyatt and Dashwoods's European Union Law* (London, Sweet & Maxwell, 2006), p. 940–942 and see the discussion on the legal basis of Directive 2001/77/EC of 27 September 2001 on the promotion of electricity produced from renewable energy sources in the internal electricity market, *OJ* L 283/33 and P. BERTOLDI, R. BOWIE, P. HODSON, J. LORENTZEN, H. VAGGEN MALVIK, A. TOTH, L. WERRING, B. YORDI AGUIRRE, in L. WERRING (ed.) *EU Energy Law, Volume III, Energy Efficiency and Renewable Energy Sources* (Leuven, Claeys & Casteels (2006)), p. 16–18.

1239 Case C-300/89, *Commission of the European Communities v Council of the European Communities* [1991] ECR I-2867.

1240 *Ibid.*, paragraph 10. See also Case 45/86, *Commission of the European Communities v Council of the European Communities* [1987] ECR 1493, paragraph 11; Case C-84/94, *United Kingdom*

1029. The Court of Justice in the *Netherlands v European Parliament and Council* case[1241] has made clear that Article 192 of the TFEU is not necessarily the correct legal basis if a measure pursues an environmental objective.[1242] It follows from case law of the Court of Justice that the main rule in this respect is a single legal basis which could be either Article 192 of the TFEU or Article 114 of the TFEU: *"[i]f examination of a Community measure reveals that it pursues a twofold purpose or that it has a twofold component and if one of those is identifiable as the main or predominant purpose or component, whereas the other is merely incidental, the act must be based on a single legal basis, namely that required by the main or predominant purpose or component"*.[1243]

1030. Only in exceptional circumstances, according to the Court Of Justice, *"if on the other hand it is established that the act simultaneously pursues a number of objectives or has several components that are indissociably linked, without one being secondary and indirect in relation to the other, such an act will have to be founded on the various corresponding legal bases"*.[1244] The Court of Justice also emphasised that a dual legal basis is only possible if the procedures for each legal basis are not incompatible with each other and the rights of the European Parliament will not be undermined due to the use of two legal bases.[1245] If different decision-making

of Great Britain and Northern Ireland v *Council of the European Union* [1996] ECR I-5755, paragraph 25; Case C-268/94, *Portuguese Republic v Council of the European Union* [1996] ECR I-6177, paragraph 22; Case C-336/00, *Republik Österreich v Martin Huber* [2002] ECR I-7699, paragraph 30; Case C-176/03, *Commission of the European Communities v Council of the European Union* [2005] ECR I-7879, paragraph 45 and Case C-411/06, *Commission of the European Communities v European Parliament and Council of the European Union*, [2007] ECR I-6199, paragraph 77.

1241 Case C-377/89, *Kingdom of the Netherlands v European Parliament and Council of the European Union* [2001] ECR I-7079.

1242 J. H. JANS and H. H. B. VEDDER, *European Environmental Law* (Groningen, Europa Law Publishing, 2008), p. 69.

1243 Case C-178/03, *Commission of the European Communities v European Parliament and Council of the European Union* [2006] ECR I-107, paragraph 42. See also Case C-36/98, *Kingdom of Spain v Council of the European Union* [2001] ECR I-779, paragraph 59; Case C-211/01, *Commission of the European Communities v Council of the European Union* [2003] ECR I-8913, paragraph 39 and Case C-338/01, *Commission of the European Communities v Council of the European Union* [2004] ECR I-4829, paragraph 55.

1244 Case C-178/03, *Commission of the European Communities v. European Parliament and Council of the European Union* [2006] ECR I-107, paragraph 43. See also Case C-336/00, *Republik Österreich v Martin Huber* [2002] ECR I-7699, paragraph 31; Case C-281/01, *Commission of the European Communities v Council of the European Union* [2002] ECR I-12049, paragraph 35; Case C-211/01, *Commission of the European Communities v Council of the European Union* [2003] ECR I-8913, paragraph 40 and Case C-94/03, *Commission of the European Communities v Council of the European Union* [2006] ECR I-1, paragraph 36.

1245 Case-300/89, *Commission of the European Communities v Council of the European Communities* (or the *Titanium dioxide* case) [1991] ECR I-2867, paragraph 10; Case C-338/01, *Commission of the European Communities v Council of the European Union* [2004] ECR I-4829, paragraph 57; Case C-178/03, *Commission of the European Communities v European Parliament and Council of the European Union* [2006] ECR I-107, paragraph 59 and Case

procedures are combined, all modalities must also be combined. For example if Article 192 of the TFEU and Article 114 of the TFEU are considered, this would mean that the Committee of the Regions must also be consulted.

1031. When a Member State in this context would like to adopt more stringent measures Article 193 of the TFEU is applicable if the measure is based upon Article 192 of the TFEU. If on the other hand, the measure is based upon Article 114 of the TFEU and a Member State would like to adopt stricter environmental standards, it needs to comply with the procedure and standards of paragraphs 4–6 of Article 114 of the TFEU.[1246]

2. CHOICE BETWEEN ARTICLE 194 OF THE TFEU AND OTHER PROVISIONS OF THE TFEU

(a) General remarks

1032. The analysis above highlights that the choice of a legal basis for a measure must be based on objective factors which include in particular the aim, object and content of the measure, as confirmed by the Court of Justice in its *Commission v Council of the European Communities* case.[1247] If a measure has a manifold objective or various aspects are involved, its legal basis in the TFEU should be determined on the basis of the Treaty Article which corresponds most with its main aim or main aspect. Thus, the European legislator's choice of the legal basis for a measure must be determined on the basis of the aim and content of the proposed measure which are amenable to judicial review.[1248]

1033. Moreover, the main object of a measure in question should be the decisive factor.[1249] A general Treaty Article is according to the Court of Justice a sufficient legal basis, although the measure in question seeks to attain in a subordinate manner an aim sought by a specific Treaty Article.[1250] Consequently, before the European legislator has its recourse to Article 194 of the TFEU, the

C-94/03, *Commission of the European Communities v Council of the European Union* [2006] ECR I-1, paragraph 52.

1246 J. H. JANS and H. H. B. VEDDER, *European Environmental Law* (Groningen, Europa Law Publishing, 2008), p. 69–70.

1247 Case C-300/89, *Commission of the European Communities v Council of the European Communities* [1991] ECR I-2867.

1248 Case 45/86, *Commission of the European Communities v Council of the European Communities* [1987] ECR 1493, paragraph 5 and Case C-300/89, *Commission of the European Communities v Council of the European Communities* [1991] ECR I-2867, paragraph 10.

1249 Case C-377/89, *Kingdom of the Netherlands v European Parliament and Council of the European Union* [2001] ECR I-7079, paragraph 27.

1250 *Ibid.*, paragraphs 27 and 28.

aim and content of the proposed measure should fall within the scope of the energy field.

1034. The main rule is a single legal basis for the adoption of measures on a European level. Only in exceptional circumstances, if the given measure simultaneously pursues different objectives or has several components that are indissociably linked, without one being secondary and indirect in relation to the other, such measures will have to be founded on the various corresponding legal bases. However, such a dual legal basis (for example Articles 194 and 114 of the TFEU) is only acceptable if the procedures for each legal basis are not incompatible with each other and the rights of the European Parliament will not be undermined due to the use of two legal bases.

1035. Whether the decision-making procedures for example prescribed in paragraph 3 of Article 194 of the TFEU and Article 114 of the TFEU or paragraph 2 of Article 192 of the TFEU and paragraph 4 of Article 207 of the TFEU[1251] could be combined is not clear. The latter combination was at issue in the *Commission v Parliament and Council* case[1252] in which the Court of Justice, after deciding that the regulation at issue should have been founded on a dual legal basis, reiterated its case law of the *Commission* v *Council* (*Titanium Dioxide*) case[1253] in which it held that "*recourse to a dual legal basis is not possible where the procedures laid down for each legal basis are incompatible with each other or*[1254] *where the use of two legal bases is liable to undermine the rights of the Parliament*". The Court of Justice stated that the consequences of the *Titanium Dioxide* case were different in the *Commission v Parliament and Council* case[1255] at issue.

1036. It first held that *recourse to Article 133 EC as an additional basis could not in this case have had any impact on the voting rules applicable within the Council because, in the same way as Article 175(1) EC, Article 133(4) EC provides*

1251 This provision of the TFEU provides for the Union's competence in respect of the common commercial policy. For analysis of this provision with respect to environmental product standards see J. H. JANS and H. H. B. VEDDER, *European Environmental Law* (Groningen, Europa Law Publishing, 2008), p. 70–76.

1252 Case C-178/03, *Commission of the European Communities v. European Parliament and Council of the European Union* [2006] ECR I-107, paragraphs 40–60 and see also Case C-94/03, *Commission of the European Communities v Council of the European Union* [2006] ECR I-1, paragraphs 34–56.

1253 Case-300/89, *Commission of the European Communities v Council of the European Communities* (or the *Titanium dioxide* case) [1991] ECR I-2867, paragraphs 17 to 21.

1254 It should be emphasised that the use of the word "or" is confusing since the Court of Justice had only the second instance (i.e. where the rights of the Parliament were undermined) refused to accept a dual legal basis.

1255 Case C-178/03, *Commission of the European Communities v European Parliament and Council of the European Union* [2006] ECR I-107.

that the Council, in exercising the powers conferred upon it by that provision, is to act by a qualified majority. Secondly, recourse to Article 133 EC jointly with Article 175(1) EC is likewise not liable to undermine the Parliament's rights because, although the first-mentioned article does not formally provide for the participation of that institution in the adoption of a measure of the kind at issue in this case, the second article, on the other hand, expressly refers to the procedure provided for in Article 251 EC. In contrast to the situation at issue in the abovementioned Titanium dioxide case, the use of a combination of legal bases does not therefore in this case involve any encroachment upon the Parliament's rights since recourse to Article 175(1) EC enables that institution to adopt the measure under the co-decision procedure.[1256]

1037. Consequently, because paragraph 1 of Article 192 of the TFEU and paragraph 4 of Article 207 of the TFEU both provide for a qualified majority vote and since the Court of Justice found that the rights of the Parliament were not undermined in view of the fact that paragraph 4 of Article 207 of the TFEU does not provide any role of the Parliament, the Court of Justice concluded that, in contrast to the situation at issue in the *Titanium dioxide* case, the dual legal basis is possible.[1257]

1038. In a more recent judgment, namely the *European Parliament v Council of the European Union case*[1258] which was handed down by the Court a few days after the *Commission v Parliament and Council* case[1259], the Court of Justice was asked to pronounce whether the dual legal basis combining Article 192 of the TFEU, which provided for the co-decision procedure, and Article 352 of the TFEU, which mandates the consultation procedure, could be allowed. The Court of Justice concluded that the contested regulation pursues objectives set out in Articles 2 EC and 3(1)(k) EC and in Title XVII of the EC Treaty, "*without that title by itself conferring on the Community the power to realize those objectives, the Community legislature ought to have had recourse to both the third paragraph of Article 159 EC and Article 308 EC (see, to that effect, Case 242/87 Commission v Council [1989] ECR 1425, paragraphs 6 and 37, and Kadi and Al Barakaat International Foundation v Council and Commission, paragraphs 211 to 214), while complying with the legislative procedures laid down therein, that is to say,*

1256 Case C-178/03, *Commission of the European Communities v. European Parliament and Council of the European Union* [2006] ECR I-107, paragraphs 58 and 59, emphasis added.

1257 Notes case law discussion of 22 September 2009, Europa Institute, Leiden University, p. 2 and 3 see http://media.leidenuniv.nl/legacy/2009-09-22.pdf.

1258 Case C-166/07, *European Parliament v Council of the European Union* [2009] ECR I-7135.

1259 Case C-411/06, *Commission of the European Communities v European Parliament and Council of the European Union*, [2007] ECR I-6199.

both the 'co-decision' procedure referred to in Article 251 EC and the requirement that the Council should act unanimously".[1260]

1039. This ruling of the Court of Justice implies that as of now different legislative procedures, resulting from a dual legal basis, which for example provide for a co-decision procedure and unanimity in the Council, is possible. Consequently, this would mean that the dual legal basis combining paragraph 3 of Article 194 of the TFEU and Article 114 of the TFEU and combining two different legislative procedures might be an acceptable option.

1040. Unfortunately, the Court of Justice in the *European Parliament v Council of the European Union case*[1261] did not explain why it accepts this approach and why the consequences of the *Titanium Dioxide* case were different in this case. With respect to the powers of the European Parliament, it could be argued that in the case of co-decision a unanimous Council cannot override the European Parliament and for this reason the Parliament's powers will not be undermined. Nonetheless, due to this lack of explanation by the Court of Justice, it remains uncertain to which extent different legislative procedures can be combined.

1041. Since the *European Parliament v Council of the European Union case*[1262] is a chamber judgment and not a Grand Chamber judgment, we might have to wait for further explanations until a Grand Chamber judgment tackles this issue. However, in the author's view the analysis above indicates that the most relevant issue is whether the rights of the European Parliament are being undermined and not whether the different legislative procedures are incompatible.[1263]

1042. Another issue in this context is whether it is possible to combine legal basis provisions where the nature of the competence of the Union (exclusive competence as opposed to shared competence) is different, for example having recourse to Article 207 of the TFEU jointly with Article 194 of the TFEU. The Court of Justice confirmed in its *Commission v Parliament and Council* case[1264] notably that it is from a legal point of view possible to combine exclusive and shared legal bases. However, how such cases should be handled in practice is unclear.

1260 Case C-166/07, *European Parliament v Council of the European Union* [2009] ECR I-7135, paragraph 69, emphasis added.

1261 Case C-166/07, *European Parliament v Council of the European Union*, [2009] ECR I-7135.

1262 *Ibid.*

1263 Note case law discussion of 22 September 2009, Europa Institute, Leiden University, p. 2–4 see http://media.leidenuniv.nl/legacy/2009-09-22.pdf.

1264 Case C-178/03, *Commission of the European Communities v European Parliament and Council of the European Union* [2006] ECR I-107, paragraph 40–60 and see also Case C-94/03, *Commission of the European Communities v Council of the European Union* [2006] ECR I-1, paragraphs 34–56.

1043. Advocate General Poiares Maduro pointed out in his opinion in the *Commission v Parliament and Council* case[1265] that such a combination could be possible if a clear distinction could be drawn between which elements of the measures concerned fall within the shared or exclusive competences. This means that in a given measure, it must be possible to isolate the provisions covered by one or the other legal basis. Advocate General Poiares Maduro argued that it is even conceivable that one and the same provision could have a dual legal basis as long as it relates to different objectives. However, one and the same issue could not fall within exclusive and shared competence.[1266] This approach of Advocate General Poiares Maduro is in theory plausible; however it will be difficult to apply in practice.[1267]

(b) Identifying the most appropriate legal basis for energy measures

1044. The main rule for the adoption of secondary legislation, as discussed above, is a single legal basis. Thus, this part provides some guidance to which extent Article 194 of the TFEU or other provisions of the TFEU could be used as a legal basis for the adoption of secondary legislation in the energy field. As discussed above, the ultimate choice between different legal bases must be based on objective factors such as the aim, object and content of the proposed legislative which are amenable for judicial review.

1045. How the Union legislator should determine in practice the appropriate legal basis for its energy measures, will be demonstrated by an analysis of the legal basis for the Directive 2010/30/EU.[1268] The initial Commission's proposal of this directive, proposed Article 114 of the TFEU as the legal basis. However, due to the adoption of the Lisbon Treaty which led to the adoption of the new energy Article 194 of the TFEU, the Council proposed to use this new energy title instead of Article 114 of the TFEU.

1046. Since both Articles 114 and 194 of the TFEU are expressly concerned with enhancing internal market integration, they do not run contrary to each other and can be regarded as complementary provisions. However, there are clearly a number of differences between both Articles. Contrary to Article 194 of

[1265] Case C-411/06, *Commission of the European Communities v European Parliament and Council of the European Union*, [2007] ECR I-6199.

[1266] Opinion of Advocate General Poiares Maduro of 26 March 2009 in the Case C-411/06, *Commission of the European Communities v European Parliament and Council of the European Union*, [2007] ECR I-6199, paragraph 12 and 13.

[1267] Notes case law discussion of 22 September 2009, Europa Institute, Leiden University see http://media.leidenuniv.nl/legacy/2009-09-22.pdf.

[1268] Directive 2010/30/EU of the European Parliament and of the Council of 19 May 2010 on the indication by labelling and standard products information of the consumption of energy and other resources by energy-related products, *OJ* L 153/1 and for further details see *Supra*.

the TFEU, Article 114 is exclusively concerned with economics and in particular the removal of barriers which distort the functioning of the internal market. Article 194 of the TFEU introduces other policy goals than Article 114 such as the protection of the environment through energy efficiency or the development of new and renewable forms of energy. Thus, it can be concluded that both provisions have similarities, but they are by no means identical.[1269]

1047. The Commission with respect to its selection of Article 114 of the TFEU as a legal basis for the proposed directive argued that: "*[t]he legal basis of this proposal for a recast of Council Directive 92/75/EEC is provided by Article 95 of the EC Treaty, which sets out the principles for the establishment of the internal market. The ELD ensures the completion of a well-functioning internal market with a level playing field excluding technical barriers to trade, the free movement of products falling within its scope that satisfies the labelling requirements laid down in the implementing measures adopted under the Directive*". This reasoning of the Commission is clearly in favour of Article 114 of the TFEU as the most appropriate legal basis for the proposed directive.

1048. However, it can be derived from the recitals of the proposed directive[1270] that the aim of this proposal is to promote energy-savings and energy efficiency, and to achieve environmental goals. Moreover, none of the recitals of the proposed directive specifically discussed or referred to the potential benefits for the internal market. Besides, Article 1 of the proposed directive, which is entitled as the scope, states that "*1. This Directive establishes a framework for the harmonization of national measures on end-user information, particularly by means of labelling and of product information, on the consumption of energy and*

1269 Opinion of the Committee on legal affairs on the legal basis of the Proposal for a directive of the European Parliament and of the Council on the indication by labelling and standard product information of the consumption of energy and other resources by energy-related products, COM(2008)778 in the Recommendation for second reading on the Council's position at first reading for adopting a directive of the European Parliament and of the Council on the indication by labelling and standard product information of the consumption of energy and other resources by energy-related products (recast), A7–0128/2010, p. 12 and 13.

1270 Recital 3 of the proposed directive states that: "*The provision of accurate, relevant and comparable information on the specific energy consumption of energy-related products may influence the end-user's choice in favour of those products which consume or indirectly result in consuming less energy and other essential resources during use, thus prompting manufacturers to take steps to reduce the consumption of energy and other essential resources of the products which they manufacture; it should also, indirectly, encourage the efficient use of these products. In the absence of this information, the operation of market forces alone will fail to promote the rational use of energy and other essential resources for these products*". Recital 8 specifies that "*only energy-related products which have a significant impact on consumption of energy or, where relevant, of essential resources during use and which afford adequate scope for increased efficiency should be covered by an implementing measure, when provision of information through labelling may stimulate end-users to purchase more efficient products.*"

of other essential resources, and supplementary information concerning energy-related products, thereby allowing end-users to choose more energy efficient products. 2. This Directive shall apply to energy-related products which have a significant impact on the consumption of energy and, where relevant, on other essential resources during use […]".

This Article with respect to the scope of the proposed directive indicates that the primary aim and objective of the directive is to promote undoubtedly energy efficiency and saving. Any additional coherence granted to the internal market is an incidental benefit to its primary aim and objective. The proposed directive is primarily directed towards the promotion of energy efficiency and saving within the meaning of Article 194(1)(c) of the TFEU.

1049. Moreover, the Court of Justice held in its case law that if a sector-specific legal regime was adopted, this regime should be preferred to residual and general provisions.[1271] Consequently, on the basis of the foregoing and with respect to the given situation, it can be concluded that if the Union legislator must choose between Article 114 of the TFEU and Article 194 of the TFEU as the legal basis for the proposed directive, preference should be given to the latter one.

1050. In the case, where for example a measure primarily pursues a certain objective and the harmonisation of national legislation is only an incidental effect, such a measure should be adopted only on the basis of the specific Treaty provision which corresponds to its principal characteristic.[1272] On the contrary, if a measure aims to harmonise national legislation and seeks in a subordinate

---

[1271] Opinion of the Committee on Legal Affairs on the legal basis of the Proposal for a directive of the European Parliament and of the Council on the indication by labelling and standard product information of the consumption of energy and other resources by energy-related products, COM(2008)778 in the Recommendation for second reading on the Council's position at first reading for adopting a directive of the European Parliament and of the Council on the indication by labelling and standard product information of the consumption of energy and other resources by energy-related products (recast), A7-0128/2010, p. 12–16 which refers to the Case C-91/05, *Commission of the European Communities v Council of the European Union* [2008] ECR I-3651.

[1272] Case C-177/78, *Pigs and Bacon Commission v Mc Carren and Company Limited* [1979] ECR 2161, paragraphs 9–11; Case C-68/86, *United Kingdom of Great Britain and Northern Ireland v Council of the European Communities* [1988] ECR 855, paragraphs 14–16; Case C-187/93, *European Parliament v Council of the European Union* [1994] ECR 2857, paragraphs 23–26; Case C-271/94, *European Parliament v Council of the European Union*, [1996] ECR, I-1689, paragraphs 28–32; Case C-84/94, *United Kingdom of Great Britain and Northern Ireland* v *Council of the European Union* [1996] ECR I-5755, paragraphs 11–12 and 22; Case C-36/98, *Kingdom of Spain v Council of the European Union* [2001] ECR I-779, paragraph 59; Case C-281/01, *Commission of the European Communities v Council of the European Union* [2002] ECR I-12049, paragraphs 33–49; Case C-338/01, *Commission of the European Communities v Council of the European Union* [2004] ECR I-4829, paragraphs 55–60; Case C-178/03, *Commission of the European Communities v. European Parliament and Council of the European Union* [2006] ECR I-107, paragraph 42; Case C-533/03, *Commission of the European Communities v Council of the European Union* [2006] ECR I-1025, paragraphs 43–47 and

manner to achieve an aim sought by specific Treaty provisions, a general Treaty provision, for example Article 114 of the TFEU, must be the legal basis.[1273, 1274]

1051. The foregoing has demonstrated that the aim, objective and content of a given measure are the decisive elements in order to determine the appropriate legal basis for a given measure. This means that if the primary objective of the proposed directive, albeit concerned with energy policy, was the establishment and functioning of the internal market, the legal basis would have been Article 114 of the TFEU and not Article 194 of the TFEU.[1275] Thus, for the adoption of secondary legislation with a similar objective as the Third Electricity and Gas Directives, namely the creation of fully operational internal electricity and gas markets, the current legal basis of Article 114 of the TFEU will not be replaced by Article 194 of the TFEU.

1052. Moreover on the basis of the analysis above and if the aim, objective and content remain the same as the Directive 2003/96/EC on the taxation of energy products and electricity, its reform must be based on Article 113 of the TFEU and not Article 194 of the TFEU.

1053. The recitals of Directive 2003/96/EC indicate that the aim of this directive is to harmonise the taxation of energy products and electricity in order to ensure the establishment and proper functioning of the internal market. This objective falls within the scope of Article 113 of the TFEU. This provision is the appropriate legal basis for the adoption of secondary legislation which aims to harmonise turnover taxes, excise duties and other forms of indirect taxation to the extent that such a harmonisation is necessary to ensure the establishment and the functioning of the internal market and to avoid distortion of competition. Article 194 of the TFEU, as discussed above, has a different aim. Therefore, this provision would not be the appropriate legal basis for the adoption of secondary legislation reforming Directive 2003/96/EC. This

---

Case C-155/07, *European Parliament v Council of the European Union* [2008] ECR I-81103, paragraphs 34 and 35.

1273 Case C-377/98, *Kingdom of the Netherlands v European Parliament and Council of the European Union* [2001] ECR I-7079, paragraphs 27–28 and Case C-491/01, *The Queen v Secretary of State for Health, ex. parte British American Tobacco ("BAT")* [2002] ECR I-11453, paragraphs 93–94.

1274 K. LENAERTS, and P. VAN NUFFEL, in BRAY, R. (ed.), *Constitutional Law of the European Union*, 2nd ed. (London: Thomson Sweet & Maxwell, 2005), p. 91.

1275 Opinion of the Committee on legal affairs on the legal basis of the Proposal for a directive of the European Parliament and of the Council on the indication by labelling and standard product information of the consumption of energy and other resources by energy-related products, COM(2008)778 in the Recommendation for second reading on the Council's position at first reading for adopting a directive of the European Parliament and of the Council on the indication by labelling and standard product information of the consumption of energy and other resources by energy-related products (recast), A7–0128/2010, p. 12–15.

approach seems to be followed by the Commission in its proposed energy taxation directive on 13 April 2011.[1276]

1054. Although, if the European legislator changes the aim, objective and content of Directive 2003/96/EC in such a way that this secondary legislation would not only aim to harmonise the taxation of energy products and electricity, but also other objectives which fall under Article 194 of TFEU or other provisions of the TFEU, it could apply a dual legal basis. However, according to the Court of Justice this is only possible in exceptional circumstances. The proposed measure must simultaneously pursue a number of objectives or must have several components that are indissociably linked, without one being secondary and indirect in relation to the other.

1055. Moreover, such a dual legal basis is only possible if the procedures for each legal basis are not incompatible with each other and the rights of the European Parliament will not be undermined due to the use of two legal bases.[1277] With respect to the decision-making procedure of Articles 113 and 194 of the TFEU, this procedure, except for the fact that under Article 194(3) of the TFEU, the Council is not required to consult the Economic and Social Committee, is the same. Both Articles require that the Council acts unanimously in accordance with the special legislative procedure. This means that a dual legal basis of Articles 113 and 194(3) of the TFEU is possible for the adoption of secondary legislation, but this would only be possible in exceptional circumstances, as discussed above.

1056. Both the energy and environmental title of the TFEU could be used as a legal basis for the adoption of measures in areas such as energy efficiency and renewable energy sources, etc. Thus, there is a certain overlap. Moreover, from a procedural point of view there is no difference between both provisions. Nonetheless, in the author's view a significant difference between these provisions remains, if the specific Union action in relation to the energy field is based on Article 192 of the TFEU. Member States are allowed on the basis of Article 193 of the TFEU to maintain or introduce more stringent national

1276 Proposal for a Council Directive amending Directive 2003/96/EC restructuring the Community framework for the taxation of energy products and electricity, COM(2011)169/3, 41 p.

1277 See *Supra* and Case-300/89, *Commission of the European Communities v Council of the European Communities* (or the *Titanium dioxide* case) [1991] ECR I-2867, paragraph 10; Case C-338/01, *Commission of the European Communities v Council of the European Union* [2004] ECR I-4829, paragraph 57; Case C-178/03, *Commission of the European Communities v European Parliament and Council of the European Union* [2006] ECR I-107, paragraph 59 and Case C-94/03, *Commission of the European Communities v Council of the European Union* [2006] ECR I-1, paragraph 52.

measures than those provided for by the Union. This is not the case if secondary legislation is adopted on the basis of Article 194 of the TFEU.

1057. One of the consequences of the foregoing, is that for example when the Union establishes a total harmonisation in one of the aspects of the energy field on the basis of Article 194 of the TFEU, contrary to Article 192 of the TFEU[1278], there will be a transfer of internal competence to the Union legislator which results in an exclusive external competence.

1058. It can be concluded, that with respect to the choice between Article 194 of the TFEU and other provisions of the TFEU such as Article 113, 114, 192, etc. merely because a proposed measure is concerned with the energy sector, this does not automatically mean that the given measure must be based on Article 194 of the TFEU. It must be asserted on a case-by-case basis which Treaty Article (energy, environmental, fiscal, etc. title) matches the most with the main aim, objective and content of a given measure on a European level.[1279] Time and experience will also tell us how the Court of Justice will interpret the new energy title of the TFEU.

E. CONCLUDING REMARKS

1059. This chapter concludes that since the adoption of the Lisbon Treaty, despite considerable political opposition[1280], "energy" has been explicitly recognised in Article 4(2), (i) of the TFEU as a new policy area among the general category of shared competences.

1060. Thus, both the European Union and its Member States are competent to regulate in the energy area. If the Union legislator has not yet exercised its shared competence in the energy field, Member States remain free to adopt legislative measures in the energy field provided that they comply with the primary Treaty provisions. Where the Union has adopted legislative measures in the energy field, Member States remain free to exercise their competence provided that they comply with the primary provisions of the Treaty and any obligation imposed by the relevant secondary energy law provisions of the Union.

1061. The Lisbon Treaty also included a new energy title, Article 194, into the TFEU. This provision provides a legal basis for the adoption of measures which

---

1278 See *Supra*.

1279 See also *Infra* the analysis between Article 192 of the TFEU and Article 114 of the TFEU.

1280 S. ANDOURA, L. HANCHER and M. VAN DER WOUDE, Policy Proposal by Jacques DELORS Towards a European Energy Community: A Policy Proposal (Notre Europe, March 2010), p. 11.

aim to ensure most of the goals of the existing European energy policy. The Lisbon negotiators adopted a carefully crafted compromise between the shared energy competence, between the Union and its Member States, and the national sovereignty over a number of energy situations or concerns referred to in the second paragraph of Article 194 of the TFEU. Hence, the new energy title in the TFEU can be seen as a mere codification of the existing practise in the energy field. Thus, the same pre-existing gaps, flaws and limitations remain in place.

1062. This chapter concluded that merely because a proposed measure is concerned with the energy sector, this does not automatically mean that the given measure must be based on Article 194 of the TFEU. Thus, in order to determine the legal basis of each measure in the energy field, it must be asserted on a case-by-case basis which Treaty Article (energy, environmental, fiscal, etc. title) matches the most with the main aim, objective and content of a given energy measure on a European level.

1063. Compared to the former situation, it can be concluded that with respect to competences in the energy field, no major changes were adopted by the negotiators of the Lisbon Treaty.[1281] It could even weaken the existing foundations of the European energy policy.[1282]

1064. This new energy title does not offer any prospect of radical reform from the former situation. Moreover, chapters 2 and 3 demonstrated that the current policy and legal framework in the energy field are insufficient. Thus, it can be concluded that the European Union with the current legal framework in the energy field will not achieve its three central goals: competitiveness, security of supply and sustainability.

1065. The overall objective of this book is to analyse whether the current EU regulatory framework for energy law and policy suffices to address the triple objectives of the EU's energy policy, or, by contrast, whether changes to that framework are needed. Chapters 2 and 3 concluded that the current EU regulatory framework for energy law and policy is insufficient to address those triple objectives. Therefore, the next part will provide some legal and institutional options to change EU regulatory framework for energy law and policy in order to overcome the current situation in the energy field.

1281 Policy Proposal by Jacques Delors, *Towards a European Energy Community: A Policy proposal*, study by S. ANDOURA, L. HANCHER and M. VAN DER WOUDE, p. 99.

1282 L. HANCHER and A. DE HAUTECLOCQUE, "Manufacturing the EU Energy Markets: The Current Dynamics of Regulatory Practise", *Competition and Regulation in Network Industries* (2010), p. 310.

CHAPTER 6

LEGAL AND INSTITUTIONAL OPTIONS FOR A EUROPEAN ENERGY FRAMEWORK

A. INTRODUCTION

1066. Since the TFEU until recently, as discussed in chapter 4, did not contain an enabling provision putting forward the objectives, goals and procedures for Union action in the energy field, the European legislator applied general provisions of the TFEU such as Articles 113, 114, 122, 192 and 352, to intervene in the energy sector. Union competence undoubtedly existed in the energy policy area and led to the adoption of a considerable amount of secondary legislation in this field.

1067. This chapter 4 has also emphasised that the energy measures very often, or even always, seem to pass the tests relating to the principles of attribution of competences, subsidiarity and proportionality. Nonetheless, it emphasised that it must be assessed on a case-by-case basis whether each proposal in the energy field passes these tests. Chapter 4 concluded that recourse to the general provisions of the TFEU for the adoption of energy measures is not the most appropriate situation because the European legislator is unable to address the specific issues of the energy sector. Thus, an effective, coherent and robust legal framework which tackles the current challenges in the energy field could not be adopted.

1068. Chapter 5 highlighted that much hope was given to the Lisbon Treaty since this Treaty explicitly recognised energy as a new policy area among the general category of shared competences. Moreover, the IGC included the new energy title, Article 194, into the TFEU. This provision provides the new legal basis for the adoption of measures in the energy field. The adoption of Article 194 of the TFEU, as discussed above does not automatically mean that merely because a proposed measure is concerned with the energy sector, this proposal should be based on Article 194 of the TFEU. Chapter 5 argued that for each measure in the energy field it must be determined which Treaty Article matches the most with the aim, objective and content of the given measure in order to determine its legal basis. It concluded that the Lisbon negotiators adopted a carefully crafted compromise between the shared energy competence and the national sovereignty over a number of energy situations. Moreover, compared to the former situation, the Lisbon Treaty did not adopt major changes and the new energy title could even weaken the existing foundations of the European energy policy.

1069. Since chapter 5 concluded that the new energy title does not offer any prospect of radical reform from the former situation and both chapters 2 and 3 have argued that the current policy and legal framework is insufficient in achieving the three key objectives, competitiveness, security of supply and sustainability of the European energy policy, urgent, decisive and immediate action is required. This action should address the short-term as well as the long-term challenges that Europe is facing in the energy field.

1070. This chapter therefore, suggests different effective institutional and legal amendments to the current European energy framework which remedy the current deficiencies and allow the Union to adopt a specific, common and effective European energy policy and law. Before discussing these options, the next part will provide some suggestions formulated in legal doctrine and current practices, to overcome these deficiencies.

B. SUGGESTIONS IN LEGAL DOCTRINE AND CURRENT PRACTICES

1. ENHANCED COOPERATION

1071. Since the adoption of the Lisbon Treaty the rules on enhanced cooperation have been regrouped into two sets of rules. The first set of rules contains the generalities with respect to enhanced cooperation and was adopted under Article 20 of the TEU. In the Articles 326 to 334 of the TFEU the second set of rules containing the details on enhanced cooperation has been provided for. Enhanced cooperation has been designed to allow some Member States, using the institutions, procedures and mechanisms of the European Union, to cooperate further with each other in areas where others do not wish to do so. These provisions allow more ambitious Member States to cooperate with each other while making use of the institutional structures of the European Union.

1072. This mechanism for enhanced cooperation, which existed on paper since 1997 with the adoption of the Amsterdam Treaty, appears to be very attractive.

However, its procedure, before the request of twelve Member States to apply enhanced cooperation in the area of unitary patent protection, has never been used.[1283] This is perhaps due to the strict conditions attached to it. However, it is more due to political reasons which avoid a two-speed Europe.[1284] The Nice[1285] and Lisbon Treaty both made some modifications to the procedure. Nonetheless, the conditions attached to the enhanced cooperation procedure remain restrictive.

1073. Some of these conditions are, *inter alia*, that enhanced cooperation must be limited to achieve the Treaty's existing policy objectives, the scope of enhanced cooperation is limited to areas of shared competences and not exclusive competences of the European Union, the minimum number of Member States required to initiate an enhanced cooperation is at least nine Member States. The coalition of Member States which opt for an enhanced cooperation must make apparent that the policy objectives of the energy title in the TFEU cannot be achieved by all the Member States. The initiative for such an enhanced cooperation lies mainly in the hands of the Commission. With respect to the content of enhanced cooperation nothing can be done that the Union itself would not be entitled to do, the consent of the European Parliament is required for launching an enhanced cooperation, etc.

1074. Given these procedural constraints of enhanced cooperation, it remains uncertain whether this mechanism will be used by the Member States in the energy field. However, this Treaty-based mechanism has in particular the advantage that the Union institutions can be used.[1286] The possibility to start up

---

1283 Proposal for a Regulation of the European Parliament and of the Council implementing enhanced cooperation in the area of the creation of unitary patent protection, COM(2011)215/3, 25 p. After the request of those twelve Member States the Commission submitted a proposal to the Council for authorising enhanced cooperation in the area of unitary patent protection. Following the adoption of this proposal 13 other Member States requested to join the cooperation. After the consent of the European Parliament the Council adopted on 10 March 2011 the authorising decision. For more details see http://ec.europa.eu/internal_market/indprop/patent/index_en.htm#enhanced.

1284 J.-C. PIRIS, *The Lisbon Treaty A Legal and Political Analysis* (Cambridge, Cambridge University Press, 2010), p. 89; K. LENAERTS, and P. VAN NUFFEL, in BRAY, R. (ed.), *Constitutional Law of the European Union* (London, Thomson Sweet & Maxwell, 2005), p. 372–377; Policy Proposal by Jacques Delors, *Towards a European Energy Community: A Policy proposal*, study by S. ANDOURA, L. HANCHER and M. VAN DER WOUDE, p. 100 and K. LENAERTS, and P. VAN NUFFEL, *Europees Recht* (Antwerpen, Intersentia, 2011), p. 508–512.

1285 The Nice Treaty introduced a qualified majority vote instead of unanimity.

1286 N. AHNER, J.-M. GLACHANT and A. DE HAUTECLOQUE, *Legal feasibility of Schengen-like Agreements in European Energy Policy: The Cases of Nuclear Cooperation and Gas Security of Supply*, (Florence, European University Institute, 2010), p. 5 and 6 and Policy Proposal by Jacques Delors, *Towards a European Energy Community: A Policy proposal*, study by S. ANDOURA, L. HANCHER and M. VAN DER WOUDE, p. 100 and 101.

an enhanced cooperation between several Member States could be an option, if the procedural constraints can be overcome, for a specific energy-related issue where insurmountable difficulties exist, making a compromise or the unanimity requirement[1287] impossible. Such a specific energy-environmental-related issue could for example be the adoption of a CO_2 tax or the adoption of a measure which requires unanimity under Article 194 of the TFEU. However, in order to establish a stronger and more coherent European energy policy which requires different amendments to the current legal framework, enhanced cooperation in the author's view appears not to be a successful option.[1288]

2. REINFORCED COOPERATION OR IN-BUILT CLOSER COOPERATION – THE SCHENGEN MODEL

1075. An alternative option to enhanced cooperation is reinforced cooperation or in-built closer cooperation outside the institutional framework of the European Union through the adoption of partial agreements between some Member States under international law. The most well-known examples of such a reinforced cooperation are the Eurozone and the Schengen Agreement. The latter was initially established between Germany, France and the Benelux countries because a consensus on abolishing the controls of persons at the intra-Union borders could not be reached.[1289]

1076. The most recent example is the adoption of the Treaty relating to the establishment of the "Functional Airspace Block Europe Central (hereinafter FABEC)" between Germany, Belgium, France, Luxembourg, the Netherlands and Switzerland. This FABEC Treaty creates the framework, general conditions and governance under which the contracting states have to ensure air traffic management and the provision of air navigation services in the airspace concerned.[1290] Such functional airspace blocks should put an end to the current fragmentation of European airspace and enable more efficient and shorter flights.[1291] This kind of reinforced cooperation outside the institutional

1287 This was the case for the establishment of an EU patent system.

1288 See *Infra.*

1289 J.-C. PIRIS, *The Lisbon Treaty A Legal and Political Analysis* (Cambridge, Cambridge University Press, 2010), p. 90 and 91.

1290 Airspace concerned means *"the airspace over the territory of the Contracting States and that under their responsibility in accordance with International Civil Aviation Organization (ICAO) rules, as defined in Article 3 of this Treaty."*

1291 IP/10/1648, Brussels of 2 December 2010, *Belgium, France, Germany, Luxembourg, the Netherlands and Switzerland sign agreement towards the Single European Sky*; Treaty relating to the establishment of the functional airspace block "Europe Central" between the Federal Republic of Germany, the Kingdom of Belgium, the French Republic, the Grand Duchy of

framework of the European Union should aim to be reintegrated into the institutional framework of the European Union.[1292]

1077. The legal criteria and procedural requirements for such a reinforced cooperation outside the institutional framework of the European Union are less onerous than those set for enhanced cooperation. However, there are still a number of legal limits which should be complied with. Member States are free to enter agreements *inter se*, but those agreements need to be in full compliance with European Union law. In particular, the principle of the primacy of Union law limits the ability of Member States to conclude a reinforced cooperation agreement outside the institutional framework of the European Union or a Schengen-like agreement.[1293] According to case law of the Court of Justice of the European Union, this principle is absolute in the sense that Union law takes precedence over any national law, including principles of a national constitutional structure.[1294]

1078. Nonetheless, it is uncertain how much European Union law must be in place and whether an actual conflict is required between existing Union law and a Schengen-like agreement in order to conclude that such an agreement is not allowed due to the principle of the primacy of Union law. On the basis of case law of the Court of Justice of the European Union and legal doctrine, the relevant criteria in order to determine whether in a specific situation a conflict exists between European Union law and a Schengen-like agreement are subsidiarity, pre-emption and primacy of Union law.[1295] In particular, the membership of the European Union imposes a number of legal constraints on the scope and content of a Schengen-like agreement.[1296] For example such agreements could not cover aspects regarding which the European Union has exclusive competences, aspects which are governed by European Union law, aspects which are not governed by

Luxembourg, the Kingdom) of The Netherlands and the Swiss Confederation. See www.fabectreaty.tv/downloads/FABEC_Treaty_English_version.pdf.

1292 Policy Proposal by Jacques Delors, *Towards a European Energy Community: A Policy proposal*, study by S. ANDOURA, L. HANCHER and M. VAN DER WOUDE, p. 105.

1293 N. AHNER, J.-M. GLACHANT and A. DE HAUTECLOQUE, *Legal feasibility of Schengen-like Agreements in European Energy Policy: The Cases of Nuclear Cooperation and Gas Security of Supply*, (Florence, European University Institute, 2010), p. 6 and 7.

1294 K. LENAERTS, and P. VAN NUFFEL, in BRAY, R. (ed.), *Constitutional Law of the European Union* (London, Thomson Sweet & Maxwell, 2005), p. 666 and 667; Case C-6/64, *Flaminio Costa v E.N.E.L* [1964] ECR 585, p. 594 and Case C-11/70, *Internationale Handelsgesellschaft mbH v Einfuhr- und Vorratsstelle für Getreide und Futtermittel* [1970] ECR 1125.

1295 For a detailed analysis of these criteria with respect to Schengen-like agreements see N. AHNER, J.-M. GLACHANT and A. DE HAUTECLOQUE, *Legal feasibility of Schengen-like Agreements in European Energy Policy: The Cases of Nuclear Cooperation and Gas Security of Supply*, (Florence, European University Institute, 2010), p. 7–13.

1296 Presentation N. AHNER and A. DE HAUTECLOQUE, *Legal feasibility of 'Schengen-like' Agreements in European Energy Policy: The Cases of Nuclear Cooperation and Gas Security of Supply*, (ELRF Conference 9 December 2010, Cambridge Law Faculty.

European Union law, but the Union has a duty to act, they cannot conflict with general principles of Union law, etc.

1079. This means that only specific relevant topics of national energy competences and policies could be covered by a reinforced cooperation agreement outside the institutional framework of the European Union or a Schengen-like agreement.[1297]

1080. In legal doctrine, the application of Schengen-like agreements has been analysed in two energy areas: nuclear policy and security of gas supply. With respect to nuclear policy, in this doctrine it is argued that an *inter se* agreement could be adopted on disposal of high-level radio-active waste and spent nuclear fuel, as well as the setting-up of regional centres of disposal, and the standardisation of reactor design. On the contrary, an *inter se* agreement on market rules for nuclear investors and operators on the design and use of long-term electricity contracts and on the creation of joint ventures, the use of open seasons mechanisms and investment in merchant lines, appeared not to be legally feasible. According to this doctrine, such an *inter se* agreement would not pass the pre-emption criterion since most market frame issues (such as long term contracts, joint ventures, etc.) fall under the competition rules which is one of the areas in which the Union has exclusive competences.

1081. With respect to security of gas supply in this doctrine it is highlighted that *inter se* agreements could not be adopted in the areas of mutual transparency commitments, harmonisation aspects of emergency plans and a common supply framework agreement since they could not fulfil the pre-emption criterion.[1298]

1082. On the basis of the foregoing, the adoption of a reinforced cooperation agreement outside the institutional framework of the European Union or a Schengen-like agreement could be an option for certain aspects of the energy area. However, due to the fact that the European legislator has been very active in the energy field, the scope of such Schengen-like agreements will be limited in the author's view. Only for a limited number of targeted areas in the energy field, as discussed above, Schengen-like agreements could serve as the right legal

---

1297 Policy Proposal by Jacques Delors, *Towards a European Energy Community: A Policy proposal*, study by S. ANDOURA, L. HANCHER and M. VAN DER WOUDE, p. 105 and 106.

1298 N. AHNER, J.-M. GLACHANT and A. DE HAUTECLOQUE, *Legal feasibility of Schengen-like Agreements in European Energy Policy: The Cases of Nuclear Cooperation and Gas Security of Supply*, (Florence, European University Institute, 2010), p. 10–27 and N. AHNER, J.-M. GLACHANT and A. DE HAUTECLOQUE, "Legal feasibility of Schengen-like Agreements in European Energy Policy: The Cases of Nuclear Cooperation and Gas Security of Supply", *Policy brief* (2010), 5 p.

framework.[1299] If the Union would become one of the contracting parties this might solve some of the legal constraints.[1300] Nonetheless, this option remains to be analysed on a case-by-case basis. Thus, in the author's view little hope should be given to reinforced cooperation or in-built closer cooperation in order to establish a stronger and more coherent European energy policy.

3. FUNCTIONAL OR REGIONAL COOPERATION

1083. A group of Member States could decide to cooperate in certain aspects of the energy field on a regional basis. In the spring of 2006, the European Regulators' Group for Electricity and Gas (ERGEG), at the request of the European Commission, set up as an interim step towards a single energy market, 7 electricity regions[1301] and 3 gas regions.[1302] These regional initiatives provide a forum for regulators, network operators and other stakeholders of neighbouring countries to discuss a variety of issues of common interest such as congestions management, balancing, transparency, security of supply, etc.[1303]

1084. So far, they followed a voluntary bottom-up approach whereby each region has set its own priorities. This allowed the regions to focus on their specific problems. Although such a voluntary bottom-up approach also has some shortcomings among which the inherent risk that those different regions implement different solutions to similar issues. This could over time impede the integration process and endanger the establishment of the internal energy market in the longer run.[1304] Recently the Commission adopted a communication on the future role of the regional initiatives in which it presented

1299 N. AHNER, J.-M. GLACHANT and A. DE HAUTECLOQUE, *Legal feasibility of Schengen-like Agreements in European Energy Policy: The Cases of Nuclear Cooperation and Gas Security of Supply*, (Florence, European University Institute, 2010), p. 10–28.

1300 Policy Proposal by Jacques Delors, *Towards a European Energy Community: A Policy proposal*, study by S. ANDOURA, L. HANCHER and M. VAN DER WOUDE, p. 105 and 106.

1301 Note that the electricity regions have been defined in the Commission Decision 2006/770/EC of 9 November 2006 amending the Annex to the Regulation No 1228/2003 on conditions for access to the network for cross-border exchanges in electricity, *OJ* L 312/59.

1302 For gas, contrary to the electricity regions, the regions have not yet been formally defined in a binding decision. At the moment, it is under discussion whether the gas regions should be formally defined.

1303 See also the Study, *"From regional markets to a single European market"* of Everis and Mercados of 28 April 2010, http://ec.europa.eu/energy/gas_electricity/studies/doc/2010_gas_electricity_markets.pdf and ERGEG Conclusions paper of 21 may 2010, *Strategy for delivering a more integrated European energy market: The role of the ERGEG Regional Initiatives*, www.energy-regulators.eu/portal/page/portal/EER_HOME/EER_PUBLICATIONS/CEER_ERGEG_PAPERS/Cross-Sectoral/2010/E10-RIG-10-04_Strategy_Conclusions_21-May-10.pdf.

1304 Communication from the Commission to the European Parliament and the Council of 7 December 2010, *The future Role of Regional Initatives*, COM(2010)721, p. 2–7.

possible directions for reviewing their role, clarifying their membership and reinforcing their effectiveness.[1305]

1085. As a start, such a voluntary form of regional cooperation between Member States could be useful to achieve a number of goals (e.g. the adoption of common standards, the coordination of investments, etc.). Although in the longer run it is uncertain whether such cooperation between Member States due to its voluntary commitment is the right track for the development of the European energy market. Moreover, the current regional initiatives have a limited scope. They mainly focus on common issues of the energy markets. If they would like to contribute to the three central goals competitiveness, security of supply and sustainability they must certainly broaden their scope.

1086. Such voluntary cooperation will be insufficient for the establishment of a specific, common and effective European energy policy and law. Nonetheless, in the author's view this form of cooperation on a regional level could be very useful as a starting point to a more formal cooperation. This could lead to the adoption of an energy Treaty between Member States which initially cooperated voluntary under the regional initiative.[1306]

1087. The suggestions discussed above in the author's view do not seem to be the most successful options. The next part therefore proposes some amendments to the existing institutional and legal framework relating to the energy field. These amendments can be divided into two options. On the one hand, the adoption of different amendments to the TFEU and on the other, the adoption of a new Treaty in the energy field.

C. AMENDMENTS TO THE PROVISIONS OF THE TREATIES

1088. The first option on a European level is to provide the Union, with the necessary tools through amendments of the Treaties in order to allow the European Union to establish a stronger and more coherent European energy policy.

1. PROCEDURE FOR AMENDING THE TREATIES

1089. According to Article 48 of the TEU the Treaties can be amended in accordance with an ordinary revision procedure and a simplified revision

1305 *Ibid.*, 10 p.
1306 See *Infra*.

procedure.[1307] The later procedure is an innovative feature of the Lisbon Treaty which allows that all or part of the provisions can be amended through a unanimous vote of the European Council. Such a decision will not enter into force until it is approved by the Member States in accordance with their respective constitutional requirements. The positive element of this procedure is that the Treaties can be amended without convening an Intergovernmental Conference which could make the adoption of proposed amendments much easier.[1308]

1090. However, Article 48 of the TEU also states that decisions taken on the basis of this simplified revision procedure shall not increase the competences conferred on the Union in the Treaties. Since the scope of the simplified revision procedure is limited, it is uncertain whether this procedure could be used to adopt the necessary amendments to the energy provisions of the TFEU.

1091. Consequently, the ordinary revision procedure appears to be the most obvious procedure to amend the treaties with respect to the energy provisions. This procedure allows the government of any Member State, the European Parliament or the Commission to submit proposals to the Council which either increase or reduce the competences conferred on the Union in the Treaties for the amendment of the Treaties. Such proposals are submitted to the European Council by the Council, and the national Parliaments are notified.[1309]

1092. Although, this ordinary revision procedure would be the most appropriate one, it must be highlighted that it remains uncertain whether at this stage all the Member States would agree to amend the energy provisions of the Treaties and to equip the European Union with more competences in this area. The next part nonetheless, will make abstraction of these realities of the current European policy, and formulates some concrete amendments to the Treaties in order to allow the Union to adopt a specific, common and effective energy policy and law on a European level.

1307 See Article 48, (1) tot (7) of the TEU.

1308 Policy Proposal by Jacques Delors, *Towards a European Energy Community: A Policy proposal*, study by S. ANDOURA, L. HANCHER and M. VAN DER WOUDE, p. 102. It should be noted that this is not the only innovation of the Lisbon Treaty, it allows for amendment of the unanimity requirement to that of a qualified majority vote and adaptation of any legislative procedure to that of the ordinary procedure in the TFEU of the aspects of Title V of the TEU dealing with the Union's external action and specific provisions on the common foreign and security policy. See D. CHALMERS, G. DAVIES and G. MONTI, *European Union Law* (Cambridge, Cambridge University Press, 2010), p. 76.

1309 Article 48(2)-(5) of the TEU and see for further details K. LENAERTS, and P. VAN NUFFEL, *Europees Recht* (Antwerpen, Intersentia, 2011), p. 53–57.

2. TREATY AMENDMENTS WITH RESPECT TO ENERGY

1093. If all Member States are convinced that the current institutional and legal framework on a European level is insufficient and lacking the necessary means to achieve the three central goals, competitiveness, security of supply and sustainability in the energy field, one of the options to overcome this situation, in the author's view, is to adopt a number of Treaty amendments. In order to create a stronger and more coherent energy title, not only the present energy competences and objectives in Article 194 of the TFEU should be amended, but also some of the Treaty provisions which empower the Union to adopt certain measures in the energy field.

(a) *Article 3 of the TFEU*

1094. Before formulating some amendments to other Treaty provisions, some might argue that it would be better to add energy as an exclusive competence of the Union in Article 3 of the TFEU, instead of a shared competence between the Union and its Member States. Irrespective of whether this is possible in practice, in the author's view this is not a good option since a number of aspects in the energy field should preferably be tackled on a Member State level (for example some energy efficiency aspects) or should remain within their sovereignty.

(b) *Article 122 of the TFEU*

1095. The first Treaty provision requiring amendment is Article 122 of the TFEU. This Article provides, since the adoption of the Lisbon Treaty, that the Council, on a proposal from the Commission, may decide, in a spirit of solidarity between Member States, upon the measures appropriate to the economic situation, in particular if severe difficulties arise in the supply of certain products, notably in the area of energy.

1096. The meaning of this explicit reference to energy in Article 122 of TFEU and whether this has any impact on the relationship between Articles 122 and 194 of the TFEU is unclear.

1097. Moreover, the decision-making procedure between both provisions varies considerably.[1310] Therefore, it could be argued that the reference to the energy area should be removed from Article 122 of the TFEU and state that the energy area is even excluded from its scope. On the contrary, Article 194 of the TFEU must contain a similar provision as Article 122 of TFEU allowing the

---

[1310] See *Supra*.

Council to adopt measures if severe difficulties arise in the supply of energy products.

(c) Articles 191 and 192 of the TFEU

1098. Beside Article 122 of the TFEU, the provisions dealing with the environmental objectives and competences on a European level in the Articles 191 and 192 of the TFEU should be amended. One of the objectives enumerated in paragraph 1 of Article 191 of the TFEU is the *"prudent and rational utilisation of natural resources"*. This provision has proven in the past to be useful for the adoption of measures in the energy field.[1311] However, since the adoption of the Lisbon Treaty, Article 194 of the TFEU contains a similar objective by stating that the Union policy shall aim *"to promote energy efficiency and energy savings and the development of new and renewable forms of energy"*. Thus, the TFEU contains two provisions allowing the adoption of energy efficiency and other energy-related measures.

1099. Moreover, Article 192(2) of the TFEU which constitutes the legal basis for the adoption of environmental measures, states that the Council shall act unanimously in accordance with a special legislative procedure for the adoption of *"measures significantly affecting a Member State's choice between different energy sources and the general structure of its energy supply"*. A similar provision, as discussed above, can be found in Article 194(2) which states that the measures in the energy field, in order to achieve the objectives referred to in Article 194 of the TFEU, *"shall not affect a Member State's right to determine the conditions for exploiting its energy resources, its choice between different energy sources and the general structure of its energy supply, without prejudice to Article 192(2)(c)"*. The meaning of the reference to Article 192(2)(c) TFEU in Article 194 of the TFEU, as discussed above, is uncertain. Nonetheless, the author argued that this reference is a matter of procedural restrictions. Thus, the right of veto for Member States provided for in accordance with Article 192(2)(c) of the TFEU could not be gotten around if the Union suggests measures which *"significantly affect a Member State's choice between different energy sources and the general structure of its energy supply"*.[1312]

1100. Both the energy and environmental titles of the TFEU, as discussed above, could be used as a legal basis for the adoption of measures in areas such as energy efficiency, renewable energy sources, energy measures which require unanimity, etc. Consequently, there is an overlap between the Articles 192 and 194 of the TFEU. Moreover, from a procedural point of view there is no

---

1311 *Ibid.*

1312 *Ibid.*

difference between both provisions. Nonetheless, in the author's view a significant difference between these provisions remains if the specific Union action in relation to the energy field is based on Article 192 of the TFEU. In that situation Member States are allowed on the basis of Article 193 of the TFEU to maintain or introduce more stringent national measures than those provided for by the Union. This is not the case if secondary legislation is adopted on the basis of Article 194 of the TFEU.

1101. Given these uncertainties and different consequences due to the use of Article 192 and/or 194 of the TFEU, it is not only recommendable to rewrite Article 194 of TFEU, but also that Article 192 of the TFEU should not contain any objectives or legal bases anymore allowing the adoption of measures in the energy field and state that the energy area is excluded from its scope.

(d) Articles 170–172 of the TFEU

1102. The fourth objective of the Union's energy policy mentioned in Article 194 of the TFEU is the promotion of interconnection of energy networks. In addition to this objective the TFEU also contains the Articles 170–172 on the promotion of TENs in the areas of transport, telecommunications and energy infrastructure. In order to avoid ambiguities between Articles 194 and 170–172 of the TFEU, it could be argued that it is better to exclude the energy infrastructure in Articles 170–172 of the TFEU and provide similar provisions of those Articles in Article 194 of the TFEU. Moreover, these provisions should be added with the objective of ensuring an optimal functioning of the grids and pipelines, guarantee a European-wide coverage of the grids and pipelines, and optimise the interconnections with external suppliers and resources.[1313]

(e) Article 194 of the TFEU

1103. With respect to the amendments to Article 194 of the TFEU it is important that the Intergovernmental Conference first clarifies the meaning of some of the guiding principles, referred to in the new energy title, *inter alia*, "*in the context of the establishment and functioning of the internal market*" and "*in a spirit of solidarity between Member States*", but also the purpose of these guiding principles. The analysis in chapter 5 highlighted a few uncertainties relating to the enumerated objectives of the European energy such as the scope of ensuring the functioning of the energy market and security of energy supply of the Union[1314] which should be clarified by the IC.

1313 Policy Proposal by Jacques Delors, *Towards a European Energy Community: A Policy proposal*, study by S. ANDOURA, L. HANCHER and M. VAN DER WOUDE, p. 87 and 88.

1314 See *Supra*.

1104. Moreover, it could be argued that Article 194 of the TFEU should be supplemented with other objectives in the energy area. The first objective which needs to be added is the promotion of measures on an international level in the energy field. This objective should strengthen Europe's position on an international level to deal with the security of energy supply of the Union and its Member States, but it should also allow the Union to speak and act collectively in the energy area on an international level.

1105. The second objective which is lacking in Article 194 of the TFEU and should be supplemented is a clear commitment to promote research and development in the energy field supported by adequate funding facilities. These funding facilities could be financed from different sources such as an energy levy on forms of energy production and on energy consuming goods. Article 194 of the TFEU should therefore contain a legal basis allowing the adoption of an energy levy. This legal basis allows the Union to adopt its own financial means which is essential to finance its policy in the energy field. Moreover, the revenues of this energy levy could also be used to establish a European Energy Fund to steer investment decisions and allow producers and grid and pipeline operators to have access to additional financial instruments such as long-term loans and credit facilities.[1315]

1106. Besides completing the objectives of the Union's energy policy in Article 194 of the TFEU, the TFEU should contain a legal basis to appoint a European energy regulator in order to ensure non-discrimination, the efficient functioning of the market and effective competition of European energy markets. Such a European energy regulator should have decision-making powers and be able to take directly binding decisions on market players such as grid and pipeline operators and their users, as well as on investment-related issues.[1316] As discussed above, the so-called *Meroni Doctrine* which entails that independent executive organs, unless a specific Treaty provision is available and the delegation remains limited to the aims and means provided for by this provision, cannot be attributed discretionary powers. Thus, unless the Treaties are amended, a European energy regulator with decision-making powers cannot be established.

1107. Therefore, it could be argued that Article 13(1) of the TEU should be supplemented by another Union institution, namely the European Energy Regulator. Moreover, with respect to the powers, decision-making procedures,

[1315] Policy Proposal by Jacques Delors, *Towards a European Energy Community: A Policy proposal*, study by S. ANDOURA, L. HANCHER and M. VAN DER WOUDE, p. 89–91.

[1316] See *Supra* and Policy Proposal by Jacques Delors, *Towards a European Energy Community: A Policy proposal*, study by S. ANDOURA, L. HANCHER and M. VAN DER WOUDE, p. 87.

bodies, etc. of the European energy regulator, the energy title of the TFEU should be supplemented with a separate and new Article 194B of the TFEU.

1108. With respect to the decision-making procedure, it has been argued above that energy markets are complex markets which often raise new issues requiring decisions to be taken sooner rather than later. However, the current decision-making process of the institutional machinery on a European level is by far sufficiently flexible to adopt the necessary measures at any given moment.[1317] Therefore, the current decision-making procedures in Article 194(2) of the TFEU should be supplemented with a decision-making procedure enabling the adoption of measures in the energy field in a much shorter time span (for example within a period of three months). However, the use of such a shorter decision-making process could not have as a consequence that an ordinary legislative procedure is replaced by a special legislative procedure.

1109. The current provisions of Article 194(2) and (3) of the TFEU require unanimity for the adoption of measures which *"affect a Member State's right to determine the conditions for exploiting its energy resources, its choice between different energy sources and the general structure of its energy supply"* and those which are *"primarily of fiscal nature"*.[1318] It could be argued that the unanimity requirement for the adoption of measures which *"affect a Member State's choice between different energy sources and the general structure of its energy supply"* should be replaced by a qualified majority vote.[1319]

(f) Article 288 of the TFEU

1110. The final suggestion relates to the adoption of regulations instead of directives for the adoption of secondary legislation in the energy field. Under Article 288 of the TFEU there are five legal instruments, regulations, directives, decisions, recommendations and opinions, the first three of which are binding and the others are non-binding instruments. With the adoption of the Lisbon Treaty a slightly modified and more straightforward definition of a decision has been introduced. It makes the decision a normative act (binding in its entirety) and contrary to the previous definition it mentions the addressee only as a possibility.[1320]

1317 See *Supra*.

1318 See *Supra* and Policy Proposal by Jacques Delors, *Towards a European Energy Community: A Policy proposal*, study by S. ANDOURA, L. HANCHER and M. VAN DER WOUDE, p. 87.

1319 See *Supra*.

1320 J.-C. PIRIS, *The Lisbon Treaty: A Legal and Political Analysis* (Cambridge, Cambridge University Press, 2010), p. 93 and 94.

1111. In the energy field most of the adopted measures are currently directives and in some exceptional cases the European legislator adopted regulations to achieve its policy goals in the energy field. According to the definition in Article 288 of the TFEU directives are *"binding, as to the result to be achieved, upon each Member State to which it is addressed, but shall leave to the national authorities the choice of form and methods"*. Since directives allow Member States themselves to determine freely the way in which the intended result is to be achieved within their national legal system, they reflect the idea of subsidiarity. Directives are not directly applicable in the domestic legal systems of the Member States, except for directives which contain directly effective provisions.[1321] They only obtain there full legislative status after they have been implemented into national law. For the implementation of directives, Member States can decide upon the form and methods which in principle leaves the competent national authorities some discretion.[1322]

1112. Such discretion is not the only problem with directives, as we have seen above in chapter 2 of this book, different Member States are still in breach with the implementation and enforcement of a number of directives in the energy field. This lack of implementation and enforcement undermines the credibility of the common EU energy policy. In practice, on a European level, when Member States agree on directives in the energy field there is a clear commitment and political will. However, a direct commitment to achieve and implement in their national jurisdictions the intended results of a directive is often missing.

1113. Unlike directives, regulations are, at the basis of Article 288 of the TFEU, directly applicable in all Member States. They automatically form part of the provisions of a Member State's legal system without it being necessary to implement them. Moreover, regulations, unlike directives, are binding in their entirety. This means that regulations are intended to subject a situation to rules which are all-embracing and, if necessary, precise. A regulation has a general application which means that, according to case law of the Court of Justice, it is *"applicable to objectively determined situations and involves legal consequences for categories of persons viewed in a general and abstract manner"*.[1323, 1324]

1114. In order to avoid implementation and interpretation issues with the implementation of directives, measures in the energy field should preferably be

1321 Please note that a directive can only have vertical, but no horizontal direct effect.

1322 K. LENAERTS, and P. VAN NUFFEL, in BRAY, R. (ed.), *Constitutional Law of the European Union* (London, Thomson Sweet & Maxwell, 2005), p. 766–769.

1323 Case C-6/68, *Zuckerfabrik Watenstedt GmbH v Council of the European Communities* [1968] ECR 409, p. 415.

1324 K. LENAERTS, and P. VAN NUFFEL, in BRAY, R. (ed.), *Constitutional Law of the European Union* (London, Thomson Sweet & Maxwell, 2005), p. 763 and 764.

adopted on a European level in a manner as detailed as possible. Moreover, these measures should be directly applicable in all the Member States without the adoption on a national level of any implementing measures. Therefore, I would recommend that the European legislator mainly, or preferably only, adopts regulations instead of directives, or any other legal instrument, for future measures in the European energy framework.

1115. In practice, this approach appears to have been applied recently with the adoption of the Gas Security Regulation No 994/2010. Instead of amending the existing Directive 2004/67/EC concerning measures to safeguard security of natural gas supply, the European legislator has opted for the adoption of a regulation. This Gas Security Regulation No 994/2010 established different measures, such as a clear definition and attribution of responsibilities, coordination of the response to concrete disruptions, etc., aiming to safeguard the security of supply of gas.

1116. The advantage of using a regulation instead of a directive in this context is that in a very short period a harmonised set of security of supply standards were directly applicable in all the Member States. A regulation is operational well before a directive. Another argument in favour of a regulation is clarity, since it establishes common standards for security of supply which are directly applicable to the market participants. Moreover, since a wide range of participants across Europe will be involved in the emergency plans under the Gas Security Regulation No 994/2010, it is essential to have a clear definition of the roles and procedures, and a clear obligation to cooperate. The Commission has therefore argued that a regulation offers advantages over a directive in terms of timing, and clear and direct application.[1325]

1117. Another example of this approach, which is currently under discussion, is the proposal for a Regulation on energy market integrity and transparency.[1326] The purpose of this Regulation on energy market integrity and transparency is to establish rules prohibiting abusive practices on wholesale energy markets coherent with those applying in financial markets. It also provides for the monitoring of the wholesale energy markets to be done by the Agency for the Cooperation of Energy Regulators. According to the proposal, the most appropriate legal instrument to establish consistent rules applying to cross-

---

1325 Commission Staff Working Document of 16 July 2009, Accompanying document to the Proposal for a Regulation of the European Parliament and of the Council concerning measures to safeguard security of gas supply and repealing Directive 2004/67/EC, SEC(2009)979, p. 31 and 32.

1326 Proposal of 8 December 2010 for a Regulation of the European Parliament and of the Council on energy market integrity and transparency, COM(2010)726, 32 p.

border energy markets with a European level monitoring function and coordinated enforcement across the Member States is a Regulation.[1327]

1118. This view is in the author's opinion the correct approach which should be encouraged when the European legislator in the future proposes measures in the energy field. Moreover, it avoids gold-plating[1328], since it removes the scope for Member States to elaborate on the EU rules. The use of directly applicable regulations enables immediate application in all the Member States and guarantees that all the actors are subject to the same provisions of EU law at the same moment. However, in practice it must still be monitored whether the adopted regulations are complied with.

1119. The enforcement of regulations and any other form of secondary legislation is another issue which requires more attention on a European level. The European Commission is by virtue of Article 17(1) TEU considered to be the guardian of the Treaties. Under this competence it checks whether other institutions, agencies, Member States, natural and legal persons comply with the provisions of the Treaties. Under Article 337 of the TFEU the European Commission, to perform this task, has the right to collect any information and carry out any checks required within the limits and on conditions laid down by the Council. The European Commission may even bring actions before the Court of Justice if institutions (such as European Parliament, Council, etc.) have infringed Union law or Member States fail to comply with their obligations under EU law.[1329]

Despite these provisions and even the possibility to impose lump sums, penalty payments and fines on natural and legal persons for infringements of competition law, there is still a lack of enforcement of secondary legislation. Beside arguing in favour of an increased use of lump sums, penalty payments and fines to make sure that Member States and market participants comply with the EU law provisions, another option could be an enhanced role for ACER with respect to the enforcement of secondary legislation. However, as discussed above, in order to allow the European legislator to assign such a task to ACER the problem of the so-called *Meroni* doctrine needs to be solved first.

1327 *Ibid.*, p. 3.

1328 In some Member States the transposing of directives into national law often leads to refinements and adds e.g. labelling obligations, deadlines, technical requirements, authorisation procedures and other administrative requirements.

1329 K. LENAERTS and P. VAN NUFFEL in R. BRAY and N. CAMBIEN (eds.), *European Union Law* (London, Sweet & Maxwell, 2011), p. 506–509.

3. CONCLUSION

1120. This part suggested a number of amendments to the Treaties which should provide the Union with the necessary tools to establish a stronger and more coherent energy policy on a European level. In order to achieve this goal, in the author's view, not only the present energy competences and objectives in Article 194 of the TFEU should be amended, but also some of the Treaty provisions which empower the Union to adopt certain measures in the energy field such as Articles 122, 191, 192, 170–172 of the TFEU and Article 13(1) of the TEU.

D. A NEW ENERGY TREATY ON A EUROPEAN LEVEL

1121. Besides the Treaty amendments discussed above, Member States could also opt for the adoption of a new energy-specific Treaty. This Treaty could create a new legal and institutional framework with a stronger and more coherent European energy regulatory space.

1122. A new energy Treaty on a European level is a very attractive option in particular for Member States which have already cooperated voluntarily under the regional initiative. The current Treaties do not provide a framework for the drafting and adoption of such a treaty between only some of the Member States.[1330] However, the possibility for Member States to adopt a treaty is allowed, otherwise it would be a drastic curtailment of the Member States' treaty-making powers.[1331]

Currently, it is uncertain whether some more ambitious Member States or all Member States would be willing to pool their competences and policies in the energy field under a supranational structure. Moreover, as discussed with the reinforced cooperation or in-built closer cooperation option outside the institutional framework of the European Union, there are some legal constraints.

1123. One of those constraints could be the non-discrimination principle which outlaws, within the scope of the application of the Treaties, unjustified

---

[1330] B. DE WITTE, "Chameleonic Member States: Differentiation by means of partial and parallel international agreements" in B. DE WITTE, D. HANF and E. VOS (eds.), *The Many faces of Differentiation in EU Law* (Intersentia, Antwerpen, 2001), p. 235.

[1331] B. DE WITTE, "Old-fashioned Flexibility: International Agreements between Member States of the Union" in G. DE BÚRCA and J. SCOTT (eds.), *Constitutional Change in the EU From Uniformity to Flexibility?* (Oxford, Hart Publishing, 2000), p. 39 and 40.

discrimination on the basis of nationality.[1332] It has been questioned in legal doctrine whether this principle implies that all reciprocal advantages conferred on nationals of two or more cooperating Member States should be extended to all citizens of the Union? Does this principle require that bilateral treaties should become multilateral?

1124. In a number of cases the Court of Justice had to consider these questions such as the *Matteucci* case[1333], *Bachmann* case[1334] and *D.* case.[1335] In the latter case, where Mr D., a German national, argued that the Dutch law on wealth tax was discriminatory, the Grand Chamber of the Court of Justice also had to address the bilateral Convention between Belgium and the Netherlands for the avoidance of double taxation of income and property and for the regulation of certain other taxation matters. This convention under Article 25(3) stated that Belgian residents would have been entitled to allowances and other tax benefits which the Netherlands granted to its own residents. Given this provision it was questioned whether Union law requires that the Dutch authorities treat a German national like a Belgian national?

The Court of Justice stated that: *"[s]Similar treatment with regard to wealth tax in the Netherlands of a taxable person, such as Mr D., resident in Germany and a taxable person resident in Belgium presupposes that those two taxable persons are regarded as being in the same situation.*

It is to be remembered that, in order to avoid the same income and assets being taxed in both the Netherlands and Belgium, Article 24 of the Belgium-Netherlands Convention allocates powers of taxation between those two Member States and Article 25(3) lays down a rule under which natural persons resident in one of those two States are entitled in the other to the personal allowances which are granted by it to its own residents.

The fact that those reciprocal rights and obligations apply only to persons resident in one of the two Contracting Member States is an inherent consequence of bilateral double taxation conventions. It follows that a taxable person resident in Belgium is not in the same situation as a taxable person resident outside Belgium so far as concerns wealth tax on real property situated in the Netherlands.

A rule such as that laid down in Article 25(3) of the Belgium-Netherlands Convention cannot be regarded as a benefit separable from the remainder of the

---

1332 Article 18 of the TFEU and see J. WOUTERS, "*Constitutional Limits of Differentiation: The Principle of Equality*" in B. DE WITTE, D. HANF and E. VOS (eds.), *The Many faces of Differentiation in EU Law* (Intersentia, Antwerpen, 2001), p. 301 – 345.

1333 Case 235/87, *Annunziata Matteucci v Communauté française of Belgium and Commissariat général aux relations internationales of the Communauté française of Belgium* [1988] ECR 5589.

1334 Case C-204/90, *Hanns-Martin Bachmann v Belgian State* [1992] ECR I-249.

1335 Case C-376/03, *D. v Inspecteur van de Belastingdienst/Particulieren/Ondernemingen buitenland te Heerlen* [2005] ECR I-5821.

Convention, but is an integral part thereof and contributes to its overall balance".[1336]

The Court of Justice took the view that there is no violation of the non-discrimination principle if the mutual benefits of a bilateral agreement are not extended to Union citizens which are not resident in either state. It demonstrates that we cannot assume that on the basis of the principle of non-discrimination all advantages under bilateral agreements must automatically be extended to all other Union nationals.[1337] There could be good reasons why those nationals with respect to the benefits are not in the same situation.

It can be concluded that differential treatment on the basis of bilateral agreements or a new energy Treaty must be justified by the Member States even when it follows from a bilateral agreement. Moreover, the adoption of a bilateral agreement or a new energy Treaty does not in itself remove the application of Union law.[1338]

1125. Another legal barrier with respect to the adoption of a new energy treaty is that Member States are not longer free to conclude international Treaties in areas for which the Union has exclusive competences or where the Union has exercised its shared powers. Moreover, according to case law of the Court of Justice "*Member States may not enter into international commitments outside the framework of the Community institutions, even if there is no contradiction between those commitments and the common rules*"[1339] when the given area has been occupied or largely covered by common provisions.[1340]

Other limits to the treaty competence of Member States are the substantive compatibility between such treaties and the rules of the EU law. The conclusion and application of these treaties must respect the principle of the primacy of EU law, but also other principles such as sincere cooperation and pre-emption.[1341] It can be concluded that intergovernmental cooperation between a limited number of Member States is allowed. However, due to their membership of the

---

1336 *Ibid.*, paragraphs 59–62.

1337 R. SCHÜTZE, "EC Law and International Agreements of the Member States – An Ambivalent Relationship?" *Cambridge Yearbook of European Legal Studies* (Oxford, Hart Publishing, 2006–2007), p. 421–422.

1338 M. CREMONA, "Defending the Community Interest: the Duties of Cooperation and Compliance" in M. CREMONA and B. DE WITTE (eds.), *EU Foreign Relations Law Constitutional Fundamentals* (Oxford, Hart Publishing, 2008), p. 135–138.

1339 Case C-467/98, *Commission of the European Communities v Kingdom of Denmark* [2002] ECR I-9519, paragraph 108.

1340 M. CREMONA, "Defending the Community Interest: the Duties of Cooperation and Compliance" in M. CREMONA and B. DE WITTE (eds.), *EU Foreign Relations Law Constitutional Fundamentals* (Oxford, Hart Publishing, 2008), p. 129.

1341 B. DE WITTE, "Chameleonic Member States: Differentiation by means of partial and parallel international agreements" in B. DE WITTE, D. HANF and E. VOS (eds.), *The Many faces of Differentiation in EU Law* (Intersentia, Antwerpen, 2001), p. 234–246.

EU there are certain legal constraints on the scope and content of such cooperation.[1342]

1126. To solve some of these constraints, it could be envisaged to adopt a new Energy Treaty between the EU and some of those more ambitious Member States or to put the new Energy Treaty under the existing Union structure as the EURATOM Treaty. Both options could resolve some of the legal constraints with the existing institutional framework of the European Union and allow the institutions of the European Union to develop and apply the new energy policy goals for the participating Member States. Moreover, they allow the institutions of the European Union to ensure consistency between the existing energy policy developed under the new energy title of the TFEU, and other Union policies, and the more ambitious energy policy developed under the new energy Treaty. Hence, a new energy Treaty on a European level would ideally be placed in the present Union structure and rely on the institutional framework of the Union or the EU should become one of the parties of such a new energy Treaty.

1127. However, a number of legal issues will remain if the Union would adopt a new energy Treaty with some of the more ambitious Member States or to put such a Treaty under the existing Union structure. There are no established rules on how to negotiate such a treaty. How will the division of powers be clarified between the Union and those Member States and, how will such a Treaty be concluded and ratified?

Once the treaty has been concluded, other questions will rise such as the presentation of the Union and those Member States, how will the implementation be monitored and who will be competent to interpret such a Treaty? What is the legal effect of such a treaty in the Union's and the Member States' legal orders? Who is bound by such a Treaty? It could be argued that the answer might be obvious: only those parties (Union and those Member States) which have acceded to the agreement are bound by the Treaty and the others not. Such an answer is not obvious from the perspective of Union law. The participation of the Union could be made subject to the participation of the majority or all of its Member States. However, the suggestion of a separate energy treaty was to offer an alternative framework for inter-state cooperation in the energy field, particularly where less than the majority or all of the Member States are possibly willing to cooperate. Moreover, this suggestion would limit the Union's autonomy of action.[1343] Other legal constraints with respect to the adoption of a new energy Treaty on a European level are for example whether such a Treaty

[1342] *Ibid.*

[1343] G. DE BAERE, *Constitutional Principles of EU External Relations* (Oxford, Oxford University Press, 2008), p. 232–246.

affects the internal market and constitutional foundations of the Union, the demarcation vis-à-vis other Union Treaties, etc.[1344]

On the basis of the discussion above and the various legal issues which the adoption of a new energy Treaty would comprise, it is important that such a treaty clarifies the commitments entered into by the Union and the participating Member States, and clearly indicates the limits of the Union's participation, as well as the responsibilities between the Union and the participating Member States.[1345]

1128. Assuming that all legal constraints with respect to the adoption of a new energy Treaty would be solved, the scope of such a treaty must at least include similar powers, decision-making procedures, objectives in the energy field, etc. as Article 194 of the TFEU and the Treaty amendments with respect to energy, discussed above.[1346] With respect to the institutional framework of the new energy Treaty some inspiration can be drawn from the institutional framework of the Energy Community Treaty which is to a large extent similar to that on a European level.

1129. The most important institution of the Energy Community Treaty is the Ministerial Council which provides the general policy guidelines, takes measures to meet the objectives of the Treaty and adopts procedural acts. The Ministers of Energy of the Contracting Parties and two representatives of the Union form the Ministerial Council.

1130. Another institution is the Permanent High Level Group which prepares the work of the Ministerial Council and consists of one representative of each party of the Treaty and comprises of representatives of the Ministers of Energy of the Contracting Parties and two representatives of the European Commission. The third institution is the Regulatory Board which advises the other institutions and issues recommendations in the event of cross-border disputes. This Board is composed of representatives of the Energy Regulatory Authorities of each Contracting Party and the Union. The Energy Community is also advised by two fora composed of representatives of all interested parties. Finally, the permanent Secretariat which has its seat in Vienna, provides, amongst other things, administrative support to all the other institutions of the Energy Community.[1347]

1344 Policy Proposal by Jacques Delors, *Towards a European Energy Community: A Policy proposal*, study by S. ANDOURA, L. HANCHER and M. VAN DER WOUDE, p. 102–105.

1345 P. EECKHOUT, *External Relations of the European Union Legal and Constitutional Foundations* (Oxford, Oxford University Press, 2004), p. 223.

1346 See *Supra*.

1347 For more details See M. HUNT and R. KAROVA, "The Energy Acquis Under the Energy Community Treaty and the Integration of South East European Electricity Markets: An

1131. However, there are not only legal constraints, but also political constraints which indicate that the adoption of a new energy Treaty on a European level will not be an easy option to follow. It will take some time before such a Treaty will be conceived, negotiated, concluded and ratified between some ambitious Member States or all the Member States.[1348]

E. CONCLUSION

1132. Given the current institutional and legal framework on a European level the previous chapters concluded that the Union will not be able to achieve its key objectives: competitiveness, security of supply and sustainability of the European energy policy. Therefore, urgent, decisive and immediate action is required on a European level. Chapter 6 first discussed some suggestions formulated in legal doctrine and current practices, such as enhanced cooperation, reinforced cooperation or in-built closer cooperation, the so-called Schengen model, and functional or regional cooperation, which should allow overcoming some of the deficiencies in the current institutional and legal framework.

1133. With respect to enhanced cooperation this chapter argued that due to the procedural constraints, it remains uncertain whether this mechanism will be used by the Member States in the energy field. However, one of the strengths of this cooperation is the fact that it is a Treaty-based mechanism which has the advantage that it is incorporated in the European institutional and legal framework and thus, the Union's institutions can be used. Consequently, this chapter states that enhanced cooperation could be an option for specific energy related issues (for example the adoption of a CO_2 tax), but it will be insufficient to establish a stronger and more coherent European energy policy which requires different amendments to the current legal framework.

This chapter stated that the adoption of a reinforced cooperation agreement outside the institutional framework of the European Union, or a Schengen-like agreement, could be an option for certain aspects of the energy area (for example on the disposal of high-level radio-active waste and spent nuclear fuel, as well as the setting-up of regional centres of disposal, and the standardisation of reactor design). However, since the Union has been very active over the past years in the energy field, the scope of such cooperation will be rather limited. Some of the legal constraints suffered by this option could be solved if the Union would

Uneasy Relationship?" in B. DELVAUX, M. HUNT and K. TALUS (eds.), *EU Energy Law and Policy Issues* (Rixensart, Euroconfidentiel, 2010), p. 51–86.

1348 Policy Proposal by Jacques Delors, *Towards a European Energy Community: A Policy proposal*, study by S. ANDOURA, L. HANCHER and M. VAN DER WOUDE, p. 107–119.

become one of the contracting parties. Nonetheless, this option is in the author's view insufficient.

1134. The latter option discussed is a voluntary form of regional cooperation between Member States which could be useful to achieve a number of objectives in the energy field (for example the adoption of common standards, the coordination of investments). Due to the voluntary commitment and the limited scope of the existing regional cooperation, chapter 6 concluded this option will be insufficient for the establishment of a specific, common and effective European energy policy and law. Nonetheless, it could be a predecessor to a more formal cooperation (for example an "Energy Treaty") between Member States.

1135. Given the limited scope of the options formulated in the legal doctrine and current practices, the next part of chapter 6 suggested two options of effective institutional and legal amendments to the current European energy framework. The first option is a number of Treaty amendments which should lead to a stronger and more coherent energy title in the TFEU. In this option it is argued that, *inter alia*, that the reference to the energy area should be removed from Article 122 of the TFEU, to rewrite Article 194 of the TFEU, to remove from Article 192 of the TFEU all the objectives or legal bases allowing the adoption of measures in the energy field and state that the energy area is excluded from its scope, etc. Moreover, this chapter suggested a number of amendments to Article 194 of the TFEU and supplements with different objectives.

1136. The second option formulated in chapter 6 is the adoption of a new energy-specific Treaty. Such a Treaty could create a new legal and institutional framework with a stronger and more coherent European energy regulatory space. However, there are some legal constraints. The first possible constraint discussed in chapter 6 is the non-discrimination principle which outlaws, within the scope of the application of the Treaties, unjustified discrimination on the basis of nationality. This chapter argued that differential treatment on the basis of bilateral agreements or a new energy Treaty must be justified by the Member States even when it follows from a bilateral agreement.

1137. Another legal barrier with respect to the adoption of a new energy treaty is that Member States are not longer free to conclude international Treaties in areas for which the Union has exclusive competences or where the Union has exercised its shared powers. The analysis in chapter 6 concluded that intergovernmental cooperation between a limited number of Member States is allowed. However, due to their membership of the EU there are certain legal constraints on the scope and content of such a cooperation. In order to solve these legal constraints chapter 6 argued that it could be envisaged to adopt a new

Energy Treaty between the EU and some of those more ambitious Member States or to put such a treaty under the existing Union structure as the EURATOM Treaty. However, these options create other legal constraints and in particular the question who is bound by such a new Energy Treaty.

1138. Chapter 6 states that the scope of this new "Energy Treaty" must include similar powers, decision-making procedures, objectives in the energy field, etc. as Article 194 of the TFEU and the Treaty amendments with respect to energy, discussed above. Moreover, with respect to the institutional framework of such a new "Energy Treaty" some inspiration can be drawn from the institutional framework of the Energy Community Treaty. Chapter 6 also emphasises that there are not only legal constraints, but also political constraints which indicate that the adoption of a new energy Treaty on a European level will not be an easy option to follow.

1139. Due to the various legal constraints with respect to the adoption of a new Energy Treaty, preference should be given to the number of Treaty amendments proposed in chapter 6 which should lead to a stronger and more coherent energy title in the TFEU.

CHAPTER 7
GENERAL CONCLUSION

1140. The overall objective of this book was to analyse the current regulatory framework for energy law and policy on a European level and to determine whether this framework is sufficient to address the triple objectives (competitive, secure and sustainable energy supply) of the EU's energy policy.

1141. The well-being of European citizens, industry and economy depends on affordable, secure and sustainable energy. However, Europe is in a new energy era which is characterised by rising energy prices at international level, the challenges of climate change and security of supply. To tackle these challenges of the new energy era, the European energy market requires an over-arching EU legal regulatory framework. Such a regulatory framework for energy law and policy will not only allow the EU and its Member States to achieve the triple objectives, but also generate the right investment signals for potential investors.

1142. This book argued that due to the fact that Member States have various challenges in common, such as their increasing reliance on fossil fuels and import dependence on oil and natural gas, climate change and other environmental constraints, rising demand for energy and higher energy prices, that an approach based on 27 individual energy policies competing with each other will likely be insufficient. Thus, instead of regulatory competition, a more centralised EU level approach should be followed to deal with these energy challenges of the Member States.

1143. Chapter 2 demonstrated that over the past years the Union has been very active in adopting a large number of policy documents in the energy field. It highlighted that although there are three central objectives in the EU energy policy, the internal energy market has been the key objective for many years, which has been overemphasised and might come at the expense of the other objectives.

Although the European Commission prioritises the creation of the internal energy market, chapter 2 argued that a truly single EU internal market for electricity and gas is far from complete, since too many barriers to competition and too many differences between Member States still exist. With respect to the external security of supply, chapter 2 stated that there is still no common approach towards partners, suppliers or transit countries. The lack of such a common approach will continue to undermine the development of a comprehensive EU energy policy.

1144. This chapter argued that the current measures on a European level which should promote a diversification of the economy away from oil are not moving fast enough. Moreover, without investments in smart, effective and competitive energy infrastructure and new generation capacity, the internal energy supply could be undermined. Chapter 2 concluded that the common EU energy policy needed for achieving the three central and ambitious goals remains insufficient in its results. Due to the fragmentation in the EU energy policy, the current European energy policy framework, although it is very ambitious, will not achieve its three central goals.

1145. With respect to the European energy legal framework and in particular the secondary legislation, chapter 3 demonstrated that the Union has a relatively well-developed set of measures in order to achieve the three central goals of its energy policy. The EU is currently a champion in setting ambitious targets for 2020 and beyond 2030. However, the requisite policy and legal instruments are seldom aligned to deliver these targets.

1146. Chapter 3 argued that the success of a sustainable, competitive and secure European legal framework in the energy field depends on the willingness of the different actors involved in the drafting and adoption of secondary legislation. Often the adoption of secondary legislation in the energy field is a political compromise which will certainly not increase its quality. Moreover, due to a number of shortcomings, such as a lack of implementation and enforcement by Member States, traditional reliance on framework directives, inefficient decision-making procedures for the adoption of energy legislation, etc., chapter 3 concluded that it is uncertain, whether the Union with its current legal framework of secondary legislation, is capable of achieving its three central goals by 2020. Since action in the energy field on a European and national level must comply with the provisions of the Treaties, chapter 4 analysed the primary law provisions which allowed Union action in the energy field before the adoption of the Lisbon Treaty.

1147. The first part of chapter 4 provided an historical overview of the Treaty context with respect to energy. It was concluded that, before the adoption of the Lisbon Treaty, the amendments in the Treaties with respect to the energy field were rather limited. The next part of chapter 4 analysed to which extent, on the basis of the principles of attribution, subsidiarity and proportionality, before the adoption of the Lisbon Treaty, the Union was competent to adopt measures in the energy field. With respect to the principle of attribution, it was argued that, although the Treaties did not confer any specific powers to the European legislator to adopt measures in the energy field, recourse to more general legal bases in the TFEU appeared to be sufficient in order to comply with the principle of attribution.

The second principle analysed in this chapter 4 was the subsidiarity principle which is designed to determine whether Union intervention is appropriate or not. All the legislative measures adopted on a European level must pass the so-called subsidiarity test. With respect to the energy field, arguments could be found which allow energy measures to pass the subsidiarity test.

The third principle analysed in chapter 4 was the principle of proportionality. This principle ensures a proportionate relationship between the means and the ends of legislative measures. It has been highlighted that the European legislator often applies a standard reference. Nonetheless, chapter 4 emphasised that it should be assessed whether each measure complies with the three-tier proportionality test.

1148. On the basis of the analysis of the principles of attribution, subsidiarity and proportionality chapter 4 concluded that these principles do not seem to constitute a barrier for the adoption of measures in the energy field. Their main objective is to ensure that action in the energy sector on a European level falls within the competences of the Union. Any shortcomings in the current European energy legal framework should not be attributed to the existence and application of these principles.

1149. The next part of chapter 4 discussed the scope and application of the Articles 106(1), (2) and 345 of the TFEU in order to determine to which extent they could hamper the application of some of the TFEU provisions and secondary legislation to the electricity and gas markets. With respect to Article 106(1) of the TFEU chapter 4 argued that it is not possible to categorise neatly which types of measures infringe this provision. Even if this would be possible the categories would not be closed. Nonetheless, on the basis of the case law discussed, chapter 4 identified some general themes.

1150. With respect to Article 106(2) of the TFEU, case law of the Court of Justice in the energy field seemed to send to the Commission the signal that the Court was unwilling to deny competence to Member States in these sensitive areas as long as the Commission itself had not acted by adopting appropriate legislative alternatives. Thus, in order to ensure that Member States would opt for a less restrictive approach to guarantee its chosen public service goals, the Commission is obliged to propose a legislative solution and not a judicial one.

The Commission realised such legislative alternatives with the adoption of the electricity and gas directives. Nonetheless, chapter 4 concluded that even with the adoption of the electricity and gas directives, Member States appear to maintain a substantial discretion with respect to the adoption of public service obligations in the energy field and they do not decide which services could be classified as services of general economic interest.

1151. The next part of chapter 4 discussed the scope of Article 345 of the TFEU and to which extent this provision could hinder the adoption of measures in the energy field. Contrary to some legal doctrine, this part stated that Article 345 of the TFEU does not prevent the Commission from adopting the ownership unbundling model in the energy sector. Moreover, this model does not exclude the freedom for Member States to choose the property ownership regime under which they wish to place the transmission activities.

1152. Chapter 4 continued with an analysis of the free movement of goods provisions. These fundamental provisions of the TFEU have set boundaries which also apply to the electricity and gas markets and must be taken into account by both the national and European legislator when they adopt measures in these markets. In this part a distinction was made between the treatment of fiscal and non-fiscal barriers to trade under Union law. With respect to the fiscal barriers, the analysis of case law of the Court of Justice demonstrated that the use, generation, transport, consumption, etc. of energy products in a number of Member States is subject to a tax, customs duties or charges having equivalent effect, which are not necessarily prohibited by the Articles 28, 30 and 110 of the TFEU. Thus, in order to determine whether such a charge on energy products is compatible with these provisions of the TFEU, a case-by-case analysis will be necessary in the light of the existing case law of the Court of Justice.

With respect to the non-fiscal barriers, chapter 4 discussed both the *Dassonville* and *Keck formula*. This part argued that, under Article 36 of the TFEU and the so-called mandatory requirements, some national measures which restricted the free movement of goods can be justified by the Court of Justice. It concluded with respect to the energy field that it is uncertain whether the Court in subsequent cases would continue to apply the approach taken by the Court in the *Campus Oil* case and allow Member States on the basis of the concept of security of supply to derogate from the basic free movement provisions of the Treaty.

The analysis of the Golden Share cases in this part of chapter 4 proved that the Court of Justice interpreted the public security and security of supply grounds strictly. It also demonstrated that although the protection of the environment is not listed in Article 36 of the TFEU, the Court of Justice recognised in its case law that certain obstacles to the free movement provisions of the TFEU can be justified on the basis of the protection of the environment. Chapter 4 concluded that both security of supply and environmental protection arguments, on certain conditions, could justify, in energy, national measures which restrict the free movement of goods.

1153. Before the adoption of the Lisbon Treaty, the Union adopted in the energy sphere secondary legislation on the basis of explicit or more general provisions of the Treaty such as Articles 113, 114, 122, 192 and 352 of the TFEU. These provisions allowed the Union to establish a legal framework in the energy field on a European level. Although, the TFEU did not contain an enabling provision putting forward the objectives, goals and procedures for Union action in the energy field, the European legislator adopted the past years an enormous amount of secondary legislation in the energy field. Chapter 4 analysed the meaning and scope of these general provisions of the TFEU.

1154. Moreover, chapter 4 discussed the Renewable Directive 2009/28/EC and Directive 2008/101/EC. With respect to the Renewable Directive 2009/28/EC chapter 4 argued that it was not necessary to add Article 114 of the TFEU as a legal basis for the adoption of this directive. Chapter 4 also argued that Article 192(1) of the TFEU is not the correct legal basis for the adoption of the Renewable Energy Directive 2009/28/EC. The fact that the Renewable Energy Directive 2009/28/EC imposes on Member States legally binding targets for the overall share of energy from renewable sources in energy consumption and in transport is a violation of Article 192(2) of the TFEU. Moreover, the Renewable Energy Directive 2009/28/EC imposed various obligations with respect to biofuels and other bioliquids which affect a Member State's land use. Chapter 4 therefore concluded that the correct legal basis for the Renewable Energy Directive 2009/28/EC is Article 192(2) of the TFEU instead of Article 192(1) of the TFEU. With respect to the provisions of the Directive 2008/101/EC, chapter 4 concluded that the provisions of this directive are beyond the scope of the jurisdiction of the Member States.

1155. The final section of chapter 4 demonstrated that not only Treaty provisions can influence the adoption of secondary legislation, but so can the Court of Justice's case law. This section concluded that, mainly due to the so-called *Meroni doctrine*, the Agency has been given an advisory role in order to handle cross-border issues on a European level in both the electricity and gas markets. It concluded that if this doctrine remains one of the obstacles to attribute discretionary powers, there is a strong need for a thorough evaluation of this doctrine.

1156. With the adoption of the Treaty of Lisbon, the Union legislature's activities in the energy field finally have a legal basis for the adoption of a European energy policy. This new energy title of the TFEU, Article 194, has been analysed in detail in chapter 5. Since Lisbon, both the European Union and its

Member States are competent to regulate in the energy area. If the Union legislator has not yet exercised its shared competence in the energy field, Member States remain free to adopt legislative measures in the energy field, provided that they comply with the primary Treaty provisions.

Chapter 5 highlighted that whenever lines are drawn, as with the adoption of the Lisbon Treaty, there are inevitably borderline disputes. Also in the energy field there might be some borderline issues as to whether the actions/provisions in the energy field on a European level fall within one of the Member States' exclusive competences or fall within the shared competences.

1157. The next section of chapter 5 argued that the Lisbon Treaty continues to proceduralise the principle of subsidiarity and concluded that this Treaty and its Protocol did not turn the principle of subsidiarity into a hard and efficient political safeguard of federalism. Chapter 5 continued with a discussion of the changes to the Articles 122, 191, 352(1) of the TFEU and the extension of the scope of the common commercial policy with the area of foreign direct investment since the adoption of the Lisbon Treaty. This discussion was followed by a detailed analysis of Article 194 of the TFEU.

1158. The three guiding principles of paragraph 1 of Article 194 of the TFEU are (a) the establishment and functioning of the internal market, (b) the need to preserve and improve the environment and (c) a spirit of solidarity between Member States. These guiding principles should be taken into account and within their framework the European legislator must unfold its energy policy. However, chapter 5 argued that their sense, contents and purpose remain uncertain. Moreover, it is questionable whether these guiding principles subordinate or could even restrict the adoption of the energy policy on the basis of Article 194 of the TFEU.

Chapter 5 argued that the meaning of some objectives of the new energy title is not always crystal clear. Moreover, whether the new legal basis of Article 194 of the TFEU with respect to the promotion of the interconnection of networks will be able to overcome the current difficulties and shortcomings with the TEN-E framework remains questionable. Chapter 5 concluded that the main significance of paragraph 1 of Article 194 TFEU is that it determines the range of Union competences in the energy field. It obtained this significance via paragraph 2 of Article 194 TFEU since this provision enables the Union to adopt the measures necessary to achieve the objectives of paragraph 1.

1159. The next section of chapter 5 discussed paragraph 2 of Article 194. In particular with respect to the reference to Article 192(2)(c) of the TFEU, chapter 6 concluded that this reference is a matter of procedural restrictions. Thus, the right of veto for Member States provided for in accordance with Article 192(2)(c) of the TFEU could not be avoided if the Union suggests measures which *"significantly affect a Member State's choice between different energy sources and the general structure of its energy supply"*. Chapter 5 argued that this provision clearly demonstrated that the new Article 194 of the TFEU is not well-written.

This section in chapter 5 was followed by an analysis of paragraph 3 of Article 194 of the TFEU, Declaration 35 on Article 194 of the TFEU and the legal basis of the Gas Regulation No 994/2010.

1160. The next sector of Chapter 5 provided some guidance to which extent Article 194 of the TFEU or other provisions of the TFEU could be used as a legal basis for the adoption of secondary legislation in the energy field. It concluded that, merely because a proposed measure is concerned with the energy sector, this does not automatically mean that the given measure must be based on Article 194 of the TFEU. Consequently, chapter 5 argued that in order to determine the legal basis of each measure in the energy field, it must be asserted on a case-by-case basis which Treaty Article (energy, environmental, fiscal, etc. title) matches the main aim, objective and content of a given energy measure on a European level best.

1161. Chapter 5 concluded that the new energy title is a carefully crafted compromise between the shared energy competence (between the Union and its Member States) and the national sovereignty over a number of energy situations or concerns referred to in the second paragraph of Article 194 of the TFEU. With respect to Article 194 of the TFEU it concluded that this provision can be seen as a mere codification of the existing practice in the energy field. Thus, the same pre-existing gaps, flaws and limitations remain in place. The new energy title will not offer prospect of radical reform from the former situation.

1162. In order to overcome the current situation and to allow the Union to achieve its three central goals in the energy field, chapter 6 suggested two main legal and institutional options. Before it discussed these 2 options it concluded that some suggestions formulated in legal doctrine and current practices, inter alia, enhanced cooperation, reinforced cooperation or inbuilt closer cooperation, the so-called Schengen model, and functional or regional cooperation are insufficient in achieving a stronger and more coherent European energy policy.

1163. Chapter 6 highlighted that it could be argued that it would be better to add energy as an exclusive competence of the Union in Article 3 of the TFEU, instead of a shared competence between the Union and its Member States. Irrespective of whether this is possible in practice, chapter 6 stated that this is not a good option since a number of aspects in the energy field should preferably be tackled on a Member State level or should remain within their sovereignty.

1164. The first option discussed in chapter 6 is a number of Treaty amendments to the TFEU ensuring a stronger and more coherent energy title in the TFEU. The main constraint in this option is a political one. It is currently uncertain whether Member States would be willing to adopt the suggested changes to the TFEU.

1165. The second option discussed in chapter 6 is the adoption of a new energy-specific Treaty allowing the more ambitious Member States to work more closely together in achieving the three central goals in the energy field. However, with this second option there are not only political constraints, but also some legal constraints. The first possible constraint analysed in chapter 6 is the non-discrimination principle which outlaws, within the scope of the application of the Treaties, unjustified discrimination on the basis of nationality. Chapter 6 argued that differential treatment on the basis of bilateral agreements or a new energy Treaty must be justified by the Member States even when it is the result of a bilateral agreement.

1166. Another legal barrier with respect to the adoption of a new energy treaty is that Member States are not longer free to conclude international Treaties in areas in which the Union has exclusive competences or where the Union has exercised its shared powers. In order to overcome this constraint, chapter 6 argued that it could be envisaged to adopt a new Energy Treaty between the EU and some of those more ambitious Member States or to put such a treaty under the existing Union structure as the EURATOM Treaty. However, these options create other legal constraints.

1167. Assuming that all legal constraints with respect to the adoption of a new energy Treaty would be solved, chapter 6 argued that the scope this new energy Treaty must include similar powers, decision-making procedures, objectives in the energy field, etc. as Article 194 of the TFEU and the Treaty amendments with respect to energy, discussed in chapter 6. Moreover, with respect to the institutional framework of such a new energy Treaty some inspiration can be drawn from the institutional framework of the Energy Community Treaty.

1168. Due to the various legal constraints with respect to the adoption of a new Energy Treaty, chapter 6 concluded that preference should be given to the number of Treaty amendments proposed in chapter 6 instead of the adoption a new energy Treaty.

BIBLIOGRAPHY

AHNER, N., and RUESTER, S., "Challenges for European natural gas supply security regulation", *Network Industries Quarterly* (2010).

AHNER, N., GLACHANT, J.,-M., and DE HAUTECLOQUE, A., "Legal feasibility of Schengen-like Agreements in European Energy Policy: The Cases of Nuclear Cooperation and Gas Security of Supply", *Policy brief* (2010).

AHNER, N., GLACHANT, J.,-M., and DE HAUTECLOQUE, A., *Legal feasibility of Schengen-like Agreements in European Energy Policy: The Cases of Nuclear Cooperation and Gas Security of Supply*, (Florence, European University Institute, 2010).

AKKERMANS, B., and RAMAEKERS, E., "Article 345 TFEU (ex Article 295 EC), Its Meaning and Interpretations", *European Law Journal* (2010).

ALTER, K., and MEUNIER-AITSAHALIA, S., "Judicial Politics in the European Community: European Integration and the Pathbreaking Cassis de Dijon Decision", Comparative Political Studies (1994).

AMTENBRINK, F., GEELHOED, A., and KINGSTON, S., "Chapter × Economic, Monetary and Social Policy" in KAPTEYN, P., J., G., MCDONNELL, A., M., MORTELMANS, K., J., M., and TIMMERMANS, C., W., A., (eds.), *The Law of the European Union and the European Communities* (The Netherlands, Kluwer Law International, 2008).

ANADUCCI, S., "Towards a Comprehensive Community Energy Policy: Some Legal Considerations", *Diritto comunitario e degli scambi internazionali* (1976).

ARNESEN, F., "Chapter XVI The EFTA Court Decision on Norwegian Hydropower and Norway's response to that decision" in ROGGENKAMP, M. M., and HAMMER, U., (eds.), European Energy Law Report VI (Antwerp, Intersentia, 2009).

ARNULL, A., *The European Union and its Court of Justice* (Oxford, Oxford University Press, 2006).

ARNULL, A., DASHWOOD, A., DOUGAN, M., ROSS, M., SPAVENTA, E., and WYATT, D., *European Union Law* (London, Sweet & Maxwell, 2006).

BARBER, N., W., "The limited Modesty of Subsidiarity", *European Law Journal* (2005).

BARENTS, R., and SLOT, P., J., in KAPTEYN, P., J., G., MCDONNELL, A., M., MORTELMANS, K., J., M., and TIMMERMANS, C., W., A., (eds.), *The Law of the European Union and the European Communities* (The Netherlands, Kluwer Law International, 2008).

BARNARD, C., "Derogations, Justifications and the Four Freedoms: Is State Interest Really Protected?" in BARNARD, C., and ODUDU, O., (eds.), *The Outer Limits of European Union Law* (Oxford, Hart Publishing, 2009).

BARNARD, C., "Filling the Remaining Pieces into the Goods and Persons Jigsaw", *European Law Review* (2001).

BARNARD, C., *The substantive Law of the EU: The Four Freedoms* (Oxford, Oxford University Press, 2007).

BARRETT, G., "The king is dead, long live the king: the recasting by the Treaty of Lisbon of the provisions of the Constitutional Treaty concerning national parliaments", *European Law Review* (2008).

BARTON, B., REDGWELL, C., RØNNE, A., and ZILLEMAN, D., N., (eds.), "Introduction" in BARTON, B., REDGWELL, C., RØNNE, A., and ZILLEMAN, D., N., (eds.), *Energy Security – Managing Risk in a Dynamic Legal and Regulatory Environment* (Oxford, Oxford University Press, 2004).

BECKER, F., "Market Regulation and the 'Right to Property' in the European Economic Constitution" *Yearbook of European Law 2007* (Oxford, Clarendon, 2007).

BELYI, A., "New dimensions of energy security of the enlarging EU and their impact on relations with Russia", *European Integration* (2004).

BERMANN, G., "Taking Subsidiarity Seriously", *Columbia Law Review* (1994).

BERNITZ, U., and NERGELIUS, J., *General Principles of European Community Law* (The Hague, Kluwer Law International, 2000).

BERTOLDI, P., BOWIE, R., HODSON, P., LORENTZEN, J., VAGGEN MALVIK, H., TOTH, A., WERRING, L., YORDI AGUIRRE, B., in WERRING, L., (ed.), *EU Energy Law, Volume III, Energy Efficiency and Renewable Energy Sources* (Leuven, Claeys & Casteels, 2006).

BJØRNEBYE, H., "The Security of Supply Directive – The Right Aims but the Wrong Means" in DELVAUX, B., HUNT, M., and TALUS, K., (eds.) *EU Energy Law and Policy Issues* (Rixensart, Euroconfidentiel, 2008).

BJØRNEBYE, H., *Investing in EU energy security – Exploring the regulatory approach to tomorrow's electricity production*, Dissertation submitted to the faculty of law, University of Oslo, 2009.

BJØRNEBYE, H., *Investing in EU Energy Security. Exploring the regulatory approach to tomorrow's electricity production* (The Netherlands, Kluwer Law International, 2010), 456 p.

BOWDEN, P., "Nuclear regulation: a brave new world?", *EELR* (1995).

BRIGHT, C., "Article 90, economic policy and the duties of Member States", *E.C.L.R.* (1993).

BRONCKERS, M., and VAN DER VLIES, R., "The European court's *PreussenElektra* judgment: tensions between EU principles and national renewable energy initiatives", *E.C.L.R.* (2001).

BRUGGEMAN, V., and DELVAUX, B., "EU Energy Policy and Legislation under Pressure since the UNFCC and the Kyoto Protocol?" in PEETERS, M., and DEKETELAERE, K., (eds.), *EU Climate Change policy – The Challenge of New Regulatory Initiatives* (London, Edward Elgar Publishing, 2006).

BRUS, M., M., T., A., RAIMOND, R., R., and DRUPSTEEN, T., G., "Balancing National and European Competence in Environmental Law", *Connecticut Journal of International Law* (1993–1994).

BUNGENBERG, M., "Going Global? The EU Common Commercial Policy After Lisbon" in HERMANN, C., and TERHECHTE, J., P., (eds.), *European Yearbook of International Economic Law* (2010).

BYWATER, M., "La Politique énergétique commune", *Revue Marché Commun* (1977).

CAMERON, P., *Competition in Energy Markets, Law and Regulation in the European Union* (Oxford, Oxford University Press, 2007).

CAMERON, P., D., "The revival of nuclear power: an analysis of the legal implications", *J. Env. L.* (2007).

CAMERON, P., D., *Competition in Energy Markets – Law and Regulation in the European Union* (Oxford, Oxford University Press, 2007).

CASS, D., Z., "The word that saves Maastricht? The principle of subsidiarity and the division of powers within the European Community", *Common Market Law Review* (1992).

CENEVSKA, I., "The European Parliament and the European Atomic Energy Community: a legitimacy crisis?", *European Law Review* (2010).

CHALMERS, D., DAVIES, G., and MONTI, G., *European Union Law* (Cambridge, Cambridge University Press, 2010).

CONSTANTINESCO, V., "Who's Afraid of Subsidiarity?", *Yearbook of European Law* (1992).

COOPER, I., "The Watchdogs of Subsidiarity: National Parliaments and the Logic of Arguing in the EU", *Journal of Common Market Studies* (2006).

CORCELLO, G., "La perspective communautaire du principe de précaution", *Revue du Marché Commun et de l'Union Européenne* (2001).

CORTHAUT, T., "Plus ça change, plus c'est la même chose? A Comparison with the Constitutional Treaty", *Maastricht Journal of European and Comparative Law* (2008).

CRAIG, P., "Competence: clarity, conferral, containment and consideration", *European Law Review* (2004).

CRAIG, P., "The Treaty of Lisbon, process, architecture and substance", *European Law Review* (2008).

CRAIG, P., and DE BÚRCA, G., *EU law text, cases and materials* (Oxford, Oxford University Press, 2003).

CRAIG, P., and DE BÚRCA, G., *EU Law Text, Cases and Materials* (Oxford, Oxford University Press, 2008).

CRAIG, P., *EU Administrative Law* (Oxford, Oxford University Press, 2006).

CREMONA, M., "Defending the Community Interest: the Duties of Cooperation and Compliance" in CREMONA, M., and DE WITTE, B., (eds.), *EU Foreign Relations Law Constitutional Fundamentals* (Oxford, Hart Publishing, 2008).

CROSS, E., D., "Pre-emption of Member State Law in the European Economic Community: A Framework for Analysis", *Common Market Law Review* (1992).

CROSS, E., D., DELVAUX, B., HANCHER, L., SLOT, P., J., VAN CALSTER, G., and VANDENBERGHE, W., "EU Energy Law (chapter 5)" in ROGGENKAMP, M., M., RØNNE, A., REGWELL, C., and DEL GUATO, I., (eds.), *Energy Law in Europe National, EU and International Regulation*, 2nd ed. (Oxford, Oxford University Press, 2007).

CROSS, E., D., HANCHER, L., and SLOT, P., J., "EC Energy Law" in ROGGENKAMP, M., M., RØNNE, A., REGWELL, C., and DEL GUATO, I., (eds.), *Energy Law in Europe, National, EU and International Law and Institutions* (Oxford, Oxford University Press, 2001).

CUSACK, T., F., "A tale of two treaties: an assessment of the Euratom Treaty in relation to the EC Treaty", *Common Market Law Review* (2003).

DAINTITH, Y., and WILLIAMS, S., F., *The legal integration of energy markets* (Berlin, W. De Gruyter, 1987).

DASHWOOD, A., "Relationship between Member States and EC/EU" in MCDONNELL, A., (ed.), *A review of forty years of Community law – Legal developments in the European Communities and the European Union* (The Hague, Kluwer Law International, 2005).

DASHWOOD, A., "States in the European Union", *European Law Review* (1998).

DASHWOOD, A., "The Limits of European Community Powers", *European Law Review* (1996).

DASHWOOD, A., "The relationship between the Member States and the European Community/European Union" in MCDONELL, A., (ed.), *A review of forty years of Community Law* (The Hague, Kluwer Law International, 2002).

DASHWOOD, A., "The relationship between the Member States and the European Union/Community", *Common Market Law Review* (2004).

DAVIES, G., "Subsidiarity: The wrong idea, in the wrong place, at the wrong moment", *Common market Review* (2006).

DAVIES, G., *Nationality Discrimination in the European Internal Market* (The Hague, Kluwer Law International, 2003).

DE BAERE, G., *Constitutional Principles of EU External Relations* (Oxford, Oxford University Press, 2008).

DE BAUW, R., "La politique énergétique" in LASOK, D., and SOLDATOS, P., (eds.) *Les Communautés Européennes en fonctionnement* (Bruxelles, Bruylant 1981).

DE BÚRCA, G., "The Principle of Proportionality and its Application in EC law", *Yearbook of European law* (Oxford, Clarendon, 1993).

DE BÚRCA, G., and DE WITTE, B., "The delimitation of powers between the EU and its Member States" in ARNULL, A., and WINCOTT, D., (eds.), *Accountability and Legitimacy in the European Union* (Oxford, Oxford University Press, 2002).

DE CENDRA DE LARRAGAN, J., "EU Climate change mitigation policy: is it adequate, is it fair?" in DELVAUX, B., HUNT, M., and TALUS, K., (eds.) *EU energy law and policy issues* (Rixensart, Euroconfidentiel, 2008).

DE CENDRA DE LARRAGÁN, J., *Distributional choices in EU climate change law and policy – Towards a principled approach*, Dissertation submitted to the faculty of law, Maastricht University, 2010.

DE SAEDELEER, N., "Environmental Principles – From Political Slogans to Legal Rules", *European Journal of International Law* (2005).

DE SAEDELEER, N., "The Precautionary Principle in EC Health and Environmental Law", *European Law Review* (2006).

DE SAEDELEER, N., *Environmental Principles From Political Slogans to Legal Rules* (Oxford, Oxford University Press, 2002).

DE WITTE, B., "Chameleonic Member States: Differentiation by means of partial and parallel international agreements" in DE WITTE, B., HANF, D., and VOS, E., (eds.), *The Many faces of Differentiation in EU Law* (Intersentia, Antwerpen, 2001).

DE WITTE, B., "Old-fashioned Flexibility: International Agreements between Member States of the Union" in DE BÚRCA, G., and SCOTT, J., (eds.), *Constitutional Change in the EU From Uniformity to Flexibility?* (Oxford, Hart Publishing, 2000).

DEHOUSSE, F., "Towards a Real New Energy Policy for the European Union? The 2007 Challenge", *Studia Diplomatica* (2007).

DEKETELAERE, K., (ed.), "European 'Environmental Law" in *International Encyclopedia of Laws* (The Hague, Kluwer Law International, 2000).

DEKETELAERE, K., "European Environmental Tax Policy: Proposal for a Policy Vision and a Legal Framework", *EELR* (1996).

DELBEKE, J., (ed.), *EU Energy Law – EU Environmental Law – The EU Greenhouse Gas Emission Trading Scheme* (Leuven, Claeys & Casteels, 2006).

DELVAUX, B., "Promoting Biofuels in Energy Supply: the European Legal Framework", *EELR* (2004)

DELVAUX, B., "Recente ontwikkelingen in het Europees energierecht en -beleid – Is de Sagrada Familia eindelijk af?" in DEKETELAERE, K., en DELVAUX, B., (eds.) *Jaarboek Energierecht 2007* (Antwerpen, Intersentia, 2008).

DELVAUX, B., "The Agency for the Cooperation of Energy Regulators: A New Beginning?" in B. DELVAUX, M. HUNT and K. TALUS (eds.), *EU Energy Law and Policy Issues* (Rixensart, Euroconfidentiel, 2010).

DELVAUX, B., "The EC State Aid Regime regarding Renewables: Opportunities and Pitfalls", *EELR* (2003).

DELVAUX, B., "The gas transmission regulation 1775/2005 – Has a genius been born?" in ROGGENKAMP, M. M., and HAMMER, U., (eds.), *European Energy Law Report IV* (Antwerpen, Intersentia, 2007).

DELVAUX, B., en VANDEN BORRE, T., "Recente ontwikkelingen in het Europese Energierecht en -beleid – Heel wat lekkers in de Europese pijplijn" in DEKETELAERE, K., en DELVAUX, B. (eds.) *Jaarboek Energierecht 2008* (Antwerpen, Intersentia, 2009).

DELVAUX, B., HUNT, M., and TALUS, K., (eds.), *EU Energy Law and Policy Issues* (Rixensart, Euroconfidentiel, 2008).

DELVAUX, B., HUNT, M., and TALUS, K., (eds.), *EU Energy Law and Policy Issues* (Rixensart, Euroconfidentiel, 2009).

DERUYTTER, T., GELDHOF, W., and VANDENDRIESSCHE, F., "Public Service Obligations in the Electricity and Gas Markets" in DELVAUX, B., HUNT, M., and TALUS, K., (eds.) *EU Energy Policy and* Law (Cambridge, Intersentia, 2011), p. 63–95.

DEVROE, W., "Privatization and Community Law: Neutrality versus Policy", *Common Market Law Review* (1997).

DHONDT, N., "Integration of Environmental Protection into the EC Energy Policy", *Yearbook of European Environmental Law* (2005).

DHONDT, N., *Integration of Environmental Protection into other EC Policies: legal theory and practice* (Groningen, Europa Law Publishing, 2003).

DOHERTY, A., "Retail market development and public service objectives" in C. JONES (ed.), *EU Energy law, The international energy market* (Leuven, Claeys & Casteels, 2010).

DOHERTY, M., G., "Judicial Use of Principles of EC Environmental Policy", *Environmental Law Review* (2000).

DOUGALL, D., S., M., "European Community Energy law in Perspective" in DOUGALL, D., S., M., and WÄLDE, T., W., (eds.), *European Community Energy Law - selected topics, International Energy and Resources law & Policy Series* (Nijhoff, London, Graham & Trotman/Martinus, 1994).

DOUGLAS-SCOTT, S., *Constitutional Law of the European Union* (Harlow, Pearson education Limited, 2002).

DOUMA, W., T., *The Precautionary Principle Its Application in International, European and Dutch Law* (PhD., Groningen, 2003).

DOYLE, A., and CARNEY, T., "Precaution and Prevention: Giving effect to Article 130r Without Direct Effect", *European Environmental Law Review* (1999).

Editorial Comments to Campus Oil, *ECLR* (1984).

EDWARD, D., and HOSKINS, M., "Article 90: Deregulation and EC Law Reflections arising from the XVI FIDE Conference", *Common Market Law Review* (1995).

EECKHOUT, P., *EU External Relations Law* (Oxford, Oxford University Press, 2011).

EECKHOUT, P., *External Relations of the European Union Legal and Constitutional Foundations* (Oxford, Oxford University Press, 2004).

EHRICKE, U., and HACKLÄNDER, D., "European Energy Policy on the Basis of the New Provisions in the Treaty of Lisbon" in BAUSCH, A., and SCHWENKER, B., (eds.), *Handbook Utility Management* (Berlin Heidelberg, Springer-Verlag, 2009).

EILMANSBERGER, T., "Bilateral Investment Treaties and EU Law" *Common Market Law Review* (2009).

EMILIOU, N., "Subsidiarity: An effective Barrier against 'The Enterprises of Ambition'", *European Law Review* (1992).

EMILIOU, N., "Subsidiarity: Panacea or Fig Leaf?" in O'KEEFFE, D., and TWOMY, P., M., (eds.) *Legal Issues of the Maastricht Treaty* (London, Chancery Press, 1994).

EMILIOU, N., *The Principle of Proportionality in European Law - A Comparative Study* (London, Kluwer Law International, 1996).

EPINEY, A., "Environmental Principles" in MACRORY, R., (ed.), *Reflections on 30 Years of EU Environmental Law: A high level of protection* (Groningen, Europa Law Publishing, 2005).

ESTELLA, A., *The EU Principle of Subsidiarity and its Critique* (Oxford, Oxford University Press, 2002).

ESTY, D., C., "Regulatory competition in focus" *Journal of International Economic Law* (2000).

FAULL, J., and NIKPAY, A., (eds.), *The EC Law of Competition* (Oxford, Oxford University Press, 2008).

FISHER, E., JONES, J., and VON SCHOMBERG, R., (eds.), *Implementing the Precautionary Principle: Perspectives and Prospects* (Cheltenham, Edward Elgar, 2006).

FOSTER, N., *Foster on EU Law* (Oxford, Oxford University Press, 2009).

GANOVA, A., *European Union Energy Supply Policy: Diversified in Unity?* (Institut Européen des Hautes Etudes Internationales, 2007).

GELDHOF, W., and VANDENDRIESSCHE, F., "European Electricity and Gas Market Liberalisation Background, Status, Developments" in DELVAUX, B., HUNT, M., and TALUS, K., (eds.), *EU Energy Law and Policy Issues* (Rixensart, Euroconfidentiel, 2008).

GENY, F., *Méthode d'interprétation et sources en droit privé positif* (Paris, Librairie générale de droit et de jurisprudence, 1954).

GERADIN, D., MUNOZ, R., and PETIT, N., *Regulation through Agencies in the EU – A new paradigm of European Governance* (Cheltenham, Elgar, 2005).

GOLDBERG, S., and BJØRNEBYE, H., "The Third Energy Package of the European Union, Introduction and comment" in DELVAUX, B., HUNT, M., and TALUS, K., (eds.), *EU Energy Law and Policy Issues* (Cambridge, Intersentia, 2011).

GORMLEY, L., W., "The Definition of Measures having Equivalent Effect" in ARNULL, A., EECKHOUT, P., and TRIDIMAS, T., (eds.), *Continuity and Change in EU Law Essays in Honour of Sir Francis Jacobs* (Oxford, Oxford University Press, 2008).

GORMLEY, L., W., *EU law of free movement of goods and customs union* (Oxford, Oxford University Press, 2009).

GOUCHA SOARES, A., "Pre-emption, Conflicts of Power and Subsidiarity", *European Law Review* (1998).

GOYDER, D., G., *EC Competition Law* (Oxford, Oxford University Press, 2003).

GREEN, N., "The legal basis of a Community Energy Policy", *European Law Review* (1983).

GUAL, M., and DEL RIO, P., "The promotion of green electricity in Europe: present and future", *European Environment* (2004).

GUNST, A., J., "Energy Trade in the European Common Market – Free Movement, Exceptions and Regulatory Inaction", *Journal of Energy & Natural Resources Law* (2003).

GYSELEN, L., "Case note on Albany, Brentjens' and Drijvende Bokken" *Common Market Law Review* (2000).

HAGHIGHI, S., S., *Energy Security The External Legal Relations of the European Union with Major Oil- and Gas Supplying Countries* (Oxford, Hart Publishing, 2007).

HALBERSTAM, D., "The Political Morality of Federal Systems", *Virginia Law Review* (2004).

HANCHER, L. "Case note on Corbeau", *Common Market law Review* (1994).

HANCHER, L., "A single European Energy Market – rhetoric or reality?", *Energy Law Journal* (1990).

HANCHER, L., "Delimitation of Energy Law Jurisdiction: the EU and its Member States: From Organisational to Regulatory Conflicts", *Journal of Energy & Natural Resources Law* (1998).

HANCHER, L., "The new EC Constitution and the European Energy market" in ROGGENKAMP M., M., and HAMMER, U., (eds.), *European Energy Law Report II* (Antwerpen, Intersentia, 2005).

HANCHER, L., and DE HAUTECLOCQUE, A., "Manufacturing the EU Energy Markets: The Current Dynamics of Regulatory Practise", *Competition and Regulation in Network Industries* (2010).

HANCHER, L., and JANSSEN, S., "Shared Competences and Multi-Faceted Concepts – European Legal Framework for Security of Supply", in BARTON, B., REDGWELL, C., RØNNE, A., and ZILLMAN, D., (eds.), *Energy Security – Managing Risk in a*

Dynamic Legal and Regulatory Environment (Oxford, Oxford University Press, 2004).

HARBO, T.,-I., "The function of the Proportionality Principle in EU Law", *European Law Journal* (2010).

HARTKAMP, A., and HONDIUS, E., *et al.*, *Towards a European Civil Code* (The Hague, Kluwer Law International, 2004).

HARTLEY, T., C., *The Foundations of Community Law An Introduction to the Constitutional and administrative law of the European Community* (Oxford, Oxford University Press, 2007).

HARTLEY, T., C., *The Foundations of European Community Law. An Introduction to the Constitutional and Administrative Law of the European Community*, (Oxford, Oxford University Press, 2010).

HENKEL, C., "The allocation of Powers in the European Union: A closer look at the principle of subsidiarity", *Berkeley Journal of International Law* (2002).

HERVEY, T., K., "Up in Smoke? Community (Anti) Tobacco Law and Policy", *El Rev.* (2001).

HUNT, M., "EU Energy Policy in 2007–2008" in HUNT, M., and TALUS, K., (eds.), *The EU Energy Directory* (Rixensart, Euroconfidentiel, 2008).

HUNT, M., "Ownership Unbundling: the Main Legal Issues in a controversial Debate" in DELVAUX, B., HUNT, M., and TALUS, K., (eds.), *EU Energy Law and Policy Issues* (Rixensart, Euroconfidentiel, 2008).

HUNT, M., and KAROVA, R., "The Energy Acquis Under the Energy Community Treaty and the Integration of South East European Electricity Markets: An Uneasy Relationship?" in DELVAUX, B., HUNT, M., and TALUS, K., (eds.) *EU Energy Law and Policy Issues* (Rixensart, Euroconfidentiel, 2010).

HUNT, M., "Closed Distribution Systems: The End of Private Networks?" in DELVAUX, B., HUNT, M., and TALUS, K., (eds.), *EU Energy Law and Policy Issues* (Cambridge, Intersentia, 2011).

IMMENGA, U., "The Development of European Energy Policy: From the ECSC Treaty to the Internal market" in MESTMÄCKER, E., J., (ed.), *Natural Gas in the Internal Market - A Review of Energy Policy* (London, Graham and Trotman, 1993).

JACOBS, F., "The role of the Court of Justice of the European Union in the Protection of the Environment", *Journal of Environmental Law* (2006).

JACOBS, F., "The Role of the Court of Justice of the European Union in the protection of the environment", *Journal of Environmental Law* (2006).

JANS, J., H., and VEDDER, H., H., B., *European Environmental Law* (Groningen, Europa Law Publishing, 2008).

JANS, J., H., *European Environmental Law* (Groningen, Europa Law Publishing, 2000).

JENSEN, W., G., *Energy in Europe: 1945–1980* (London, GT Foulis, 1967).

JOERGES, C., and DEHOUSSE, R., (eds.), *Good Governance in Europe's Integrated Market* (Oxford, Oxford University Press, 2002).

JOERGES, C., and NEYER, J., *Deliberative Supranationalism Revisited*, EUI Working Paper No. 2006/20 (Florence, European University Institute, 2006).

JOHNSTON, A., "Maintaining the Balance of Power: Liberalisation, Reciprocity and Electricity in the European Community", *Journal of Energy and Natural Resources Law* (1999).

JOHNSTON, A., "Ownership Unbundling: Prolegomenon to a Legal Analysis" in M. BULTERMAN, L. HANCHER, A. MCDONNELL and H. SEVENSTER (eds.), *Views of European Law from the Mountain – Liber Amicorum Piet Jan Slot* (The Netherlands, Kluwer Law International, 2009).

JOHNSTON, A., "Take-or-pay contracts for Renewables: An analysis of European Legal issues", in B. DELVAUX, M. HUNT and K. TALUS (eds.), *EU energy law and policy issues* (Rixensart, Euroconfidentiel, 2008).

JOHNSTON, A., NEUHOFF, K., FOUQUET, D., RAGWITZ, M., and RESCH, G., "The proposed new EU renewables directive: interpretation, problems and prospects", *EEELR* (2008).

JONES, A., and SUFRIN, B., *EC Competition Law – Text, cases and materials* (Oxford, Oxford University Press, 2008).

JONES, C., W., and WEBSTER, W., *EU Energy Law – Volume I The Internal Energy Markets* (Leuven, Claeys & Casteels, 2006).

KELLY, G., H., "An Evaluation of the European Union's Emissions Trading Scheme in Practice", *EELR* (2006).

KHANNA, D., "The Defeat of the European Tobacco Advertising Directive: A blow for Health", *YEL* (2001).

KIIVER, P., "The Treaty of Lisbon, the National Parliaments and the principle of subsidiarity", *Maastricht Journal* (2008).

KRÄMER, L., *EC Environmental Law* (London, Sweet & Maxwell, 2007).

KROES, N., "Improving Competition in European Energy Markets through Effective Unbundling", *Fordham International Law Journal* (2008).

KUMM, M., "Constitutionalising Subsidiarity in Integrated Markets: The Case of Tobacco Regulation in the European Union", *European Law Journal* (2006).

KURCZ, M., "La répartition des compétences au sein de l'Union européenne", *Revue du Droit de l'Union Européenne* (2005).

LAPIE, P., O., "La Politique énergétique commune", *Revue Marché Commun* (1964).

LEAL ARCAS, R., "Is the Kyoto Protocol an adequate environmental agreement to resolve the climate change problem", *EELR* (2002).

LEE, M., "The environmental implications of the Lisbon Treaty", *Environmental Law Review* (2008).

LENAERTS, K., "Regulating the regulatory process: Delegation of powers in the European Union Community", *European Law Review* (1993).

LENAERTS, K., "The Principle of Subsidiarity and the Environment in the European Union: Keeping the Balance of Federalism", *Fordham International Law Journal* (1993).

LENAERTS, K., and VAN NUFFEL, P., *Europees Recht* (Antwerpen, Intersentia, 2011).

LENAERTS, K., and VAN NUFFEL, P., *Europees recht in hoofdlijnen* (Antwerpen, Maklu, 2008).

LENAERTS, K., and VAN NUFFEL, P., in BRAY, R., (ed.), *Constitutional Law of the European Union* (London, Thomson Sweet & Maxwell, 2005).

LENAERTS, K., and VAN NUFFEL, P., in BRAY, R., and CAMBIEN, N., (eds.), *European Union Law* (London, Sweet & Maxwell, 2011).

LORD MACKENZIE-STUART, "Subsidiarity – A Busted Flush?", in CURTIN, D., and O'KEEFE, D., (eds.), *Constitutional Adjudication in European and National Law: essays for T. F. O'Higgins* (London, Butterworths, 1992).

LUCAS, N., J., D., *Energy and the European Communities* (London, Europa Publications, 1977).

LYONS, T., *EC customs law* (Oxford, Oxford University Press, 2008).

MACCORNICK, N., "Democracy and Subsidiarity in the European Commonwealth", in MACCORNICK, N., (ed.), *Questioning Sovereignty* (Oxford, Oxford University Press, 1999).

MAHMOUDI, S., "Integration of Environmental Considerations into Transport" in MACRORY, R., (ed.), *Reflections on 30 Years of EU Environmental Law: A high level of protection* (Groningen, Europa Law Publishing, 2005).

MALMENDIER, B., and SCHENDEL, J., "Unbundling Germany's Energy Networks", *Journal of Energy & Natural Resources Law* (1998).

MARENCO, G., "Public Sector and Community Law", *Common Market Law Review* (1983).

MARIN DURAN, G., and MORGERA, E., "Towards Environmental Integration in EC External Relations? A comparative analysis of selected association agreements", *Yearbook of European Environmental Law* (2006).

MEIER-WEIGT, B., "Die Vorhaben von europäischem Interesse nach den Transeuropäischen Energienetze – TEN-E – Leitlinien und ihre Umsetzung in den europäischen Mitgliedsstaaten", *IR* (1/2007).

MICHAELIS, H., *Energiemarkt und Energiepolitik in einer Europäischen Union* (Frankfurt am Main, A. Metzner, 1976).

MORRIS, J., (ed.), *Rethinking Risk and the Precautionary Principle* (Oxford, Buttherworth-Heinmann, 2000).

MORTELMANS, K., J., M., "Case note to Campus Oil", *Common Market Law Review* (1984).

MORTELMANS, K., J., M., "Chapter VIII The Functioning of the Internal Market: The Freedoms" in KAPTEYN, P., J., G., MCDONNELL, A., M., MORTELMANS, K., J., M., and TIMMERMANS, C., W., A., (eds.), *The Law of the European Union and the European Communities* (The Netherlands, Kluwer Law International, 2008).

MÜLLER-MICHAELIS, W., *Die Integration der westeuropäischen Energiemärkte* (Göttingen, Vandenhoeck u. Ruprecht, 1968).

NEERGAARD, U., "Chapter 1 Services of General Economic Interest: The Nature of the Beast" in KRAJEWSKI, M., NEERGARD, U., and VAN DE GRONDON, J., (eds.) *The Changing Legal Framework for Services of General Interest in Europe Between Competition and Solidarity* (The Hague, T.M.C. Asser Press, 2009).

NEVELING, S., "Europäiserung der Energieaufsicht? – Vorschläge von Kommission und ERGEG zur Neuordnung (Teil 1)", *Heft – IR Energie, Verkehr, Abfall,* Wasser (2007).

NICOLAIDIS, K., and SHAFFER, G., "Managed Mutual Recognition Regimes: Governance Without Global Government", *Law and Contemporary Problems* (2005).

NISTOR, L., N., *Public Services and the European Union A Study of the Impact of Community Law on Health Care, Health Insurance and Education Services,*

Proefschrift ter verkrijging van het doctoraat in de Rechtsgeleerdheid aan de Rijksunversiteit Groningen, 2009.

NOËL, P., "Ensuring success for the EU Regulation on gas supply security", University of Cambridge (EPRG) & European Council on Foreign Relations (ECFR), 9 February 2010.

NOËL, P., and FINDLATER, S., "A comment on the draft Regulation on security of gas supply", EPRG, University of Cambridge, 3 July 2009.

O'RIORDAN, T., and CAMERON, J., (eds.), *Interpreting the Precautionary principle* (London, Earthscan, 1994).

O'RIORDAN, T., CAMERON, J., and JORDAN, A., (eds.), *Reinterpreting the Precautionary principle* (London, Cameron May, 2001).

OEEC, *Europe's need for oil Implications and Lessons of Suez Crisis* (Paris, OEEC, 1958).

OLE GRAM MORTENSEN, B., "The EU Emission Trading Directive", *EELR* (2004).

OLIVER, P. "Some Further Reflections on the Scope of Articles 28–30 (ex 30–36) EC", *Common Market Law Review* (1999).

OLIVER, P., "A review of the Case law of the Court of Justice on Articles 30 to 36 EEC in 1984", *Common Market Law Review* (1985).

OLIVIER, P., and JARVIS, M., *Free Movement of Goods in the European Community* (London, Sweet & Maxwell, 2003).

OPPERMANN, T., *Europarecht* (Munchen, Beck, 1991).

PAGH, P., "The Battle on Environmental Policy Competences – Challenging the Stricter Approach: Stricter Might Lead to Weaker Protection" in MACRORY, R., *Reflections on 30 years of EU Environmental law – a high level of protection* (Groningen, Europa Law Publishing, 2006).

PEETERS, M., "Emissions Trading a New Dimension to European Environmental law: The Political Agreement of the European Council on Greenhouse Gas Allowance Trading", *EELR* (2003).

PEETERS, M., and DEKETELAERE, K., (eds.), *EU Climate Change policy – The Challenge of New Regulatory Initiatives* (London, Edward Elgar Publishing, 2006).

PELKMANS, J., "Mutual Recognition in Goods: On Promises and Disillusions", *Journal of European Public Policy* (2007).

PETERSON, J., "Subsidiarity: A definition to Suit Any Vision?", *Parliamentary Affairs* (1994).

PIRANI, S., STERN, J., and YAFIMAVA, K. *The Russo-Ukrainian Gas Dispute of January 2009: A comprehensive assessment* (Oxford, Oxford University Studies, 2009).

PIRIS, J., C., *The Lisbon Treaty – A Legal and Political Analysis* (Cambridge, Cambridge University Press, 2010).

POCKLINGTON, D., "European Emission Trading – the Business Perspective", *EELR* (2002).

RASHBROOKE, G., "Clarification or Complication? The New Energy Title in the Draft Constitution for Europe", *Journal of Energy & Natural Resources Law* (2004).

RODRIGUES, S., "Towards a General EC Framework Instrument Related to SGEI? Political Considerations and Legal Constraints" in KRAJEWSKI, M., NEERGARD, U., VAN DE GRONDON, J., (eds.), *The Changing Legal Framework for Services of General Interest in Europe Between Competition and Solidarity* (The Hague, T.M.C. Asser Press, 2009).

ROGGENKAMP, M., M., RØNNE, A., REDGWELL, C., and DEL GAUYO, I., (eds.), *Energy law in Europe, National, EU and International Law and Institutions* (Oxford, Oxford University Press, 2001).

ROSS, M., "A healthy approach to services of general economic interest? The BUPA judgment of the General Court", *European Law Review* (2009).

ROSS, M., "Article 16 E.C. and services of general interest: from derogation to obligation?", *European Law Review* (2000).

SACHDEVA, A., M., "Regulatory competition in European company law", *European Journal of Law & Economics* (2010), p. 137–139.

SAGA, B., "Introduction of Competition in Gas Markets: Effects on Contract Structures and Security of Supply" in HANCHER, L., (ed.), *The European Energy Market: Reconciling Competition and Security of Supply* (Köln, Bundesanzeiger, 1995).

SAUTER, W., "Services of general economic interest and universal service in EU law", *European Law Review* (2008).

SCHILLING, T., "A new dimension of Subsidiarity: Subsidiarity as a Rule and a Principle", *Yearbook of European Law* (1994).

SCHMIDT, S., K., (ed.), "Mutual Recognition as a New Mode of Governance", *Journal of European Public Policy* (2007).

SCHRIJVER, N., *Sovereignty over natural resources. Balancing rights and duties* (Cambridge, Cambridge University Press, 1997).

SCHUMACHER, E., F., "The Struggle for a European Energy Policy", *Journal of Common Market Studies* (1964).

SCHUMACHER, T., "The environmental integration clause in Article 6 of the EC Treaty: Prioritising Environmental Protection", *Environmental Law Review* (2001).

SCHÜTZE, R., "Co-operative Federalism Constitutionalised: The Emergence of Complementary Competences in the EC Legal Order", *European Law Review* (2006).

SCHÜTZE, R., "EC Law and International Agreements of the Member States – An Ambivalent Relationship?" *Cambridge Yearbook of European Legal Studies* (Oxford, Hart Publishing, 2006–2007).

SCHÜTZE, R., "Subsidiarity after Lisbon: reinforcing the safeguards of federalism?", *Cambridge Law Journal* (2009).

SCHÜTZE, R., "Supremacy without Pre-emption? The Very Slowly Emergent Doctrine of Community Pre-emption", *Common Market Law Review* (2006).

SCHWARZE, J., "European Energy Policy in Community Law" in MESTMÄCKER, E., J., (ed.), *Natural Gas in the Internal Market – A Review of Energy Policy* (London, Graham and Trotman, 1993).

SCHWARZE, J., *European Administrative Law* (London, Sweet and Maxwell, 2006).

SCOTT, J., and VOS, E., "The Juridification of Uncertainty: Observations of the Ambivalence of the Precautionary Principle within the EU and the WTO" in JOERGES, C., and DEHOUSSE, R., (eds.), *Good Governance in Europe's Integrated Market* (Oxford, Oxford University Press, 2002).

SEVENSTER, H., G., *Milieubeleid en Gemeenschapsrecht: het interne juridische kader en de praktijk* (Deventer, Kluwer, 1992).

SHAN, W., and ZHANG, S., "The Treaty of Lisbon: Half Way towards a Common Investment Policy" *The European Journal of International Law* (2011).

SHAW, D., NADIN, V., and SEATON, K., "The application of subsidiarity in the making of European environmental law", *European Environment* (2000).

SHUIBHNE, N., N., (ed.), *Regulating the Internal Market* (Cheltenham, Edward Elgar, 2006).

SLOT, P., J., "Sectoral Policies" in KAPTEYN, P., J., G., MCDONNELL, A., M., MORTELMANS, K., J., M., and TIMMERMANS, C., W., A., (eds.), *The Law of the European Union and the European Communities* (The Netherlands, Kluwer Law International, 2008).

SNELL, J., "Who Has Got the Power: Free Movement and Allocation of Competences", *YEL* (2003).

SNELL, J., *Goods and Services in EC Law A Study of the Relationship Between the Freedoms* (Oxford, Oxford University Press, 2002).

SOLE, D., "Services of General Economic Interest in the EU: A 50-Year-Battle between Liberalization, Deregulation and Subsidiarity", *Eur. J.L. Reform* (2003).

SOMSEN, H., "Discretion in European Community Environmental Law: An analysis of ECJ Case Law", *Common Market Law Review* (2003).

STEINER, J., "Subsidiarity under the Maastricht Treaty" in O'KEEFFE, D. and TWOMY, P., M., (eds.) *Legal Issues of the Maastricht Treaty* (London, Chancery Press, 1994).

STERN, J., "Security of European Natural Gas Supplies: the Impact of Import Dependence and Liberalization", *RIIA Working Paper* (Royal Institute of International Affairs, London, 2002).

STRAETMANS, G., "The Outer Limits of Free Movement of Goods" in BULTERMAN, M., HANCHER, L., MCDONNELL, A., and SEVENSTER, H. (eds.), *Views of European Law from the Mountain – Liber Amicorum Piet Jan Slot* (The Netherlands, Kluwer Law International, 2009).

SYKES, A., O., "Regulatory Competition or regulatory harmonization? A Silly Question", *Journal of International Economic Law* (2000).

TALUS, K., "Energy Security in European Energy Law" in HUNT, M., and TALUS, K., (eds.), *The EU Energy Directory* (Rixensart, Euroconfidentiel, 2008).

TALUS, K., and JOHNSTON, A., "Comment on Pielow, Brunekreeft and Ehlers on ownership unbundling", *Journal of World Energy Law & Business* (2009).

TALUS, K., *Vertical Natural Gas Transportation capacity, upstream commodity contracts and EU Competition law*, Dissertation submitted to the faculty of law, University of Helsinki, 2010.

TALUS, K., *Vertical Natural Gas Transportation Capacity, Upstream Commodity Contracts and EU Competition Law* (The Netherlands, Kluwer Law International, 2010), 321 p.

TEMPLE LANG, J., "Community Constitutional Law: Article 5 EEC Treaty", *Common Market Law Review* (1990).

TERRA, B., J., M., and WATTEL, P., J., *European Tax Law* (Alphen aan den Rijn, Kluwer Law International, 2008).

TOTH, A. G., "The Legal status of the Declarations Annexed to the Single European Act", *Common Market Law Review* (1986).

TOTH, A., G., "Is subsidiarity justifiable?", *European Law Review* (1994).

TOTH, A., G., "The principle of Subsidiarity in the Maastricht Treaty", *Common Market Law Review* (1992).

TRIDIMAS, T., *The General Principles of EU Law* (Oxford, Oxford University Press, 2006).

TROUWBORST, A., "International nature conservation law and the adaptation of biodiversity to climate change: a mismatch?", *Journal of Environmental Law* (2009).

TRUE, C., "Legislative competences of Euratom and the European Community in the energy sector: the Nuclear Package of the Commission", *European Law Review* (2003).

VAN CALSTER, G., "Against Harmonisation – Regulatory Competition in Climate Change Law", *Carbon & Climate Law Review* (2008).

VAN CALSTER, G., and DEKETELAERE, K., "Amsterdam, the Intergouvernmental Conference and Greening the EU Treaty", *EELR* (1998).

VAN DEN BERGH, P., "Forced Harmonisation of Contract Law in Europe: Not to be Continued" in GRUNDMANN, S., and STUYCK, J., (eds.), *An Academic Green Paper on European Contract Law* (The Hague, Kluwer, 2002).

VAN GERVEN, W., "The Effect of Proportionality on the Actions of Member States of the European Community: National Viewpoints from Continental Europe", in ELLIS, E., (ed.), *The Principle of Proportionality* (Oxford, Hart Publishing, 1999).

VANDEN BORRE, T., "Derde pakket voor de vrijmaking van de elektriciteitsmarkt: Quo Vadis?" in "Het Europese Energie – en Klimaatpakket", DEKETELAERE, K., en DELVAUX, B., (eds.), *Milieu – en Energierecht* (2008).

VANDEN BORRE, T., "Via nieuwe en fragmentarische maatregelen in het Federale energiebeleid naar een globale visie?" in DEKETELAERE, K., (ed.), *Jaarboek Energierecht 2005–2006* (Antwerp, Intersentia, 2007).

VANDENDORPE, W., "Recent developments on Energy Efficiency on the European Union level", in DELVAUX, B., HUNT, M., and TALUS, K., (eds.) *EU energy law and policy issues* (Rixensart, Euroconfidentiel, 2008).

VAN DER LINDE, C., "External energy policy: Old fears and new dilemmas in a larger Union" in SAPIR, A., (ed.), *Fragmented Power: Europe and the global economy* (Brussels, Bruegel, 2007).

VANDERMEERSCH, D., "Restrictions on the Movement of Oil In and Out the European Community: The Campus Oil and Bulk Oil Cases", *Journal of Energy and Natural Resources Law* (1987).

VANDERMEERSCH, D., "The Single European Act and the Environmental Policy of the European Community", *European Law Review* (1987).

VEDDER, H., H., B., *Competition Law and Environmental Protection in Europe – Towards Sustainability?* (Groningen, Europa Law Publishing, 2003).

VERHAEGEN, K., MEEUS, L., DELVAUX, B., and BELMANS, R., "Electricity produced from renewable energy sources – what target are we aiming for?", *Energy Policy* (2007).

VERHOEVEN, A., *The European Union in Search of a Democratic and Constitutional Theory* (The Hague, Kluwer Law International, 2002).

VINOIS, J., A., "Les réseaux transeuropéens: une nouvelle dimension donée au Marché unique", *Revue du Marché Unique Européen* (1993).

VON BOGDANDY, A., and BAST, J., "The European Union's Vertical Order of Competences: The Current Law and Proposals for its Reform", *Common Market Law Review* (2002).

VON DER GROEBEN, H., VON BOECKH, H., THIESING, J., EHLERMANN, C. D., (eds.), *Kommentar zum EWG-Vertrag* (Baden-Baden, Nomos, 1983).

VON ROSENBERG, H., "No ACE to win the trick – the proposed ACER and its influence on EC Competition law", *E.C.L.R.* (2008).

WANG, X., and WISER, G., "The Implementation and Compliance Regimes under the Climate Change Convention and its Kyoto Protocol", *RECIEL* (2002).

WEATHERILL, S., "Competence Creep and Competence Control", *YBEL* (2004).

WEATHERILL, S., "Using National Parliaments to improve Scrutiny of the limits of EU Action", *European Law Review* (2003).

WEATHERILL, S., "Why Harmonise?" in TRIDIMAS, T., and NEBBIA, P., (eds.) *European Union Law for the Twenty-First Century* (Oxford and Portland, Hart Publishing, 2004).

WEATHERILL, S., "Why Object to the Harmonisation of Private Law by the EC?", *European Review of Private Law* (2004).

WEATHERILL, S., and BEAUMONT, P., *EU law – The essential guide to legal workings of the European Union* (London, Penguin Books, 1993).

WEATHERILL, S., and VOGENAUER, S., (eds.), *The Harmonisation of European Contract Law: Implications for European Private Laws, Business and Legal Practise* (Oxford, Hart Publishing, 2006).

WEBER, D., *Tax avoidance and the EC treaty freedom: a study on the limitations under European law to the prevention of tax avoidance* (The Hague, Kluwer Law International, 2005).

WEILER, J., H., H., "A Constitution for Europe", *Journal of Common Market Studies* (2002).

WERNICKE, S., "Chapter 3 Taking Stock: The EU Institutions and Services of General Economic Interest" in KRAJEWSKI, M., NEERGARD, U., and VAN DE GRONDON, J. (eds.), *The Changing Legal Framework for Services of General Interest in Europe Between Competition and Solidarity* (The Hague, T.M.C. Asser Press, 2009).

WILLIAMS, R., "Community Development Cooperation Law, Sustainable Development and the Convention on Europe: From Dislocation to Consistency?", *Yearbook of European Environmental Law* (2005).

WOUTERS, J., "Constitutional Limits of Differentiation: The Principle of Equality" in DE WITTE, B., HANF, D., and VOS, E., (eds.), *The Many faces of Differentiation in EU Law* (Intersentia, Antwerpen, 2001).

WYATT, D., "Community Competence to Regulate the Internal Market" in DOUGAN, M., and CURRIE, S., (eds.), *50 years of the European Treaties looking back and thinking forward* (Oxford, Hart Publishing, 2009).

WYATT, D., "Constitutional significance of the Tobacco Advertising Judgement of the Court of Justice of the European Union" and DE BÚRCA, G., "The Political Will versus Constitutional Limits", *The ECJ's Tobacco Advertising Judgment*, ECLS Occasional Paper, No 5 (Cambridge, Cambridge University Press, 2001).

ZACKER, C., "Environmental Law of the European Economic Community: New Powers Under the Single European Act", *International and Comparative Law Review* (1991).

ZAFIROVA, Z., "Unbundling the network: the case for ownership unbundling?", *International Energy Law & Taxation Review* (2007).
ZIMMERMANN, C., F., and TALUS, K., "Regulation of Electricity Markets at EU Level", *EEELR* (2008).

LIST OF ECJ CASES

Case 9/56, *Meroni/High Authority* [1957–8] ECR 133.
Joined Cases 27–58, 28–58 and 29–58, *Compagnie des hauts fourneaux et fonderies de Givors and others v High Authority of the European Coal and Steel Community* [1960] ECR 241.
Case 26/62, *NV Algemene Transport- en Expeditie Onderneming van Gend & Loos v Netherlands Inland Revenue Administration* [1963] ECR 1.
Case 6/64, *Flaminio Costa v E.N.E.L.* [1964] ECR 1251.
Case 57/65, *Alfons Lütticke GmbH v Hauptzollamt Sarrelouis* [1966] ECR 205.
Case 7/68, *Commission of the European Communities v Italian Republic* [1968] ECR 423.
Case 24/68, *Commission of the European Communities v Italian Republic* [1969] ECR 193.
Joined Cases 2 and 3/69, *Sociaal Fonds voor de Diamantarbeiders v S.A. Ch. Brachfeld & Sons and Chougol Diamond Co* [1969] ECR 211.
Case 41/69, *ACF Chemiefarma/Commissie* [1970] ECR 221.
Case 11/70, *Internationale Handelsgesellschaft mbH v Einfuhr- und Vorratsstelle für Getreide und Futtermittel* [1970] ECR 1125.
Case 22/70, *Commission of the European Communities v Council of the European Communities – European Agreement on Road Transport* [1971] ECR 263.
Case 10/71, *Ministère public luxembourgeois v Madeleine Muller, Veuve J.P. Hein and others* [1971] ECR 723.
Case 43/71, *Politi s.a.s. v Ministry for Finance of the Italian Republic* [1971] ECR 1039.
Case 39/72, *Commission of the European Communities v Italian Republic* [1973] ECR 101.
Case 77/72, *Carmine Capolongo v Azienda Agricole Maya* [1973] ECR 611.
Case 2/73, *Riseria Luigi Geddo v Ente Nazionale Risi* [1973] ECR 865.
Case 5/73, *Balkan-Import-Export GmbH v Hauptzollamt Berlin-Packhof* [1973] ECR 1091.
Case 127/73, *Belgische Radio en Televisie v SV SABAM and NV Fonior* [1974] ECR 313.
Case 155/73, *Giuseppe Sacchi* [1974] ECR 409.
Case 8–74, *Procureur du Roi v Benoît and Gustave Dassonville* [1974] ECR 837.
Case 63/74, *W. Cadsky SpA v Istituto nazionale per il Commercio Estero* [1975] ECR 281.
Case 23/75, *Rey Soda/Cassa Conguaglio Zucchero* [1975] ECR 1279.
Joined Cases 3, 4 and 6/76, *Cornelis Kramer and others* [1976] ECR 1279.
Case 35/76, *Simmenthal SpA v Ministère des finances italien* [1976] ECR 1871.
Case 46/76, *W. J. G. Bauhuis v The Netherlands State* [1977] ECR 5.
Case 74/76, *Iannelli & Volpi SpA v Ditta Paolo Meroni* [1977] ECR 557.
Case 105/76, *Interzuccheri S.p.A. v Societa Rezzano e Cavassa* [1977] ECR 1029.
Case 5/77, *Carlo Tedeschi v Denkavit Commerciale s.r.l.* [1977] ECR 1555.
Case 61/77, *Commission of the European Communities v Ireland* [1978] ECR 417.
Case 106/77, *Amministrazione delle Finanze dello Stato v Simmenthal SpA* [1978] ECR 629.

Case 83/78, *Pigs Marketing Board v Raymond Redmond* [1978] ECR 2374.
Case 120/78, *Rewe-Zentral AG v Bundesmonopolverwaltung für Branntwein* [1979] ECR 649.
Case 128/78, *Commission of the European Communities v United Kingdom of Great Britain and Northern Ireland* [1979] ECR 419.
Case 168/78, *Commission of the European Communities v French Republic* [1980] ECR 347.
Case 169/78, *Commission of the European Communities v Italian Republic* [1980] ECR 385.
Case 171/78, *Commission of the European Communities v Kingdom of Denmark* [1980] ECR 447.
Case 177/78, *Pigs and Bacon Commission v Mc Carren and Company Limited* [1979] ECR 2161.
Case 15/79, *P.B. Groenveld BV v Produktschap voor Vee en Vlees* [1979] ECR 3409.
Case 34/79, *Regina v Maurice Donald Henn and John Frederick Ernest Darby* [1979] ECR 3795.
Case 804/79, *Commission of the European Communities v United Kingdom of Great Britain and Northern Ireland* [1981] ECR 1045.
Case 113/80, *Commission of the European Communities v Ireland* [1981] ECR 1625.
Case 155/80, *Sergius Oebel* [1981] ECR 1993.
Joined Cases 188 to 190/80, *French Republic, Italian Republic and United Kingdom of Great Britain and Northern Ireland v Commission of the European Communities* [1982] ECR 2545.
Case 272/80, *Criminal proceedings against Frans-Nederlandse Maatschappij voor Biologische Producten BV* [1981] ECR 3277.
Case 75/81, *Joseph Henri Thomas Blesgen v Belgian State* [1982] ECR 1211.
Case 283/81, *Srl CILFIT and Lanificio di Gavardo SpA v Ministry of Health* [1982] ECR 3415.
Case 319/81, *Commission of the European Communities v Italian Republic* [1983] ECR 601.
Case 66/82, *Fromançais SA v Fonds d'orientation et de régularisation des marchés agricoles (FORMA)* [1983] ECR 395.
Joined Cases 96–102, 104, 105, 108 and 110/82, *NV IAZ International Belgium and others v Commission of the European Communities* [1983] ECR 3369.
Case 132/82, *Commission of the European Communities v Kingdom of Belgium* [1983] ECR 1649.
Case 158/82, *Commission of the European Communities v Kingdom of Denmark* [1983] ECR 3573.
Case 72/83, *Campus Oil Limited and others v Minister for Industry and Energy and others* [1984] ECR 2727.
Case 182/83, *Robert Fearon & Company Limited v Irish Land Commission* [1984] 3677.
Case 207/83, *Commission of the European Communities v United Kingdom of Great Britain and Northern Ireland* [1985] ECR 1201.
Case 240/83, *Procureur de la République v Association de défense des brûleurs d'huiles usagées* [1985] ECR 531.
Case 293/83, *Gravier v City of Liège* [1985] ECR 593.

Case 106/84, *Commission of the European Communities v Kingdom of Denmark* [1986] ECR 833.

Joined Cases 279/84, 280/84, 285/84 and 286/84, *Walter Rau Lebensmittelwerke and others v Commission of the European Communities* [1987] ECR 1069.

Case 148/85, *Direction générale des impôts et procureur de la République v Marie-Louise Forest, née Sangoy, and Minoterie Forest SA* [1986] ECR 3349.

Case 218/85, *Association comité économique agricole régional fruits et légumes de Bretagne v A. Le Campion* [1986] ECR 3513.

Case 45/86, *Commission of the European Communities v Council of the European Communities* [1987] ECR 1493.

Case 60/86, *Commission of the European Communities v United Kingdom of Great Britain and Northern Ireland* [1988] ECR 3921.

Case 66/86, *Ahmed Saeed Flugreisen and Silver Line Reisebüro GmbH v Zentrale zur Bekämpfung unlauteren Wettbewerbs e.V.* [1989] ECR 803.

Case 68/86, *United Kingdom of Great Britain and Northern Ireland v Council of the European Communities* [1988] ECR 855.

Case 291/86, *Central-Import Münster/Münster* [1988] 3679.

Case 302/86, *Commission of the European Communities v Kingdom of Denmark* [1988] ECR 4607.

Case 165/87, *Commission of the European Communities v Council of the European Communities* [1988] ECR 5545.

Case 235/87, *Annunziata Matteucci v Communauté française of Belgium and Commissariat général aux relations internationales of the Communauté française of Belgium* [1988] ECR 5589.

Case C-2/88, *Imm. J. J. Zwartveld and others* [1990] ECR I-3365.

Case C-18/88, *RTT v GB-INNO-BM SA* [1991] ECR I-5941.

Case C-21/88, *Du Pont de Nemours Italiana SpA v Unità sanitaria locale N° 2 di Carrara* [1990] ECR 889.

Case C-62/88, *Hellenic Republic v Council of the European Communities* [1990] ECR I-1527.

Case C-202/88, *France v Commission* [1991] ECR I-1223.

Case C-331/88, *The Queen v Minister of Agriculture, Fisheries and Food and Secretary of State for Health, ex parte: Fedesa and others* [1990] ECR 4023.

Case C-347/88, *Commission of the European Communities v Hellenic Republic* [1990] ECR I-4747.

Case C-23/89, *Quietlynn Limited and Brian James Richards v Southend Borough Council* [1990] ECR I-3059.

Case C-118/89, *Lingenfelser v Federal Republic of Germany* [1990] ECR I-2637.

Case C-152/89, *Commission of the European Communities v Grand Duchy of Luxembourg* [1991] ECR 3141.

Case C-155/89, *Belgian State v Philipp Brothers SA.* [1990] ECR 3265.

Case C-260/89, *Elliniki Radiophonia Tileorassi Anonimi Etaira v Dimotiki Etairia Pliroforissis* [1991] ECR I-2925.

Case C-288/89, *Stichting Collectieve Antennevoorziening Gouda and others v Commissariaat voor de Media* [1991] ECR I-4007.

Case C-300/89, *Commission of the European Communities v Council of the European Communities* [1991] ECR I-2867.

Joined Cases C-1/90 and C-176/90, *Aragonesa de Publicidad Exterior SA and Publivía SAE v Departamento de Sanidad y Seguridad Social de la Generalitat de Cataluña* [1991] ECR I-4151.

Case C-2/90, *Commission of the European Communities v Kingdom of Belgium* [1992] ECR 4431.

Case C-41/90, *Höfner v Macroton* [1991] ECR I-1979.

Joined Cases C-78/90, C-79/90, C-80/90, C-81/90, C-82/90 and C-83/90, *Compagnie Commerciale de l'Ouest and others v Receveur Principal des Douanes de La Pallice Port* [1992] ECR 1874.

Case C-179/90, *Merci Convenzionali Porto di Genova v Siderurgica Gabrielli SpA* [1991] ECR I-5889.

Case C-204/90, *Hanns-Martin Bachmann v Belgian State* [1992] ECR I-249.

Case C-286/90, *Anklagemyndigheden v Peter Michael Poulsen and Diva Navigation Corp.* [1992] ECR I-6019.

Case C-158/91, *Criminal proceedings against Jean-Claude Levy* [1993] ECR I-4287.

Joined Cases C-267/91 and C-268/91, *Criminal proceedings against Bernard Keck and Daniel Mithouard* [1993] ECR I 06097.

Case C-320/91, *Criminal proceedings against Paul Corbeau* [1994] ECR I-1783.

Case C-344/91, *Federal Republic of Germany v Commission of the European Communities* [2004] ECR I-2081.

Joined Cases C-92/92 and C-326/92, *Phil Collins v Imtrat Handelsgesellschaft mbH and Patricia Im- und Export Verwaltungsgesellschaft mbH and Leif Emanuel Kraul v EMI Electrola GmbH* [1993] I-5145.

Case C-350/92, *Kingdom of Spain v Council of the European Union* [1995] ECR I-1985.

Case C-379/92, *Peralta* [1994] ECR I-3453.

Case C-391/92, *Commission of the European Communities v Hellenic Republic* [1995] ECR I-1621.

Case C-393/92, *Municipality of Almelo and others v NV Energiebedrijf Ijsselmij* [1994] ECR I 1477.

Joined Cases C-401/92 and 402/92, *Criminal proceedings against Tankstation 't Heukske vof and J. B. E. Boermans* [1994] ECR I-2199.

Case C-405/92, *Etablissements Armand Mondiet SA v Armement Islais SARL* [1993] ECR I-6133.

Case C-18/93, *Corsica Ferries Italia Srl v Corpo dei Piloti del Porto di Genova* [1994] ECR I-1783.

Case C-19/93 P, NV, *Centraal Overijsselse Nutsbedrijven NV and Regionaal Energiebedrijf Salland NV v Commission of the European Communities* [1995] ECR I-3319.

Case C-130/93, *Lamaire NV v Nationale Dienst voor Afzet van Land- en Tuinbouwprodukten* [1994] ECR I-3215.

Joined Cases C-133/93, C-300/93 and C-362/93, *Antonio Crispoltoni v Fattoria Autonoma Tabacchi and Giuseppe Natale and Antonio Pontillo v Donatab Srl* [1994] ECR 4863.

Case C-187/93, *European Parliament v Council of the European Union* [1994] ECR 2857.

Case C-323/93, *Société Civile Agricole du Centre d'Insémination de la Crespelle v Coopérative d'Elevage et d'Insémination Artificielle du Département de la Mayenne* [1997] ECR I-5077.

Case C-412/93, *Société d'Importation Edouard Leclerc-Siplec v TF1 Publicité SA and M6 Publicité SA* [1995] ECR I-179.

Joined Cases C-418/93, C-419/93, C-420/93, C-421/93, C-460/93, C-461/93, C-462/93, C-464/93, C-9/94, C-10/94, C-11/94, C-14/94, C-15/94, C-23/94, C-24/94 and C-332/94, *Semeraro Casa Uno Srl v Sindaco del Comune di Erbusco (C-418/93), Semeraro Mobili SpA v Sindaco del Comune di Erbusco (C-419/93), RB* [1996] ECR I-02975.

Case C-426/93, *Federal Republic of Germany v Council of the European Union* [1995] ECR I-3723.

Case C-18/94, *Barbara Hopkins and others v National Power plc and Powergen plc.* [1996] ECR 2281.

Case C-84/94, *United Kingdom of Great Britain and Northern Ireland* v *Council of the European Union* [1996] ECR I-5755.

Case C-90/94, *Haahr Petroleum Ltd v Åbenrå Havn, Ålborg Havn, Horsens Havn, Kastrup Havn NKE A/S, Næstved Havn, Odense Havn, Struer Havn and Vejle Havn, and Trafikministeriet* [1997] ECR I-4085.

Case C-157/94, *Commission of the European Communities v Kingdom of the Netherlands* [1997] ECR I-5699.

Case C-158/94, *Commission of the European Communities v Italian Republic,* [1997] ECR I-5789.

Case C-159/94, *Commission of the European Communities v French Republic* [1997] ECR I-5815.

Case C-233/94, *Federal Republic of Germany v European Parliament and Council of the European Union* [1997] ECR I-2405.

Case C-268/94, *Portuguese Republic v Council of the European Union* [1996] ECR I-6177.

Case C-271/94, *European Parliament v Council of the European Union* [1996] ECR I-1689.

Case C-120/95, *Nicolas Decker v Caisse de maladie des employés privés* [1998] ECR I 1831.

Case C-124/95, *The Queen, ex parte Centro-Com Srl v HM Treasury and Bank of England* [1997] ECR I 81.

Case C-183/95, *Affish BV v Rijksdienst voor de keuring van Vee en Vlees* [1997] ECR I-4315.

Case C-265/95, *Commission of the European Communities v French Republic* [1997] ECR I-6959.

Case C-284/95, *Safety Hi-Tech Srl v S. & T. Srl.* [1998] ECR I-4301.

Case C-341/95, *Gianni Bettati v Safety Hi-Tech Srl* [1998] ECR I-4355.

Case C-375/95, *Commission of the European Communities v Hellenic Republic* [1997] ECR I-5981.

Case C-388/95, *Kingdom of Belgium v Kingdom of Spain* [2000] ECR 3123.

Case C-1/96, *The Queen v Minister of Agriculture, Fisheries and Food, ex parte Compassion in World Farming Ltd* [1998] ECR I-1251.

Case C-55/96, *Job Centre coop. arl.* [1997] ECR I-7119.

Case C-67/96, *Albany International BV v Stichting Bedrijfspensioenfonds Textielindustrie* [1999] ECR 5751.

Case C-157/96, *The Queen v Ministry of Agriculture, Fisheries and Food, Commissioners of Customs & Excise, ex parte National Farmers' Union, David Burnett and Sons Ltd, R. S. and E. Wright Ltd, Anglo Beef Processors Ltd, United Kingdom Genetics, Wyjac Calves Ltd, International Traders Ferry Ltd, MFP International Ltd, Interstate Truck Rental Ltd and Vian Exports Ltd* [1998] ECR I-2211.

Case C-163/96, *Criminal proceedings against Silvano Raso and Others* [1998] ECR-I-533.

Case C-180/96, *United Kingdom of Great Britain and Northern Ireland v Commission of the European Communities* [1998] ECR I-2265.

Case C-203/96, *Chemische Afvalstoffen Dusseldorp BV and Others v Minister van Volkshuisvesting, Ruimtelijke Ordening en Milieubeheer* [1998] ECR I-4075.

Case C-213/96, *Outokumpu Oy* [1998] ECR I-1777.

Case C-266/96, *Corsica Ferries France SA v Gruppo Antichi Ormeggiatori del porto di Genova Coop. arl, Gruppo Ormeggiatori del Golfo di La Spezia Coop. arl and Ministero dei Trasporti e della Navigazione* [1998] ECR I-3949.

Case C-309/96, *Daniele Annibaldi v Sindaco del Comune di Guidonia and Presidente Regione Lazio* [1997] ECR I-7493.

Case C-389/96, *Aher-Waggon GmbH v Bundesrepublik Deutschland* [1998] ECR 4473.

Case C-112/97, *Commission of the European Communities v Italian Republic* [1999] ECR 1821.

Case C-147–8/97, *Deutsche Post AG v Gesellschaft für Zahlungssysteme mbH GZS) (C-147/97) and Citicorp Kartenservice GmbH (C-148/97)* [2000] ECR I-825.

Case C-149/97, *Institute of the Motor Industry* [1998] ECR I-7053.

Case C-232/97, *L. Nederhoff & Zn. v Dijkgraaf en hoogheemraden van het Hoogheemraadschap Rijnland* [1999] ECR I-6385.

Case C-269/97, *Commission of the European Communities v Council of the European Union* [2000] ECR I-2257.

Case C-437/97, *EKW and Wein & Co* [2000] ECR I-1157.

Case C-293/97, *The Queen v Secretary of State for the Environment and Ministry of Agriculture, Fisheries and Food, ex parte H.A. Standley and Others and D.G.D. Metson and Others* [1999] ECR I-2603.

Case C-302/97, *Klaus Konle v Republik Österreich* [1999] ECR I-3099.

Case C-414/97, *Commission of the European Communities v Kingdom of Spain* [1999] ECR I-5585.

Case C-36/98, *Kingdom of Spain v Council of the European Union* [2001] ECR I-779.

Joined Cases C-175/98 and C-177/98, *Criminal proceedings against Paolo Lirussi (C-175/98) and Francesca Bizzaro (C-177/98)* [1991] ECR I-6881.

Case C-209/98, *Entreprenørforeningens Affalds/Miljøsektion (FFAD) v Københavns Kommune* [2000] 3743.

Case C-228/98, *Charalampos Dounias v Ypourgio Oikonomikon* [2000] ECR I-577.

Case C-318/98, *Criminal proceedings against Giancarlo Fornasar, Andrea Strizzolo, Giancarlo Toso, Lucio Mucchino, Enzo Peressutti and Sante Chiarcosso* [2000], ECR 4785.

Case C-367/98, *Commission of the European Communities v Portuguese Republic* [2002] ECR I-4731.

Case C-377/98, *Kingdom of the Netherlands v European Parliament and Council of the European Union* [2001] ECR I-7079.

Case C-379/98, *PreussenElektra AG v Schhleswag AG, in the presence of Windpark Reußenköge III GmbH and Land Schleswig-Holstein* [2001] ECR I-2099.
Case C-393/98, *Ministério Público and António Gomes Valente v Fazenda Pública* [2001] ECR I-1327.
Case C-398/98, *Commission of the European Communities v Hellenic Republic* [2001] ECR I-7915.
Case C-467/98, *Commission of the European Communities v Kingdom of Denmark* [2002] ECR I-9519.
Case C-54/99, *Association Eglise de scientologie de Paris and Scientology International Reserves Trust v The Prime Minister* [2000] ECR I-1335.
Case C-74/99, *The Queen v Secretary of State for Health and Others, ex parte Imperial Tobacco Ltd and Others* [2000] ECR I – 8599.
Case C-163/99, *Portuguese Republic v Commission of the European Communities* [2001] ECR I-2613.
Case C-234/99, *Niels Nygård v Svineafgiftsfonden, and Ministeriet for Fødevarer, Landbrug og Fiskeri* [2002] ECR I-3657.
Case C-258/99, *Commission of the European Communities v Kingdom of Spain* [1991] ECR 3977.
Case C-314/99, *Kingdom of the Netherlands v Commission of the European Communities* [2002] ECR I-5521.
Case C-475/99, *Firma Ambulanz Glöckner v Landkreis Südwestpfalz* [2001] ECR I-8089.
Case C-483/99, *Commission of the European Communities v French Republic* [2002] ECR I-4781.
Case C-503/99, *Commission of the European Communities v Kingdom of Belgium* [2002] ECR I-4809.
Case C-513/99, *Concordia Bus Finland Oy Ab, formerly Stagecoach Finland Oy Ab v Helsingin kaupunki and HKL-Bussiliikenne* [2002] ECR I-7213.
Joined Cases C-515/99 and C-527/99 to C-540/99 and Joined Cases C-519/99 to C-524/99 and C-526/99), *Hans Reisch and Others (joined cases C-515/99 and C-527/99 to C-540/99) v Bürgermeister der Landeshauptstadt Salzburg and Grundverkehrsbeauftragter des Landes Salzburg and Anton Lassacher and Others (joined cases C-519/99 to C-524/99 and C-526/99) v Grundverkehrsbeauftragter des Landes Salzburg and Grundverkehrslandeskommission des Landes Salzburg* [2002] ECR I-2157.
Case C-3/00, *Kingdom of Denmark v Commission of the European Communities* (or the so-called *Danish Foodstuff* case) [2003] ECR I-2643.
Case C-275/00, *European Community, represented by the Commission of the European Communities v First NV and Franex NV* [2002] ECR I-10943.
Case C-302/00, *Commission of the European Communities v French Republic* [2002] ECR I-2055.
Case C-336/00, *Republik Österreich v Martin Huber* [2002] ECR I-7699.
Case C-339/00, *Ireland v. Commission of the European Communities* [2003] ECR I-1175.
Case C-463/00, *Commission of the European Communities v Kingdom of Spain* [2003] ECR I-4581.

Joined Cases C-465/00, C-138/01 and C-139/01, *Rechnungshof v Österreichischer Rundfunk and Others and Christa Neukomm and Joseph Lauermann v Österreichischer Rundfunk* [2003] ECR I-4989.

Case C-98/01, *United Kingdom of Great Britain and Northern Ireland* [2003] ECR I-4641.

Case C-189/01, *H. Jippes, Afdeling Groningen van de Nederlandse Vereniging tot Bescherming van Dieren and Afdeling Assen en omstreken van de Nederlandse Vereniging tot Bescherming van Dieren v Minister van Landbouw, Natuurbeheer en Visserij* [2001] ECR I-5689.

Case C-211/01, *Commission of the European Communities v Council of the European Union* [2003] ECR I-8913.

Case C-236/01, *Monsanto Agricoltura Italia SpA and Others v Presidenza del Consiglio dei Ministri and Others* [2003] ECR I-8105.

Case C-281/01, *Commission of the European Communities v Council of the European Union* [2002] ECR I-12049.

Case C-300/01, *Doris Salzmann* [2003] ECR I-4899.

Case C-338/01, *Commission of the European Communities v Council of the European Union* [2004] ECR I-4829.

Case C-383/01, *De Danske Bilimportører v Skatteministeriet, Told- og Skattestyrelsen* [2003] ECR I-6065.

Case C-387/01, *Harald Weigel and Ingrid Weigel v Finanzlandesdirektion für Vorarlberg* [2004] ECR I-4981.

Case C-452/01, *Margarethe Ospelt and Schlössle Weissenberg Familienstiftung* [2003] ECR I-9743.

Case C-463/01, *Commission of the European Communities v Federal Republic of Germany* [2004] ECR I-11705.

Case C-491/01, The *Queen v Secretary of State for Health, ex parte British American Tobacco (Investments) Ltd and Imperial Tobacco Ltd* [2002] ECR I-11453.

Case C-1/02, *Borgmann* [2004] ECR I-3219.

Case C-309/02, *Radlberger Getränkegesellschaft mbH & Co. and S. Spitz KG v Land Baden-Württemberg* [2004] ECR I-11763.

Case C-6/03, *Deponiezweckverband Eiterköpfe v Land Rheinland-Pfalz* [2005] ECR I-2753.

Case C-86/03, *Greece* v *Commission* [2005] ECR I-10979.

Case C-94/03, *Commission of the European Communities v Council of the European Union* [2006] ECR I-1.

Case C-176/03, *Commission of the European Communities v Council of the European Union* [2005] ECR I-7879.

Case C-178/03, *Commission of the European Communities v European Parliament and Council of the European Union* [2006] ECR I-107.

Case C-210/03, *Swedish Match AB and Swedish Match UK Ltd v Secretary of State for Health* [2004] ECR I-11893.

Case C-320/03, *Commission of the European Communities v Republic of Austria* [2005] ECR I-9871.

Case C-376/03, *D. v Inspecteur van de Belastingdienst/Particulieren/Ondernemingen buitenland te Heerlen* [2005] ECR I-5821.

Case C-380/03, *Federal Republic of Germany v European Parliament and Council of the European Union* [2006] ECR I-11573.

Case C-436/03, *European Parliament v Council of the European Union* [2006] ECR I-3733.

Case C-459/03, *Commission of the European Communities v Ireland* [2006] ECR I-4635.

Case C-533/03, *Commission of the European Communities v Council of the European Union* [2006] ECR I-1025.

Case C-66/04, *United Kingdom of Great Britain and Northern Ireland v European Parliament and Council of the European Union* [2005] ECR I-10553.

Case C-156/04, *Commission of the European Communities v Hellenic Republic* [2007] ECR I-4129.

Joined Cases C-158/04 and C-159/04, *Alfa Vita Vassilopoulos AE, formerly Trofo Super-Markets AE and Carrefour Marinopoulos AE v Elliniko Dimosio, Nomarchiaki Aftodioikisi Ioanninon* [2006] ECR I-8135.

Case C-168/04, *Commission of the European Communities v French Republic* [1980] ECR 347.

Case C-174/04, *Commission of the European Communities v Italian Republic* [2005] ECR I-4933.

Case C-217/04, *United Kingdom of Great Britain and Northern Ireland v European Parliament and Council of the European Union (hereinafter ENISA case)* [2006] ECR I-3771.

Joined Cases C-282/04 and C-283/04, *Commission of the European Communities v Kingdom of the Netherlands* [2006] ECR I-9141.

Joined Cases C-393/04 and C-41/05, *Air Liquide Industries Belgium SA v Ville de Seraing and Province de Liège* [2006] ECR I- 05293.

Cases C-413/04 and C-414/04, *European Parliament v Council of the European Union*, [2006] ECR I-11221.

Case C-446/04, *Test Claimants in the FII Group Litigation v Commissioners of Inland Revenue* [2006] ECR I-11753.

Joined Cases C-463/04 and C-464/04, *Federconsumatori, Adiconsum, ADOC, Ercole Pietro Zucca (C-463/04) and Associazione Azionariato Diffuso dell'AEM SpA, Filippo Cuccia, Giacomo Fragapane Pietro Angelo Puggioni, Annamaria Sanchirico, Sandro Sartorio (C-464/04) v Comune di Milano* [2007] ECR I-10419.

Case C-517/04, *Visserijbedrijf D.J. Koornstra & Zn. Vof v Productschap Vis* [2006] ECR I-5015.

Case C-91/05, *Commission of the European Communities v Council of the European Union* [2008] ECR I-3651.

Case C-110/05, *Commission of the European Communities v Italian Republic* [2009] ECR I-519.

Case C-112/05, *Commission of the European Communities v Federal Republic of Germany* [2007] ECR I-8995.

Case C-142/05, *Åklagaren v Percy Mickelsson and Roos* [2008] ECR I-7841.

Case C-157/05, *Winfried L. Holböck v Finanzamt Salzburg-Land* [2007] ECR I-4051.

Case C-173/05, *Commission of the European Communities v Italian Republic* [2007] ECR I-4917.

Case C-313/05, *Maciej Brzeziński v Dyrektor Izby Celnej w Warszawie* [2007] ECR I-513.

Case C-319/05, *Commission of the European Communities v* Federal Republic of Germany [2007] ECR I-9811.

Joined Cases C-402/05P and 415/05, *Yassin Abdullah Kadi and Al Barakaat International Foundation v Council of the European Union and Commission of the European Communities* [2008] ECR I-6351.

Joined Cases C-439/05P and C-454/05P, *Land Oberösterreich and Republic of Austria v Commission of the European Communities* [2007] ECR I-7141.

Case C-265/06, *Commission of the European Communities v Portuguese Republic* [2008] ECR I-2245.

Case C-274/06, *Commission of the European Communities v Kingdom of Spain* [2008] ECR I-26.

Case C-411/06, *Commission of the European Communities v European Parliament and Council of the European Union* [2007] ECR I-6199.

Case C-439/06, *Citiworks AG v Flughafen Leipzig/Halle GmbH, Bundesnetzagentur* [2008] ECR I-3919.

Case C-88/07, *Commission of the European Communities v Kingdom of Spain* [2009] ECR I-1353.

Case C-155/07, *European Parliament v Council of the European Union* [2008] ECR I-81103.

Case C-166/07, *European Parliament v Council of the European Union* [2009] ECR I-7135.

Case C-187/07, *Endendijk*, [2008] ECR I-2115.

Case C-239/07, *Julius Sabatauskas and Others* [2008] ECR I- 7523.

Case C-326/07, *Commission of the European Communities v Italian Republic* [2009] ECR I-2291.

Case C-171/08, *European Commission v Portuguese Republic* not yet reported.

Case C-265/08, *Federutility, Assogas, Libarna Gas SpA, Collino Commercio SpA, Sadori Gas Srl, Egea Commerciale Srl, E.On Vendita Srl, Sorgenia SpA v Autorità per l'energia elettrica e il gas*, not yet reported.

Case C-543/08, *European Commission v Portuguese Republic*, not yet reported.

Case C-366/10, Reference for a preliminary ruling from High Court of Justice Queen's Bench Division (Administrative Court) (United Kingdom) made on 22 July 2010 – *The Air Transport Association of America, American Airlines, Inc., Continental Airlines, Inc., United Airlines, Inc. v The Secretary of State for Energy and Climate Change*, *OJ* C260/6.

LIST OF GENERAL COURT CASES AND OTHER EUROPEAN CASES

Case T-76/96 R, *The National Farmers' Union, International Traders Ferry Ltd, UK Genetics, RS & EM Wright Ltd and Prosper De Mulder Ltd v Commission of the European Communities* [1996] ECR II-815.

Joined Cases T-125/96 and T-152/96, *Boehringer Ingelheim Vetmedica GmbH and C.H. Boehringer Sohn v Council of the European Union and Commission of the European Communities* [1999] ECR II-3427.

Case T-199/96, *Laboratoires pharmaceutiques Bergaderm SA and Jean-Jacques Goupil v Commission of the European Communities* [1998] ECR II-2805.

Case T-13/99, *Pfizer Animal Health SA v Council of the European Union* [2002] ECR II-3305.

Case T-70/99 R, *Alpharma Inc. v Council of the European Union* [1999] ECR II-2027.

Case T-70/99, *Alpharma Inc. v Council of the European Union* [2002] ECR II-3495.

Case T-392/02, *Solvay Pharmaceuticals* v *Council* [2003] ECR II-4555.

Joined Cases T-366/03 and T-235/04, *Land Oberösterreich and Republic of Austria v Commission of the European Communities* [2005] ECR II-4005.

Case T-229/04, *Kingdom of Sweden v Commission of the European Communities* [2007] ECR II-2437.

Case T-182/06, *Kingdom of the Netherlands v Commission of the European Communities* [2007] ECR II-1983.

LIST OF ECJ OPINIONS

Opinion 1/75 Draft OECD Understanding on Local Cost Standard [1975] ECR 1255.

Opinion 1/76 of the Court of Justice of 26 April 1977 on the *Draft Agreement establishing a European laying-up fund for inland waterway vessels* [1977] ECR 741.

Opinion 1/94, *Competence of the Community to conclude international agreements concerning services and the protection of intellectual property* [1994] ECR I-5267.

Opinion 2/94 of the Court of 28 March 1996, *Accession by the Community to the European Convention for the Protection of Human Rights and Fundamental Freedoms* [1996] ECR I-1759.

Opinion 2/00 of the Court of Justice of 6 December 2001 on the *Cartagena Protocol* [2001] ECR 9713.

LIST OF ADVOCATE GENERAL OPINIONS

Opinion of Advocate General Sir Gordon Slynn in Case 72/83, *Campus Oil Limited and others v Minister for Industry and Energy and others.*

Opinion of Advocate General Tesauro in Case C-300/89, *Commission of the European Communities v Council of the European Communities.*

Reasoned submission by Advocate General Tizzano in Case C-339/00, *Ireland v Commission of the European Communities.*

Opinion of Advocate General Jacobs in Case C-350/92, *Kingdom of Spain v Council of the European Union.*

Opinion of Advocate General Jacobs in Case C-412/93, *Société d'Importation Edouard Leclerc-Siplec v TF1 Publicité SA and M6 Publicité SA.*

Opinion of Advocate General Cosmans in Joined Cases C-157/94 *Commission* v *Netherlands*, C-158/94 *Commission* v *Italy*, C-159/94 *Commission* v *France* and C-160/94 *Commission* v *Spain.*

Opinion of Advocate General Léger in Case C-233/94, *Federal Republic of Germany v European Parliament and Council of the European Union.*

Opinion of Advocate General Fennelly in Case C-376/98, *Federal Republic of Germany v European Parliament and Council of the European Union.*

Opinion Advocate General Léger in Case C-36/98, *Kingdom of Spain v Council of the European Union.*

Opinion Advocate General Jacobs in Case C-379/98, *PreussenElektra AG v Schhleswag AG, in the presence of Windpark Reußenköge III GmbH and Land Schleswig-Holstein.*

Opinion of Advocate General Léger in Case C-438/02, *Krister Hanner.*

Opinion of the Advocate General Ruiz-Jarabo Colomer in Case C-6/03, *Deponiezweckverband Eiterköpfe v Land Rheinland-Pfalz.*

Opinion of Advocate General Poiares Maduro in Case C-273/04, *Republic of Poland v Council of the European Union.*

Opinion of Advocate General Ruiz-Jarabo Colomer in Case C-265/08, *Federutility, Assogas, Libarna Gas SpA, Collino Commercio SpA, Sadori Gas Srl, Egea Commerciale Srl, E.On Vendita Srl, Sorgenia SpA v Autorità per l'energia elettrica e il gas.*

Opinion of Advocate General Poiares Maduro Case C-411/06, *Commission of the European Communities v European Parliament and Council of the European Union.*

Opinion Advocate General Kokott of 6 October 2011 in the Case C-366/10, Reference for a preliminary ruling from High Court of Justice Queen's Bench Division (Administrative Court) (United Kingdom) made on 22 July 2010 – *The Air Transport Association of America, American Airlines, Inc., Continental Airlines, Inc., United Airlines, Inc. v The Secretary of State for Energy and Climate Change.*

LIST OF JUDGMENTS OF NATIONAL COURTS

Brunner v European Union Treaty, German Constitutional Court (Bundesverfassungsgericht) BverfGE 89, 155 (October 12,1993), translated in English at *Common Market Law Review* (1994) 57 and 33 *I.L.M.* (1994) p. 388.

LIST OF UNION TREATIES

Treaty establishing the European Coal and Steel Community.

Treaty establishing the European Economic Community.

Treaty establishing the European Atomic Energy Community.

Single European Act.

Treaty of Amsterdam amending the Treaty on European Union, the Treaties establishing the European Communities and certain related acts.

Treaty of Nice amending the Treaty on European Union, the Treaties establishing the European Communities and certain related acts.

Treaty of Lisbon amending the Treaty on European Union and the Treaty establishing the European Community.

LIST OF TREATY PROTOCOLS AND DECLARATIONS

Declaration 41 on Article 308 of the TFEU.

Declaration of the role of national parliaments in the EU annexed to the Final Act of the Maastricht Treaty and the Protocols annexed to the Amsterdam and Nice Treaties.

Protocol (No 1) to the Treaty on the Functioning of the European Union on the role of national parliaments in the European Union.

Protocol (No 2) to the Treaty on the Functioning of the European Union on the application of the principles of subsidiarity and proportionality.

Protocol (No 2) Amending the Treaty establishing the European Atomic Energy Community.

LIST OF UNION REGULATIONS

Regulation of the Supply Agency of the European Atomic Energy Community amending the rules of the Supply Agency of 5 May 1960 determining the manner in which demand is to be balanced against the supply of ores, source materials and special fissile materials, *OJ* L 193/37.

Regulation No 293/74 of the Council of 30 January 1974 on information for the establishment of comprehensive energy balance sheets for the Community, *OJ* L 32/1.

Council Regulation (EC) No 2236/95 of 18 September 1995 laying down general rules for the granting of Community financial aid in the field of trans-European networks, *OJ* L 228/1.

Council Regulation No 736/96 of 22 April 1996 on notifying the Commission of investment projects of interest to the Community in the petroleum, natural gas and electricity sectors, *OJ* L 102/1.

Regulation No 2422/2001 of the European Parliament and of the Council of 6 November 2001 on a Community energy efficiency labelling programme for office equipment, *OJ* L 332/1.

Council Regulation (EC) No 1/2003 of 16 December 2002 on the implementation of the rules on competition laid down in Articles 81 and 82 of the Treaty, *OJ* L1/1.

Regulation No 1228/2003 of 26 June 2003 on conditions for access to the network for cross-border exchanges in electricity, *OJ* L 20/24.

Regulation (EC) No 1159/2005 of the European Parliament and of the Council of 6 July 2005 amending Council Regulation (EC) No 2236/95 laying down general rules for the granting of Community financial aid in the field of trans-European networks, *OJ* L 191/16.

Regulation 1775/2005 of the European Parliament and of the Council of 28 September 2005 on conditions for access to the natural gas transmission networks, *OJ* L 289/1.

Regulation (EC) No 680/2007 of the European Parliament and of the Council of 20 June 2007 laying down general rules for the granting of Community financial aid in the field of the trans-European transport and energy networks, *OJ* L 162/1.

Regulation (EC) No 106/2008 of the European Parliament and of the Council of 15 January 2008 on a Community energy-efficiency labelling programme for office equipment, *OJ* L 39/1.

Regulation (EC) No 663/2009 of the European Parliament and of the Council of 13 July 2009 establishing a programme to aid economic recovery by granting Community financial assistance to projects in the field of energy, *OJ* L 200/31.

Regulation (EC) No 713/2009 of the European Parliament and of the Council of 13 July 2009 establishing an Agency for the Cooperation of Energy Regulators, *OJ* L 211/1.

Regulation (EC) No 714/2009 of the European Parliament and of the Council of 13 July 2009 on conditions for access to the network for cross-border exchanges in electricity and repealing Regulation (EC) No 1228/2003, *OJ* L 211/15.

Regulation (EC) No 715/2009 of the European Parliament and of the Council of 13 July 2009 on conditions for access to the natural gas transmission networks and repealing Regulation (EC) No 1775/2005, *OJ* L 211/36.

Commission Regulation (EC) No 859/2009 of 18 September 2009 amending Regulation (EC) No 244/2009 as regards the ecodesign requirements on ultraviolet radiation of non-directional household lamps, *OJ* L 247/3.

Regulation (EC) No 1222/2009 of the European Parliament and of the Council of 25 November 2009 on the labelling of tyres with respect to fuel efficiency and other essential parameters, *OJ* L 342/46.

Regulation (EU) No 539/2010 of the European Parliament and of the Council of 16 June 2010 amending Council Regulation (EC) No 1083/2006 laying down general provisions on the European Regional Development Fund, the European Social Fund and the Cohesion Fund as regards simplification of certain requirements and as regards certain provisions relating to financial management, *OJ* L 158/1.

Council Regulation (EU, Euratom) No 617/2010 of 24 June 2010 concerning the notification to the Commission of investment projects in energy infrastructure within the European Union and repealing Regulation No 736/96, *OJ* L 180/7.

Regulation No 994/2010 of the European Parliament and of the Council of 20 October 2010 concerning measures to safeguard security of gas supply and repealing Council Directive 2004/67/EC, *OJ* L 295/1.

Commission Regulation (EU) No 1015/2010 of 10 November 2010 implementing Directive 2009/125/EC of the European Parliament and of the Council with regard to ecodesign requirements for household washing machines, *OJ* L 293/21.

Commission Regulation (EU) No 1016/2010 of 10 November 2010 implementing Directive 2009/125/EC of the European Parliament and of the Council with regard to ecodesign requirements for household dishwashers, *OJ* L 293/31.

LIST OF UNION DIRECTIVES

Directive 68/414/EEC of 20 December 1968 imposing an obligation on Member States of the EEC to maintain minimum stocks of crude oil, *OJ* L 308/14.

Directive 73/238/EEC of 24 July 1973 on measures to mitigate the effects of difficulties in the supply of crude oil and petroleum products, *OJ* L 228/1.

Directive 75/339/EEC of 20 May 1975 obliging the Member States to maintain minimum stocks of fossil fuel at thermal power stations, *OJ* L 153/35.

Council Directive 75/442/EEC of 15 July 1975 on waste, *OJ* L 194/39.

Council Directive 76/769/EEC of 27 July 1976 on the approximation of the laws, regulations and administrative provisions of the Member States relating to restrictions on the marketing and use of certain dangerous substances and preparations, *OJ* L 262/201.

Council Directive 85/337/EEC of 27 June 1985 on the assessment of the effects of certain public and private projects on the environment, *OJ* L 216/40.

Council Directive 85/536/EEC of 5 December 1985 on crude-oil savings through the use of substitute fuel components in petrol, *OJ* L 34/20.

Council Directive 90/377/EEC of 29 June 1990 concerning a Community procedure to improve the transparency of gas and electricity prices charged to industrial end-users, *OJ* L 185/16.

Council Directive 90/547/EEC of 29 October 1990 on the transit of electricity through transmission grids, *OJ* L 313/30.

Council Directive 91/296/EEC of 31 May 1991 on the transit of natural gas through grids, *OJ* L 147/37

Council Directive 91/414/EEC of 15 July 1991 concerning the placing of plant protection products on the market, *OJ* L 230/1.

Council Directive 92/43/EEC of 21 May 1992 on the conservation of natural habitats and of wild fauna and flora, *OJ* L 206/7.

Council Directive 92/75/EEC of 22 September 1992 on the indication by labelling and standard product information of the consumption of energy and other resources by household appliances and its amendments and implementing measures, *OJ* L 297/16.

Council Directive 93/76/EEC of 13 September 1993 to limit carbon dioxide emissions by improving energy efficiency (SAVE), *OJ* L 237/28.

Directive 94/62/EC of the European Parliament and Council of 20 December 1994 on packaging and packaging waste, *OJ* L 365/10.

Commission Directive 95/12/EC of 23 May 1995 implementing Council Directive 92/75/EEC with regard to energy labelling of household washing machines, *OJ* L 136/35, as amended by Directive 96/89/EC, *OJ* L 388/85.

Council Directive 96/48/EC of 23 July 1996 on the interoperability of the trans-European high-speed rail system, *OJ* L 235/6.

Directive 96/57/EC of 3 September 1996 on energy efficiency requirements for household electric refrigerators, freezers and combinations thereof, *OJ* L 236/36.

Council Directive 96/82/EC of 9 December 1996 on the control of major-accident hazards involving dangerous substances, *OJ* L 10/13.

Directive 96/92/EC of the European Parliament and the Council of 19 December 1996 concerning common rules for the internal market in electricity, *OJ* L 27/10.

Directive 98/30/EC of the European Parliament and the Council of 22 June 1998 concerning common rules for the internal market in natural gas, *OJ* L 204/1.

Directive 98/8/EC of the European Parliament and of the Council of 16 February 1998 concerning the placing of biocidal products on the market, *OJ* L 123/1.

Directive 98/93/EC of 14 December 1998 amending Directive 68/414/EEC imposing an obligation on Member States of the EEC to maintain minimum stocks of crude oil and/or petroleum products, *OJ* L 358/100.

Commission Directive 1999/9 of 26 February 1999 amending Directive 97/17/EC implementing Council Directive 92/75/EEC with regard to energy labelling of household dishwashers, *OJ* L 56/46.

Council Directive 1999/22/EC of 29 March 1999 relating to the keeping of wild animals in zoos, *OJ* L 94/24.

Directive 1999/94/EC of the European Parliament and of the Council of 13 December 1999 relating to the availability of consumer information on fuel economy and CO_2 emissions in respect of the marketing of new passenger cars, *OJ* L 12/16.

Directive 2000/60/EC of the European Parliament and of the Council of 23 October 2000 establishing a framework for Community action in the field of water policy, *OJ* L 327/1.

Directive 2001/16/EC of the European Parliament and of the Council of 19 March 2001 on the interoperability of the trans-European conventional rail system, *OJ* L 110/1.

Directive 2001/42/EC of the European Parliament and of the Council of 27 June 2001 on the assessment of the effects of certain plans and programmes on the environment *OJ* L 197/30.

Directive 2001/77/EC of the European Parliament and of the Council of 27 September 2001 on the Promotion of Electricity Produced from Renewable Energy Sources in the Internal Electricity Market, *OJ* L 283/33.

Directive 2002/3/EC of the European Parliament and of the Council of 12 February 2002 relating to ozone in ambient air, *OJ* L 67/14.

Directive 2002/47EC of the European Parliament and the Council of 6 June 2002 on financial collateral arrangements, *OJ* L 168/43.

Directive 2002/91/EC of the European Parliament and of the Council of 16 December 2002 on the energy performance of buildings, *OJ* L 1/65.

Directive 2003/30/EC of the European Parliament and of the Council of 8 May 2003 on the promotion of the use of biofuels or other renewable fuels for transport, *OJ* L 123/42.

Directive 2003/54/EC of 26 June 2003 concerning the common rules for the internal market in electricity and repealing Directive 96/92/EC, *OJ* L 176/37.

Directive 2003/55/EC of the European Parliament and the Council of 26 June 2003 concerning common rules for the internal market in natural gas and repealing Directive 98/30/EC, *OJ* L 176/57.

Commission Directive 2003/66/EC of 3 July 2003 amending Directive 94/2/EC implementing Council Directive 92/75/EEC with regard to energy labelling of household electric refrigerators, freezers and their combinations, *OJ* L 170/10.

European Parliament and Council Directive 2003/87/EC of 13 October 2003 establishing a scheme for greenhouse gas emission allowance trading within the Community and amending Council Directive 96/61/EC *OJ* L 275/32.

Council Directive 2003/96/EC of 27 October 2003 restructuring the Community framework for the taxation of energy products and electricity, *OJ* L 283/51, last amended by Directives 2004/74/EC and 2004/75/EC, *OJ* L 157/87.

Directive 2004/8/EC of the European Parliament and of the Council of 11 February 2004 on the promotion of cogeneration based on a useful heat demand in the internal energy market and amending Directive 92/42/EEC, *OJ* L 52/50.

Council Directive 2004/67 of 26 April 2004 concerning measures to safeguard security of natural gas supply in 2004, *OJ* L 127/92.

Directive 2004/107/EC of the European Parliament and of the Council of 15 December 2004 relating to arsenic, cadmium, mercury, nickel and polycyclic aromatic hydrocarbons in ambient air, *OJ* L 23/3.

Directive 2005/32/EC of the European Parliament and of the Council of 6 July 2005 establishing a framework for the setting of ecodesign requirements for energy-using products and amending Council Directive 92/42/EEC and Directives 96/57/EC and 2000/55/EC of the European Parliament and of the Council, *OJ* L 191/29.

Directive 2005/89/EC of the European Parliament and of the Council of 18 January 2006 concerning measures to safeguard security of electricity supply and infrastructure investment, *OJ* L 33/22.

Directive 2006/7/EC of the European Parliament and of the Council of 15 February 2006 concerning the management of bathing water quality and repealing Directive 76/160/EEC, *OJ* L 64/37.

Directive 2006/32/EC on the promotion of end-use efficiency and energy efficiency, *OJ* L 114/64.

Council Directive 2006/67/EC of 24 July 2006 imposing an obligation on Member States to maintain minimum stocks of crude oil and/or petroleum products, *OJ* L 217/209.

Directive 2008/1/EC of the European Parliament and of the Council of 15 January 2008 concerning integrated pollution prevention and control (codified version), *OJ* L 24/8.

Directive 2008/92/EC of the European Parliament and of the Council of 22 October 2008 concerning a Community procedure to improve the transparency of gas and electricity prices charged to industrial end-users, *OJ* L 298/9.

Directive 2008/98/EC of 19 November 2008 on waste and repealing certain Directives, *OJ* L 312/3.

Directive 2009/28/EC of the European Parliament and of the Council of 23 April 2009 on the promotion of the use of energy from renewable sources and amending and subsequently repealing Directives 2001/77/EC and 2003/30/EC, *OJ* L 140/16.

Directive 2009/29/EC of the European Parliament and of the Council of 23 April 2009 amending Directive 2003/87/EC so as to improve and extend the greenhouse gas emission allowance trading scheme of the Community, *OJ* L 140/63.

Directive 2009/31/EC of the European Parliament and of the Council of 23 April 2009 on the geological storage of carbon dioxide and amending Council Directive 85/337/EEC, European Parliament and Council Directives 2000/60/EC, 2001/80/EC, 2004/35/EC, 2006/12/EC, 2008/1/EC and Regulation (EC) No 1013/2006, *OJ* L 140/114.

Directive 2009/72/EC of the European Parliament and the Council of 13 July 2009 concerning common rules for the internal market in electricity and repealing Directive 2003/54/EC, *OJ* L 211/55.

Directive 2009/73/EC of the European Parliament and of the Council of 13 July 2009 concerning common rules for the internal market in natural gas and repealing Directive 2003/55/EC, *OJ* L 211/94.

Council Directive 2009/119/EC of 14 September 2009 imposing an obligation on Member States to maintain minimum stocks of crude oil and/or petroleum products (from 2012), *OJ* L 265/9.

Directive 2009/125/EC of the European Parliament and of the Council of 21 October 2009 establishing a framework for the setting of ecodesign requirements for energy-using products, *OJ* L 285/10.

Directive 2010/30/EU of the European Parliament and of the Council of 19 May 2010 on the indication by labelling and standard products information of the consumption of energy and other resources by energy-related products, *OJ* L 153/1.

Directive 2010/31/EU of the European Parliament and of the Council of 19 May 2010 on the energy performance of buildings, *OJ* L 153/13.

LIST OF UNION DECISIONS

Decision 77/706/EEC on the setting of a community target for a reduction in the consumption of primary energy sources of energy in the event difficulties in the supply of crude oil and petroleum products, *OJ* L 292/9.

Council Decision 93/500/EEC of 13 September 1993 concerning the promotion of renewable energy sources in the Community (ALTENER programme), *OJ* L 235/41.

Council Decision of 28 March 1996 laying down a series of measures aimed at creating a more favourable context for the development of trans-European networks in the energy sector, *OJ* L 161/154.

Council and Commission Decision 98/191/EC, ECSC, EURATOM of 23 September 1997 on the conclusion, by the European Communities, of the Energy Charter Treaty and the Energy Charter Protocol on energy efficiency and related environmental aspects, *OJ* L 69/1.

Council Decision 98/352/EC of 18 May 1998 concerning a multiannual programme for the promotion of renewable energy sources in the Community (ALTENER II programme), *OJ* L 159/53.

Council Decision no 647/2000/EC of the European Parliament and of the Council of 28 February 2000 adopting a multiannual programme for the promotion of energy efficiency (SAVE) (1998 to 2002), *OJ* L 79/6.

Council Decision 2001/469/EC of 14 May 2001 concerning the conclusion on behalf of the European Community of the Agreement between the Government of the United States of America and the European Community on the coordination of energy-efficient labelling programs for office equipment, *OJ* L 172/1.

Council Decision 2002/358/EC of 25 April 2002 concerning the approval, on behalf of the European Community, of the Kyoto Protocol to the United Nations Framework

Convention on Climate Change and the joint fulfilment of commitments thereunder, *OJ* L 130/1.

Decision 1600/2002/EC of the European of the European Parliament and of the Council of 22 July 2002 laying down the Sixth Community Environment Action Programme, *OJ* L 242/1.

Commission Decision of 11 March 2003 established a European Community Energy Star Board (ECESB) to manage and coordinate the introduction of the Energy Star in Europe: Commission Decision 2003/168/EC of 11 March 2003 establishing the European Community Energy Star Board, *OJ* L 67/22.

Commission Decision 2003/367/EC of 15 May 2003 establishing the Rules of Procedure of the European Community Energy Star Board, *OJ* L 125/9.

Decision No 1229/2003/EC of the European Parliament and of the Council of 26 June 2003 laying down a series of guidelines for trans-European energy networks and repealing Decision No 1254/96/EC, *OJ* L 176/11.

Decision No 1230/2003/EC of the European Parliament and of the Council of 26 June 2003 adopting a multiannual programme for action in the field of energy: "Intelligent Energy – Europe", *OJ* L 176/29.

Council Decision 2005/905/EC of 17 October 2005 on the signing by the European Community of the Energy Community Treaty, *OJ* L 329/30.

Council Decision 2006/500/EC of 29 May 2006 on the conclusion by the European Community of the Energy Community Treaty, *OJ* L 198/15.

Commission Decision 2006/770/EC of 9 November 2006 amending the Annex to the Regulation No 1228/2003 on conditions for access to the network for cross-border exchanges in electricity, *OJ* L 312/59.

Decision 1364/2006/EC of the European Parliament and of the Council of 6 September 2006 laying down guidelines for trans-European energy networks and repealing Decision 96/391/EC, *OJ* L 262/1.

Council Decision 2006/1005/EC of 18 December 2006 concerning the conclusion on behalf of the European Community of the Agreement between the Government of the United States of America and the European Community on the coordination of energy-efficient labelling programs for office equipment, *OJ* L 381/24.

Decision No 406/2009/EC of the European Parliament and of the Council of 23 April 2009 on the effort of Member States to reduce their greenhouse gas emissions to meet the Community's greenhouse gas emission reduction commitments up to 2020, *OJ* L 140/136.

LIST OF COMMISSION DOCUMENTS

Commission of the European Communities of 14 June 1985, *Completing the Internal Market: White Paper from the Commission to the European Council*, COM(1985)310.

Commission working document of 2 May 1988, *The internal energy market*, COM(1988)238.

Communication from the Commission to the Council and the European Parliament, *The principle of subsidiarity*, SEC(92)1990.

Commission of the European Communities, Proposal for a Council Directive concerning common rules for the internal market in electricity and Proposal for a Council Directive concerning common rules for the internal market in natural gas, COM(1991)548.

Commission Report to the Council, Parliament and ESC of 23 July 2002, *Second Biennial Report on the Application of the Principle of Mutual Recognition in the Single Market*, COM(1992)419.

White Paper of the European Commission of 5 December 1993 on Growth, Competitiveness, Employment – The Challenges and Ways Forward into the 21st Century, COM(93)700.

Green Paper, *For a European Union Energy Policy*, COM(94)659.

White Paper of the European Commission of 13 December 1995, *An energy policy for the European Union*, COM(1995)628.

White Paper on an Energy Policy for the European Union, COM(95)682.

Communication of the Commission of 26 November 1997 on Energy for the Future: Renewable Sources of Energy, White Paper for a Community Strategy and Action Plan, COM (97)599.

White paper of the European Commission of 18 December 2007 on the integration of EU Mortgage Credit Markets, COM(2007)807.

Report from the Commission to the Council on Civil Protection, Tourism and Energy, SEC(1996)496.

Commission of the European Communities of 3 November 1999, *Commission report to the European Council – Better lawmaking 1999*, COM(99)562.

Communication of 2 February 2000 from the Commission on the precautionary principle, COM(2000)1.

Committee and Committee of the Regions: Action Plan to Improve Energy Efficiency in the European Community of 26 April 2000, COM(2000)247.

European Commission Green Paper towards a European strategy for the security of the energy supply of 29 November 2000, COM (2000)769.

European Commission, *Guide to the concept and practical application of Articles 28–30 EC*, January 2001.

Communication from the Commission of 2001 – *Services of general interest in Europe* (2001/C 17/04).

Communication from the Commission of 7 November 2001 to the European Parliament, the Council, the Economic and Social Committee and the Committee of the Regions on alternative fuels for road transportation and on a set of measures to promote the use of biofuels, COM(2001)547.

Communication from the Commission to the Council and the European Parliament, *Report on the Green Paper Towards a European Strategy for the Security of Energy Supply*, COM(2002)321.

Proposal for a Directive of the European Parliament and of the Council on the promotion of the use of biofuels for transport presented by the Commission pursuant to Art. 250 (2) of the EC-Treaty, COM(2002)508.

Proposal for a Directive of the European Parliament and of the Council on energy end-use efficiency and energy services, COM(2003)739.

Communication from the Commission to the Council and the European Parliament on the share of renewable energy in the EU, Commission Report in accordance with Article 3 of Directive 2001/77/EC, evaluation of the effect of legislative instruments and other Community policies on the development of the contribution of renewable energy sources in the EU and proposals for concrete actions of 26 May 2004, COM(2004)366.

Green paper of 1 March 2005 of the European Commission on successions and wills, COM(2005)65.

Proposal for a Directive of the European Parliament and of the Council on energy end-use efficiency and energy services, 19 May 2005, Council of the European Union, interinstitutional file 2003/0300 (COD).

European Commission Competition DG, *Energy sector inquiry – issues paper non-confidential version of 15 November 2005.*

Green paper of 8 march 2006, *A European Strategy for Sustainable, Competitive and Secure Energy* of the Commission, COM(2006)105.

Commission staff working document, *Annex to the Green Paper, A European strategy for sustainable, competitive and secure energy – what is at stake – background document*, COM(2006)105.

Communication from the Commission of 19 October 2006, *Action Plan for Energy Efficiency: Realising the Potential*, COM(2006)545.

Proposal for a Directive of the European Parliament and of the Council amending Directive 2003/87/EC so as to include aviation in the scheme for greenhouse gas emission allowance trading within the Community, COM(2006)818.

Commission staff working document of 20 December 2006, *Accompanying document to the Proposal for a Directive of the European Parliament and of the Council amending Directive 2003/87/EC so as to include aviation in the scheme for greenhouse gas emission allowance trading within the Community Impact Assessment of the inclusion of aviation in the scheme for greenhouse allowance trading within the Community*, SEC(2006)1684.

Communication from the Commission to the Council and the European Parliament, *Renewable Energy Road Map Renewable energies in the 21st century: building a more sustainable future*, COM(2006)848.

DG Competition Report on Energy Sector Inquiry of 10 January 2007, SEC(2007)1724.

Communication from the Commission of 10 January 2007 to the Council and the European Parliament, *Prospects for the internal gas and electricity market*, COM(2006)841.

Communication from the Commission to the Council and the European Parliament of 10 January 2007, *Priority interconnection plan*, COM(2006)846.

Communication of the Commission of 10 January 2007, *Renewable Energy Road Map – Renewable energies in the 21st century: building a more sustainable future*, COM(2006)848.

Communication from the Commission of 10 January 2007, *Inquiry pursuant to Article 17 of Regulation (EC) No 1/2003 into the European gas and electricity sectors (Final Report)*, COM(2006)851.

Communication from the Commission to the European Council and the European Parliament of 10 January 2007, *An energy policy for Europe*, COM(2007)1.

Proposal for a Directive of the European Parliament and of the Council amending Directive 2003/54/EC of the European Parliament and of the Council of 26 June 2003 concerning common rules for the internal market in electricity, COM(2007)528.

Proposal for a Directive of the European Parliament and of the Council amending Directive 2003/55/EC of the European Parliament and of the Council of 26 June 2003 concerning common rules for the internal market in natural gas, COM(2007)529.

Proposal for a Regulation of the European Parliament and of the Council establishing an Agency for the Cooperation of Energy Regulators, COM(2007)530.

Proposal for a Regulation of the European Parliament and of the Council amending Regulation (EC) No 1228/30 on conditions for access to the network for cross-border exchanges in electricity, COM(2007)531.

Proposal for a Regulation of the European Parliament and of the Council amending regulation (EC) No 1775/2005 on conditions for access to the natural gas transmission networks, COM(2007)532.

Communication from the Commission of 22 November 2007 to the Council, the European Parliament, the European Economic and Social Committee and the Committee of the Regions, *A European strategic energy technology plan (SET-plan) – Towards a low carbon future*, COM(2007)723.

Communication from the Commission to the Council and the European Parliament on a first assessment of national energy efficiency action plans as required by Directive 2006/32/EC on energy end-use efficiency and energy services – Moving forward together on energy efficiency, COM(2008)11

Communication from the Commission to the European Parliament, the Council, the European Economic and Social Committee and the Committee of the Regions – Supporting early demonstration of sustainable power generation from fossil fuels, COM(2008)13.

Proposal for a Directive of the European Parliament and of the Council amending Directive 2003/87/EC so as to improve and extend the greenhouse gas emission allowance trading system of the Community, COM(2008)16.

Proposal for a Decision of the European Parliament and of the Council on the effort of Member States to reduce their greenhouse gas emissions to meet the Community's greenhouse gas emission reduction commitments up to 2020, COM(2008)17.

Proposal for a Directive of the European Parliament and of the Council on the promotion of the use of energy from renewable sources, COM(2008)19.

Communication from the Commission to the European Parliament, the Council, the European Economic and Social Committee and the Committee of the Regions – *2020 by 2020 – Europe's climate change opportunity*, COM(2008)30.

Communication from the Commission to the European Parliament, the Council, the European Economic and Social Committee and the Committee of Regions of 13 November 2008 on the Directive 2004/76/EC of 26 April 2004 concerning measures to safeguard security of natural gas supply, COM(2008)769.

Communication from the Commission of 13 November 2008 to the European Parliament, the Council, the European Economic and Social Committee and the

Committee of the Regions, *Second Strategic Energy Review: an EU energy security and solidarity action plan*, COM(2008)781.

Communication from the Commission to the European Council – *A European Economic Recovery Plan*, COM(2008)800.

Impact Assessment Guidelines, SEC(2009)92 of 15 January 2009.

White paper of 1 April 2009, *Adapting to climate change: Towards a European framework for action*, COM(2009)147.

MEMO/09/176 of 21 April 2009 of the European Commission, "Q&A on the European Parliament's vote on the Third Internal Energy Market Legislative Package".

Commission staff working document of 16 July 2009 accompanying document to the Proposal for a Regulation of the European Parliament and of the Council concerning measures to safeguard security of gas supply and repealing Directive 2004/67/EC, COM(2009)978.

Proposal for a Council Regulation concerning the notification to the Commission of investment projects in energy infrastructure within the European Community and repealing Regulation (EC) No 736/96, COM(2009)361.

Proposal of 16 July 2009 for a Regulation of the European Parliament and of the Council concerning measures to safeguard security of gas supply and repealing Directive 2004/67/EC, COM(2009)363.

Commission staff working document, Free Movement of Goods – Guide to the application of Treaty provisions governing Free Movement of Goods (Articles 28–30 EC) of 12 May 2009, SEC(2009)673.

Commission staff working document of 23 June 2009, *Synthesis of the complete assessment of all 27 National Energy Efficiency Action Plans as required by the Directive 2006/32/EC on energy end-use efficiency and energy services, Moving forward together on saving energy*, SEC(2009)889.

Commission staff working document of 16 July 2009 accompanying document to the proposal for a regulation of the European Parliament and of the Council concerning measures to safeguard security of supply and repealing Directive 2004/67/EC, The January 2009 Gas supply disruption to the EU: an assessment, SEC(2009)977.

Commission staff working paper of 22 January 2010 – *Interpretative note on Directive 2009/72/EC concerning common rules for the internal market in electricity and Directive 2009/73/EC concerning common rules for the internal market in natural gas – The unbundling Regime.*

Communication from the Commission of 11 March 2010 to the Council and the European Parliament, report on progress in creating the internal gas and electricity market, COM(2010)84.

Report from the Commission to the Council and the European Parliament on the implementation of the European Energy Programme for Recovery, COM(2010)191.

Report from the Commission of 4 May 2010 to the European Parliament, the Council, the European Economic and Social Committee and the Committee of the Regions on the implementation of the trans-European Energy Networks in the period 2007–2009, Pursuant to Article 17 of Regulation (EC) 680/2007 and Articles 9(2) and 15 of Decision 1364/2006/EC, COM(2010)203.

Communication from the Commission to the Council, the European Parliament, the European Economic and Social Committee and the Committee of the Regions, *Towards a comprehensive European international investment policy*, COM(2010)343.

Proposal for a Regulation of the European Parliament and of the Council establishing transitional arrangements for bilateral investment agreements between Member States and third countries, COM(2010)344.

Report of 8 October 2010 from the Commission on subsidiarity and proportionality COM(2010)504.

Commission staff working document of 10 November 2010, *State of play in the EU energy policy* accompanying document in the Communication from the Commission to the Council, the European Parliament, the European Economic and Social Committee and the Committee of the Regions, *Energy 2020 A Strategy for competitive, sustainable and secure energy*, SEC(2010)1346.

Communication from the Commission of 10 November 2010 to the Council, the European Parliament, the European Economic and Social Committee and the Committee of the Regions, *Energy 2020 A Strategy for competitive, sustainable and secure energy*, COM(2010)639.

Commission staff working paper of 11 November 2010, *An energy policy for consumers*, SEC(2010)1407.

Commission staff working document of 11 November 2010, *The functioning of the retail electricity markets for consumers in the European Union*, SEC(2010)1409.

Report of the Committee on Economic and Monetary Affairs of 12 November 2010 on the proposal for a Council regulation on State aid to facilitate the closure of uncompetitive coal mines, A7–0324/2010.

Communication from the Commission of 17 November 2010 to the European Parliament, the Council, the European Economic and Social Committee and the Committee of the Regions, *Energy infrastructure priorities for 2020 and beyond – A Blueprint for an integrated European energy network*, COM(2010)677.

Commission staff working document of impact assessment accompanying document to the Communication from the Commission to the European Parliament, the Council, the European Economic and Social Committee and the Committee of the Regions, *Energy infrastructure priorities for 2020 and beyond – A Blueprint for an integrated European energy network*, SEC(2010)1395.

Communication from the Commission of 7 December 2010 to the European Parliament and the Council, *The future Role of Regional Initiatives*, COM(2010)721.

Communication from the Commission to the European Parliament, the Council, the European Economic and Social Committee of the Regions, *A Resource-efficient Europe – Flagship initiative under the Europe 2020 Strategy*, COM(2011)21.

Communication from the Commission to the European Parliament and the Council of 31 January 2011, *Renewable Energy: Progressing towards the 2020 target*, COM(2011)31.

Communication of the Commission to the European Parliament, the Council, the European Economic and Social Committee and the Committee of the regions of 8 March 2011, *Energy Efficiency Plan 2011*, COM(2011)109.

Commission staff working document of 8 March 2011, Impact Assessment, Accompanying document to the Communication of the Commission to the European Parliament, the Council, the European Economic and Social Committee and the Committee of the Regions, *Energy Efficiency Plan 2011*, SEC(2011)277.

Communication from the Commission of 8 March 2011 to the European Parliament, the Council, the European Economic and Social Committee and the Committee of the Regions, *A Roadmap for moving to a competitive low carbon economy in 2050*, COM(2011)112.

White Paper of 28 March 2011, *Roadmap to a Single European Transport Area – Towards a competitive and resource transport efficient transport system*, COM(2011)144.

Proposal for a Council Directive amending Directive 2003/96/EC restructuring the Community framework for the taxation of energy products and electricity, COM(2011)169/3.

Proposal for a Regulation of the European Parliament and of the Council implementing enhanced cooperation in the area of the creation of unitary patent protection, COM(2011)215/3.

Communication from the Commission to the European Parliament, the Council, the European Economic and Social Committee and the Committee of the Regions, *Energy Roadmap 2050*, COM(2011)885.

Communication from the Commission to the European Parliament, the Council, the European Economic and Social Committee and the Committee of the Regions, *Renewable Energy: a major player in the European energy market*, COM(2012)271.

Commission staff working document Accompanying the document Communication from the Commission to the European Parliament, the Council, the European Economic and Social Committee and the Committee of the Regions, *Renewable Energy: a major player in the European energy market*, SWD(2012)164.

PRESS RELEASES

Europe could save 20% of its energy by 2020", *Rapid Press Releases* 22 June 2005, following the adoption of the Green Paper on Energy Efficiency.

Press release of 22 April 2009 of the European Parliament, "MEPs give green light to further liberalisation of EU electricity and gas markets".

IP/09/1490 of 8 October 2009, The Commission opens infringement proceedings against Belgium concerning its gas transit system.

IP/10/836 of 24 June 2010, Commission requests 20 Member States to implement and apply Single Market rules without delay.

Press Release (Committee on Industry, Research and Energy (ITRE) of 9 November 2010, *Energy Efficiency Action Plan: focusing on efficient buildings and eco-design.*

IP/10/1648, Brussels of 2 December 2010, *Belgium, France, Germany, Luxembourg, the Netherlands and Switzerland sign agreement towards the Single European Sky*; Treaty relating to the establishment of the functional airspace block "Europe Central" between the Federal Republic of Germany, the Kingdom of Belgium, the French Republic, the Grand Duchy of Luxembourg, the Kingdom of the Netherlands and the Swiss Confederation.

Press release of the Council of the European Union of 10 December 2010, *Council extends state aid regime for the coal sector*, PRESSE 344, 17766/10.
European Union welcomes Cancún Agreement as important step towards global framework for climate action, Cancún 11 December 2010, IP/10/1699.
Press release of the European Parliament of 23 November 2010, *Parliament asks for state aid for coal mines to be allowed until 2018.*
Press release UN Climate Change Conference in Cancún delivers balanced package of decisions, restores faith in multilateral process, UNFCCC secretariat.
IP/11/468 of 13 April 2011, *Energy taxation: Commission promotes energy efficiency and more environmental friendly products.*

VARIOUS DOCUMENTS

Report on the European Union of 7 January 1976 by Mr Leo Tindemans, Prime Minister of Belgium, to the European Council. Bulletin of the European Communities, Supplement 1/76.
Report of 1977 of the Study Group on the Role of the Public Finance in European Integration.
Third Environmental Action Programme focused on the prevention principle, *OJ* C 46/1.
Council Resolution of 16 September 1986 concerning new Community energy policy objectives for 1995 and convergence of the policies of the policies of the Member States, *OJ* C 241/1.
Council Resolution of 8 July 1996 on the White Paper An energy policy for the European Union, *OJ* C 224/1.
EAEC Council: The Statutes of the Euratom Supply Agency, *OJ* L 27/534.
EAEC Supply Agency: Rules of the Supply Agency of the European Atomic Energy Community determining the manner in which demand is to be balanced against the supply of ores, source materials and special fissile materials, *OJ* L 32/777.
EVERSON, M., MAJONE, G., "Chapter VII: The problem of Delegation", in EVERSON, M., MAJONE, G., METCALFE, L., and SCHOUT, A., "*The Role of specialized Agencies in Decentralising EU Governance*", Report presented to the Commission.
Conference of the Representatives of the Governments of the Member States of 22 February 2000, CONFER 4711/00.
European Parliament Resolution on the Commission Green Paper Towards a European Strategy of Energy Supply, COM(2000)769 – C5-0145/2001-2001/2071(COS).
Climate Change Secretariat, *Guide to the Climate Change Convention and its Kyoto Protocol* (2002).
Energy Star – European Community – Energy Efficiency labelling programme for office equipment – Working Plan, 9 October 2003.
Eurelectric, *Legal Note on the Energy Chapter of the Constitution of the European Union*, Legal affairs working group (February 2005).
CEER response to the Energy Green Paper of 11 July 2006.
Study on the energy savings Potentials in EU Member States, Candidate Countries and EEA Countries Final Report for the European Commission Directorate-General

Energy and Transport, EC Service Contract Number TREN/D1/239–2006/S07.6664.

AGREEMENT between the Government of the United States of America and the European Community on the coordination of energy-efficiency labelling programs for office equipment, *OJ* L 381/26.

Council of the European Union of Brussels, Brussels European Council of 8/9 March 2007, 2 May 2007, 7224/1/07 REV 1.

European Parliament resolution of 10 July 2007 on prospects for the internal gas and electricity market (2007/2089(INI)).

Community guidelines of 1 April 2008 on State aid for environmental protection, *OJ* C 82/1.

United Nations Framework Convention on Climate Change – Declarations, *OJ* L 33/13.

Legal service of the Council of the European Union, draft written observations to the president and the members of the Court of Justice of the European Union in the Case C-366/10.

European Parliament adopted on 3 February 2009 a Resolution on the Second Strategic Energy Review (2008/2239(INI).

Policy Proposal by Jacques Delors, *Towards a European Energy Community: A Policy proposal*, study by S. ANDOURA, L. HANCHER and M. VAN DER WOUDE.

Study, *"From regional markets to a single European market"* of Everis and Mercados of 28 April 2010.

ERGEG Conclusions paper of 21 may 2010, *Strategy for delivering a more integrated European energy market: The role of the ERGEG Regional Initiatives.*

ENTSO-E, *Ten-Year Network Development Plan 2010–2020* (28 June 2010).

IEA, World Energy Outlook 2010, November 2010.

Study of November 2010, *The functioning of the retail electricity markets for consumers in the European Union*, ECME Consortium, EAHC/FWC/2009 86 01.

Presentation N. AHNER and A. DE HAUTECLOQUE, *Legal feasibility of 'Schengen-like' Agreements in European Energy Policy: The Cases of Nuclear Cooperation and Gas Security of Supply*, (ELRF Conference 9 December 2010, Cambridge Law Faculty.

European Council conclusions of 4 February 2011.

Opinion of the Committee on legal affairs on the legal basis of the proposal for a directive of the European Parliament and of the Council on the promotion of the use of energy from renewable sources (COM(2008)0019 – C6–0046/2008 – 2008/0016(COD)).

Opinion of the Committee on legal affairs on the legal basis of the Proposal for a directive of the European Parliament and of the Council on the indication by labelling and standard product information of the consumption of energy and other resources by energy-related products, COM(2008)778 in the Recommendation for second reading on the Council's position at first reading for adopting a directive of the European Parliament and of the Council on the indication by labelling and standard product information of the consumption of energy and other resources by energy-related products (recast), A7–0128/2010.

Stock-taking document, *Towards a new Energy Strategy for Europe 2011–2020.*

INTERNET RESOURCES

www.cels.law.cam.ac.uk/publications/occasional%20papers/Paper_5.pdf
http://ec.europa.eu/competition/sectors/energy/inquiry/index.html
https://www.entsoe.eu/fileadmin/user_upload/_library/SDC/TYNDP/TYNDPfinal_document.pdf
http://ec.europa.eu/clima/news/index_en.htm
http://ec.europa.eu/clima/policies/brief/eu/index_en.htm
http://ec.europa.eu/clima/policies/transport/aviation/index_en.htm
http://ec.europa.eu/climateaction/index_en.htm
http://ec.europa.eu/comm/competition/sectors/energy/inquiry/index.html
http://ec.europa.eu/comm/competition/sectors/energy/inquiry/issues_paper15112005.pdf.
http://ec.europa.eu/competition/sectors/energy/inquiry/index.html
http://ec.europa.eu/energy/energy_policy/index_en.htm
http://ec.europa.eu/energy/gas_electricity/studies/doc/2010_gas_electricity_markets.pdf
http://ec.europa.eu/energy/gas_electricity/third_legislative_package_en.htm
http://ec.europa.eu/energy/index_en.htm
http://ec.europa.eu/energy/observatory/gas/gas_en.htm
http://ec.europa.eu/energy/observatory/trends_2030/doc/trends_to_2030_update_2009.pdf
http://ec.europa.eu/internal_market/indprop/patent/index_en.htm#enhanced
http://europa.eu/abc/treaties/archives/en/entr14b.htm
http://media.leidenuniv.nl/legacy/2009-09-22.pdf
http://unfccc.int/2860.php
www.consilium.europa.eu/ueDocs/cms_Data/docs/pressData/en/ec/93135.pdf
www.consilium.europa.eu/ueDocs/cms_Data/docs/pressData/en/ec/94932.pdf
www.energyregulators.eu/portal/page/portal/EER_HOME/EER_PUBLICATIONS/CEER_ERGEG_PAPERS/Cross-Sectoral/2010/E10-RIG-10-04_Strategy_Conclusions_21-May-10.pdf
www.europarl.europa.eu/summits/edinburgh/default_en.htm
www.fabectreaty.tv/downloads/FABEC_Treaty_English_version.pdf
www.unep.org/Documents.Multilingual/Default.asp?documentid=97&articleid=1503

INDEX

References are made to paragraph numbers.

A

access 167
ACER 172, 174-176, 811, 812, 820, 831-852, 1119
Agency 21, 47, 69, 112, 173-176, 234, 246, 289, 575, 811, 813-817, 819, 827-831, 833-852, 854, 860, 1117, 1155
Agency for the Cooperation of Energy Regulators 21, 69, 173, 174, 176, 811, 813, 1117
allowances 220, 221, 683, 684, 690, 692, 1124
Amsterdam Protocol 356-358, 361, 366-368, 383, 895, 916, 917
Article 5 of the TEU 301, 327, 341, 348-350, 352, 917, 919
Article 14 of the TFEU 428
Article 18 of the TFEU 404, 408, 410, 510, 1123
Article 101 of the TFEU 404
Article 106(1) of the TFEU 404, 407, 409, 410, 412, 416, 419, 421-423, 425, 431, 1149
Article 106(2) of the TFEU 399, 401, 417, 419, 424, 427, 431, 434, 442, 447-451, 456, 458, 468, 477-479, 1150
Article 106 of the TFEU 394, 407, 411, 415, 416, 418, 425, 435, 472, 480
Article 109 of the TFEU 404
Article 113 of the TFEU 395, 653, 655, 657, 733, 855, 997, 1052, 1053
Article 114 of TFEU 336
Article 125 of the EEA 485, 487, 488, 503
Article 192 of the TFEU 297, 332, 628, 663, 673, 717, 720, 726, 728-734, 736, 739, 740, 742-745, 750, 762, 763, 766, 855, 983, 986, 994, 996, 1025-1027, 1029-1031, 1035, 1037, 1038, 1056-1058, 1100, 1101, 1135
Article 194 of the TFEU 23-26, 892, 894, 936-942, 945, 947-950, 952, 957, 959, 960, 968, 969, 973-976, 978, 983, 986-998, 1000, 1016-1019, 1023, 1025, 1032, 1033, 1035, 1039, 1042, 1044-1046, 1049, 1051-1053, 1056-1058, 1061, 1062, 1068, 1074, 1093, 1097-1100, 1102-1106, 1120, 1128, 1135, 1138, 1157-1161, 1167
Article 258 of the TFEU 400
Article 345 of the TFEU 404, 480-483, 485, 490, 499-502, 504, 857, 1151
Article 352 of the TFEU 332, 333, 337, 338, 340, 792, 828, 933-936, 1038
Articles 34–37 of the TFEU 394, 453, 566, 567, 633
Articles 49–66 of the TFEU 394
Articles 101–102 of the TFEU 394
Articles 107–108 of the TFEU 394
Articles 170–172 of the TFEU 20, 395, 793, 859, 973, 1102
Articles 191–193 of the TFEU 395

Articles 191–193 of the TFEU 405
attribution 18, 22, 275, 278, 279, 319, 320, 324-327, 330, 332, 333, 335, 338, 340, 350, 353, 390, 393, 719, 818, 856, 919, 1004, 1067, 1115, 1147, 1148
aviation sector 679-681

B

balancing 62, 135, 168, 170, 363, 666, 668, 669, 672, 703, 963, 1083
biofuels 66, 183-185, 190, 365, 368, 669, 760, 763, 765, 774, 782, 783, 786, 1154
bioliquids 190, 368, 760, 763, 765, 782, 786, 1154
black-out 46

C

capacity allocation 168, 847
CDM credits 225
certificates of origin 617
charges having equivalent effect 514-516, 521, 529, 532, 1152
choice between different energy sources 297, 311, 313, 731, 736, 737, 741, 776-780, 782, 894, 922, 937, 980, 982, 984, 986, 987, 989, 991, 993, 1016, 1099, 1109, 1159
choice of energy 757
climate change 3, 4, 7, 9, 14, 49, 51, 66, 71, 92-100, 133, 135, 137, 142, 145, 180, 194, 210, 215, 220, 222, 225, 274, 624, 626, 662, 674, 679, 681, 682, 685-687, 691, 693, 697, 698, 720, 757, 805, 922-925, 1141, 1142
coal 4, 8, 30, 31, 45, 48, 95, 106, 108, 191, 282-285, 372, 737, 738, 741, 778, 820, 943, 946, 959, 966, 985
cogeneration 191-193, 203, 855
common market 27, 280, 283, 288, 298, 321, 332, 346, 347, 349, 353, 354, 363, 409, 411, 417, 472, 481, 505, 514, 534, 569, 636, 637, 650, 876, 882, 895, 904, 911, 926, 928, 933, 943
competitiveness 3, 51, 52, 87, 114, 122, 135, 138-140, 144, 145, 158, 184, 270, 304, 756, 757, 795, 805, 1064, 1069, 1085, 1093, 1132
compliance 94, 204, 252, 350, 428, 431, 472, 498, 505, 640, 836, 903, 907, 910, 914, 915, 918, 1077, 1124, 1125
conditions for access 69, 160, 163, 167, 170, 335, 811, 835, 844, 847, 855, 1083
congestion 161, 166, 168, 171, 847
Constitutional Treaty of Europe 310, 311, 317, 863-865, 926
consumer protection 68, 69, 89, 151, 300, 311, 555, 872
consumption 30, 31, 33-35, 43, 47, 48, 53, 54, 57, 82, 90, 91, 103, 106-108, 110, 115, 127, 131, 180-182, 187, 194, 198, 201, 204, 206-208, 210-215, 217, 218, 228, 233, 235, 236, 263, 335, 368, 440, 526, 530, 532, 684, 717, 756, 759, 768, 770, 774, 776, 780-782, 784, 786, 942, 960, 970, 1045, 1046, 1048, 1049, 1051, 1152, 1154
cooperation of TSOs 835, 838
cross-border exchanges 69, 160, 163, 171, 335, 811, 855, 972, 1083
cross-border infrastructure 176, 834, 835, 844, 847
cross-border issues 67, 172, 175, 813, 819, 833, 851, 852, 860, 958, 1155

D

Dassonville formula 542, 543, 549, 614
decision procedure 306, 907, 1026, 1036, 1038, 1039
Declaration 35 998, 1159
Declaration of the EU Treaty 302, 303
delegation of powers 822-826, 828, 831
delimitation of powers 322, 323, 346
derogation clauses 745
direct investment 926-928, 931, 1157
discretionary powers 812, 818, 822-824, 826, 829, 853, 854, 860, 1106, 1155
discrimination 88, 153, 306, 404, 408, 410, 445, 510, 517, 547, 551-553, 563, 564, 611, 683, 839, 841, 850, 935, 952, 954, 1106, 1123, 1124, 1136, 1165
discriminatory 67, 86, 152, 161, 166-169, 171, 257, 414, 446, 474, 478, 487, 509, 516, 518, 521, 524, 526, 527, 543, 544, 551, 556, 558, 603, 611, 612, 621, 627, 630, 631, 653, 699, 707, 1124
distortion of competition 356, 367, 643, 654, 710, 1053
distribution 77, 103, 131, 147, 150, 155, 156, 159, 180, 189, 196, 197, 208, 238, 365, 438, 439, 459, 463, 493, 495, 496, 530, 582, 583, 593, 617, 717, 794, 911, 942, 970
distribution system operators 155, 493, 495
doctrine of pre-emption 876-878
dual legal basis 761, 762, 766, 767, 1030, 1034, 1035, 1037-1039, 1043, 1054, 1055

E

ecodesign 53, 206, 208, 209
ECSC Treaty 17, 31, 32, 36, 282-285, 820-822, 825
EC Treaty 278, 295, 297, 300-303, 306, 307, 309, 319, 327, 329, 341, 349, 370, 380, 425, 428, 513, 516, 521, 549, 625, 626, 640, 650, 664, 665, 672, 696, 704, 717, 718, 730, 732, 733, 744, 748, 753, 771, 855, 884, 925, 933, 934, 943, 944, 1038, 1047
EEC Treaty 291-293, 297, 298, 300, 301, 348, 354, 380, 571, 739, 809, 882
EFTA 484-489, 503
EFTA Court 484-489, 503
Electricity Regulatory Forum 172, 813
Emissions 2, 53, 56, 64, 71, 80, 83, 93, 94, 96, 99, 100, 106, 191, 192, 194, 201, 208, 218, 220-222, 224-226, 229, 338, 366, 613, 616, 669, 679-682, 684, 687, 705, 716, 720, 733, 741, 755, 805
emissions of greenhouse gases 56, 741
emission trading 71, 180, 220-224, 693
emission trading scheme (ETS) 71, 220-224, 687, 693
Energy Charter Treaty 112, 248, 249, 958
Energy Community Treaty 112, 250-254, 958, 1128-1130, 1138, 1167
energy consumption 30, 31, 34, 35, 43, 47, 48, 54, 106-108, 115, 127, 181, 187, 194, 198, 201, 204, 207, 210, 368, 756, 770, 780, 781, 786, 960, 970, 1048, 1154
energy demand 1, 41, 47, 115, 180, 194, 197
energy efficiency 45, 51, 53-60, 71, 72, 75, 77, 80, 95, 96, 103, 104, 106, 112, 115, 116, 135, 152, 180, 191, 194, 195, 197, 199-202, 206, 207, 210, 212, 215, 217, 225, 229, 248, 249,

251, 265, 311, 312, 338, 462, 465-467, 669, 770, 855, 866, 937, 938, 969, 970, 996, 1027, 1046, 1048, 1056, 1094, 1098, 1100
energy efficiency plan 55-57, 80, 116, 970
energy efficiency tax 996
energy end-use efficiency 71, 104, 196-198, 200
energy infrastructure 48, 74-79, 101, 107, 124, 128, 129, 138, 177, 178, 263, 395, 793, 799, 801-803, 805, 806, 955, 965, 973, 985, 1102, 1144
energy markets 4, 10, 19, 27, 30, 34, 36, 39, 127, 131, 147, 163, 176, 252, 265, 268, 281, 287, 290-293, 298, 299, 302, 303, 309, 316, 335, 366, 395, 491, 501, 575, 590, 636, 811, 852, 859, 957-959, 964, 978, 1024, 1063, 1085, 1106, 1108, 1117
energy mix 8, 48, 51, 106, 232, 757, 768, 951, 985
energy performance certificate 205
energy performance of buildings 53, 201-204, 855, 905
energy production 35, 41, 92, 1105
energy products 40, 41, 85, 226, 228, 229, 532, 533, 591, 593, 601, 655, 657, 855, 963, 1052-1054, 1097, 1152
energy resources 10, 35, 298, 314, 594, 672, 894, 937, 980, 982, 983, 986, 989, 994, 1016, 1099, 1109
energy saving 32, 36, 42, 194, 198, 201, 670, 770, 937, 938, 969, 970
energy sector inquiry 19, 60, 61, 63, 120, 151, 493
energy security 30, 32, 33, 40-42, 44, 72, 73, 79, 123, 127, 149, 231, 240, 247, 249, 255, 428, 434, 555, 561, 562, 569, 572, 574-578, 580-582, 590, 591, 593, 601, 603, 616, 619, 951, 958, 962, 964-967, 1002
energy services 71, 104, 196, 198-200, 965
energy sources 8, 17, 27, 33, 34, 36, 45, 51, 77, 100, 107, 131, 135, 181-183, 189, 251, 263, 297, 311-313, 365, 368, 386, 605, 608, 609, 614, 616, 672, 731, 736-739, 741, 755-757, 759, 767-769, 774-780, 782, 804, 805, 836, 855, 894, 905, 922, 937, 946, 971, 977, 980, 982-987, 989, 991, 993, 1016, 1021, 1027, 1056, 1099, 1100, 1109, 1159
energy supply 1, 4-6, 9, 13, 27, 32-34, 40-44, 46, 51, 56, 83, 100, 103, 105, 122, 123, 127, 129, 137, 181, 183, 185, 194, 196, 230-232, 240, 247, 248, 251, 280, 297, 311-315, 599, 602, 606, 731, 736, 738, 741, 755, 776-780, 782, 866, 894, 937, 938, 960, 961, 963, 965, 966, 968, 980, 982, 984-987, 989, 991, 993, 998, 999, 1016-1018, 1021, 1099, 1103, 1104, 1109, 1140, 1144, 1159
energy taxation 180, 229, 313, 996, 1053
energy-using products 206, 208
enhanced cooperation 112, 240, 1071-1075, 1077, 1132, 1133, 1162
environmental guarantee clause 745, 749, 750
environmental protection 71, 152, 307, 462, 464-466, 555, 558, 611-613, 616, 621, 622, 624-626, 628-631, 635, 692, 696, 710, 711, 727, 728, 749, 763, 858, 962, 966, 1152
EURATOM Treaty 17, 286, 288-291, 943-946, 959, 1126, 1137, 1166
European Coal and Steel Community 4, 31, 282, 372, 820, 943
European Network of Transmission System Operators for Electricity (ENTSO-E) 165, 166, 171, 804, 838-840, 849
European Network of Transmission System Operators for Gas (ENTSO-G) 171, 246, 838-840, 849
European Regulators' Group for Electricity and Gas (ERGEG) 172, 813, 819, 852, 1083

EU's external energy policy 806
EU Treaty 300-303, 306, 394, 696
exclusive competence 19, 350, 354, 360, 725, 867, 872, 877, 931, 992, 994, 1042, 1094, 1163
exclusive rights 398, 400, 404, 407, 409, 411-413, 415-417, 419, 423, 426, 429, 447, 453, 454, 456, 458, 583
executive powers 823, 824, 826, 828, 830, 831, 933
exploitation 35, 36, 100, 284, 298, 426, 672, 739, 741, 983, 1016
external aspects of energy supply 247
external competences 44, 718, 719, 721
external costs 709
external energy policy 51, 806, 931, 1002
external security of supply 125, 126, 1143

F
fiscal nature 266, 508, 515, 534, 565, 730-733, 894, 922, 937, 995, 996, 1109
fiscal obstacles 653, 654
flexibility clause 788
force majeure 967
foreign direct investment 926-928, 931, 1157
fossil fuels 9, 71, 84, 95, 108, 737, 946, 971, 977, 1142
free movement 19, 39, 87, 293, 334, 354, 363, 367, 394, 399, 403, 404, 408, 410, 414, 480, 484, 486, 489, 499, 505-507, 510, 513, 516-518, 530, 534-536, 540, 542, 547, 549, 554-556, 559-561, 564, 565, 568, 569, 586, 587, 589, 591 602, 604, 610, 613, 614, 622, 623, 626, 632-635, 637, 640, 647, 751, 858, 885, 927, 935, 1047, 1152
free movement of establishment 592, 594
free movement of goods 19, 293, 334, 354, 367, 394, 399, 414, 505-507, 510, 513, 516-518, 530, 534-536, 540, 542, 547, 554, 556, 559, 560, 564, 565, 568, 586, 587, 592, 595, 626, 632, 635, 858, 885, 1152
functional or regional cooperation 1083, 1132, 1162

G
gas 2, 9, 14, 19, 20, 27, 30-33, 36, 38, 40-42, 44, 46-48, 51, 60, 61, 63, 64, 67-69, 71, 72, 77, 78, 80, 81, 86-88, 90, 91, 93, 95, 96, 99, 100, 105, 106, 108-113, 117-120, 123, 128, 130, 145, 147-149, 158-160, 167, 169-172, 174, 176-178, 186, 191, 194, 220, 222-225, 227, 233, 237, 240-247, 249, 250, 257, 258, 260, 261, 264, 265, 280, 285, 291, 292, 334, 335, 368, 385, 387, 388, 397, 398, 400, 401, 403-406, 437, 456, 457, 460, 461, 463-475, 477-480, 490-496, 498, 501, 504, 506, 525-529, 565-567, 576, 581, 582, 586, 588, 590, 592, 593, 597, 601, 602, 613, 616, 633, 651, 670, 679-682, 686, 737-739, 741, 755, 758, 776, 778, 794, 799, 803-805, 811, 813-815, 819, 832, 834, 836, 838-845, 847-852, 855, 857, 858, 860, 931, 946, 951, 953, 955, 959-965, 967, 971, 982, 985, 1000-1017, 1019-1022, 1051, 1074, 1077, 1078, 1080-1083, 1115, 1116, 1142, 1143, 1149, 1150, 1152, 1155, 1159
Gas Regulatory Forum 172, 813
gas supply 109, 110, 237, 240-242, 245, 246, 1000, 1002-1008, 1011-1013, 1015-1017, 1019-1021, 1080, 1081, 1115, 1116
gas supply disruption 109, 110, 245, 1002, 1007

general economic interest 307, 416, 424, 425, 427-430, 432, 433, 438, 443, 448, 454, 462, 465-468, 472-474, 478, 857, 1150
general legal bases 390, 856, 1147
general non-energy provisions 658
general principle of EU law 374
general structure of its energy supply 297, 311, 313, 731, 736, 738, 741, 776-778, 780, 782, 894, 937, 980, 982, 984-987, 989, 991, 993, 1016, 1021, 1099, 1109, 1159
generation 51, 66, 68, 69, 71, 73, 78, 106, 107, 129, 147, 150, 153, 157, 165, 180, 237, 239, 439, 440, 444, 491, 494, 496, 520, 522, 532, 609, 613, 618, 619, 942, 951, 970, 1144, 1152
golden share 591, 592, 602, 633, 1152
greenhouse gas emissions 2, 64, 71, 80, 93, 96, 99, 100, 194, 220, 222, 224, 225, 613, 616, 679-681, 755, 805
gross energy consumption 35, 181
gross final consumption of energy 759, 768, 776, 784
guarantees of origin 189, 193, 760

H

harmonisation 7, 66, 67, 163, 167, 252, 261, 368, 536, 545, 574, 576, 623, 640, 642, 643, 648-650, 654-657, 711, 724, 725, 746, 747, 753, 764, 841, 849, 874, 877, 934, 997, 1027, 1050, 1053, 1057, 1081
horizontal division 321
human health 576, 662, 666-668, 700, 702, 763, 885

I

imperative requirements 558, 559, 604, 611, 707
implied powers doctrine 718, 719
incumbent 120, 490
independent entity 814
independent system operator (ISO) 496-498
independent transmission operator (ITO) 497, 498
Industry, Research and Energy (ITRE) 58, 174, 832
infrastructure 16, 48, 51, 72, 74-79, 98, 101, 106, 107, 113, 120, 124, 128, 129, 138, 160, 171, 176-178, 237, 243-245, 247, 263, 278, 335, 395, 398, 491, 496, 734, 779, 793, 795, 797, 799-806, 834, 835, 842, 844-847, 855, 955, 962, 965, 972, 973, 985, 1007, 1008, 1011, 1015, 1019, 1102, 1144
infringement 256-258, 400, 410, 415, 423, 535, 609, 925
integration principle 307, 622-629
interconnection 51, 68, 78, 119, 128, 158, 166, 237, 257, 794, 839, 841, 937, 938, 972-974, 978, 1102, 1158
interconnection of energy networks 937, 938, 972, 1102
interconnectors 166, 815, 842, 844, 847, 972
internal electricity and gas market 77
Internal Energy Agency (IEA) 47, 112
internal energy market 37, 39, 86, 120, 122, 141, 174, 191, 240, 256, 299, 366, 398, 400, 401, 804, 832, 855, 957, 958, 978, 1084, 1143

internal market 14, 16, 27, 31, 32, 36, 40, 69, 87, 90, 112, 117, 121, 123-125, 131, 136, 141, 145, 147, 149-151, 159, 163, 166, 226, 237, 250, 264, 278, 280, 285, 291-295, 311, 332-336, 340, 365, 395, 397, 398, 422, 479, 490, 495, 496, 498, 510, 513, 514, 516, 518, 519, 547, 628, 636-638, 640, 642, 643, 646, 647, 649, 650, 652, 654, 658, 717, 730, 756, 764, 765, 794, 799, 811, 842, 855, 866, 871, 872, 874, 876, 935, 937, 938, 940, 947, 948, 956, 961, 976, 998, 1027, 1046-1048, 1051, 1053, 1103, 1127, 1143, 1158
internal taxation 509, 510, 517-519, 521-523, 527, 531, 618
international environmental agreements 718, 728

K
Keck formula 549, 632, 1152
Kyoto Protocol 93-95, 99, 210, 680, 681, 687, 692

L
labelling 53, 210-213, 215-219, 335, 367, 669, 1045-1049, 1051, 1118
legally binding target 58, 386
legislative package 174, 758, 811, 832
liberalisation 19, 63, 77, 86, 89, 147, 174, 367, 398, 399, 466, 473, 490, 794, 832, 926
Lisbon Treaty 16, 17, 22, 269, 272, 275-278, 281, 305, 318, 323, 327, 333, 354, 380, 389, 390, 428, 652, 856, 859, 862-864, 866, 871, 877, 881, 882, 887, 888, 891, 897-902, 907, 908, 911, 913, 920, 921, 923, 926, 928, 930, 933-935, 937, 943, 944, 972, 975, 979, 989, 994, 1023, 1024, 1045, 1059, 1061, 1063, 1068, 1071, 1072, 1075, 1089, 1095, 1098, 1110, 1146, 1147, 1153, 1156, 1157
low-carbon 70, 80-82, 137, 142, 222, 265, 802, 805

M
Maastricht Treaty 327, 341, 345, 346, 349, 354, 899
mandatory national targets 759, 769, 770, 784
mandatory requirement 555, 604, 622
mandatory targets 187, 770, 775, 776, 780
Meroni doctrine 812, 817, 820, 825, 827-830, 853, 854, 860, 1106, 1119, 1155
minimum stocks 33, 232-234, 590, 855
monopoly 120, 121, 400, 401, 410, 414, 415, 420, 422-424, 426, 444, 450, 491, 492, 582

N
national action plans 188
National Energy Efficiency Action Plans (NEEAP's) 71, 104, 115, 200
national Parliaments 350, 895-897, 899-904, 906-912, 934, 1091
national regulators 66, 67, 69, 172, 173, 175, 368, 813, 814, 835
National Renewable Energy Action Plans (NREAP's) 101, 102, 107
natural resources 206, 283, 297, 303, 313, 315, 394, 398, 404, 436, 446, 451, 455, 458, 461, 476, 491, 569, 572, 578, 662, 669-672, 739, 741, 763, 1098
necessity test 376, 387, 561, 592
network codes 176, 834, 839-841, 849
network connection 839
network development plan 165, 498, 804, 839

networks 4, 68, 69, 76-79, 123, 131, 153, 155, 161, 166, 167, 170, 237, 267, 301, 311, 335, 395, 459, 490, 491, 493, 793-796, 798-800, 804, 805, 811, 814, 855, 872, 937, 938, 972-974, 978, 1102, 1158
new and renewable forms of energy 311, 937, 938, 969, 971, 1046, 1098
new energy era 4, 13, 46, 51, 951, 1141
new energy title 22, 23, 25, 278, 303, 315, 394, 652, 859, 866, 937, 943, 976, 978, 1014, 1023-1025, 1045, 1058, 1061, 1064, 1068, 1069, 1103, 1126, 1156, 1158, 1161
new energy Treaty 26, 1121, 1122, 1124-1128, 1131, 1136-1139, 1165-1168
non-discrimination principle 1123, 1124, 1136, 1165
non-discriminatory 67, 86, 152, 161, 166-169, 171, 257, 478, 487, 518, 699
non-renewable 670, 671, 741
nuclear 8, 30, 36, 47, 48, 92, 108, 286-290, 737, 738, 741, 776, 778, 943, 945, 946, 965, 966, 977, 985, 1074, 1077, 1078, 1080-1082, 1133
nuclear sources 737, 946, 977

O

oil 9, 30-33, 35, 40-42, 44, 46-49, 56, 72, 73, 78, 91, 95, 105, 106, 108, 112, 113, 123, 130, 178, 183, 191, 194, 227, 232-235, 249, 287, 315, 554, 565-570, 572-584, 589, 590, 595, 597, 609, 610, 614, 633, 670, 739, 741, 751, 804, 809, 855, 931, 960, 965, 982, 1142, 1144, 1152
ordinary legislative procedure 312, 428, 639, 729, 731, 735, 742, 743, 777, 797, 908, 921, 937, 980, 981, 988, 995, 1014, 1021, 1108
ordinary revision procedure 1089, 1091, 1092
ownership unbundling 67, 267, 483, 490, 491, 494-502, 857, 1151

P

package 19, 63-65, 69, 71, 96, 132, 157, 163, 170, 174, 186, 256, 259, 289, 388, 490, 494, 496, 497, 500, 758, 811, 832, 962
permits 416, 466, 841, 948
polluter pays principle 694, 709, 710
precautionary principle 529, 694, 697-700, 702, 713
preserve and improve the environment 311, 937, 938, 940, 947, 949, 976, 1158
prevention principle 694, 703, 705, 706
primary energy 33, 47, 48, 51, 54, 106, 108, 191, 757, 958
primary energy consumption 48, 106, 108
primary energy sources 33, 757
primary law 12, 14, 19, 20, 22, 275, 276, 975, 1146
principle of attribution of powers 324, 325, 327, 332, 333, 350, 353
principle of conferral of competences 331, 339
principle of conferral of powers 341, 640
principle of cooperation or loyal cooperation 882
principle of precedence or primacy 882
principle of proportionality 369-375, 380, 382, 384, 385, 388, 474, 478, 560, 716, 752, 856, 1147
procedural restrictions 987, 993, 1099, 1159
projects of common interest 796, 797, 800

property ownership 404, 481-483, 486-488, 501-503, 1151
proportionality 18, 22, 275, 278, 279, 319, 320, 323, 350, 356, 369-380, 382-386, 388, 390, 392, 393, 429, 443, 474, 478, 499, 560-562, 584, 592, 594, 600-603, 613, 626, 716, 752, 856, 895, 903-905, 917-919, 1067, 1147, 1148
proportionality test 376-380, 384, 386, 392, 443, 560-562, 600, 601, 613, 856, 1147
proportionality test *sensu stricto* 561
proportionate 363, 365, 384, 392, 487, 525, 580, 600, 699, 1147
protection of the environment 304, 604, 611, 615, 616, 621, 626, 627, 630, 631, 634, 699, 708, 751, 804, 965, 1027, 1046, 1152
prudent use 671
public security 553, 569, 573, 577-580, 582-584, 588, 591, 592, 594-599, 601, 602, 610, 633, 1152
public service 152, 159, 399, 400, 429, 448-450, 458, 460-462, 464-469, 473, 474, 476, 478, 479, 857, 1150
public service obligations 152, 159, 399, 400, 462, 464-468, 473, 474, 476, 478, 479, 857, 1150

Q

qualified majority 294, 306, 309, 312, 315, 399, 781, 783, 786, 925, 1036, 1037, 1072, 1089, 1109
quantitative restrictions 284, 445, 507, 508, 537, 541, 542

R

rational use 266, 671, 763, 1048
reinforced cooperation 1075-1077, 1079, 1082, 1122, 1132, 1133, 1162
renewable electricity (RES-E) 107, 182, 605
renewable energy 8, 45, 48, 51, 66, 68, 71, 77, 95, 100-102, 107, 135, 181-183, 186-190, 229, 251, 263, 312, 365, 368, 386, 605, 607-609, 614, 616, 670n 737, 738, 755, 756, 758-761, 763, 765-776, 778-780, 782, 784-787, 804, 805, 836, 855, 905, 946, 971, 985, 996, 1027, 1056, 1100, 1154

S

Safeguard clause 694, 711
SEA 97, 128, 255, 293-296, 298-300, 347, 676, 725, 800, 870
secondary legislation 14-16, 19, 21, 24, 87, 143, 144, 146, 256, 259, 261, 267, 269-272, 276, 278, 292, 395, 402, 406, 437, 461, 477, 479, 504, 505, 625, 636, 651, 652, 659, 665, 705, 758, 787, 792, 799, 811, 812, 855, 857, 859-861, 890, 906, 953, 961, 971, 1044, 1051, 1053-1056, 1066, 1100, 1110, 1119, 1145, 1146, 1149, 1153, 1155, 1160
security of energy supply 5, 40-44, 56, 83, 103, 105, 122, 123, 137, 183, 194, 230-232, 240, 247, 248, 251, 311, 312, 599, 602, 755, 866, 937, 938, 960, 961, 963, 965, 966, 968, 1017, 1018, 1103, 1104
security of gas supply 240-242, 246, 1004-1006, 1012, 1015, 1017, 1019-1021, 1080, 1081, 1116
security of supply 4, 10, 14, 49, 51, 52, 77, 85, 87, 105, 109, 110, 112, 114, 121-126, 135, 136, 138-141, 144, 145, 152, 154, 201, 239, 240, 245, 251, 252, 270, 283, 290, 299, 304, 314, 428, 462, 464-467, 555, 578, 580, 587, 589-592, 595, 601-603, 609, 610, 633, 635,

757, 799, 805, 846, 858, 945, 955, 961-967, 1002, 1010, 1017, 1064, 1069, 1074, 1077, 1078, 1081-1083, 1085, 1093, 1115, 1116, 1132, 1141, 1143, 1152
separation 66, 153, 557
services of general economic interest 307, 416, 424, 425, 427-430, 448, 468, 478, 857, 1150
shared competence 22, 311, 660, 722, 867, 881, 888, 891, 1042, 1043, 1060, 1094, 1156, 1163
shared regulatory competence 873
simplified revision procedure 1089, 1090
Single European Act 293, 298, 668, 672, 706, 713, 718, 723, 739, 983
single legal basis 761, 1029, 1034, 1044
single supplier 109
solidarity 51, 72, 73, 234, 243, 428, 808, 920, 935, 937, 938, 940, 950-956, 972, 976, 982, 1004, 1095, 1103, 1158
source principle 694, 706-708, 715
sovereignty 8, 298, 313, 329, 342, 411, 672, 689, 757, 1023, 1061, 1068, 1094, 1161, 1163
special legislative procedure 654, 731, 733, 735, 740, 742, 777, 781, 786, 932, 937, 986, 989, 991, 995, 996, 1021, 1055, 1099, 1108
spirit of solidarity 243, 808, 920, 937, 938, 940, 950, 976, 1004, 1095, 1103, 1158
Strategic Energy Technology (SET) 3, 7, 15, 33, 48, 50, 70, 73, 83, 85, 104, 106-108, 168, 169, 172, 181-183, 185, 188, 198, 199, 224, 227, 235, 236, 238, 257, 263, 265, 270, 291, 292, 326, 327, 348, 350, 365, 371, 394, 422, 428, 505, 506, 554, 558, 568, 605, 622, 684, 692, 717, 720, 738, 744, 778, 780, 797, 805, 813, 820, 821, 835, 839, 845, 858, 882, 917, 932, 934, 985, 1038, 1071, 1077, 1083, 1084, 1116, 1145, 1152
subsidiarity 18, 22, 275, 278, 279, 301, 304, 307, 319, 320, 322, 323, 340-344, 346-365, 368, 369, 380, 384-386, 388, 390-393, 428, 640, 646, 650, 667, 856, 876, 895, 899, 903-905, 907-911, 913-919, 1067, 1078, 1111, 1147, 1148, 1157
subsidiarity principle 341, 342, 358-361, 363, 365, 368, 385, 391, 392, 856, 899, 915, 917, 1147
subsidiarity test 354, 357, 360, 364, 369, 391, 856, 1147
suitability test 376, 388, 561
supply 1, 4-6, 9, 10, 13, 14, 27, 32-34, 40-44, 46, 49, 51, 52, 56, 66, 69, 77, 83, 85, 87, 90, 91, 97, 100, 103, 105, 109, 110, 112-114, 121-127, 129, 131, 135-141, 144, 145, 147, 150, 152-154, 157, 159, 166, 180, 181, 183, 185, 194, 196, 201, 203, 217, 230-234, 237-248, 251, 252, 258, 270, 280, 283, 284, 287-290, 297, 299, 304, 311-315, 335, 366, 423, 428, 438, 440, 442, 444, 450, 451, 454, 457-459, 462, 464-467, 469-471, 473, 491, 493, 494, 496, 505, 525, 555, 570, 578-581, 583, 587, 589-592, 595, 599, 601-603, 605, 606, 609, 610, 614, 633, 635, 731, 736, 738, 741, 755, 757, 776-780, 782, 799, 805, 808, 809, 846, 855, 858, 866, 894, 920, 937, 938, 942, 945, 955, 958, 960-968, 971, 975, 980, 982, 984-987, 989, 991, 993, 998-1000, 1002-1008, 1010-1013, 1015-1021, 1064, 1069, 1074, 1077, 1078, 1080-1083, 1085, 1093, 1095, 1097, 1099, 1103, 1104, 1109, 1115, 1116, 1132, 1140, 1141, 1143, 1144, 1152, 1159
sustainability 52, 85, 92, 103, 114, 121, 122, 135-141, 144, 145, 180, 266, 270, 283, 365, 368, 629, 757, 760, 763-765, 782, 1064, 1069, 1085, 1093, 1132
sustainability criteria 368, 760, 763-765

T

tariffs 68, 89, 156, 169, 257, 771
taxation 84, 180, 226-229, 313, 395, 499, 509-511, 517-519, 521-524, 527, 531, 618, 653-655, 657, 732, 751, 855, 996, 997, 1052-1054, 1124
TEN-E funding programme 801, 802
TEN-E Guidelines 800, 801, 806, 973
TEN Financing Regulation 800, 801, 973
third package 69, 163, 170, 186, 388, 494, 496, 497, 500, 962
third-party access 257, 491, 839
three-tier proportionality test 376, 384, 386, 392, 856, 1147
Trans-European Networks (TEN) 85, 117, 165, 250, 301, 311, 395, 498, 793, 794, 796, 799-807, 838, 839, 849, 872, 973, 974, 978, 1158
transmission 51, 67, 69, 77, 103, 119, 131, 147, 149, 150, 153-156, 159-162, 164-167, 169-171, 180, 196, 237, 238, 257, 267, 335, 397, 444, 490-494, 496-499, 501, 502, 613, 617, 794, 804, 811, 838, 841, 855, 904, 942, 972, 1151
transmission system operators (TSO) 51, 69, 154-156, 160, 161, 164-167, 171, 238, 257, 497, 838
transparency 31, 62, 66, 67, 69, 120, 149, 155, 157, 159, 170, 177, 257, 358, 361, 397, 429, 1003, 1081, 1083, 1117
Treaty of Nice 308, 309

U

unanimously 636, 654, 673, 731, 733, 740, 742, 777, 787, 932, 937, 986, 989, 991, 994-996, 1038, 1055, 1099
unbundling 4, 67, 69, 153, 157, 159, 267, 368, 483, 490-502, 857, 1151
Union competence 313, 725, 855, 867, 869, 885, 982, 1066
Union policy on energy 311, 312, 866, 937, 938, 941, 942
Union Treaties 275, 281, 318, 333, 1127

V

vertical division 320-322
vertical integration 490
vertically integrated 147, 153, 491, 493, 496, 498
veto right 781, 911, 993